90 Springer Series in Solid-State Sciences

Edited by Peter Fulde

Springer Series in Solid-State Sciences

Editors: M. Cardona   P. Fulde   K. von Klitzing   H.-J. Queisser

Managing Editor: H. K. V. Lotsch

Volumes 1–49 are listed at the end of the book

50 **Multiple Diffraction of X-Rays in Crystals**
By Shih-Lin Chang

51 **Phonon Scattering in Condensed Matter**
Editors: W. Eisenmenger, K. Laßmann, and S. Döttinger

52 **Superconductivity in Magnetic and Exotic Materials**   Editors: T. Matsubara and A. Kotani

53 **Two-Dimensional Systems, Heterostructures, and Superlattices**
Editors: G. Bauer, F. Kuchar, and H. Heinrich

54 **Magnetic Excitations and Fluctuations**
Editors: S. Lovesey, U. Balucani, F. Borsa, and V. Tognetti

55 **The Theory of Magnetism II**   Thermodynamics and Statistical Mechanics   By D. C. Mattis

56 **Spin Fluctuations in Itinerant Electron Magnetism**   By T. Moriya

57 **Polycrystalline Semiconductors,** Physical Properties and Applications
Editor: G. Harbeke

58 **The Recursion Method and Its Applications**
Editors: D. Pettifor and D. Weaire

59 **Dynamical Processes and Ordering on Solid Surfaces**   Editors: A. Yoshimori and M. Tsukada

60 **Excitonic Processes in Solids**
By M. Ueta, H. Kanzaki, K. Kobayashi, Y. Toyozawa, and E. Hanamura

61 **Localization, Interaction, and Transport Phenomena**   Editors: B. Kramer, G. Bergmann, and Y. Bruynseraede

62 **Theory of Heavy Fermions and Valence Fluctuations**   Editors: T. Kasuya and T. Saso

63 **Electronic Properties of Polymers and Related Compounds**
Editors: H. Kuzmany, M. Mehring, and S. Roth

64 **Symmetries in Physics**   Group Theory Applied to Physical Problems
By W. Ludwig and C. Falter

65 **Phonons: Theory and Experiments II** Experiments and Interpretation of Experimental Results   By P. Brüesch

66 **Phonons: Theory and Experiments III** Phenomena Related to Phonons
By P. Brüesch

67 **Two-Dimensional Systems: Physics and New Devices**
Editors: G. Bauer, F. Kuchar, and H. Heinrich

68 **Phonon Scattering in Condensed Matter V**
Editors: A. C. Anderson and J. P. Wolfe

69 **Nonlinearity in Condensed Matter**
Editors: A. R. Bishop, D. K. Campbell, P. Kumar, and S. E. Trullinger

70 **From Hamiltonians to Phase Diagrams**   The Electronic and Statistical-Mechanical Theory of sp-Bonded Metals and Alloys   By J. Hafner

71 **High Magnetic Fields in Semiconductor Physics**
Editor: G. Landwehr

72 **One-Dimensional Conductors**
By S. Kagoshima, H. Nagasawa, and T. Sambongi

73 **Quantum Solid-State Physics**
Editors: S. V. Vonsovsky and M. I. Katsnelson

74 **Quantum Monte Carlo Methods** in Equilibrium and Nonequilibrium Systems   Editor: M. Suzuki

75 **Electronic Structure and Optical Properties of Semiconductors**   Second Edition
By M. L. Cohen and J. R. Chelikowsky

76 **Electronic Properties of Conjugated Polymers**
Editors: H. Kuzmany, M. Mehring, and S. Roth

77 **Fermi Surface Effects**
Editors: J. Kondo and A. Yoshimori

78 **Group Theory and Its Applications in Physics**
By T. Inui, Y. Tanabe, and Y. Onodera

79 **Elementary Excitations in Quantum Fluids**
Editors: K. Ohbayashi and M. Watabe

80 **Monte Carlo Simulation in Statistical Physics** An Introduction
By K. Binder and D. W. Heermann

81 **Core-Level Spectroscopy in Condensed Systems**
Editors: J. Kanamori and A. Kotani

82 **Introduction to Photoemission Spectroscopy**
By S. Hüfner

83 **Physics and Technology of Submicron Structures**
Editors: H. Heinrich, G. Bauer, and F. Kuchar

84 **Beyond the Crystalline State**   An Emerging Perspective   By G. Venkataraman, D. Sahoo, and V. Balakrishnan

85 **The Fractional Quantum Hall Effect** Properties of an Incompressible Quantum Fluid
By T. Chakraborty and P. Pietiläinen

86 **The Quantum Statistics of Dynamic Processes**
By E. Fick and G. Sauermann

87 **High Magnetic Fields in Semiconductor Physics II** Transport and Optics   Editor: G. Landwehr

88 **Organic Superconductors**
By T. Ishiguro and K. Yamaji

89 **Strong Correlation and Superconductivity**
Editors: H. Fukuyama, S. Maekawa, and A. P. Malozemoff

90 **Earlier and Recent Aspects of Superconductivity**
Editors: J. G. Bednorz and K. A. Müller

91 **Electronic Properties of Conjugated Polymers III**   Basic Models and Applications
Editors: H. Kuzmany, M. Mehring, and S. Roth

J. G. Bednorz   K. A. Müller (Eds.)

Earlier and Recent Aspects of
Superconductivity

Lectures from the International School,
Erice, Trapani, Sicily, July 4–16, 1989

With 281 Figures

Springer-Verlag Berlin Heidelberg New York
London Paris Tokyo Hong Kong

Dr. J. Georg Bednorz
Professor Dr. Dr. h.c. mult. K. Alex Müller

IBM Forschungslaboratorium, Säumerstrasse 4,
CH-8803 Rüschlikon, Switzerland

Series Editors:

Professor Dr., Dres. h. c. Manuel Cardona
Professor Dr., Dr. h. c. Peter Fulde
Professor Dr., Dr. h. c. Klaus von Klitzing
Professor Dr. Hans-Joachim Queisser

Max-Planck-Institut für Festkörperforschung, Heisenbergstrasse 1,
D-7000 Stuttgart 80, Fed. Rep. of Germany

Managing Editor:
Dr. Helmut K. V. Lotsch

Springer-Verlag, Tiergartenstrasse 17, D-6900 Heidelberg, Fed. Rep. of Germany

ISBN 3-540-52156-9 Springer-Verlag Berlin Heidelberg New York
ISBN 0-387-52156-9 Springer-Verlag New York Berlin Heidelberg

This work is subject to copyright. All rights are reserved, whether the whole or part of the material is concerned, specifically the rights of translation, reprinting, reuse of illustrations, recitation, broadcasting, reproduction on microfilms or in other ways, and storage in data banks. Duplication of this publication or parts thereof is only permitted under the provisions of the German Copyright Law of September 9, 1965, in its version of June 24, 1985, and a copyright fee must always be paid. Violations fall under the prosecution act of the German Copyright Law.

© Springer-Verlag Berlin Heidelberg 1990
Printed in Germany

The use of registered names, trademarks, etc. in this publication does not imply, even in the absence of a specific statement, that such names are exempt from the relevant protective laws and regulations and therefore free for general use.

Offset printing: Druckhaus Beltz, D-6944 Hemsbach/Bergstr.
Bookbinding: J. Schäffer GmbH & Co. KG., D-6718 Grünstadt
2154/3150-543210 – Printed on acid-free paper

Preface

This volume contains the proceedings of "Earlier and Recent Aspects of Superconductivity", the 18th course of the International School of Materials Science and Technology, which was held at the Ettore Majorana Centre for Scientific Culture in Erice (Sicily), Italy, July 4-16, 1989.

In the wake of the discovery of high-T_c oxides many international conferences have been held on the subject, mainly intended for those already working in high-T_c research. Hence, the idea evolved of organizing a school dedicated to students and young scientists entering the field of superconductivity. The intention was to give a broad perspective of the field with its eight decades of research history and thus to link existing fields of knowledge and gain continuity. The new high-T_c cuprates were integrated, both historically and by subject, into the field of classical superconductivity, as well as into its newer branches of heavy-fermion, organic and chalcogenide superconductors. Even the organizers and the lecturers were surprised how many cross-links became apparent between the different branches of superconductivity.

The lectures have been grouped into four parts, namely:

– Fundamental properties of superconductors
– Coherence-length-related properties
– Electronic and magnetic properties
– Theoretical models

The number of applications from potential participants greatly exceeded the capacity of the facilities available. Ultimately, 100 students from 25 countries were selected to attend the course. The remarkable surroundings of the medieval town of Erice promoted a most stimulating but nevertheless relaxed atmosphere. Its good restaurants and pleasant parks encouraged the formation of small discussion groups, thus furthering personal contacts and helping to deepen various aspects. To promote such interaction, the program was arranged with sufficient intermission times so as not to overburden the participants.

The course was held under the auspices of the European Physical Society and of UNESCO. The latter institution, through the Regional Office for Science and Technology in Europe (ROSTE), provided special grants for participants from Eastern Europe and, under the EPS-UNESCO Agreement on Energy Storage and Saving, for the publication of the proceedings as well. The sponsorship of IBM Italy, the IBM Zurich Research Laboratory, and CNR

Italy (Physical Science Committee) is gratefully acknowledged. The NATO Scientific Affairs Division kindly provided special travel grants to participants from Turkey and Portugal and, through the National Science Foundation, to participants from the United States. Finally, the support provided by the institutions acting through the Majorana Centre, namely, the Italian Ministries of Education and Technological Research and the Regional Government of Sicily, is gratefully acknowledged.

Last but not least, the professional and dedicated help of the staff of the Ettore Majorana Centre, especially when required at odd times, added to the school's success and is sincerely appreciated. In the organizational phase of the school and throughout the period of finalizing the proceedings the untiring help of our secretary in Rüschlikon, Mrs. Mann, was of immeasurable value, as on so many earlier occasions.

Zurich
February 1990

J.G. Bednorz
K.A. Müller

Contents

Part I   Fundamental Properties of Superconductors

Some Remarks on the History of Superconductivity
By H. Thomas (With 21 Figures) 2

Electronic Structures of Oxide Superconductors – Development of Concepts
By A. Kitazawa (With 16 Figures) 45

Oxygen Nonstoichiometry and Valence States in Superconductive Cuprates
By B. Raveau, C. Michel, M. Hervieu, J. Provost, and F. Studer
(With 22 Figures) 66

Superconductivity and Magnetism in Chevrel Phases
By Ø. Fischer (With 10 Figures) 96

Organic Superconductivity: The Role of Low Dimensionality and Magnetism
By D. Jérome (With 19 Figures) 113

Flux Quantisation and Quantum Coherence in Conventional and HTC Superconductors and Their Application to SQUID Magnetometry
By C.E. Gough (With 13 Figures) 141

Similarities and Differences Between Conventional and High-T_c Superconductors
By A. Barone (With 2 Figures) 163

Part II   Coherence-Length-Related Properties

Short Coherence Length and Granular Effects in Conventional and High T_c Superconductors
By G. Deutscher (With 11 Figures) 174

Critical Currents in Single-Crystal and Bicrystal Films
By P. Chaudhari, D. Dimos, and J. Mannhart (With 1 Figure) 201

What Limits the Critical Current Density in High-T_c Superconductors?
By J. Mannhart (With 11 Figures) 208

Muon Spin Rotation Experiments in High-T_c Superconductors
By H. Keller (With 7 Figures) 222

(Spin) Glass Behavior in High-T_c Superconductors
By I. Morgenstern (With 20 Figures) 240

Microwave Absorption in Granular Superconductors
By K.W. Blazey (With 8 Figures) 262

Microwaves and Superconductivity: Processes in the Intergranular Medium
By A.M. Portis 278

Part III Electronic and Magnetic Properties

Superconductivity of Strongly Correlated Electrons: Heavy-Fermion Systems
By F. Steglich (With 10 Figures) 306

Pair Breaking in Superconductors
By P. Fulde and G. Zwicknagl (With 13 Figures) 326

On the Electronic Structure and Related Physical Properties of 3d Transition Metal Compounds
By G.A. Sawatzky (With 16 Figures) 345

The Electronic Structure of Previous and Present High-T_c Superconductors – Investigations with High-Energy Spectroscopies
By J. Fink, J. Pflüger, Th. Müller-Heinzerling, N. Nücker, B. Scheerer, H. Romberg, M. Alexander, R. Manzke, T. Buslaps, R. Claessen, and M. Skibowski (With 20 Figures) 377

A Positive Experimental Test for Pairing Mechanisms Including $3d_{z^2}$ Hole Symmetry: Correlation of the Relative Weight of $3d_{z^2}$ vs $3d_{x^2-y^2}$ Hole States with the Critical Temperature
By A. Bianconi, P. Castrucci, A. Fabrizi, M. Pompa, A.M. Flank, P. Lagarde, H. Katayama-Yoshida, and G. Calestani (With 9 Figures) . 407

Magnetic and Electronic Correlations in $YBa_2Cu_3O_{6+x}$
By J.M. Tranquada (With 13 Figures) 422

Magnetic Correlations and Spin Dynamics in $La_{2-x}Sr_xCuO_4$ from NQR Relaxation
By A. Rigamonti, F. Borsa, M. Corti, T. Rega, J. Ziolo, and F. Waldner (With 14 Figures) 441

A Credo for a Spy: Nuclear Spin Interactions in Superconductors
By M. Mehring (With 6 Figures) 467

Part IV Theoretical Models

Phonons and Charge-Transfer Excitations in High-Temperature Superconductors
By A.R. Bishop (With 5 Figures) 482

Experimental Constraints and Theory of Layered High-Temperature Superconductors
By T. Schneider and M. Frick (With 5 Figures) 501

Generalized Hubbard Models for Cu-O-Based Superconductors: Field-Theoretical and Monte-Carlo Results
By J. Wagner, R. Putz, G. Dopf, B. Ehlers, L. Lilly, A. Muramatsu, and W. Hanke (With 9 Figures) 518

Index of Contributors 529

Part I

Fundamental Properties of Superconductors

Some Remarks on the History of Superconductivity

H. Thomas

Institut für Physik, Universität Basel, Klingelbergstrasse 82,
CH-4056 Basel, Switzerland

This lecture is intended to give a brief review of the history of superconductivity. Since it is impossible to treat all aspects with equal weight, I have put the main emphasis on the earlier developments of the thermodynamic and electromagnetic properties of superconductors, but have been rather selective with respect to more recent aspects. Although I shall present the various subjects in an introductory manner, I assume a certain elementary knowledge of the field. For an introduction into the field of superconductivity, I refer to a selection of books [1-7].

In the earlier history, one may distinguish two periods of high activity: A first one comprising the discovery of the phenomenon of superconductivity and of the existence of a critical current and a critical magnetic field, and the first persistent-current experiments by H.K.Onnes during the years 1911-1914. A detailed account of this period has been given in Ref.[8]. The second period starts with the discoveries of the specific-heat anomaly in 1932 and of the Meissner-Ochsenfeld effect in 1933, leading to the development of the Gorter-Casimir theory of the thermodynamics of superconductors in 1934, and of the London theory of electromagnetic behaviour in 1935. A discussion [9] which took place in 1935 gives a vivid picture of the activity in this period. A lot of interesting historical details may be found in the personal recollections by C.J.Gorter and K.Mendelssohn of work on superconductivity until 1940 in Leiden and in Oxford, respectively [10,11]. Of high interest is further the historical review on superconductivity by H.B.G.Casimir [12].

To the landmarks of the more recent history belong: on the theoretical side, after the development of the Ginzburg-Landau theory in 1950 and of Pippard's non-local electrodynamics in 1953, finally, after many attempts, the elucidation of the microscopic mechanism by the BCS theory in 1957, further the prediction of the Josephson effect in 1962; on the experimental side the discovery of the isotope effect in 1950, the demonstration of an energy gap in the electronic excitation spectrum by various experiments in 1953-1960, the discovery of flux quantization in 1961, the observation of the flux-line lattice in 1967, and of macroscopic quantum interference phenomena since 1964; on the materials side the work of B.T.Matthias on intermetallic compounds, and, most recently, the discovery of oxidic superconductors leading to the breakthrough towards higher transition temperatures by G.Bednorz and K.A.Müller in 1986.

References to historical key papers have been included as part of the presentation in the main text; other references, in particular to books and review articles, are placed as usual at the end.

1. Basic Phenomena of Superconductivity

1.1 What had happened before

The phenomenon of conduction of electricity by certain materials was discovered still in the era of static electricity by S.Gray (1729). The properties of metallic conductivity were investigated using galvanic electricity by G.S.Ohm (1827).

The first important step towards a microscopic understanding of metallic conduction was the development of the (classical) electron theory of metals by E.Riecke, P.Drude, and H.A.Lorentz (1898 - 1905):

E.Riecke: Zur Theorie des Galvanismus und der Wärme.
Annalen der Physik und Chemie(3) 66 (1898) 353-389, 545-581,

E.Riecke: Ueber das Verhalten der Leitfähigkeiten der Metalle für Wärme und für Elektrizität.
Annalen der Physik (4) 2 (1900) 835-842,

P.Drude: Zur Elektronentheorie der Metalle.
Annalen der Physik(4) 1 (1900) 566-613, 3 (1900) 369-402,

H.A.Lorentz: The motion of electrons in metallic bodies.
Proc.Acad.Sci.Amsterdam 7 (1904-05) 438, 585, 684.

It was based on the concept of a gas of free electrons scattered by the atoms, and led to the famous formula for the electrical conductivity

$$\sigma = ne^2 l/mv, \qquad (1.1)$$

where n is the density of free electrons, e and m are their charge and their mass, respectively, v is their mean thermal velocity, and l their mean free path.

The prediction for the behaviour of the resistivity $\rho = 1/\sigma$ at low temperatures based on this formula depended on the assumption about the temperature dependence of n: If the electrons would condense on the atoms at low temperatures, then the metal should become an insulator at T = 0 (Kelvin 1902); if no condensation of the electron gas would occur (n = const), then ρ should vanish as $v(T) \propto \sqrt{T}$.

In 1908, H.K.Onnes had succeeded in liquefying He (Leiden held the monopoly for the production of liquid helium until 1923!), and it was with the above background that he and his coworkers studied the temperature dependence of the resistivity of pure metals like Au and Pt at low temperatures.

The resistivity was found neither to diverge nor to vanish, but rather to approach a residual resistance for T → 0, which depended strongly on the impurity content as already found by Matthiessen (1864). Moreover, an ideal resistivity vs temperature curve constructed by subtracting the residual resistance did not show at all a $\sqrt{T}$ behaviour (Fig.1). It is remarkable that H.K.Onnes associated the observed temperature dependence already with the thermal motion of the atoms. He even deduced

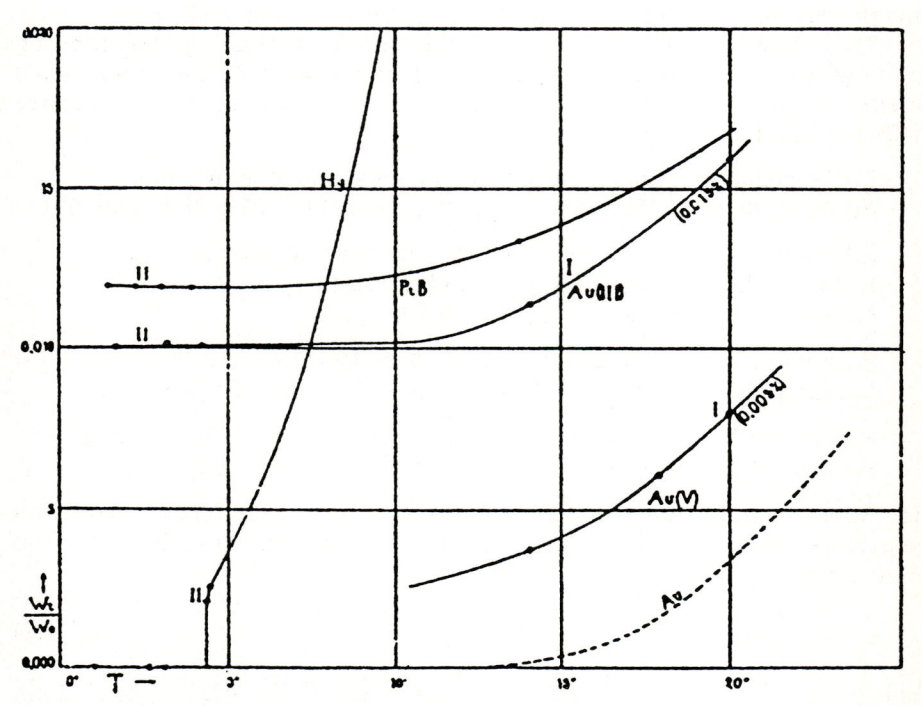

Fig.1. Resistance of various samples of Au and Pt at low temperatures. (Measurements by H.K.Onnes, from [8])

a conductivity formula under the assumption that the mean free path was determined by the amplitude of Einstein oscillators, which yielded an exponential temperature dependence for T → 0.

1.2 Discovery of Superconductivity

Disappearance of Resistance. Since impurities seemed to prevent the resistance from vanishing at T = 0 even in pure Au and Pt, the experiments were extended to solid Hg which had been distilled repeatedly in vacuum in order to reduce the impurity content. During these investigations, H.K.Onnes together with his assistant G.Holst discovered in 1911 that the resistivity of Hg disappeared suddenly at 4.2 K (Fig.2):

> H.K.Onnes: The resistance of pure mercury at helium temperatures. Comm.Leiden 120b (28 April 1911),

> H.K.Onnes: The disappearance of the resistivity of mercury. Comm.Leiden 122b (27 May 1911),

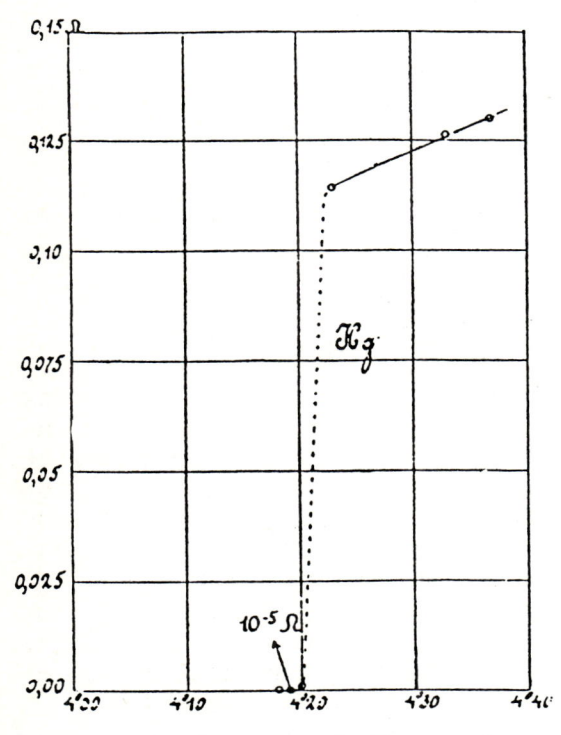

Fig.2. Resistance of Hg at low temperatures showing the transition to superconductivity. (Measurement by H.K.Onnes, from [8])

> H.K.Onnes: On the sudden change in the rate at which the resistance of mercury disappears.
> Comm.Leiden 124c (25 Nov 1911).

In the "supraconductive state", no resistance could be detected, the upper limit being of the order of 10^{-9} of the resistance at the melting point.

Critical Current Density. Further experiments showed that the current density could be increased to a certain threshold value until a voltage occurred. The threshold value was found to increase with decreasing temperature, and current densities of up to 10^3 A/mm^2 were obtained:

> H.K.Onnes: The potential difference necessary for the electric current through mercury below 4.19 K.
> Comm.Leiden 133a (1913).

When superconductivity was found to occur also in Sn and Pb,

> H.K.Onnes: The sudden disappearance of the ordinary resistance of tin, and the superconductive state of lead.
> Comm.Leiden 133d (1913),

the construction of non-resistive superconducting magnets became a practical possibility, and a facility producing fields of up to 10^5 G seemed to be a realistic project.

Critical Magnetic Field. In 1914, it was discovered that at a sharply defined threshold value of the magnetic field the resistance reappeared:

> H.K.Onnes: The appearance of resistance in supraconductors, which are brought into a magnetic field, at a threshold value of the field. Comm.Leiden 139f (1914).

The critical field was found to increase with decreasing temperature. Its relatively low value for the known superconductors put, for the time being, an end to the possibility of superconducting high-field magnets. Later experiments by W.Tuyn showed that the critical field vs temperature curve ends with finite slope at the critical temperature, and, in accordance with Nernst's theorem, with zero slope at $T = 0$.

According to a hypothesis put forward by Silsbee (1916), the threshold values of current and magnetic field are simply related: The magnetic field produced by the critical current at the surface of the superconductor equals the critical field.

Persistent Currents. The persistence of a current induced in a closed loop formed by a superconductor is a striking demonstration of its vanishing resistance, and may serve to push the upper limit of any residual resistance further down. Such an experiment was performed by H.K.Onnes already in 1914:

> H.K.Onnes: The imitation of an Ampere molecular current or of a permanent magnet by means of a supraconductor. Comm.Leiden 140b (1914).

In a closed coil of lead wire with an L/R relaxation time of ca 10^{-5} sec at room temperature, no decrease of the induced current could be observed in the superconductive state for the duration of the experiment (1 hour).

1.3 Meissner-Ochsenfeld Effect

The next important step in the development was the discovery by W.Meissner and R.Ochsenfeld in 1933, that the magnetic flux was expelled from the superconductor, independent of whether the field was applied in the superconductive state ("zero-field-cooled") or already in the normal state ("field-cooled") (Fig.3, 4a,c):

> W.Meissner, R.Ochsenfeld: Ein neuer Effekt bei Eintritt der Supraleitfähigkeit. Naturwiss.21 (1933) 787-788.

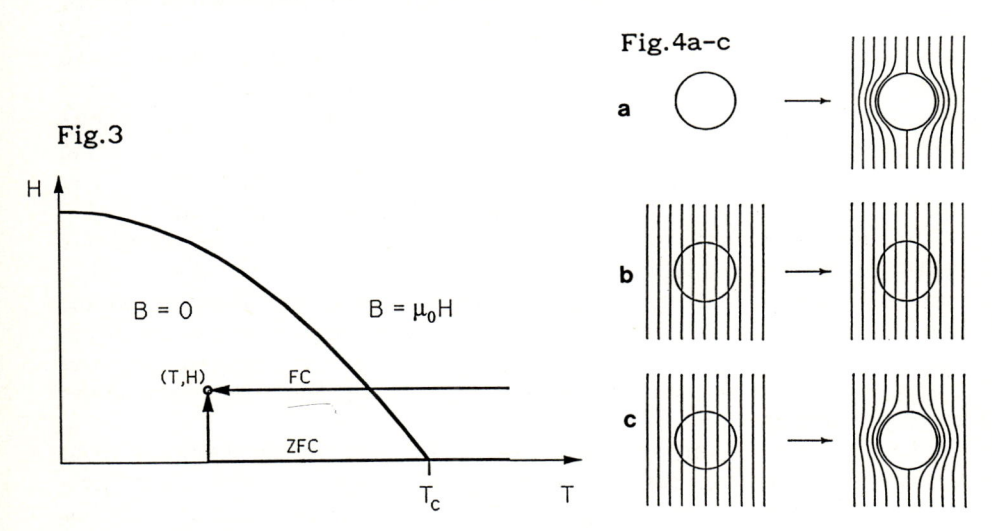

Fig.3. Paths in the (T,H)-plane for zero-field cooled (ZFC) and field-cooled (FC) experiment.

Fig.4a-c. Magnetic field distribution in the superconductive state; a) for ZFC experiment, b) for FC experiment as expected for a perfect conductor, c) for FC experiment as found by Meissner and Ochsenfeld.

A comprehensive presentation of the results of Meissner and coworkers may be found in Ref.[13]. Their results were confirmed in the years 1934-35 by K.Mendelssohn and J.D.Babbitt, G.N.Rjabinin and L.W.Shubnikow, W.J.de Haas and J.M.Casimir-Jonker, and others. In almost all field-cooled experiments, flux expulsion was found to be incomplete: Even in single crystals of pure metals a small amount of flux remained in the sample, and this amount increased with increasing impurity content. Extension of the experiments to superconducting alloys led to the discovery of type II superconductors (see Section 3).

In spite of this non-ideal behaviour, C.J.Gorter suggested already in 1933 that $B = 0$ is a general characteristic of the superconductive state (not endorsed by Meissner, see Ref.[10]), and took this as the basis for the thermodynamic theory (Section 2.1). Any remaining flux is assumed to be confined to enclosed non-superconducting regions.

The behaviour observed by Meissner and Ochsenfeld is significantly different from that which would occur at the transition to a state of perfect conduction, in which the magnetic state depends on history: In the field-cooled case, the magnetic flux present in the normal state would be frozen in (Fig.4b), and this would represent the equilibrium state. In the zero-field-cooled case, on the other hand, a magnetic field switched on in the perfectly conducting state would give rise to induction currents keeping the interior flux-free (Fig.4c), but this would represent a metastable state; consequently, at the transition back to the normal state, the induction currents would die out with Joule dissipation, i.e. this transition would in general be an irreversible process.

In contrast, according to the Meissner effect, the magnetic state of a superconductor is uniquely determined by the applied magnetic field. The flux-free state is the equilibrium state, and the screening currents vanish without any dissipation when the magnetic flux reenters the bulk of the sample at the transition back to the normal state, which thus becomes completely reversible.

Further, a clear distinction should be made between the Meissner screening current keeping the bulk of the superconductor flux-free, which is an equilibrium phenomenon, and the persistent current associated with a non-vanishing flux through a superconducting loop, which represents a metastable state, no matter for how long it persists. The older literature does often not distinguish between these two cases.

1.4 Specific Heat and Latent Heat

W.H.Keesom and coworkers discovered that at the transition to the superconductive state there occurs a jump in the specific heat,

> W.H.Keesom, J.N.van den Ende: Measurements of atomic heats of tin and zinc. Proc.Acad.Sci.Amsterdam 35 (1932) 143-155 (Comm.Leiden 219b),
>
> W.H.Keesom, J.A.Kok: On the change of the specific heat of tin when becoming supraconductive. Proc.Acad.Sci.Amsterdam 35 (1932) 743-748 (Comm. Leiden 221e),
>
> W.H.Keesom, J.A.Kok: Measurements of the Specific Heat of Thallium at Liquid Helium Temperatures. Physica 1 (1934) 175-181 (Comm. Leiden 230c),

(Fig.5), and that the transition in non-zero magnetic field is accompanied by a latent heat:

> W.H.Keesom, J.A.Kok: Measurements of the Latent Heat of Thallium Connected with the Transition, in a Constant External Magnetic Field, from the Supraconductive to the Non-Supraconductive State. Physica 1 (1934) 503-512 (Comm. Leiden 230e),
>
> W.H.Keesom, J.A.Kok: Further Calorimetric Experiments on Thallium. Physica 1 (1934) 595-608 (Comm.Leiden 232a),

(Fig.6). The existence of a latent heat had been demonstrated

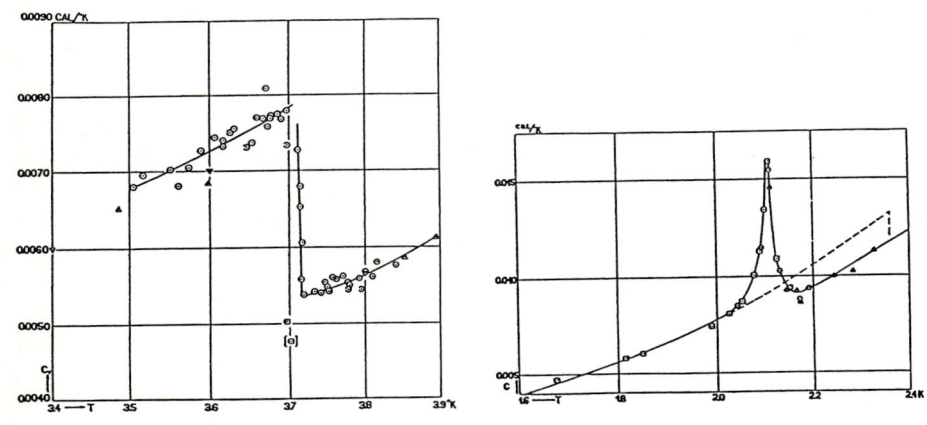

Fig.5 (left). Specific heat of Sn in zero field. (Measurement by W.H.Keesom and J.A.Kok, from [14])

Fig.6 (right). Specific heat of Tl at H = 33.6 Gauss. (Measurement by W.H.Keesom and J.A.Kok, from [14])

shortly before by the observation of a magnetocaloric effect associated with the transition:

> K.Mendelssohn, J.R.Moore: Magneto-Caloric Effect in Supraconducting Tin. Nature 133 (1934) 413.

The results by Keesom and coworkers were found to be in satisfactory agreement with expressions which had been derived by A.J.Rutgers before the discovery of the Meissner effect by applying equilibrium thermodynamics to the transition (see Section 2.1), in spite of the expected irreversibility. Moreover, an experimental test confirmed the hypothesis that the transition from the superconducting to the non-superconducting state is a reversible process. As F.London [1] remarks, this result came very close to predicting the Meissner effect.

2. Thermodynamics of Superconductors

2.1 The Superconductive State as a Thermodynamic Phase

The independence of the superconductive state of previous history and the reversibility of the transition inferred from the Meissner effect paved the way to regarding the superconductive state as a thermodynamic phase. On the basis of this concept, C.J.Gorter and H.Casimir developed a thermodynamic theory of superconductivity:

> C.J.Gorter: Theory of Supraconductivity. Nature 132 (1933) 931,
>
> C.J.Gorter, H.Casimir: On Supraconductivity I. Physica 1 (1934) 306-320.

The idea of a superconducting phase had been suggested previously by P.Langevin (1911) and W.H.Keesom (1924) but without real experimental justification.

The superconducting phase is characterized by a thermodynamic potential $\mathcal{G}_s$ as a function of temperature T and external magnetic field $\mathbf{H}_{ext}$

$$\mathcal{G}_s(T,\mathbf{H}_{ext}) = \mathcal{F}_s + \mathcal{W}, \tag{2.1}$$

where $\mathcal{F}_s$ is the free energy,

$$\mathcal{F}_s = \int f_s(T)dV + \tfrac{1}{2}\int(\mathbf{B}\cdot\mathbf{H} - \mu_o H^2_{ext})dV, \tag{2.2}$$

consisting of an intrinsic part with density $f_s(T)$ and the change in magnetic field energy due to the presence of the superconductor, and W is the interaction energy with the external magnetic circuit,

$$W = -\mu_o \int \mathbf{M} \cdot \mathbf{H}_{ext} dV. \tag{2.3}$$

Here, $\mathbf{B}$ and $\mathbf{H}$ are the magnetic induction and magnetic field in the presence of the superconductor, and $\mathbf{M} = \mu_o^{-1}\mathbf{B} - \mathbf{H}$ is the magnetization due to the Meissner screening current. By using the identity

$$\int (\mathbf{B} \cdot \mathbf{H} - \mu_o H_{ext}^2) dV = \mu_o \int \mathbf{M} \cdot \mathbf{H}_{ext} dV, \tag{2.4}$$

the thermodynamic potential may be written

$$\mathcal{G}_s(T, H_{ext}) = \int f_s(T) dV - \tfrac{1}{2}\mu_o \int \mathbf{M} \cdot \mathbf{H}_{ext} dV. \tag{2.5}$$

The total magnetic moment $\mathcal{M} = \int \mathbf{M} dV$ of the superconductor is given by

$$\mathcal{M} = -\mu_o^{-1} \partial \mathcal{G}_s / \partial \mathbf{H}_{ext} \tag{2.6}$$

(magnetic equation of state).

In the normal state, one may neglect the small diamagnetic contribution and take

$$\mathcal{G}_n(T) = \int f_n(T) dV \tag{2.7}$$

to be independent of H_{ext}.

In the case of a complete Meissner effect, the magnetic fields $\mathbf{B}$ and $\mathbf{H}$ have to be determined by solving the magnetostatic problem with $\mathbf{B} = 0$ in the superconducting phase, $\mathbf{B} = \mu_o \mathbf{H}$ outside, and the usual boundary conditions. Because of the long-range character of the magnetostatic fields, the thermodynamic properties of superconductors become shape-dependent.

2.2 Superconducting Cylinder in Parallel Field

For a superconducting cylinder of arbitrary cross section in a magnetic field H_{ext} (>0) parallel to the axis,

$$B = 0, \quad H = H_{ext}, \quad B = -H_{ext} \text{ in the superconductor}, \tag{2.8}$$

one finds

$$\mathcal{G}_s(T, H_{ext}) = (f_s(T) + \tfrac{1}{2}\mu_o H_{ext}^2) V, \tag{2.9}$$

where V is the superconducting volume.

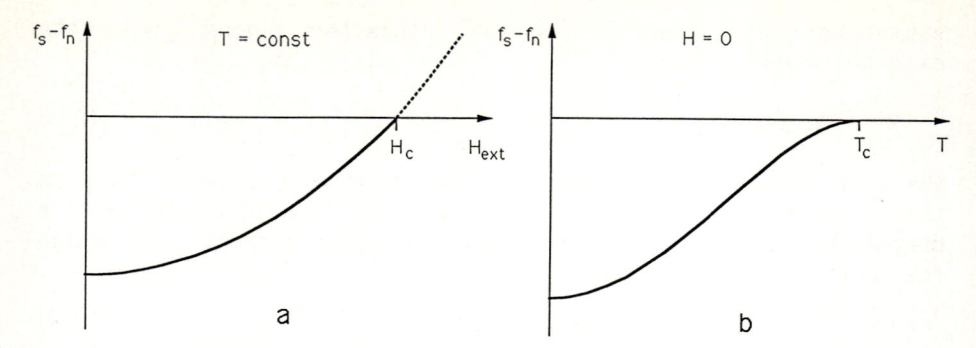

Fig.7. a) Difference of thermodynamic potential for constant temperature as function of magnetic field (Eq.(2.9)), b) difference of free energy densities for zero field as function of temperature (Eq.(2.10)).

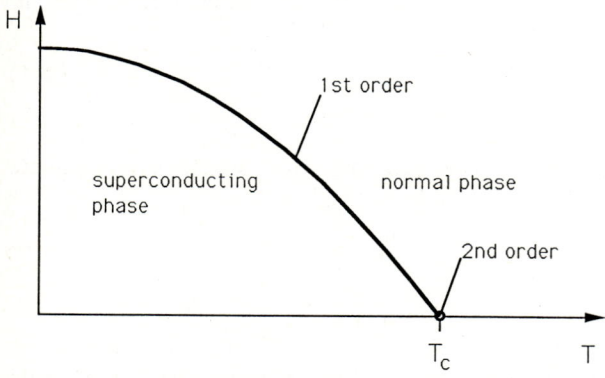

Fig.8. Phase diagram of type I superconductors.

Eq (2.9) shows that superconductivity is destroyed by a magnetic field $H_c(T)$ determined by

$$f_s(T) + \tfrac{1}{2}\mu_0 H_c^2(T) = f_n(T) \qquad (2.10)$$

(see Fig.7a). Measurement of the critical field as function of temperature thus yields directly the free-energy difference between the normal and the superconductive state (Fig.7b).

The phase diagram in the (T,H) plane shows the superconducting phase bounded by the line representing the critical field $H_c(T)$ (Fig.8).

The magnetic equation of state,

$$\begin{aligned} M &= -H_{ext} \quad (H_{ext} < H_c) \\ &= 0 \quad\quad\ \ (H_{ext} > H_c), \end{aligned} \qquad (2.11)$$

Fig.9. Magnetic equation of state of a superconducting cylinder.

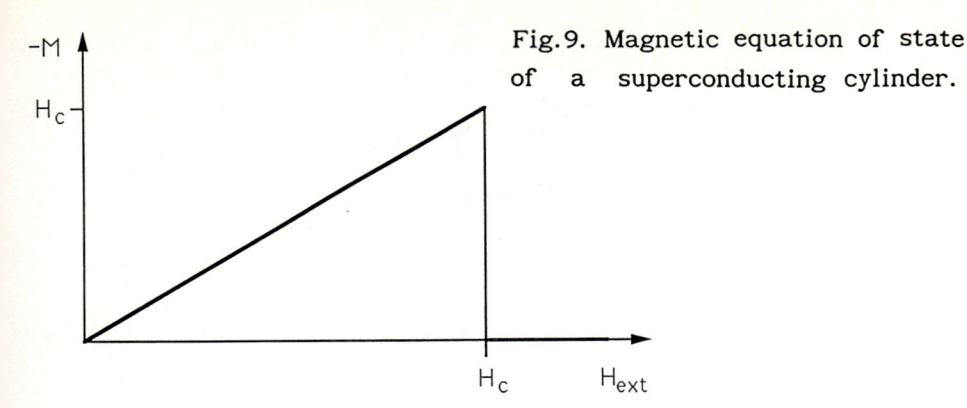

describes the Meissner effect in the superconducting phase (Fig.9).

According to Eqs (2.9,11), the thermodynamic potential of the superconductive phase may be written in the form

$$\mathcal{G}_s(T, H_{ext}) = \mathcal{G}_n(T) - \tfrac{1}{2}\mu_o(H_c^2(T) - H_{ext}^2)V. \tag{2.12}$$

The first and second derivatives with respect to T yield the latent heat

$$r = -\tfrac{1}{2}\mu_o T \partial(H_c^2)/\partial T \tag{2.13}$$

(Clapeyron's equation), and the jump in the specific heat

$$c_s - c_n = \tfrac{1}{2}\mu_o T \partial^2(H_c^2)/\partial T^2, \tag{2.14}$$

respectively. These expressions had already been derived by A.J.Rutgers by applying the first and second law of thermodynamics to the transition, prior to the discovery of the Meissner effect, see

A.J.Rutgers: Note on Supraconductivity.
Physica 1 (1934) 1055-1058,

and were found to be in good agreement with the measurements by W.K.Keesom and coworkers (see Section 1.4).

Since the critical-field curve $H_c(T)$ has a finite slope at T_c, the phase transition in zero magnetic field is of second order; for $H_{ext} \neq 0$, on the other hand, it becomes weakly first order with a small but finite latent heat (Fig.8). Actually, it was the jump in the specific heat of liquid helium and of superconductors which led P.Ehrenfest to the notion of second-order phase transitions.

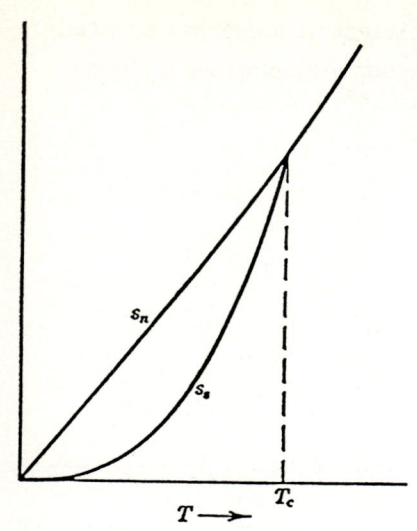

Fig.10. Entropy as a function of temperature in the normal and in the superconducting phase [1].

For T → 0, the entropy of the superconducting phase drops rapidly to zero (Fig.10), indicating the occurrence of some kind of condensation phenomenon.

2.3 The Intermediate State

In the general case, flux expulsion due to the Meissner effect leads to a deformation of the magnetic field outside the superconductor, such that in those regions of the surface where the field is compressed it reaches the critical value H_c when the external field is still below threshold. This gives rise to the appearance of an "intermediate state" consisting of alternating domains of superconducting phase with $B = 0$ and normal phase with tangential field $H = H_c$.

For a sample of ellipsoidal shape in an external field H_{ext} (>0) parallel to one of the axes, the field at the surface assumes its maximum H_{max} at the equator; in the case of complete flux expulsion,

$$H_{max} = H_{ext}/(1-N), \qquad (2.15)$$

where N is the demagnetization factor along the field direction. Thus, H_{max} reaches the critical value H_c for $H_{ext} = (1-N)H_c$.

W.J.de Haas and J.Voogd had observed already in 1931 that superconductivity in tin wires in a field *perpendicular* to the axis was partially destroyed when the field strength exceeded about half the critical value in a *parallel* field:

W.J.de Haas, J.Voogd: The magnetic disturbance of the supraconductivity of single-crystal wires of tin. Proc.Acad.Sci.Amsterdam 34 (1931) 63-69 (Comm.Leiden 212c).

The interpretation of this observation suggested by M.v.Laue,

M.v.Laue: Zur Deutung einiger Versuche über Supraleitung. Phys.Zs.33 (1932) 793-796,

was the deformation of the magnetic field due to the induced surface currents in a perfect conductor.

Gorter and Casimir showed in their 1934 thermodynamics paper that the superconducting state does in fact become unstable at $H_{ext} = (1-N)H_c$. The thermodynamic treatment of the intermediate state by F.London:

F.London: Zur Theorie magnetischer Felder im Supraleiter. Physica 3 (1936) 450-462,

and independently by R.Peierls:

R.Peierls: Magnetic Transition Curves of Supraconductors. Proc.Roy.Soc.(London)A 155 (1936) 613,

yielded the following results:

For $H_{ext} < (1-N)H_c$, the whole sample is in the superconducting state with

$$B = 0, \quad M = -H = -H_{ext}/(1-N) \quad \text{inside the sample} \tag{2.16}$$

and a thermodynamic potential

$$\mathcal{G}_s(T,H_{ext}) = \mathcal{G}_n(T) - \tfrac{1}{2}\mu_o(H_c^2(T) - H_{ext}^2/(1-N))V. \tag{2.17}$$

For values in the interval

$$(1-N)H_c \leq H_{ext} \leq H_c, \tag{2.18}$$

the sample is in the intermediate state characterized by average fields

$$\overline{H} = H_{ext} - N\overline{M} = H_c, \quad \overline{M} = -(H_c - H_{ext})/N,$$
$$\overline{B} = \mu_o(H_{ext} - (1-N)H_c)/N \quad \text{inside the sample,} \tag{2.19}$$

and a thermodynamic potential

$$\mathcal{G}_s(T,H_{ext}) = \mathcal{G}_n(T) - \tfrac{1}{2}\mu_o N^{-1}(H_c(T) - H_{ext})^2 V, \tag{2.20}$$

with volume fractions

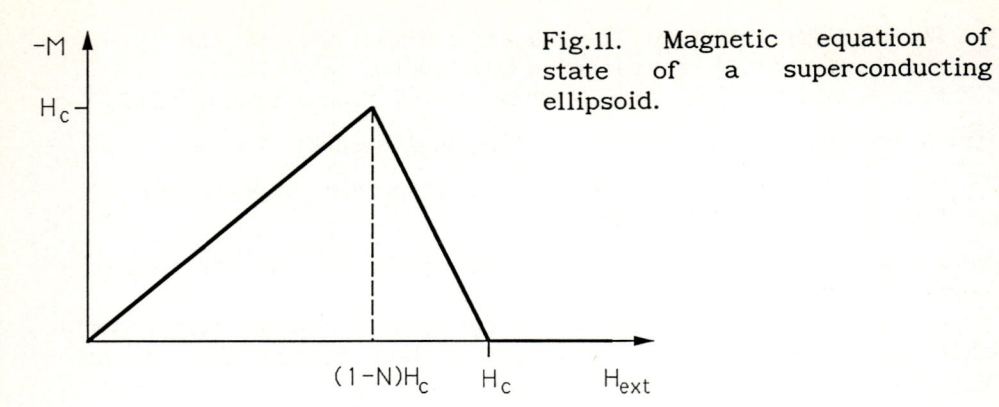

Fig.11. Magnetic equation of state of a superconducting ellipsoid.

$$x_s = |\overline{M}|/H_c, \quad x_n = 1-x_s = \overline{B}/\mu_0 H_c \qquad (2.21)$$

in the superconducting and in the normal state, respectively.

The predicted magnetic equation of state (Eqs.(2.16), (2.19) and Fig.11) was confirmed in several experiments,

W.J.de Haas, O.A.Guinau: On the Transition of a Monocrystalline Tin Sphere from the Supraconductive into the Non-Supraconductive State. Physica 3 (1936) 182-192 (Comm.Leiden 241a),

W.J.de Haas, O.A.Guinau: On the Transition of a Tin Sphere from the Non-Supraconductive State to the Supraconductive State. Physica 3 (1936) 534-542 (Comm.Leiden 241b),

K.Mendelssohn: The Transition Between the Supraconductive and the Normal State. I-Magnetic Induction in Mercury. Proc.Roy.Soc.(London)A 155 (1936) 558-570

D.Shoenberg: The Magnetization Curves of a Supraconducting Sphere and Ring. Proc.Roy.Soc.(London)A 155 (1936) 712-726,

all performed within the same year.

The structure of the intermediate state was studied by tracing the magnetic field distribution at the surface of the superconductor with a bismuth probe (A.Meshkovsky and A.Shalnikov 1947), by powder techniques (A.L.Schawlow el al 1954), and by the Faraday effect in a thin layer of a paramagnetic solution covering the surface (P.B.Alers 1957). Powder-technique experiments with sufficient resolution:

B.M.Balashova, Yu.V.Sharvin: Structure of the Intermediate State of Superconductors. Zh.eksp.teor.Fiz.31 (1956) 40-44,

T.E.Faber: The Intermediate State in Superconducting Plates. Proc.Roy.Soc.(London)A 248 (1958) 460-481,

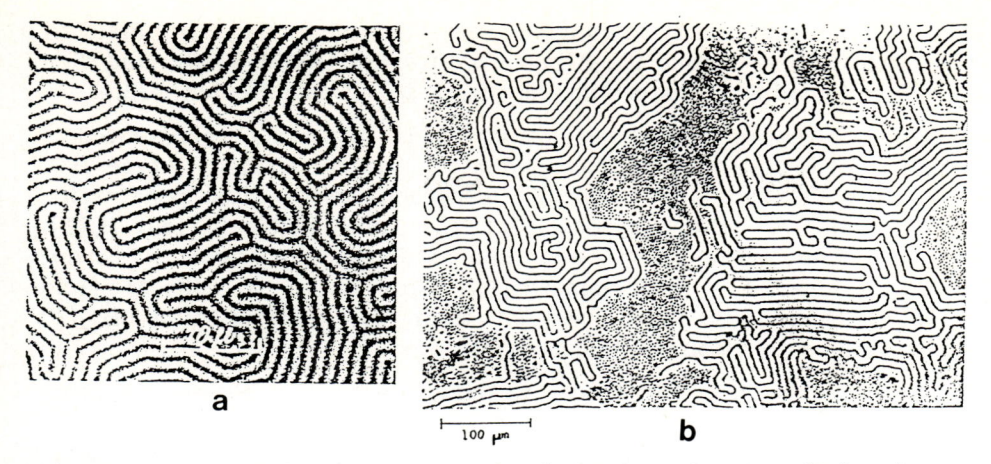

Fig.12a,b. Domain structure of a) barium ferrite [15], b) a single-crystal Pb foil in the intermediate state. (U.Essmann in Ref.[16])

revealed a corrugated structure of the domains, markedly similar to the domain structure of a ferromagnetic plate with the easy axis perpendicular to the plane (Fig.12a,b). This is really not too surprising, because the domain structure is determined in both cases by the balance between magnetostatic energy and wall energy.

A theory developed by L.D.Landau (1937, 1938, 1943) assumed a laminar structure of the intermediate state and is therefore not well applicable to these observations.

Valuable information about magnetic flux structures in superconductors may be found in the summary of an international conference held in 1973 [16].

3. Type II Superconductors

Certain superconducting alloys (in particular Pb-Bi and PbTl$_2$) were found to have very high critical fields, but did not show the large jump in the specific heat predicted by Eq.2.14. This was immediately associated with the observation that magnetic fields penetrated into these materials much below threshold, thus violating the condition for the validity of the theory:

L.W.Schubnikow, W.J.Chotkewitch: Spezifische Wärme von supraleitenden Legierungen.
Phys.Zs.Sowjetunion 6 (1934) 605-607,

T.C.Keeley, K.Mendelssohn, J.R.Moore: Experiments on Supraconductors. Nature 134 (1934) 773-774,

> W.J.de Haas, J.M.Casimir: Penetration of a Magnetic Field into Supra-Conductive Alloys. Nature 135 (1935) 30-31, and Proc.Acad.Sci.Amsterdam 38 (1935) 2-7 (Comm.Leiden 233c).

Closer investigation showed that these materials represented a different type of superconductor (later called "type II"). Whereas in ordinary ("type I") superconductors flux penetration and resistive behaviour occur at the *same* critical field $H_c(T)$, type II superconductors are characterized by *two different* critical fields: Flux penetration sets in at a critical field $H_{c1}(T)$ which is much smaller than the critical field $H_{c2}(T)$ for full restoration of resistivity:

> K.Mendelssohn, J.R.Moore: Specific Heat of a Superconducting Alloy. Proc.Roy.Soc.(London)A 151 (1935) 334-341,
>
> G.N.Rjabinin, L.W.Shubnikow: Magnetic Properties and Critical Currents of Supraconducting Alloys.
> Nature 135 (1935) 581-582, and
> Phys.Zs.Sowjetunion 7 (1935) 122-125.

The notation for the two critical fields (as $H_{k1,2}$) was introduced by L.W.Shubnikow, whose work came to an abrupt end in 1937 [11].

In these experiments, Silsbee's hypothesis was found to be strongly violated. As already observed by W.H.Keesom:

> W.H.Keesom: On the Disturbance of Supraconductivity of an Alloy by an Electric Current. Physica 2 (1935) 35-36 (Comm.Leiden 234f),

the magnetic field produced by the critical current is of the order of H_{c1} rather than H_{c2}.

It was realized only gradually that in type II materials there exist actually two different phases: The "Meissner phase" with complete flux expulsion in the region $0 < H < H_{c1}$, and the "mixed phase" or "Shubnikow phase" [4] with partial flux penetration in the region $H_{c1} < H < H_{c2}$, as shown in the phase diagram, Fig.13. (The names "intermediate state" and "mixed phase" have been chosen in a rather unfortunate way: The *intermediate* state represents a phase *mixture*, the *mixed* phase a genuine new phase *intermediate* between the Meissner phase and the normal state.)

Two different models were proposed in 1935 to explain the behaviour of type II superconductors: K.Mendelssohn and J.R.Moore assumed that the alloy consists of a network or "sponge" of high-H_c material which percolates a low-H_c matrix, such that above the lower threshold field, flux may penetrate through the "meshes" of the sponge:

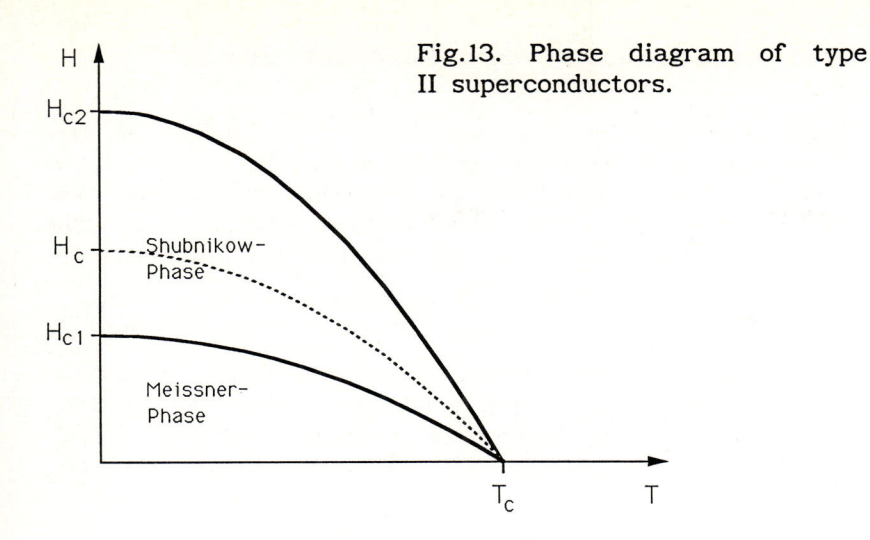

Fig.13. Phase diagram of type II superconductors.

K.Mendelssohn, J.R.Moore: Supraconducting Alloys.
Nature 135 (1935) 826-7

(see also K.Mendelssohn in Ref.[11]).

In contrast to this concept, C.J.Gorter considered the alloy as a *uniform* material characterized by two characteristic lengths, the minimum size δ of superconducting regions and the penetration depth λ of the magnetic field (see Section 4). If $\delta > \lambda$, there is no tendency to form small superconductive regions, and one expects type-I behaviour; if, however, $\delta < \lambda$, there may occur a fine partition into superconducting and normal regions in the form of thin films or filaments, which exists up to fields of the order $(\lambda/\delta)H_c$:

C.J.Gorter: Note on the Superconductivity of Alloys.
Physica 2 (1935) 449-452

It is remarkable that this criterion agrees essentially with the result of Ginzburg-Landau theory, and that the suggested fine partition into filaments almost anticipates Abrikosov's flux-line lattice (Section 5.2).

The question was taken up again by C.J.Gorter 27 years later:

C.J.Gorter: Note on the Superconductivity of Alloys.
Phys.Lett.1 (1962) 69-70.

He remarks that Mendelssohn's sponge model is appropriate for hard and cold-worked alloys, while Abrikosov's model of a flux-line lattice fits to the more homogeneous and soft ones.

In the same year, a quantitative theory of the magnetic behaviour of Mendelssohn's model was developed by C.P.Bean:

C.P.Bean: Magnetization of Hard Superconductors.
Phys.Rev.Lett.8 (1962) 250-253,

which has been found very useful for the discussion of the magnetic behaviour of hard superconductors.

An ideal material in a state with partial flux penetration will show resistive behaviour in an applied electric field due to dissipative flux motion, as first pointed out by C.J.Gorter:

C.J.Gorter: On the Possibility of a Dynamic Variety of the Intermediate Superconductive State. Physica 23 (1957) 45-56

for the case of the intermediate state. In order to obtain non-resistive current flow in the presence of flux penetration, imperfections have to be introduced into the material which prevent flux motion by pinning the domain structure or flux-line lattice.

4. Electrodynamics of Superconductors

4.1 Acceleration Theory

Complete flux exclusion from the superconductor would require the induction of diamagnetic *surface* currents without any spatial extent, which is obviously an unphysical concept.

R.Becker, G.Heller and F.Sauter studied in 1933 a model of a perfect conductor including an acceleration term

$$m\partial j_s/\partial t = n_s e^2 E, \tag{4.1}$$

where j_s is the current density, and n_s is the density of the superconducting electrons with mass m and charge e:

R.Becker, G.Heller, F.Sauter: Über die Stromverteilung in einer supraleitenden Kugel. Z.Physik 85 (1933) 772-787.

They showed that the current induced by switching on a magnetic field in the perfectly conducting state would flow as a *volume* current in a surface layer of thickness λ_L given by

$$\mu_o \lambda_L^2 = m/(n_s e^2). \tag{4.2}$$

i.e. the magnetic field would penetrate the superconductor to a depth λ_L ("penetration depth"), which, depending on the value assumed for n_s, was found to be of the order of $10^{-6} - 10^{-4}$ cm.

The same length had occurred already in 1925 in a note concerned with the different magnetic behaviour of superconductors and of molecular currents of paramagnetic atoms:

G.L.de Haas-Lorentz: Iets over het mechanisme van inductieverschijnselen. Physica 5 (1925) 384-388.

It was found that in a rotating sphere filled with electrons, the kinetic energy of the electrons (which is usually neglected in macroscopic considerations) becomes equal to the magnetic energy for a sphere of radius λ_L.

4.2 London Theory

London Equations. In the above model, one may derive from Faraday's induction law

$$\text{rot}\mathbf{E} = -\partial \mathbf{B}/\partial t \tag{4.3}$$

a conservation law

$$\mu_o \lambda_L^2 \, \text{rot}\mathbf{j} + \mathbf{B} = \mathbf{B}_o(x) \tag{4.4}$$

describing the freezing-in of an arbitrary magnetic flux $\mathbf{B}_o(x)$ in the field-cooled case, in contradiction to the Meissner effect.

In view of the Meissner effect, F. and H.London put forward the concept of the superconductor as a single macroscopic diamagnetic atom, and suggested describing its magnetic behaviour by the equation

$$\Lambda \, \text{rot}\mathbf{j}_s = -\mathbf{B}, \tag{4.5}$$

with $\Lambda > 0$ a new material parameter:

F. and H.London: The Electromagnetic Equations of the Supraconductor. Proc.Roy.Soc.(London)A 149 (1935) 71-88,

F.und H.London: Supraleitung und Diamagnetismus. Physica 2 (1935) 341-354.

In fact, when combined with Maxwell's equations for stationary fields

$$\text{div}\mathbf{B} = 0, \quad \text{rot}\mathbf{B} = \mu_o \mathbf{j}_s, \tag{4.6}$$

the London equation (4.5) yields solutions which decay exponentially in the superconductor,

$$B(z) = B_o \, e^{-z/\lambda_L}, \tag{4.7}$$

(Fig.14), with a decay length λ_L given by

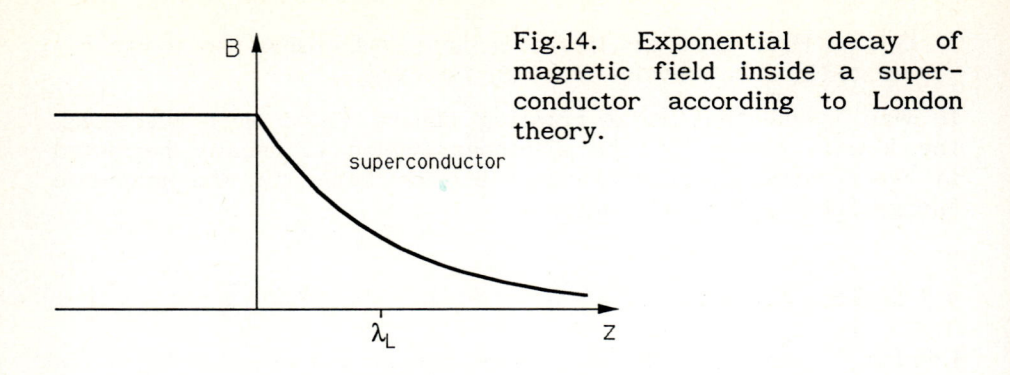

Fig.14. Exponential decay of magnetic field inside a superconductor according to London theory.

$$\mu_o \lambda_L^2 = \Lambda. \tag{4.8}$$

In the Coulomb gauge div**A** = 0, the London equation (4.5) for a singly connected superconductor may be written in terms of the vector potential **A** in the form

$$\Lambda \mathbf{j}_s = -\mathbf{A}. \tag{4.9}$$

A heuristic argument, comparing this equation with the diamagnetic part of the quantum-mechanical expression for the current,

$$\mathbf{j}_{dia} = -(e^2/m) \mathbf{A} |\psi|^2, \tag{4.10}$$

gave for the London constant the expression

$$\Lambda = m/(n_s e^2), \tag{4.11}$$

which yields the same value for the penetration depth λ_L as the acceleration theory (see Eq.(4.2)).

It is highly interesting that the London equation may be understood in terms of a condensation of the superconducting electrons in momentum space, as has been pointed out by F.London [1]. In fact, assuming **p** = m**v** + e**A** = 0 and using $\mathbf{j}_s = en_s \mathbf{v}$ for the superconducting electrons yields directly Eq.(4.9) with Λ given by Eq.(4.11).

Originally, F. and H.London proposed to abandon the acceleration equation (4.1) altogether. However, the London equation (4.5) alone does not give a complete description of the electromagnetic behaviour of superconductors: By combining the London equation (4.5) with Faraday's law (4.3), one obtains an equation of the form

$$\Lambda \partial \mathbf{j}/\partial t - \mathbf{E} = \text{grad } \mu, \tag{4.12}$$

with an arbitrary scalar function μ, which has to be fixed by an additional postulate. A symmetry argument seemed to suggest a relation $\mu = -\Lambda c^2 \rho$ where $\rho = \varepsilon_0 \mathrm{div} \mathbf{E}$ is the charge density. This choice has the interesting consequence that *electric* fields penetrate into superconductors to the same depth λ_L as magnetic fields. A measurement of the resulting capacity change when the plates of the condenser become superconducting gave, however, a negative result:

> H.London: An Experimental Examination of the Electrostatic Behaviour of Supraconductors.
> Proc.Roy.Soc.(London)A 155 (1936) 102-110.

This finally suggested the choice $\mu = 0$, which is equivalent to adopting the acceleration equation (4.1) as the second London equation.

Free Energy, Phase Equilibrium. From a consideration of the electromagnetic energy balance, F. and H.London obtained for the free energy of the superconducting phase an expression of the form

$$\mathcal{F}_s = \int (f_s(T) + \tfrac{1}{2}\Lambda j_s^2)dV + \tfrac{1}{2}\int (\mu_0^{-1}\mathbf{B}^2 - \mu_0 H_{ext}^2)dV, \qquad (4.13)$$

containing a kind of kinetic-energy term $\tfrac{1}{2}\Lambda j_s^2 = \tfrac{1}{2}n_s m \overline{v}^2$, where $\overline{v}$ is the drift velocity of the superconducting electrons. The thermodynamic potential $\mathcal{G}_s$ is obtained as in Eq.(2.1) by adding the interaction energy W with the external magnetic circuit given by Eq.(2.3). The London equation (4.5) may be obtained by substituting $\mathbf{j}$ from Maxwell's equation (4.4) and varying $\mathcal{F}_s$ with respect to $\mathbf{B}$.

The condition for phase equilibrium was derived by H.London from the energy balance for a virtual displacement of the phase boundary between the superconducting and the normal phase:

> H.London: Phase-Equilibrium in Supraconductors in a Magnetic Field. Proc.Roy.Soc.(London)A 152 (1935) 650-663.

Instead of Eq.(2.10), one finds the condition

$$f_s - f_n = -\tfrac{1}{2}\Lambda j_c^2, \qquad (4.14)$$

where j_c is the threshold value of the current density at the phase boundary. For thick superconductors,

$$\lambda_L j_s = H \qquad (4.15)$$

at the boundary of the superconducting phase, such that in this case Eq.(4.14) reduces to the equilibrium condition Eq.(2.10) of the Gorter-Casimir theory.

Size Effects. In thin superconductors with dimensions smaller than the penetration depth, the Meissner effect becomes incomplete, i.e. $\lambda_L j_s < H$ at the surface. Thus, Eq.(4.14) leads to an enhancement of the critical field.
Resistance measurements on thin wires:

> R.B.Pontius: Supraconductors of Small Dimensions. Nature 139 (1937) 1065-1066,

and on thin films:

> A.Shalnikov: Superconducting Thin Films. Nature 142 (1938) 74,

> E.T.S.Appleyard, J.R.Bristow, H.London: Variation of Field Penetration with Temperature in a Superconductor. Nature 143 (1939) 433-434,

> E.T.S.Appleyard, A.D.Misener: Superconductivity of Thin Films of Mercury. Nature 142 (1938) 474,

showed a size dependence of the critical field of the expected form.

A much more direct method is to measure the magnetic susceptibility of small particles, which yields information on both flux penetration and H_c enhancement. Such an experiment was carried out by D.Shoenberg on Hg:

> D.Shoenberg: Superconducting Colloidal Mercury. Nature 143 (1939) 434-435,

> D.Shoenberg: Properties of Superconducting Colloids and Emulsions: Proc.Roy.Soc.(London)A 175 (1940) 49-70.

The results of these investigations are in good qualitative agreement with the London theory. They show that the penetration depth is temperature dependent and diverges at T_c (Fig.15a,b).

Interphase Boundary Energy. A serious defect of the London theory lies in the fact that it gives a *negative* value

$$\sigma_L = -\tfrac{1}{2}\mu_o H_c^2 \lambda_L \qquad (4.16)$$

for the energy of the interphase boundary between the superconducting and the normal phase. Such a negative surface energy would make the superconducting phase unstable, causing a splitting-up into thin superconducting films or fibres separated by normal regions, which would remain stable for fields far above H_c.

In his paper on phase equilibrium, H.London therefore postulated the existence of an additional surface energy $\sigma' > 0$

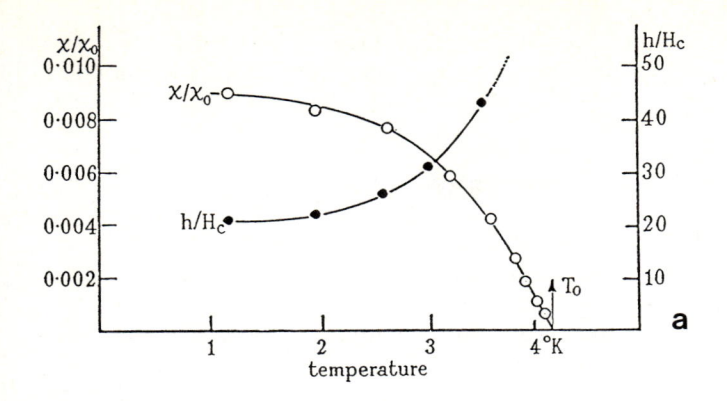

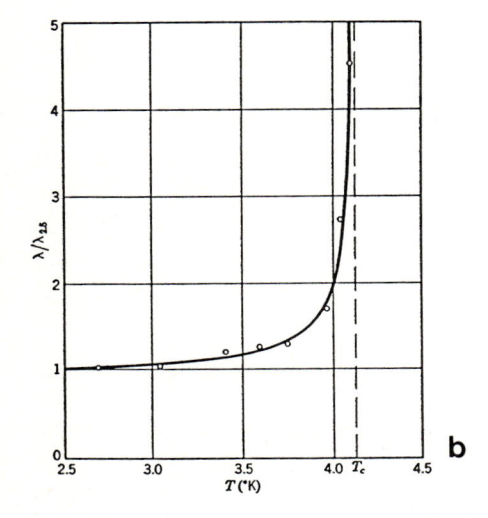

Fig.15a. Temperature variation of magnetic susceptibility χ and critical magnetic field h of colloidal mercury. (Measurements by D.Shoenberg)

Fig.15b. Temperature variation of the penetration depth derived by D.Shoenberg from the results shown in Fig.15a.

of nonmagnetic origin. The superconducting state is stable if $\sigma'+\sigma_L > 0$, but London points out that the opposite case $\sigma'+\sigma_L < 0$ offers the possibility of describing the penetration of magnetic fields in magnetic alloys. This criterion is essentially equivalent to that of Gorter (Section 3), as may be seen by defining through the relation

$$\sigma' = \tfrac{1}{2}\mu_o H_c^2 \, \delta, \tag{4.17}$$

25

a characteristic length δ signifying the thickness of the transition layer between the superconducting and the normal phase.

4.3 Pippard's Non-Local Theory

On the basis of Eqs.(4.8) and (4.11) of London theory, the penetration depth λ_o at zero temperature is expected be a characteristic constant of the metal, insensitive against the addition of small amounts of impurities. Contrary to this expectation, A.B.Pippard found in 1953 a sharp increase of λ_o when the electronic mean free path was reduced to the order of λ_o by doping Sn with small amounts of In:

> A.B.Pippard: An Experimental and Theoretical Study of the Relation between Magnetic Field and Current in a Superconductor. Proc.Roy.Soc.(London)A 216 (1953) 547-568.

He concluded that the local relation (4.9) between supercurrent and vector potential has to be replaced by a *nonlocal* relation, and, led by an analogy with the theory of the anomalous skin effect, proposed a relation of the form

$$\Lambda j_s(x) = -\xi_o^{-1} \int \mathbb{K}(x-x') \cdot A(x') dV, \qquad (4.18)$$

where Λ is London's constant, ξ_o is a new characteristic length of the superconductor of the order of 10^{-4} cm at T = 0, and $\mathbb{K}(x-x')$ is a kernel whose width ξ is assumed to decrease with decreasing mean free path l and to approach ξ_o for $l \to \infty$. The analysis yields an effective penetration depth λ which grows for $\xi \to 0$ as

$$\lambda \sim \lambda_L (\xi_o/\xi)^{1/2}, \qquad (4.19)$$

where λ_L is the London penetration depth defined by Eq.(4.8).

Microscopically, the width ξ is interpreted as the range of coherence of the superconducting wave function. This concept is used to provide a picture of the origin of the interphase boundary energy σ': Since the interphase boundary will be spread over a distance of the order of the coherence range ξ, σ' is given by Eq.(4.17) with $\delta \simeq \xi$. The total boundary energy

$$\sigma_{ns} = \tfrac{1}{2}\mu_o H_c^2 (\delta-\lambda) \qquad (4.20)$$

may become negative for very impure superconductors in which $\xi < \lambda$. Except for the difference between λ and λ_L, this is the same criterion as found previously by C.J.Gorter and by

H.London, and coincides up to a numerical factor with the criterion of Ginzburg-Landau theory.

5. Ginzburg-Landau Theory

5.1 Superconducting Order Parameter

An important step in the phenomenological description of superconductors was the introduction of an *order parameter* distinguishing the superconducting phase from the normal phase by L.D.Landau and V.L.Ginzburg in 1950:

> V.L.Ginzburg, L.D.Landau: On the Theory of Superconductivity. Collected Papers of L.D.Landau, ed. by D.ter Haar. Gordon & Breach 1967, nr 73
> (Zh.Eksp.Teor.Fiz.20 (1950) 1064-1082).

This order parameter was assumed to be a complex field

$$\Psi = |\Psi| e^{i\theta} \tag{5.1}$$

representing a kind of macroscopic wave function, such that the quantity

$$n_s = |\Psi|^2 \tag{5.2}$$

may be interpreted as the density of superconducting carriers of mass m^* and charge e^*, and the phase θ describes the macroscopic coherence properties of the superconductive state.

In terms of this order parameter, L.D.Landau and V.L.Ginzburg constructed an expression for the free energy,

$$\mathcal{F}_s[\Psi,\Psi^*;T,\mathbf{B}] = \int f_s(\Psi,\Psi^*,T)dV + \tfrac{1}{2}\int(\mu_o^{-1}\mathbf{B}^2 - \mu_o H_{ext}^2)dV, \tag{5.3}$$

where, based on arguments of gauge invariance, the intrinsic free-energy density is assumed to have the form

$$f_s = a(T)|\Psi|^2 + \tfrac{1}{2}b(T)|\Psi|^4 + \frac{1}{2m^*}|-i\hbar\nabla\Psi - e^*\mathbf{A}\Psi|^2 \tag{5.4}$$

with $a(T) <(>) 0$ for $T <(>) T_c$ and $b(T) > 0$. The vector potential $\mathbf{A}$ is defined by $\mathbf{B} = \text{rot}\mathbf{A}$ and $\text{div}\mathbf{A} = 0$. Again, the thermodynamic potential $\mathcal{G}_s = \mathcal{F}_s + W$ is obtained as in Eq.(2.1) by adding the interaction energy W with the external magnetic circuit given by Eq.(2.3).

The formulation of this free-energy expression with a gauge-invariant gradient term represents a true masterpiece of intuition. The concept of an order parameter had been introduced in 1937 by L.D.Landau in his general theory of second-order phase transitions. Already in 1934, C.J.Gorter and H.Casimir, in their two-fluid model,

C.J.Gorter, H.Casimir: Zur Thermodynamik des supraleitenden Zustandes. Phys.Zs.35 (1934) 963-966,

had constructed a thermodynamic potential for superconductors (without a gradient term) in terms of what would be called today a real order parameter, indicating the fraction of super-conducting electrons. The important step in the Ginzburg-Landau theory is the realization that gauge invariance requires the introduction of a complex order parameter.

Variation of $\mathcal{F}_s$ with respect to Ψ^* and $\mathbf{A}$ yields the Ginzburg-Landau equations

$$\frac{1}{2m^*}(-i\hbar\nabla - e^*\mathbf{A})^2\Psi + a\Psi + b|\Psi|^2\Psi = 0 \qquad (5.5)$$

for the order parameter Ψ, and

$$\mathbf{j}_s = -i\frac{\hbar e^*}{2m^*}(\Psi^*\nabla\Psi - \Psi\nabla\Psi^*) - \frac{e^{*2}}{m^*}\Psi^*\Psi\,\mathbf{A} \qquad (5.6)$$

for the supercurrent density $\mathbf{j}_s = \mu_0^{-1}\mathrm{rot}\mathbf{B} = -\mu_0^{-1}\nabla^2\mathbf{A}$.

L.P.Gorkov showed in 1959 that the Ginzburg-Landau equations with $e^* = 2e$ and $m^* = 2m$ follow for $|T-T_c| \ll T_c$ approximately from microscopic theory. It should be noted that the value of m^* is of only formal significance, and may be changed arbitrarily by rescaling (many authors take $m^* = m$); but the value of e^* may be related to observable quantities. V.L.Ginzburg [17] reports that Landau, for reasons of gauge invariance, was opposed to considering e^* as an effective charge even when Ginzburg found, shortly before the appearance of BCS theory, that by choosing $e^* \simeq (2-3)\,e$ the agreement between theory and experiment could be significantly improved.

The uniform superconducting phase in zero magnetic field is characterized by an order parameter given by

$$|\Psi|^2 = n_o = -a/b, \qquad (5.7)$$

and a free energy density difference

$$f_n - f_s = \tfrac{1}{2}\mu_0 H_c^2 = a^2/2b, \qquad (5.8)$$

which according to Eq.(2.10) defines a thermodynamic critical field $H_c(T)$.

The theory contains the London theory as a limiting case: For small magnetic fields, Ψ remains approximately constant, and one finds a decay of the field inside the superconductor with a decay length λ given by the London expression

$$\mu_o\lambda^2 = m^*/(n_o e^{*2}). \tag{5.9}$$

A second characteristic length ξ defined by

$$\xi^2 = \hbar^2/(2m^*|a|), \tag{5.10}$$

gives the length scale for variations of Ψ in zero magnetic field, and has subsequently been called coherence length. The three constants H_c, ξ and λ are related by

$$\mu_o H_c = \Phi_o/(2\pi\sqrt{2}\,\xi\lambda), \tag{5.11}$$

where $\Phi_o = 2\pi\hbar/e^*$ is the flux quantum (see Sec.7).

The behaviour of the solutions of the Ginzburg-Landau equation (5.5) depends in an important way on the ratio $\kappa = \lambda/\xi$ of these two lengths (In the original paper, only the penetration depth λ and the dimensionless parameter κ are introduced; the coherence length ξ is not mentioned explicitly.)

The motivation for the development of the Ginzburg-Landau theory, and especially for introducing the gradient term, was the need of a theory yielding a positive interphase boundary energy σ_{ns}. In fact, for $\lambda \ll \xi$, the calculation yielded an interphase boundary with a thickness of the order of ξ and a positive energy

$$\sigma_{ns} = \tfrac{4}{3}\sqrt{2}\,\tfrac{1}{2}\mu_o H_c^2\,\xi \quad (\lambda \ll \xi). \tag{5.12}$$

Landau and Ginzburg noted that for $\lambda \gtrsim \xi/\sqrt{2}$ there occurs an instability of the normal phase with respect to the formation of thin layers of superconducting phase, connected with the fact that $\sigma_{ns} < 0$ for $\lambda > \xi/\sqrt{2}$. However, they did not consider it necessary to investigate the nature of the state occurring in this case, since the experiments available to them seemed to indicate that $\lambda \ll \xi$.

5.2 The Abrikosov Flux-Line Lattice

The question of the behaviour of a superconductor for $\lambda > \xi/\sqrt{2}$ was taken up in 1957 by A.A.Abrikosov:

A.A.Abrikosov: On the Magnetic Properties of Superconductors of the Second Type.
Zh.Eksp.Teor.Fiz.32 (1957) 1442-1452 [JETP 5 (1957) 1174].

He showed that at an upper critical field

$$H_{c2} = \sqrt{2}\,\kappa H_c, \quad \text{i.e.} \quad \mu_o H_{c2} = \Phi_o/(2\pi\xi^2), \tag{5.13}$$

there occurs a second-order phase transition from the normal phase to a flux-line lattice with exactly one flux quantum Φ_o through each lattice cell, and at a lower critical field H_{c1}, which for $\kappa \gg 1$ is given by

$$H_{c1} \simeq H_c \ln(1.08\,\kappa)/(\sqrt{2}\,\kappa) \quad (\kappa \gg 1), \tag{5.14}$$

a transition into the Meissner phase.

The flux-line lattice was first observed ten years later by U.Essmann and H.Träuble:

> U.Essmann, H.Träuble: The Direct Observation of Individual Flux Lines in Type II Superconductors.
> Phys.Lett.24a (1967) 526-527.

Fig.16 shows a more recent observation obtained by scanning tunneling microscopy [18]. Experiment and theory of the vortex structure in superconductors up to 1973 is reviewed in [16].

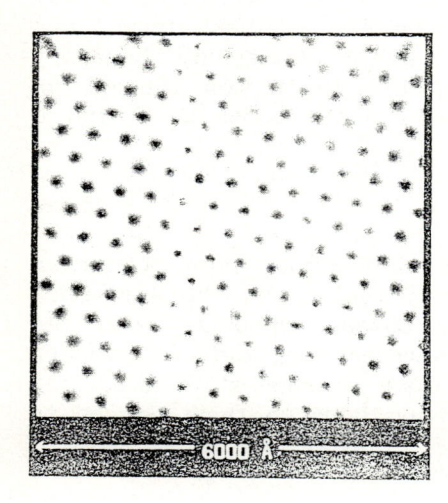

Fig.16. Flux-line lattice in $NbSe_2$, observed by scanning tunneling microscopy [18].

6. The Long Way to a Microscopic Theory

6.1 Earlier Attempts

In the meantime, the electron theory of metals had been formulated in the framework of quantum theory [19]. It was rather clear from the beginning that superconductivity could not be understood as a property of independent electrons, but represented some kind of cooperative phenomenon. However, the actual mechanism was identified only after many unsuccessful attempts. An assessment of the state of the theory in 1935 is vividly reflected in a quotation from F.London in Ref.[9]:

> The present theoretical situation may be characterized in such a way that it is rigorously demonstrable that, on the basis of the recognized conceptions of the electron theory of metals, a theory of supraconductivity is impossible - provided that the phenomenon is interpreted in the usual way.

Electron Lattice. R.de L.Kronig (1932/33) proposed that the Coulomb repulsion leads to the formation of a three-dimensional electron lattice with rather high vibration frequencies which are not excited at low temperatures, a concept which had been introduced already by F.A.Lindemann (1915) and J.J.Thomson (1922). Although the electron lattice (assumed to be commensurate with the ionic lattice) cannot move as a whole, Kronig argued that one-dimensional chains are mobile, and considered this as a possible mechanism for superconductivity. This is an interesting model, which is in fact an early version of charge density wave theory; but it has not been possible to demonstrate that it really shows superconductive behaviour.

Spontaneous Currents. A mechanism which has been suggested repeatedly is the formation of spontaneous currents, in analogy to the formation of ferromagnetic domains. Such a model was presented in 1933 by L.Landau in a paper which was later considered by him as erroneous [17], and has not been included in his Collected Papers. A general theorem due to F.Bloch (see Note ajoutée sur épreuves of Ref.[20]) stating that the ground state of a metal in zero magnetic field is always current-free seemed to rule out any spontaneous currents. I quote again F.London in Ref.[9]:

> So Bloch concluded that the only theorem about supraconductivity which can be proved is that any theory of supraconductivity is refutable, and until now experience has always verified this theorem

to demonstrate the impact of Bloch's theorem. It should be noted, however, that strictly speaking the theorem refers only to the *total* current, and does not exclude the existence of disordered current loops enclosing a large number of atoms, as suggested by W.Meissner (1934).

The concept of spontaneous currents was revived by W.Heisenberg and H.Koppe (1947/48) who argued that the Coulomb repulsion between the electrons leads to an instability of the occupied Fermi sphere with respect to a formation of packets of electrons travelling in the same direction.

Energy Gap. In 1938, H. Welker showed that the assumption of a gap Δ of the order of kT_c in the electronic excitation spectrum of a superconductor would explain the perfect diamagnetism, the behaviour of the specific heat, the temperature dependence of the critical field $H_c(T)$, and to some extent also the infinite conductivity:

>H. Welker: Ueber ein elektronentheoretisches Modell des Supraleiters. Phys. Zs. 39 (1938) 920-925.

The existence of such an energy gap had been suggested already by F. London [9].

Welker pointed out that the magnetic moment of a metal produced by an applied magnetic field consists of two contributions of opposite sign: A large diamagnetic moment due to the induced current which is given by Eq. (4.10), and a large paramagnetic moment which may be understood as originating from reversals of individual orbital moments. Ordinarily, the two contributions almost cancel each other, leaving only the small diamagnetism of normal metals. However, if there exists a gap Δ for electronic excitations, because of the Pauli exclusion principle the reversal of an orbital moment requires an energy $\geq \Delta$, and is therefore exponentially suppressed at low temperatures $kT \leq \Delta$ and sufficiently small fields. Therefore, in this case there remains only the diamagnetic contribution, and one obtains immediately the London equation (4.9). Superconductivity will be destroyed by a magnetic field large enough for the Zeeman energy gained by the reversal of an orbital moment to overcome the gap energy Δ (for this estimate it is important to take the screening of the magnetic field in the superconductor into account).

Thus, the concept of an energy gap for electronic excitations represents an important step in the understanding of the phenomenon of superconductivity.

Subsequently, Welker proposed that such a gap may arise from the magnetic exchange interactions in an electron liquid with strong short-range correlations,

>H. Welker: Supraleitung und magnetische Austauschwechselwirkung. Z. Physik 114 (1939) 525-551,

but this has not been confirmed.

6.2 Experimental Hints

Energy Gap. The first experimental indication of a gap in the electronic excitation spectrum was derived from a measurement of the Thomson effect:

> J.G.Daunt, K.Mendelssohn: An Experiment on the Mechanism of Superconductivity. Proc.Roy.Soc.(London)A 185 (1946) 225-239

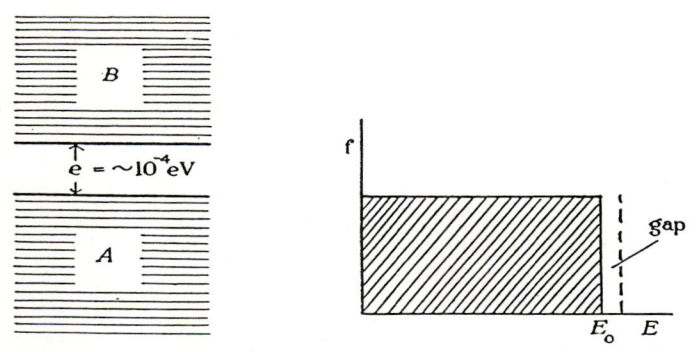

Fig.17. Energy gap in the electronic excitation spectrum of superconducting Pb derived from a measurement of the Thomson effect by J.G.Daunt and K.Mendelssohn.

(Fig.17). The experiment was actually performed already in 1939, but the results could be published only after the war [11]. Further evidence for an energy gap was obtained from a variety of experiments: Thermal conductivity,

> B.B.Goodman: The Thermal Conductivity of Superconducting Tin below 1°K. Proc.Phys.Soc.A66 (1953) 217-227,

specific heat,

> W.S.Corak, B.B.Goodman, C.B.Satterthwaite, A.Wexler: Exponential Temperature Dependence of the Electronic Specific Heat of Superconducting Vanadium. Phys.Rev.96 (1954) 1442-1444,

and infrared transmission,

> M.Tinkham: Energy Gap Interpretation of Experiments on Infrared Transmission through Superconducting Films. Phys.Rev.104 (1956) 845-846.

Ultrasonic attenuation experiments performed shortly after the development of the BCS theory,

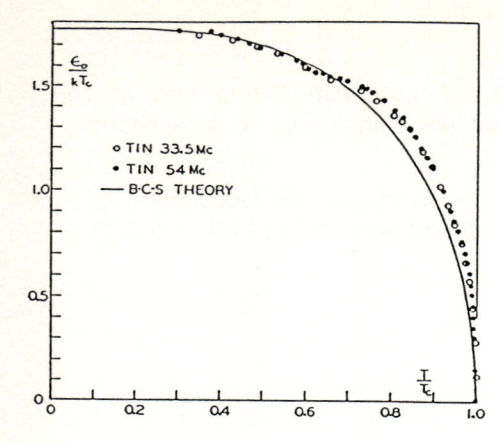

Fig.18. Electronic energy gap of Sn as function of temperature from ultrasonic attenuation experiments, compared to the prediction of BCS theory. (R.W.Morse, H.V.Bohm)

R.W.Morse, H.V.Bohm: Superconducting Energy Gap from Ultrasonic Attenuation Measurements. Phys.Rev.108 (1957) 1094-1096,

yielded a first confirmation of its predictions (Fig.18). The most direct experimental proof of an electronic energy gap comes from tunneling experiments,

I.Giaever: Energy Gap in Superconductors Measured by Electron Tunneling. Phys.Rev.Lett.5 (1960) 147-148,

(Fig.19),

J.Nicol, S.Shapiro, P.H.Smith: Direct Measurement of the Superconducting Energy Gap. Phys.Rev.Lett.5 (1960) 461-464,

I.Giaever: Electron Tunneling Between Two Superconductors. Phys.Rev.Lett.5 (1960) 464-66.

Isotope Effect. The discovery of the isotope effect in 1950:

E. Maxwell: Isotope Effect in the Superconductivity of Mercury. Phys.Rev.78 (1950) 477,

C.A.Reynolds, B.Serin, W.H.Wright, L.B.Nesbitt: Superconductivity of Isotopes of Mercury. Phys.Rev.78 (1950) 487,

indicated that the electron-lattice interaction is involved in the superconductive mechanism, and gave thus an important hint for the theory.

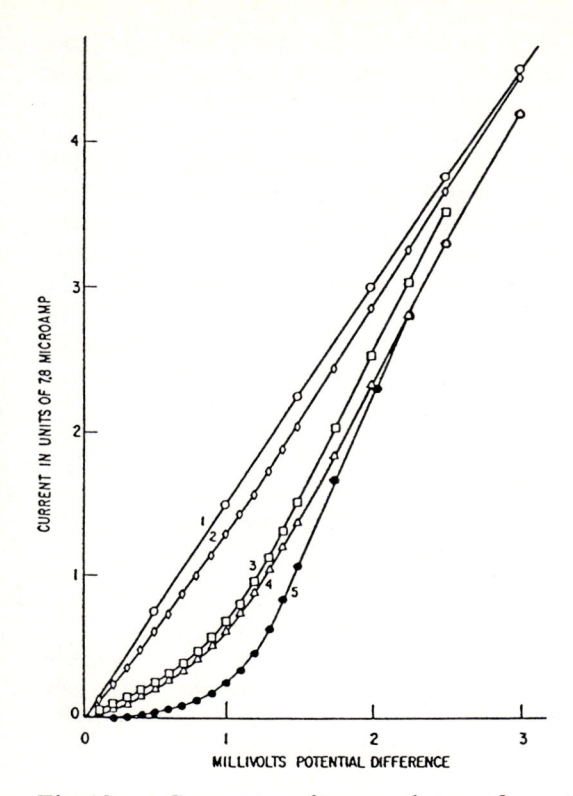

Fig.19. Current-voltage plots of a tunnel junction between Pb and Al at various temperatures showing electronic energy gap. (Measurements by I.Giaever)

6.3 Getting Close

Electron-Lattice Interaction. H.Fröhlich pointed out in 1950 that the same electron-lattice interaction which describes the scattering of conduction electrons by lattice vibrations gives rise to an indirect interaction between the electrons. He proposed this to be the interaction responsible for superconductivity:

> H.Fröhlich: Theory of the Superconducting State. I.
> The Ground State at the Absolute Zero of Temperature.
> Phys.Rev.79 (1950) 845-856.

The theory got strong support from the just discovered isotope effect. But the specific model considered, which was based on the self-interaction of electrons rather than on their mutual interaction, has not turned out to be the correct one. Nonetheless, this paper represents a very important contribution, and has played a decisive role in establishing the correct mechanism.

Bose Condensation of Electron Pairs. M.R.Schafroth (1954) noted that an ideal gas of charged bosons below its condensation point shows superconductive behaviour. He concluded that superconductivity should occur in a metal if the total effective interaction between the electrons, including the indirect interaction via lattice vibrations, gives rise to the formation of resonant states of electron pairs. Chemical equilibrium between these pairs and single electrons was proposed to lead to a Bose condensation:

> M.R.Schafroth: Theory of Superconductivity.
> Phys.Rev.96 (1954) 1442,

> M.R.Schafroth, S.T.Butler, J.M.Blatt: Quasichemical Equilibrium Approach to Superconductivity.
> Helv.Phys.Acta 30 (1957) 93-134.

A similar idea had been suggested in 1952 by V.L.Ginzburg.

Although this mechanism is not applicable to ordinary superconductors because it is incompatible with the large value of their correlation length, it has been suggested that high-temperature superconductors are possible candidates.

6.4 The Final Steps

Cooper Pairs. In 1956, L.N.Cooper discovered that two electrons with an attractive interaction in the presence of a filled Fermi sphere always form a bound pair (and not just a resonance), no matter how small the interaction, with a wave function decaying algebraically ($\propto r^{-2}$) at large distances:

> L.N.Cooper: Bound Electron Pairs in a Degenerate Electron Gas. Phys.Rev.104 (1956) 1189-90.

The binding energy Δ of a pair was found to depend exponentially on the parameter $N_F V$ where N_F is the density of states at the Fermi surface and V is a typical matrix element of the interaction; for $\Delta \simeq kT_c$, the extension of the wave function turned out to be of the order of 10^{-4} cm. The properties of a non-interacting system of such bound pairs were considered as very suggestive of those which could produce a superconducting state.

BCS Theory. Because the size of a Cooper pair is large compared to the average electron distance, the electronic state cannot realistically be described as a gas of Cooper pairs, but involves complicated many-body correlations. In 1957, J.Bardeen, L.N.Cooper and J.R.Schrieffer succeeded in constructing a variational wave function for the superconducting ground state which takes these many-body correlations into account, and which is separated from the band of single-particle excitations by an energy gap:

J.Bardeen, L.N.Cooper, J.R.Schrieffer: Microscopic Theory of Superconductivity. Phys.Rev.108 (1957)162-164,

J.Bardeen, L.N.Cooper, J.R.Schrieffer: Theory of Superconductivity: Phys.Rev.108 (1957) 1175-1204.

Their theory has been found to describe satisfactorily all superconductive phenomena in weak-coupling superconductors.

7. Macroscopic Quantum Phenomena

Perhaps the most fascinating phenomena in the field of superconductivity are those which are caused by the macroscopic quantum coherence of the superconductive state, in particular flux quantization, Josephson and quantum interference effects.

7.1 Flux Quantization

In his contribution to a discussion on superconductivity [9] in 1935, F.London considered a superconducting ring enclosing a magnetic flux Φ. Even if the flux is completely confined to the hole, there exists a vector potential $\mathbf{A}$ in the superconductor, such that

$$\oint \mathbf{A} \cdot d\mathbf{s} = \Phi \tag{7.1}$$

for any closed loop surrounding the hole. This vector potential may be represented as the gradient of a multiply-valued gauge field χ, $\mathbf{A} = \nabla\chi$, increasing by

$$\Delta\chi = \Phi \tag{7.2}$$

for each positive turn around the hole.

The wave function ψ of a carrier of charge e^* in the ring is related to the wave function ψ_0 in the absence of the flux by the gauge transformation

$$\psi = \psi_0 \, e^{-ie^*\chi/\hbar} \,. \tag{7.3}$$

If ψ is required to be single-valued, the increment $\Delta\chi$ and therefore the flux Φ must be an integer multiple

$$\Phi = n\Phi_0 \tag{7.4}$$

of the flux quantum

$$\Phi_0 = 2\pi\hbar/e^* \simeq 4\cdot 10^{-15} \; (e/e^*) \; \text{Vsec.} \tag{7.5}$$

If field penetration into the superconductor is taken into account, it follows from the London equation (4.5) that Φ has to be replaced by the "fluxoid"

$$\tilde{\Phi} = \Phi + \oint \Lambda \mathbf{j}_s \cdot d\mathbf{s}, \tag{7.6}$$

where Λ is London's constant.

London was apparently very cautious in drawing the conclusion expressed in Eqs.(7.4,5). In Ref.[9], which contains the above consideration including Eq.(7.3), no mention is made of flux quantization; even in his book [1], the statement is confined to a footnote.

In 1961, there appeared four consecutive papers in Physical Review Letters, two reporting on the observation of flux quantization:

B.S.Deaver Jr, W.M.Fairbank: Experimental Evidence for Quantized Flux in Superconducting Cylinders.
Phys.Rev.Lett.7 (1961) 43-46 (received June 16),

R.Doll, M.Näbauer: Experimental Proof of Magnetic Flux Quantization in a Superconducting Ring.
Phys.Rev.Lett.7 (1961) 51-52 (received June 19),

with a flux quantum corresponding to $e^* = 2e$ (Fig.20), the other two (N.Byers and C.N.Yang, L.Onsager) as well as three other papers in the same volume (J.Bardeen, J.B.Keller and B.Zumino, W.Brenig) commenting on the theory.

The observed equilibrium states with $\Phi = n\Phi_0$ of a superconducting ring are the minima of a free energy which is a periodic function of the enclosed flux with period Φ_0. Since the free energy in the normal state is independent of the flux, the transition temperature shows the same periodicity:

W.A.Little, R.D.Parks: Observation of Quantum Periodicity in the Transition Temperature of a Superconducting Cylinder.
Phys.Rev.Lett.9 (1962) 9-12.

The observation of a flux quantum Φ_0 with $e^* = 2e$ in all these experiments gives clear evidence for electron pairing in the superconductive state.

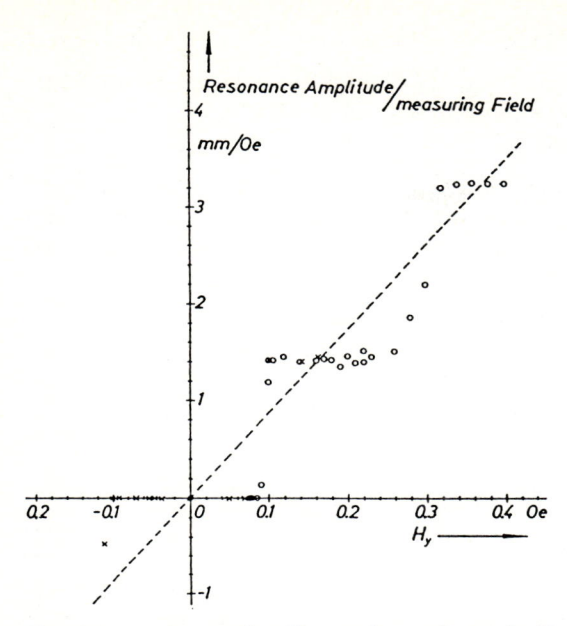

Fig.20. Magnetic flux through a hollow cylinder as function of the applied field showing flux quantization. (Measurements by R.Doll and M.Näbauer)

7.2 Josephson Effects

In 1962, B.D.Josephson pointed out that the tunnelling current between two superconductors contains, in addition to the ordinary single-particle tunnelling terms, previously neglected contributions due to the tunnelling of Cooper pairs:

> B.D.Josephson: Possible New Effects in Superconductive Tunnelling. Phys.Lett. 1 (1962) 251-253.

These terms give rise to very interesting properties: At a voltage V applied across the junction there occurs, in addition to the usual dc current, an ac supercurrent of frequency $2eV/h$ with an amplitude $|I_1|$ proportional to an effective matrix element for pair transfer; at zero applied voltage, there occurs a dc supercurrent of up to a maximum of $|I_1|$. Theoretically, the value of I_1 is equal to the current flowing in the normal state at a voltage of $\pi/2$ times the energy gap.

These predictions were soon experimentally confirmed:

Zero-voltage Josephson current,

> P.W.Anderson, J.M.Rowell: Probable Observation of the Josephson Superconducting Tunneling Effect. Phys.Rev.Lett.10 (1963) 230-232,

and finite-voltage ac Josephson current,

> S.Shapiro: Josephson Currents in Superconducting Tunneling: The Effect of Microwaves and other Observations. Phys.Rev.Lett.11 (1963) 80-82.

The tunnelling of Cooper pairs gives rise to a coupling of the *phases* of the superconducting wave functions on the two sides of the junction, as pointed out by P.W.Anderson, and worked out in more detail by B.D.Josephson:

> B.D.Josephson: Coupled Superconductors. Rev.Mod.Phys.36 (1964) 216-220.

The coupling energy depends on the relative phase φ in the form

$$f = -(\hbar/2e) I_1 \cos\varphi. \tag{7.7}$$

Associated with $f(\varphi)$ is the dc Josephson current

$$I = (2e/\hbar) \partial f/\partial \varphi = I_1 \sin\varphi. \tag{7.8}$$

The ac properties are determined by an equation of motion relating the rate of phase change to the applied voltage V,

$$d\varphi/dt = 2eV/\hbar. \tag{7.9}$$

In the presence of a vector potential **A** in the junction, the relative phase φ must be augmented by an integral

$$\eta = -(2e/\hbar) \int_1^2 \mathbf{A}\cdot d\mathbf{s} \tag{7.10}$$

along a line joining the two superconductors, in order to achieve gauge invariance. In narrow junctions, the magnetic field is nearly constant, and variations of the phase along the barrier may be neglected.

The term (7.10) gives rise to diffraction-like phenomena in a magnetic field: The maximal dc Josephson current depends on the flux Φ_B contained in the junction in the form of a Fraunhofer diffraction formula,

$$I = I_1 \left|\frac{\sin(\pi\Phi_B/\Phi_o)}{\pi\Phi_B/\Phi_o}\right| \sin\varphi, \tag{7.11}$$

with $\Phi_o = 2\pi\hbar/2e$. The observation of such a pattern,

> J.M.Rowell: Magnetic Field Dependence of the Josephson Tunnel Current. Phys.Rev.Lett.11 (1963) 200-202,

gave clear evidence that the observed supercurrent was in fact due to the Josephson effect, and not to superconducting leaks.

7.3 Macroscopic Quantum Interference

A striking demonstration of macroscopic phase coherence are the quantum interference phenomena occurring in multiply connected superconductors containing two Josephson junctions in parallel and enclosing a magnetic flux. In such circuits, there occurs, in addition to the diffraction due to the flux Φ_B contained within the junctions, an interference due to the total enclosed flux Φ_T:

R.C.Jaklevic, J.Lambe, A.H.Silver, J.E.Mercereau:
Quantum Interference Effects in Josephson Tunnelling.
Phys.Rev.Lett.12 (1964) 159-160,

R.C.Jaklevic, J.Lambe, J.E.Mercereau, A.H.Silver:
Macroscopic Quantum Interference in Superconductors.
Phys.Rev.140 (1965) A1628-1637.

The theory yields for the total Josephson current through two identical junctions in parallel the expression

$$I = I_1 \left| \frac{\sin(\pi\Phi_B/\Phi_0)}{\pi\Phi_B/\Phi_0} \right| \sin(\varphi - \pi\Phi_T/\Phi_0). \tag{7.12}$$

In the experiment, the fluxes Φ_T and Φ_B were produced by a uniform magnetic field. The observation shows clearly the interference pattern due to Φ_T and the diffraction envelope due to Φ_B (Fig.21).

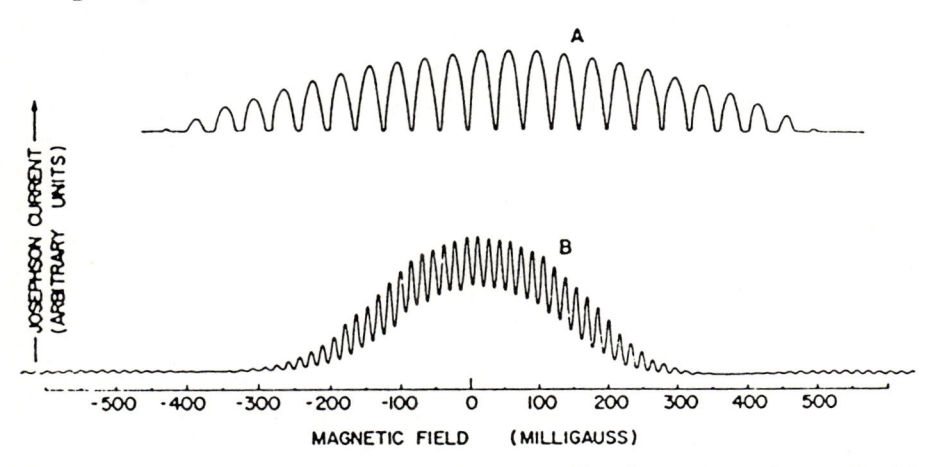

Fig.21. Maximum Josephson current flowing through a doubly connected circuit with two Josephson junctions in parallel as function of the applied magnetic field, showing interference and diffraction effects (two different samples). (Measurements by R.C.Jaklevic, J.Lambe, J.E.Mercereau, and A.H.Silver)

The interference effect depends only on the existence of an enclosed flux; it is not necessary that the superconductor itself actually experiences a magnetic field. This prediction was confirmed in an experiment where the magnetic flux Φ_T was produced by a small solenoid embedded in the space between two superconducting films connected at both ends by Josephson contacts (Aharonov-Bohm experiment):

R.C.Jaklevic, J.Lambe, A.H.Silver, J.E.Mercereau:
Quantum Interference from a Static Vector Potential in a Field-Free Region. Phys.Rev.Lett.12 (1964) 159-160.

The observed phase shift may be understood as arising from the electric field induced during the change of the flux.

8. Materials

So far, I have only discussed *phenomena* in this lecture. However, the preparation of *materials* is at least as important, both for research and for applications.

Early work on superconductive materials was carried out in a number of low-temperature laboratories: In Berlin, W.Meissner (called by C.J.Gorter "the Matthias of the late twenties and early thirties" [10]) discovered superconductivity in several transition metals, compounds and alloys. Many alloys were investigated by J.C.McLennan and coworkers in Toronto and by E.van Aubel, W.J.de Haas and J.Voogd in Leiden. The work on alloys by L.W.Shubnikow in Charkov, K.Mendelssohn in Oxford and W.J.de Haas in Leiden led to the discovery of type II superconductivity as described in Section 3. It is remarkable that already in 1941, E.Justi and coworkers in Berlin reached a critical temperature of 16 K with the compound NbN (although their original claim of persistence of superconductivity to temperatures well above 20 K could not be substantiated).

In a more recent period, B.T.Matthias with remarkable intuition and knowledge of materials found a large number of new superconducting intermetallic compounds, and proposed empirical rules for their occurrence. The discovery of Nb_3Sn in 1956 with T_c = 18.1 K and high critical field and current values was the starting point for the development of superconducting magnets. However, the progress towards higher transition temperatures was slow: The record holder in this class of materials is Nb_3Ge with T_c = 23.3 K, and a T_c of ~30 K was speculated to represent an upper limit.

The breakthrough to considerably higher transition temperatures was achieved - as you well know - by G.Bednorz and K.A.Müller in 1986 with superconducting oxides. The history of superconductivity research in oxides is reviewed by K.A.Müller [21]. The anticipation that oxide materials, so distinct from ordinary metals, might be promising candidates for high-T_c superconductivity proves an extraordinary amount of intuitive insight into the electronic peculiarities of this class of materials. In addition, the realization of such an unconventional research project certainly required a good portion of courage.

References

1. F.London: Superfluids, Vol 1. John Wiley & Sons 1950
2. E.A.Lynton: Superconductivity. Methuen 1964 (B.I-Hochschultaschenbücher 74)
3. M.Tinkham: Superconductivity. Gordon & Breach 1965
4. P.G.de Gennes: Superconductivity of Metals and Alloys. Benjamin, N.Y. 1966
5. A.Rose-Innes, E.Rhoderick: Introduction to Superconductivity. Pergamon Press 1969
6. R.D.Parks: Superconductivity, Vol.1 and 2. Marcel Dekker Inc. New York 1969
7. W.Buckel: Supraleitung. Physik Verlag 1984
8. R.de Bruyn Ouboter: Superconductivity: Discoveries During the Early Years of Low Temperature Research at Leiden 1908 - 1914. IEEE Transactions on Magnetics MAG-23 (1987) 355-370
9. A Discussion on Supraconductivity and other Low Temperature Phenomena. Proc.Roy.Soc.(London)A 152 (1935) 1-46 (Opening adress by J.C.McLennan, further contributions by D.Shoenberg, W.H.Keesom, W.Meissner, R.de L.Kronig, L.Brillouin, N.Kürti and F.Simon, R.Peierls, F.London, K.Mendelssohn, J.D.Bernal, N.F.Mott, M.Blackman)
10. C.J.Gorter: Superconductivity until 1940 In Leiden and As Seen From There. Rev.Mod.Phys.36 (1964) 3-7
11. K.Mendelssohn: Prewar Work on Superconductivity as Seen from Oxford. Rev.Mod.Phys.36 (1964) 7-11
12. H.B.G.Casimir: Superconductivity. In: Proceedings of the International School of Physics Enrico Fermi Course LVII "History of Twentieth Century Physics". Ed. by C.Weiner, Academic Press 1977; p. 170-181
13. W.Meissner, Fr.Heidenreich (Teilweise nach Messungen von W.Meissner und R.Ochsenfeld.): Über die Änderung der Stromverteilung und der magnetischen Induktion beim Eintritt der Supraleitfähigkeit. Phys.Zs.37 (1936) 449-470
14. W.H.Keesom: Das kalorische Verhalten von Metallen bei den tiefsten Temperaturen. Phys.Zs.35 (1934) 939-944

15. C. Kooy, U. Enz: Experimental and Theoretical Study of the Domain Configuration in Thin Layers of $BaFe_{12}O_{19}$.
Philips Res.Repts.15 (1960) 7-29
16. R.P.Huebener, J.R.Clem: Magnetic Flux Structure in Superconductors - A Conference Summary.
Rev.Mod.Phys.46 (1974) 409-422
17. V.L.Ginzburg: Landau's Attitude Towards Physics and Physicists. Physics Today May 1989
18. H.F.Hess, R.B.Robinson, R.C.Dynes, J.M.Valles,Jr., J.V.Waszczak: Scanning-Tunneling-Microscope Observation of the Abrikosov Flux Line Lattice and the Density of States near and inside a Fluxoid. Phys.Rev.Lett. 62 (1989) 214-216.
19. A.Sommerfeld, H.Bethe: Elektronentheorie der Metalle. In: Handbuch der Physik, 2.Auflage, herausgegeben von H.Geiger und K.Scheel, Band XXIV/2. Springer Berlin 1933; p.333-622
20. L.Brillouin: Le champ self-consistent, pour des électrons liés; la supraconductibilité. J.Phys.Radium 4 (1933) 333-361
21. K.A.Müller: The Development of Superconductivity Research in Oxides. In: Proceedings of the Second International Conference on Superconductivity (ISS '89), Tsukuba, Japan, November 14-17 (1989).

Electronic Structures of Oxide Superconductors – Development of Concepts

A. Kitazawa

Department of Industrial Chemistry, University of Tokyo,
Hongo, Bunkyo-ku, Tokyo 113, Japan

Abstract. An understanding of the basic electronic structures of the high T_c superconducting oxides has been developed based on various observations of the superconducting compositional regions as well as of the adjacent semiconducting and metallic regions [1, 2]. This review mainly emphasizes that high T_c superconductivity manifests itself in the vicinity of the semiconductor-to-metal transition, which is triggered by high-level carrier doping in a semiconductor with a nonconventional energy gap at the half-filled position. The energy gap involved has its origin in multi-body interactions: charge density waves (BaBiO$_3$) or electronic correlation (cuprates). The Fermi surface state in the cuprates is mainly composed of O$_{2p}$ and, to a lesser degree, Cu$_{3d}$ orbitals in the CuO$_2$ layer, but it seems to be strongly influenced by the presence of the local Cu spins in the background, which make the various normal-state properties of the cuprates complex. The change in the electronic structure is symmetrical on doping in the hole- and electron-doped cuprate superconductors.

1. Introduction

In order to clarify the mechanism of the recently discovered high T_c superconductivity (HTSC) in the cuprates [3, 4] and bismuthates [5, 6, 7], it is essential to understand their unique electronic structures in the normal state because the superconductivity is obtained when the electronic structure is slightly modified through weak interactions on the energy scale $k_B T_c$. The normal-state electronic structure, determined by stronger interactions, is essentially preserved in the superconducting state.

It has been observed that the superconductivity in BaPb$_{1-x}$Bi$_x$O$_3$ (BPBO, T_c = 12 K) occurs over a narrow compositional range near $x = 0.25$ where the energy gap due to the formation of the charge density wave in BaBiO$_3$ is destroyed by doping of Pb to result in a metallic state [8]. The recently discovered higher T_c superconductor Ba$_{1-x}$K$_x$BiO$_3$ [6] appears to be able to be understood along the same lines [9].

On the other hand, cuprate superconductors are obtained by doping the divalent Cu compounds, e.g. La$_2$CuO$_4$, which is a semiconductor because of the opening of an energy gap at the half-filled position in the conduction band due to the strong electronic correlation. In this system also it has been observed that HTSC occurs in the narrow compositional range near the semiconductor-to-metal transition. Although the origin of the semiconductivity is different in the bismuthates and cuprates, this similarity seems to be significant. Furthermore, the energy gaps in the two groups of materials have their origins in nonconventional mechanisms which have not been encountered in typical semiconductors.

This review first describes the nonconventional gap formation mechanisms and then the processes by which the gap is destroyed on doping to result in the metallic state. Superconductivity is seen in this transition range.

2. Charge Density Waves and Bismuthate Superconductors

2.1 BaBiO₃ as a CDW Insulator. $BaBiO_3$, a monoclinic perovskite, is a semiconductor. Electrical measurements indicate an energy gap of 0.2 eV [5], while optical measurements indicate a gap as large as 2 eV. The major interest here is the fact that this is a semiconductor, contrary to expectation. It is expected to be a metal with the conduction band half-filled because Bi^{4+} has one electron in the valence state and hence the Bi_{6s} and O_{2p} orbitals would give a half-filled conduction band in the hybridization.

The opening up of the energy gap at the half-filled position was explained by the formation of the charge density wave (CDW) according to band calculations [10], neutron diffraction [11], infrared [12] and Raman measurements [13]. CDWs had been known to occur in low-dimensional metals such as polyacetylene (1D) and $NbSe_2$ (2D), because the Fermi surface can readily be influenced by the charge modulation with a certain wavelength. $BaBiO_3$ turned out to be the first CDW insulator of 3D electronic structure, reflecting the unique feature of the perovskite structure which enables the 3D nesting of the Fermi surface [8, 10].

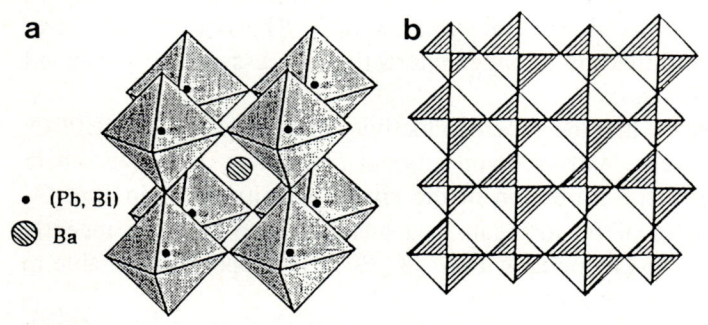

Fig. 1. Octahedron network of perovskite (a) and schematic alternate expansion and contraction of the octahedra – breathing mode deformation (b). The same applies perpendicular to the sheet

The "breathing mode phonon" of the perovskite is schematically shown in Fig. 1, which involves alternate contraction and expansion of the oxygen coordination octahedra and makes the unit cell double, resulting in the opening of the CDW gap of about 2 eV [8].

2.2 Doping of Pb into BaBiO$_3$: BaBi$_{1-x}$Pb$_x$O$_3$. The doping into BaBiO$_3$ of Pb on Bi sites is thought to create holes in the lower of the two split bands. But the activation energy of resistivity remains at 0.2 eV for $x < 0.4$ [14] and the peak position in the optical conductivity spectrum corresponding to the CDW gap remains unchanged up to $x = 0.2$ and then shows only a gradual decrease, with an indication of pseudo-gap structure surviving even in the superconducting compositional range ($x > 0.65$) [15].

This behavior is strange if the origin of CDWs, driven by Fermi-surface instability, is taken into account. A low-level hole doping would cause lowering of E_F and readily destroy the CDW. The persistent semiconducting structure could then be explained by assuming an extreme local nature of the CDW unique to this system [8]. This local CDW seems to be sustained in BaBi$_{1-x}$Pb$_x$O$_3$ due to the special electronic configuration of Bi ions: i.e., the bistable Bi^{3+}-Bi^{5+} states, and the difference in the site energy between Bi and Pb, about 0.3 eV.

The density of states model [16] is illustrated in Fig. 2. On doping, we obtain the Pb states as the mid gap acceptor levels [14] while the two peaks created by the CDW are essentially preserved unchanged [15]. The Pb states hence create the growing tail structure of the main peak in Fig. 2. The gap finally is embedded in the continuous distribution of the doped Pb levels at about $x = 0.6$. But the pseudo-gap structure survives even in the superconduct-

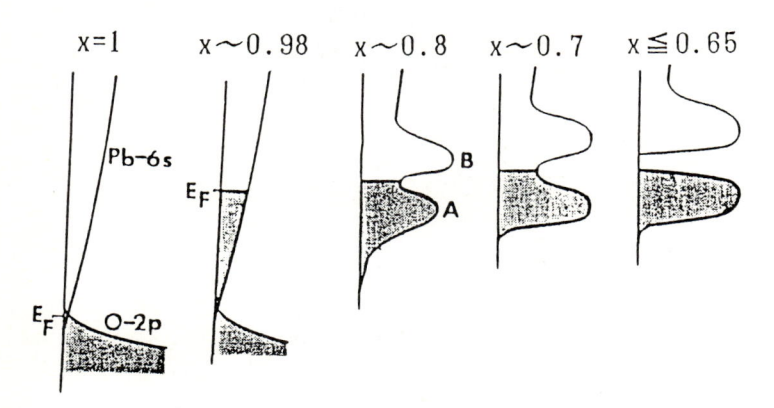

Fig. 2. A density of states model of BaBi$_{1-x}$Pb$_x$O$_3$ for different doping levels. The two peaks for $x < 0.65$ are due to CDWs. The pseudo-gap structure is retained in the superconducting region $x > 0.65$. BaPbO$_3$ ($x = 1$) is a semi-metal because of the overlapping of Pb$_{6s}$ and O$_{2p}$ bands

ing region, $x > 0.65$, with the highest T_c composition $x = 0.75$ ($T_c = 12\,\mathrm{K}$) [15, 16]. The gap energy starts decreasing from $x = 0.3$ and disappears completely at $x = 0.15$, resulting in typical metallic behavior. The superconducting T_c increases from $x = 0.65$ but decreases towards the typical metallic region beyond $x = 0.75$.

It is interesting to note that the Hall coefficient changes its sign from positive to negative as the Pb doping increases. The difference in the excitation energy observed optically (2 eV) and electrically (0.2 eV) has been discussed in terms of one- and two-particle excitation in the CDW-induced bipolaronic lattice [8].

2.3 Doping of K into BaBiO$_3$: Ba$_{1-x}$K$_x$BiO$_3$. It was often pointed out that the localized nature of the carriers in BaBi$_{1-x}$Pb$_x$O$_3$ would cause the

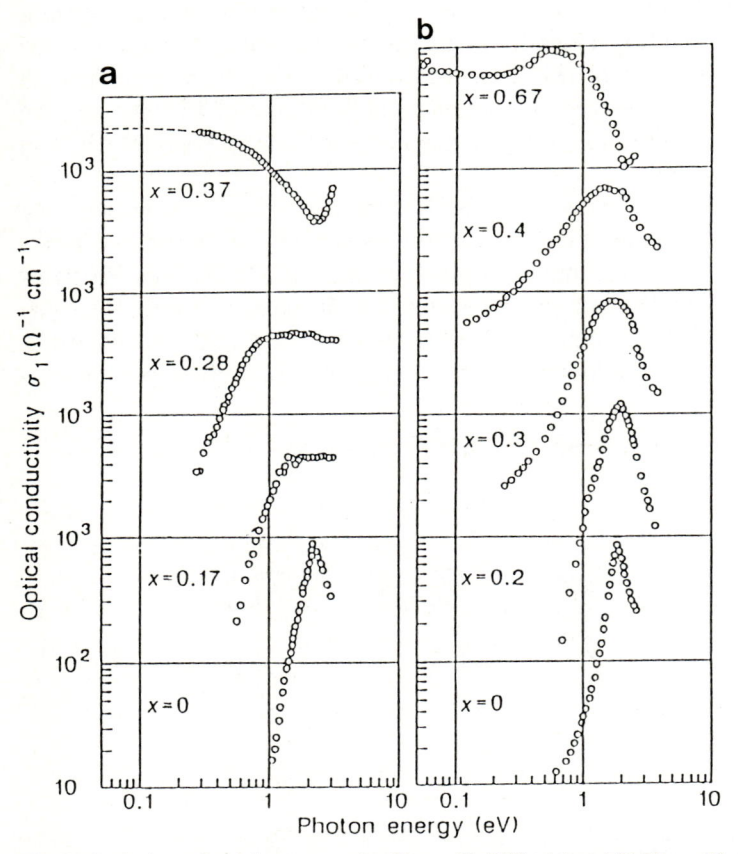

Fig. 3. Optical conductivity spectra for Ba$_{1-x}$K$_x$BiO$_3$ (a) and BaBi$_{1-x}$Pb$_x$O$_3$ (b). The peak at 2 eV is the excitation across the CDW gap. In BKBO, the CDW gap narrows with doping and disappears at $x \simeq 0.4$, while in BPBO the gap is essentially retained but mid gap states grow

suppression of T_c due to their reduced screening efficiency. Therefore doping on Ba sites which were not directly located on the current path was attempted. Due to the difficulty of the synthesis, it was only after the discovery of the cuprate superconductors that $Ba_{1-x}K_xBiO_3$ was found to be a superconductor with T_c up to 30 K at $x = 0.4$ [6, 7].

Substitution of K on Ba sites is expected to introduce holes. It is indeed p-type at the beginning according to Hall measurements, but a crossover to n-type takes place as the doping level is raised, resulting in the superconducting region [9]. Figure 3 shows that the optical spectra of BKBO and BPBO are quite similar [9].

2.4 Superconductivity of Bismuthates. Since a strong electron–phonon interaction is one of the essential factors for inducing CDW instability, one may presume the strong interaction to be preserved in the metallic region in the vicinity of the S-M transition. A large isotope effect has been observed in T_c in both BPBO [17] and BKBO [18], indicating phonon-mediated superconductivity. Another important feature is the relatively high frequency of the breathing mode phonon because of the light mass of oxygen ions, which may contribute to raising the T_c. Tunneling spectroscopy has indicated the contribution of relatively high-energy phonons in both BPBO [19] and BKBO [20].

Therefore, we may summarize the electronic structure of the bismuthates by saying that the CDW gap is destroyed by doping of acceptors in $BaBiO_3$ and the strong electron–phonon interaction in the high phonon frequency range is preserved in the vicinity of the S-M transition where the high T_c manifests itself.

3. Electronic Structure in Cuprate Superconductors

3.1 Covalent Bonding in CuO_2 Plane and Two-Dimensional Current Path. A remarkable structural feature commonly encountered in the cuprate superconductors is the presence of the two-dimensional network of Cu-O bonds or the CuO_2 square lattice. Because the typical ionic bond length of Cu-O is 0.21 nm, the significantly shortened Cu-O inter-atomic spacing, about 0.19 nm, indicates the enhanced contribution from covalent bonding. Due to the square-planar symmetry around the Cu atom, the degeneracy of the $3d$ orbitals is lifted as shown in Fig. 4. Therefore, the most likely covalent bond is obtained from the filling of electrons in the bonding orbital of $Cu_{3d(x^2-y^2)}$-O_{2p} and the half-filling of the anti-bonding orbital, σ_{dp}^*. A conceptual σ_{dp}^* bond network expected to give the conduction band is illustrated in Fig. 5.

Since the Cu-O spacing along the z-axis is as great as 0.22 nm, covalent bonding along this direction should be insignificant. The electrical conduction

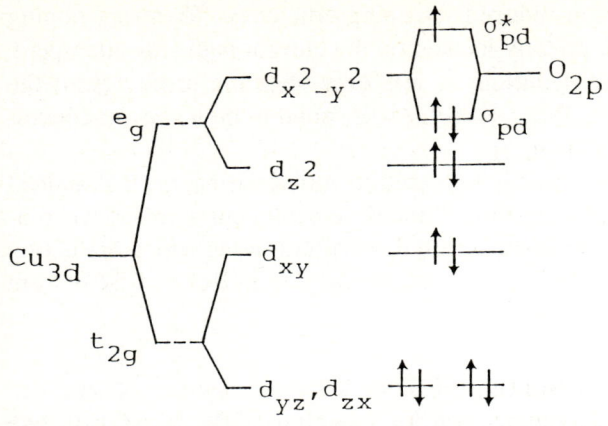

Fig. 4. Splitting of $3d$ orbital energy in the square coordinated environment. Since Cu(II) has nine $3d$ electrons, a hole is created at the $d(x^2 - y^2)$ level. This forms a σ bond with the oxygen $2p$ orbital extended toward the Cu atom, leaving a hole in the anti-bonding $\sigma^*(dp)$ orbital

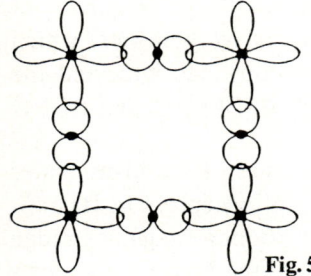

Fig. 5. A schematic network of $\sigma(dp)$ bonds in the basal CuO_2 plane

in these cuprates therefore is expected to be essentially two-dimensional, which might be modeled by assuming a pile of alternating ultra-thin metallic and insulating layers. Whether the electron transfer in the c-direction is metallic or semiconductive has been a matter of controversy [21]. But recent results on a large high-quality single crystal of $(La, Sr)_2CuO_4$ (several millimeters along the c-axis) have clearly indicated that the conduction is semiconductive along the c-axis below the tetragonal-to-orthorhombic phase transition temperature and is metallic above the transition [22]. Therefore, the transfer probability from a metallic plane to the adjacent one across the intervening insulating ionic layer seems to be a sensitive function of the nature of the ionic layer. Important superconducting parameters such as ξ_c, the coherence length along the c-axis, are dependent upon this interlayer coupling. The upper limit of the critical current could be estimated from the normal state resistivity along the c-axis by assuming Josephson coupling between the 2D metallic layers [22].

3.2 Opening of an Energy Gap in the Conduction Band. When T_c was found to depend sharply on the composition, the first question encountered was why $(La_{1-x}M_x)_2CuO_4$ became insulating upon a decrease in x [23]. As a matter of fact, this was already a subject of interest well before the discovery of HTSC in this material [24]. Figure 6 clearly shows that the oxides become insulating as the average Cu valence approaches 2+, i.e., when the conduction band is expected to become half-filled [25].

Therefore, one is led to assume that an energy gap opens at the half-filled position in the conduction band for some reason. There are essentially two conceivable mechanisms.

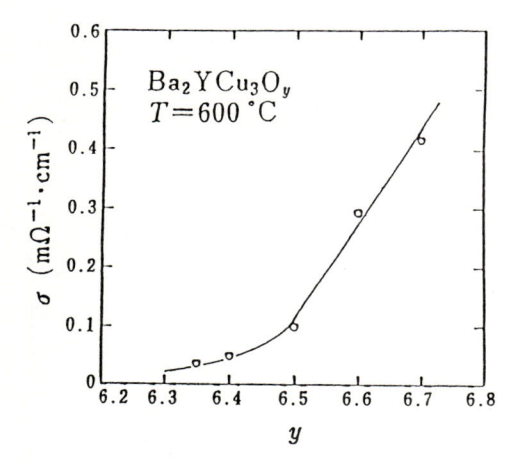

Fig. 6. Dependence of electrical conductivity of $Ba_2YCu_3O_y$ at 600°C on oxygen content, indicating that the system becomes insulating at $y = 6.5$, where Cu becomes nominally divalent. Due to the presence of two non-equivalent sites for Cu, a finite conductivity is still observed below $y = 6.5$

(a) Possibility of Fermi Surface Instability: CDW or SDW. At the earliest stage of the studies, the possibility of the Fermi surface instability was considered because of the 2D square shape of the Fermi surface expected for the half-filled σ_{dp}^* [26]. Figure 7 shows the tight-binding band of the CuO_2 square lattice. As can be seen from the figure, the band gives a free-electron-like circular Fermi surface when E_F is low. But there are four van Hove singularity points at $k = (\pm\pi/a, 0)$ and $k = (0, \pm\pi/a)$. As E_F is raised, the Fermi surface is distorted from a circle and becomes a square at the half-filled position, as shown in Fig. 8a. The density of electronic states (DOS) for this band is shown in Fig. 8b. The divergent DOS character at half-filling is due to the van Hove singularity.

If the Fermi surface is square-shaped as shown in Fig. 8a with the edges A, B, C and D, a superlattice modulation of the wave vector $|k| = \sqrt{2}\pi/2a$ along the $\langle 1, 1 \rangle$ direction, if present, can diffract electrons on line A onto the corresponding states on line B, or vice versa. In other words, each energy state on line A is hybridized with the corresponding one on line B. The same applies

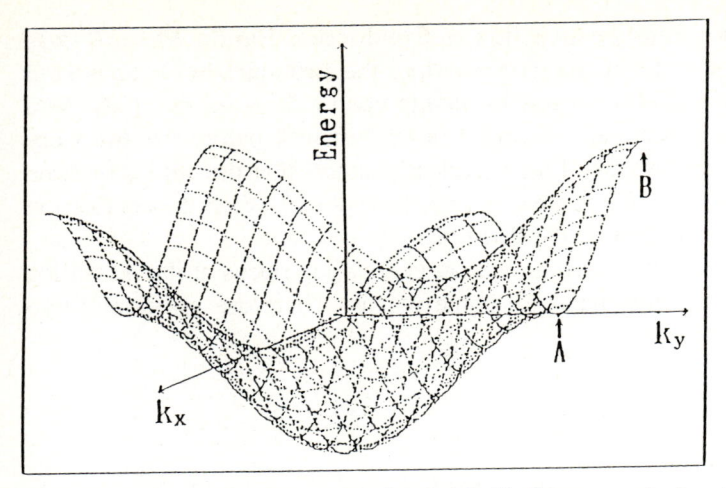

Fig. 7. Dispersion of the two-dimensional tight-binding square lattice. At the zone boundary $A(k_y = \pi/a)$ the van Hove singularity is observed. When the band is half-filled, the Fermi line becomes square, connecting the four equivalent points of A

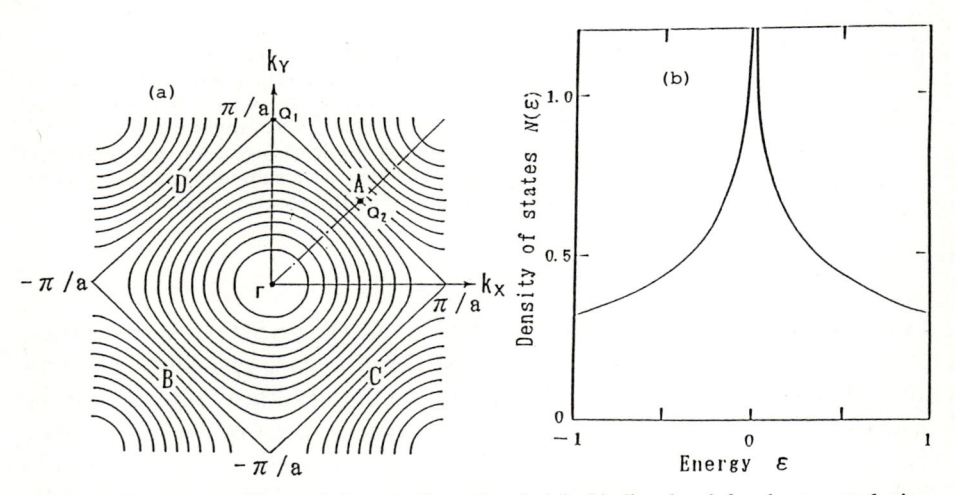

Fig. 8. (a) Equi-energy lines of the two-dimensional tight-binding band for the square lattice, showing the Fermi line to be free-electron-like for low carrier concentration but to become a square when the band is half-filled. (b) Density of states in units of $1/4t$ vs energy in units of $4t$, showing the divergence that occurs at half-filling of the band

for modulation along $\langle 1, \bar{1} \rangle$ to hybridize the states on line C with those on D. If the energy needed to create such modulation is over-compensated by the stabilization of the electronic energy due to the hybridization, the modulation should take place spontaneously and cause the Fermi surface instability. This opens up the energy gap.

The modulation could be in either charge or spin density, as long as it couples strongly with electrons, resulting in the formation of a charge density wave (CDW) or spin density wave (SDW), respectively. In the former case, the lattice is also distorted because of the coulomb interaction between the lattice and charge, while there is no direct interaction between the lattice and spins.

The possibility of CDWs, however, was discarded because no such superlattice distortion was detected [27]. SDWs, on the other hand, appeared attractive because Cu^{2+} is commonly known to show magnetism in ionic crystals.

However, the involvement of SDWs also turned out to be unlikely as first indicated by Hall measurements [28]. As shown in Fig. 9, the Hall coefficient of $(La_{1-x}Sr_x)_2CuO_4$ falls on the curve expected by assuming that each substitution of La for Sr provides one hole in the valence band up to quite a high doping level. The gap seems to persist up to as high as $x = 0.05$, where the Fermi surface shape should become closer to circular than to square-like, as can be seen from Fig. 8a. Then one cannot expect any single modulation to effectively hybridize the major part of the electronic states on the Fermi surface. Hence, it is hard to explain by the SDW mechanism the persistence of the energy gap to such a high doping level.

Therefore, one is forced to consider another possibility. But it should be remembered that the system is located in the close neighborhood of the half-filled state of the band where Fermi surface instability tends to occur and the DOS becomes divergent.

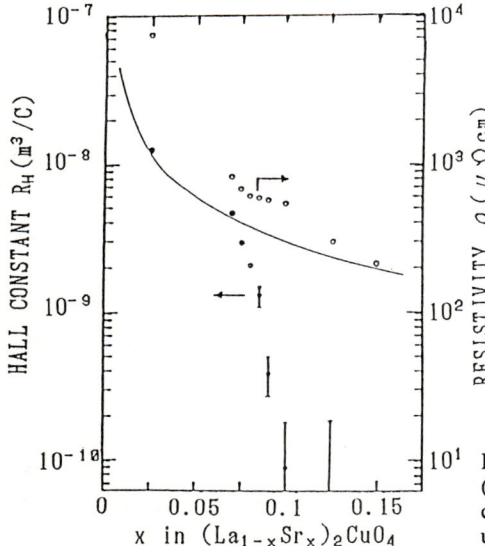

Fig. 9. The Hall constant vs doping level in $(La_{1-x}Sr_x)_2CuO_4$ at 77 K (*dots*). The solid curve is the relation $1/R_H e = x/$(unit cell volume). The circles show the resistivity at 77 K

(b) Electronic Correlation Gap (Mott-Hubbard Gap) [29]. When the overlap of the orbitals of valence electrons is large, each of the valence electrons tends to delocalize over the whole volume of the solid in order to stabilize their kinetic energy. This is the transfer energy, t, which corresponds to the band width. On the other hand, a high probability is expected for two of the delocalized electrons to come on the same unit cell, resulting in a strong repulsive interaction between them. This correlation is the interaction which is neglected in the concept of the band. When the correlation energy U becomes larger than t, the total energy may be lower if each electron is localized in each unit cell so that it does not trespass on the territory of the other electrons.

After filling every unit cell with one electron, i.e., half-filling of the conduction band, introduction of another electron to the system requires a finite energy U. In other words, there is an energy gap opened at this degree of electron filling as shown in Fig. 10. This is the concept of the Mott insulator [29].

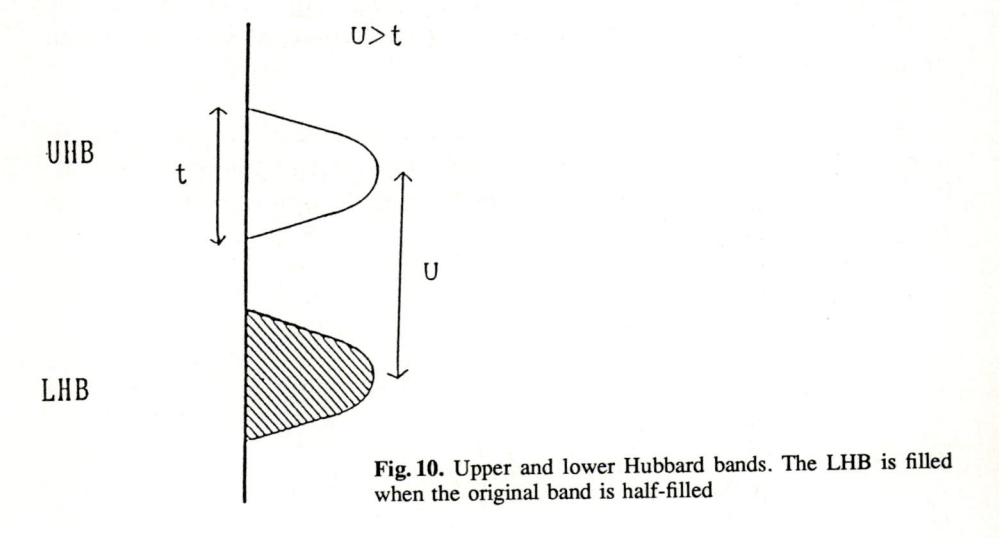

Fig. 10. Upper and lower Hubbard bands. The LHB is filled when the original band is half-filled

In the case of cuprates, this picture may hold when Cu is divalent because the $3d$ orbital is relatively small in its spatial extension and the Cu_{3d}-O_{2p} hybridization can be weaker than the correlation interaction. By the summer of 1987, the dominance of the electronic correlation had become the widely accepted mechanism for understanding the insulative nature of La_2CuO_4 and $Ba_2YCu_3O_{7-y}$ ($y > 0.5$) [28]. The strong correlation among electrons was first noticed by XPS observations of the valence band, which revealed a remarkable shift in the energy of the valence band peak away from that expected by the band calculations [30].

It should be noted that the correlation effect is of multi-body origin and hence the density of states model shown in Fig. 10 requires certain precautions regarding when it it utilized for examining the excitation spectrum.

3.3 Antiferromagnetic Ordering. In the localized electronic system, the orientation of the electron spin on a site may be determined by the spin configurations of the adjacent sites. Because of the exchange interaction, the two adjacent spins tend to orient anti-parallel to each other. In the square lattice, this requirement can be fully satisfied by arranging the spins in alternating directions.

Compared with the exchange interaction in the CuO_2 plane, the interaction across the plane is expected to be much smaller. The three-dimensional antiferromagnetic spin ordering was first suggested in La_2CuO_4 by the observation of a cusp in the magnetic susceptibility–T curve [31] and more clearly established by NOR of Cu ions [32, 33], μSR [34] and neutron diffraction [35, 36]. The 3D long range ordering was found to be lost above 240 K.

3.4 Hole Doping in the Mott Insulator. It was found that doping such as $(La_{1-x}Sr_x)_2CuO_4$, $Ba_2YCu_3O_{6+y}$, $Bi_2Sr_2(Y_{1-x}Ca_x)CuO_y$ decreased the resistivity, Hall coefficient (positive), and Seebeck coefficient [1, 2], suggesting that these changes can be understood in terms of the increased hole concentration in the highest valence band [37]. As the hole concentration increased, the AF transition temperature T_N was found to go down sharply. The phase diagram of $(La_{1-x}Ba_x)_2CuO_4$ is shown in Fig. 11, indicating the occurrence of super-

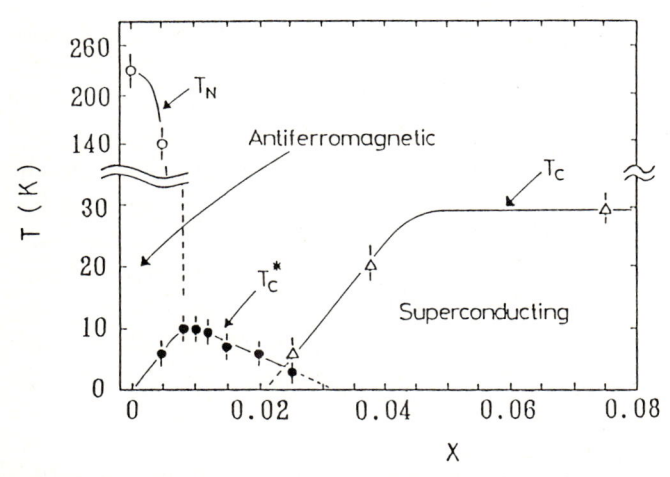

Fig. 11. Phase diagram of $(La_{1-x}Ba_x)_2CuO_4$. The metal–semiconductor transition takes place near $x = 0.025$. The magnetic phase boundaries were determined by NQR. The region below T_c^* is considered to be the spin glass phase

conductivity only after the complete destruction of the three-dimensional AF order. This has been observed commonly in the other cuprate superconductors as well [34].

In the simple Hubbard scheme, this may be understood as follows. The "hole" is doped in the lower Hubbard band as a vacancy of the localized electrons. It should be mentioned that the term "hole" carries a different meaning here from that in the band concept; here it rather means a "vacancy" for an electron on the lattice site. Then because of the absence of the electron on this particular site, the electrons on the adjacent sites can penetrate onto this site in order to stabilize their kinetic energy, i.e., due to the transfer energy t. But the long-range AF ordering based on the exchange energy J would act to prevent this, as can be seen in Fig. 12. In order to compromise the two tendencies, the rigid AF ordering may be at least locally destroyed by creating the local spin fluctuation depending on the relative size of t/J. If this brings about the stabilization of the total energy, the spin fluctuation may be induced in the ground state in association with the vacancy of an electron. This picture can be used to interpret the increase of the p-type carriers in proportion to x as well as the sharp drop of T_N.

↑ ↓ ↑ ↓ ↑ ↓ ↑ ↓ ↑ ↓   ↑ ↓ ↑ ↓ ↑ ↓ ↑ ↓ ↑ ↓
↓ ↑ ↓ ↑ ↓ ↑ ↓ ↑ ↓ ↑   ↓ ↑ ↓ ↑ ↓ ↑ ↓ ↑ ↓ ↑
↑ ↓ ↑ ↓ □ ↓ ↑ ↓ ↑ ↓   ↑ ↓ ↑ ↓ ↓ □ ↑ ↓ ↑ ↓
↓ ↑ ↓ ↑ ↓ ↑ ↓ ↑ ↓ ↑   ↓ ↑ ↓ ↑ ↓ ↑ ↓ ↑ ↓ ↑
↑ ↓ ↑ ↓ ↑ ↓ ↑ ↓ ↑ ↓   ↑ ↓ ↑ ↓ ↑ ↓ ↑ ↓ ↑ ↓
↓ ↑ ↓ ↑ ↓ ↑ ↓ ↑ ↓ ↑   ↓ ↑ ↓ ↑ ↓ ↑ ↓ ↑ ↓ ↑
              a   b

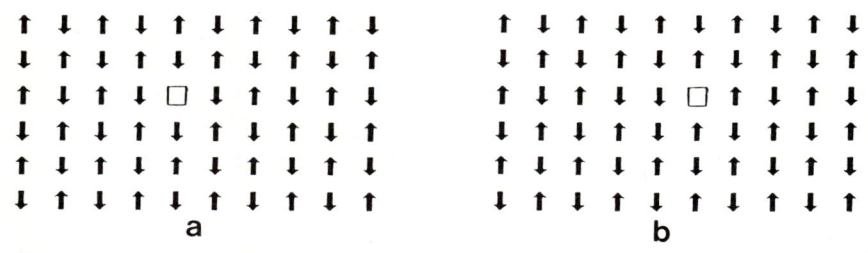

Fig. 12. (a) The antiferromagnetic spin arrangement with a "hole" in the two-dimensional square lattice. When the hole is transferred to the adjacent site, the antiferromagnetic order is locally broken around the spin (b), hence preventing the delocalization of the hole

Anderson's RVB picture is similar but assumes a more extreme case in which the ground state is composed of singlet pairs of electron spins [38].

If the above pictures are to hold in the superconducting region, the electronic structure of cuprates should be regarded as completely different from that of the band metal, because the cuprates then do not possess a Fermi surface in the literal sense. The electronic states at the Fermi level are not itinerant extended single-electron states but are rather localized and strongly correlated.

The electrons become increasingly mobile as holes are introduced only because of the presence of more vacant sites of electrons. To be more specific, the "hole" in the band concept represents an electronic state whose electron has a certain momentum and is delocalized in space. The "hole" here, on the other hand, literally means a certain unit cell where the electron is missing, i.e., a vacancy created in the lattice of the highly correlated localized electrons.

Since the BCS model is based on the band concept, the superconducting mechanism that includes the above picture of the electronic structure, i.e., superconductivity in the Hubbard band, should be essentially different from the BCS one.

3.5 Persistence of Two-Dimensional Spin Ordering on Doping. Although long-range AF spin ordering is sharply lost upon doping with holes, neutron scattering detected short-range AF domains surviving on the basal plane even in the superconducting compositional range, as shown in Fig. 13 [39]. The domains are dynamic, i.e., the AF domain boundaries are mobile, and the size of the domains (the correlation length) decreases with doping and is about equal to the average spacing between the doped holes.

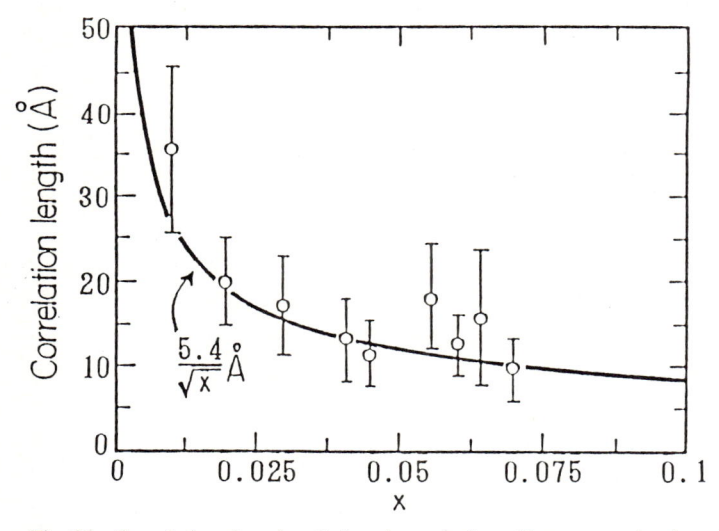

Fig. 13. Correlation length of the dynamical antiferromagnetic domains vs composition in $(La_{1-x}Sr_x)_2CuO_4$, indicating that the magnetic ordering survives in the highest T_c superconducting region, $x \simeq 0.075$. The solid line is the relation $5.4/\sqrt{x}$ [Å] which is just the average separation between the holes introduced by the Sr doping

It is amazing that the AF ordering of correlation length about 1 nm is persistent, even in the highest T_c region of $(La, Sr)_2CuO_4$, because it means that the carriers move against the background of the localized spin lattice. It had been believed that superconductivity does not get along well with the presence of localized spins in conventional superconductors. Therefore this observation tempts one to assume the positive involvement of some magnetic interaction in the mechanism of high T_c superconductivity.

3.6 O_{2p} Nature of the Fermi Surface State.

The above simple Hubbard picture has been challenged by observations, mostly by electron spectroscopy. Angle resolved UPS [40] and ELS [41] indicated a small but finite dispersion (up to 0.5 eV) of electronic states in the k-space near the Fermi edge and that one of the bands was crossing the Fermi level. Also, the electronic state involved was shown to be mostly of O_{2p} nature by using the resonant incident beam technique [40]. Reverse UPS (BIS) has recently shown that the band crossing the Fermi level is connected to the one above E_F [42].

In the early period, a certain ambiguity was involved in electron spectroscopy because of possible surface degradation and the low resolution of the energy spectrum. But improvements have been made by cleaving the specimens in the UHV chamber at low temperature, by utilizing the transmission method [41] rather than reflection and by improving the resolution limit to 30–15 meV. Recent results have further supported the conclusions.

Therefore, it seems to be clear enough that the simple Hubbard picture does not apply to the cuprates. But the gap is at least a charge transfer gap in which the lowest excitation occurs from O_{2p} to Cu_{3d} states. The question whether the Fermi level state is band-like or not may be settled as follows. If the interaction between the Fermi level O_{2p} states and the Cu_{3d} localized spin states is weak, we may obtain the state well approximated by the Fermi liquid. But if the interaction is strong enough that the O_{2p} hole is coupled with the Cu_{3d} localized spin, then the situation is rather similar to the case when E_F is situated in the Hubbard band.

At this moment it seems to be too early to judge conclusively which of the two pictures are correct: Fermi liquid or non-Fermi liquid.

Based mainly on the results from electron and optical spectroscopies, two models have been proposed to describe the Fermi level states as shown in Fig. 14. One model assumes the O_{2p} band to be located in the Hubbard gap, while the other model assumes the creation of a new modified O_{2p} state as each dopant atom is introduced. The latter model [43] was recently proposed based on UPS, XAS [44] and optical absorption spectroscopy [45]. In the former model, the holes are supposed to be itinerant ones from the very beginning and hence the semiconductive nature in the low doping region is explained by assuming the localization of carriers in the low concentration range due, for instance, to Anderson localization. In the latter, the Fermi state levels are essentially local levels, being similar to donor levels in the usual semiconductor and they form an itinerant band as the concentration increases.

So far, the charge transfer gap energy, Δ has been assigned to be about 2 eV from UPS [40], and optical absorption data [44], corresponding to excitation from the O_{2p} valence band to the Cu_{3d} upper Hubbard band. This is smaller than the correlation gap energy on the Cu site, U_{dd}, about 7 eV according to UPS data [40]. The Fermi level states in Fig. 14b are considered

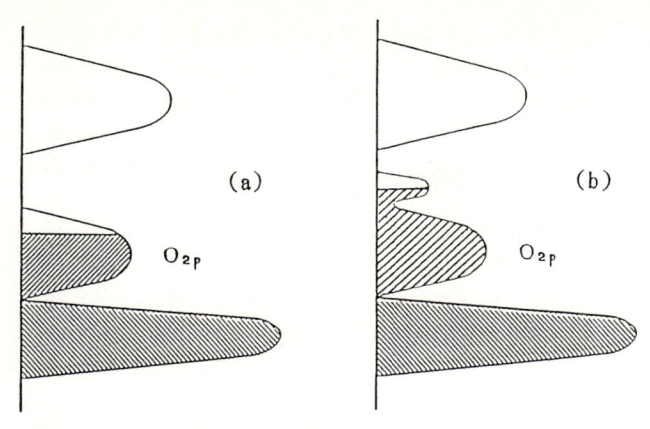

Fig. 14. Two possible extended Hubbard band schemes. (a) The Fermi level is assumed to be located in the O_{2p} itinerant band. (b) New Fermi level states are assumed to be created on doping

to be formed 1.5 eV (LSCO) below the upper Hubbard gap [44], but this is still controversial.

Furthermore, although the electron spectroscopy measurements agree on the strong O_{2p} nature of the Fermi surface state, there has been an observation that Cu_{3d} is also contributing up to 20% [46], indicating the strong hybridization of the Cu_{3d} and O_{2p} orbitals in the Fermi level states.

Concerning the position of the oxygen atoms (and Cu atoms) involved, there have been indications that they are the ones in the CuO_2 layer. This has been shown by observation by polarized incident light, which shows that the excitation across the gap has its dipole vector along the basal plane [44].

3.7 Phenomena Associated with Fermi Level States. From the transport experiments, it has been indicated that the assumption of just a simple itinerant band picture cannot explain the experimental results. Therefore, although electron spectroscopy suggests the Fermi liquid state, it should be noted that this state is significantly distorted from the one in a typical metal.

First of all, a difficulty was pointed out in understanding the too-simple linear relationship between the resistivity and temperature [47]. In typical d-band superconductors such as Nb or Nb_3Sn, the resistivity shows a tendency to saturate in both the low- and high-T regions. The simple linear relationship in the high T-region is thought to indicate a lack of strong electron–phonon coupling. But this seems to contradict the observation that the thermal conductivity shows a decrease above the superconductivity transition temperature [48], which seems to suggest strong scattering of phonons by electrons in the normal state. Recent single-crystal resistivity data have indicated a T^m relationship (m = 1.3–1.5) in LSCO rather than the simple T-linear [22] one, and hence this problem has still to be clarified.

The Seebeck coefficient, on the other hand, showed a rather flat dependence on T [49]. This is strange, considering the linear dependence observed in a typical metal. The flat dependence has been suggested for a strongly correlated electronic system in organic conductors [50].

The Hall coefficient is expected to be nearly independent of T in a simple metal because of the constant carrier concentration with T. However, a strong dependence has been consistently observed commonly among the various cuprate superconductors. Often it was reported to change almost linearly with T^{-1}, but this seems to be rather fortuitous and to depend on the composition. Any simple modification of the metallic band model, such as two-band models, cannot explain the complex but consistently observed temperature dependence of the Hall coefficient on T in the various HTSC cuprates [51].

It is often considered that the hole concentration is the key parameter determining T_c [52]. However, there must be at least one additional key parameter, such as the lattice constant, because it has been pointed out that the application of a high pressure can increase the T_c of LSCO significantly while it leaves the Hall coefficient constant [53].

No magneto-electric phenomena, such as de Haas–van Alphen and Shubnikov–de Haas effects have been observed, perhaps because of the too low mobility of the carriers or of the unusual electronic structure of the cuprates.

Therefore, the transport experiments so far may be summarized as indicating the Fermi surface to be essentially a metallic one but considerably modified to show the strange dependences of the transport properties on the temperature.

4. Electron-Doped Cuprate Superconductors

So far discussions have been confined to the hole-doped cuprate superconductors. Until the end of 1988, because of the lack of electron-doped superconductors, the special role of holes had been discussed in relation to HTSC of cuprates. The discovery of $(Nd_{1-x}Ce_x)_2CuO_4$, which showed the highest T_c about 27 K at $x = 0.075$ when heat-treated in a reducing atmosphere, made a great impact in this respect [54, 55]. It was soon found that the non-rare-earth tetravalent ion Th could replace Ce, and Nd could be replaced by Pr, Sm or Eu [56]. Also $Nd_2Cu(O_{1-x}F_x)_4$ was found to become superconducting [57]. Since all these substitutions indicate the direction towards electron-doping from the half-filled band state of Ln_2CuO_4, it is clear that the Fermi level is shifted above the half-filled level, while it is below that for hole-doped superconductors.

4.1 Symmetry Between Electron- and Hole-Doping. There have been controversies concerning whether the superconductivity composition of $(Nd_{1-x}Ce_x)_2CuO_4$ is n-type or p-type based on the sign of Hall and Seebeck coefficients. These arguments, however, seem to be unfruitful if the following facts are considered. The sign of the Hall and Seebeck coefficients of $(Nd_{1-x}Ce_x)_2CuO_4$ is negative in the low doping level, but it becomes positive as the doping level is raised, as shown in Fig. 15 [60]. HTSC is obtained near the crossover region of the sign. Consequently the observed sign of those coefficients in the HTSC region should depend on the subtle differences of the electronic structures and material conditions of the particular specimens utilized for measurements, and hence does not seem to reflect the really important features.

The important point to be stressed here should rather be the fact that the electron- and hole-doped superconducting cuprates are symmetrical with regard to the change in the conduction mechanism with composition, as can be seen from Fig. 15. In the initial stage of doping they exhibit the characteristics of a half-filled band insulator doped with electrons and holes, respectively. The carrier concentration is just that expected from the dopant level. But as the doping becomes heavier, they become metallic and HTSC sets in while

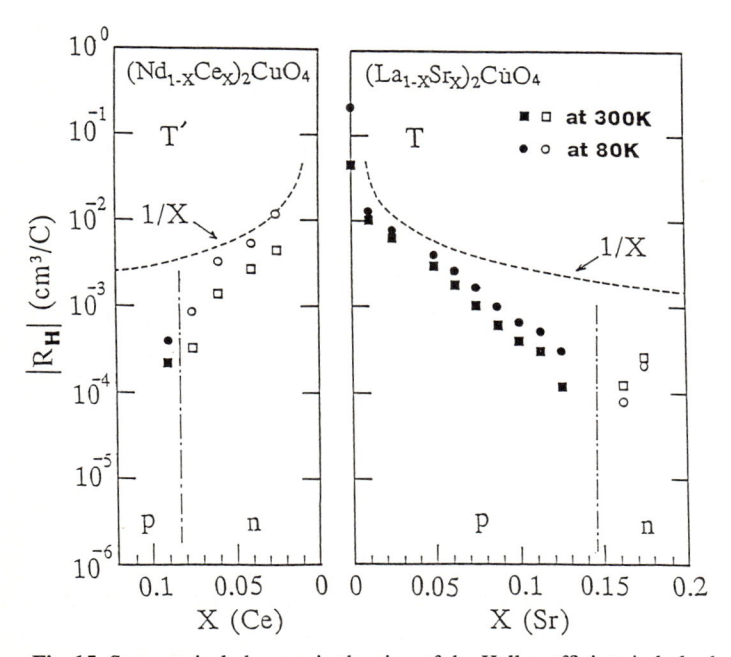

Fig. 15. Symmetrical changes in the sign of the Hall coefficient in hole-doped $(La_{1-x}Sr_x)_2CuO_4$ and electron-doped $(Nd_{1-x}Ce_x)_2CuO_4$. The highest T_c composition is $x = 0.075$ in both of the systems, while the crossover composition of the Hall coefficient is higher in LSCO (0.15) than in NCCO (0.08)

the Hall coefficient deviates from the expected curve. The metallic nature is further intensified with an increase in the conductivity and the sign of the Hall coefficient is reversed. The superconducting transition temperature, in association, goes up (this region is lacking in electron-doped superconductors so far), then goes down, and finally they become nonsuperconducting. Namely, HTSC is obtained just in the intermediate range of the two extreme cases; the one is approximated well by the extended Hubbard scheme and the other is better understood by the metallic band scheme.

4.2 HTSC Family Derived from Ln_2CuO_4. Figure 16 shows the crystal structures of the three HTSC systems, T, T' and T* derived from Ln_2CuO_4 by doping. T is the structure of the original single-layered superconductor consisting of CuO_6 octahedra, while T' is that of the electron-doped one consisting of CuO_4 square-coordinated units. T* is a hybrid of the two structures with the upper half of the unit cell equivalent to T and the lower half to T', although this was discovered [58, 59] prior to T'. Since both T and T* are known to be hole-doped superconductors, one may presume that the feasibility boundary of hole- and electron-doping is situated between the pyramidal and square planar coordination of Cu ions. This seems to be understandable because the CuO_2 layer is surrounded by the same number of cations but fewer anions for T' than T* and hence than T. Therefore the CuO_2 plane may become more electronegative in the T' structure.

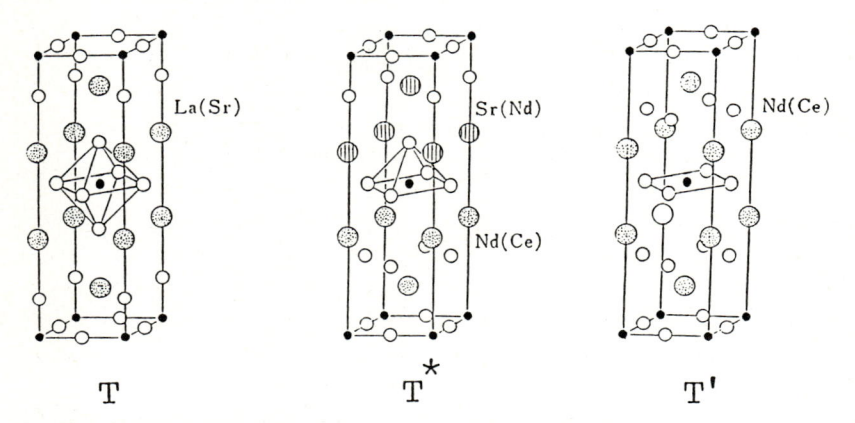

Fig. 16. Crystal structures of cuprate superconductors derived from Ln_2CuO_4

4.3 Significance of the Electron-Doped Superconductor. The significance of the electron-doped superconductor could be as follows, in addition to its impact on the effort to identify the mechanism of HTSC. First of all, the preparation can be done in a reducing atmosphere, which seems to ease the high temperature process and the vacuum process for preparation of the materials.

Secondly, the possible area of search for new materials has been doubled and hence one may expect further discoveries of new HTSC systems. Thirdly, one of the most serious drawbacks of the oxide superconductors, i.e., the extremely short coherence length, may be improved if the simple BCS relationship is considered:

$$\xi_{BCS} = \hbar v_F / \pi \Delta_0$$

where ξ_{BCS} is the BCS coherence length, v_F the Fermi velocity, and Δ_0 the superconducting energy gap at $T = 0$. Since Δ_0 is dominantly determined by T_c, a larger value is desired for v_F, which seems to be more plausible for electron-doping.

From H_{c2} measurements on a $(Nd_{0.925}Ce_{0.075})_2CuO_4$ single crystal, Δ_0 values have been reported as 7 nm along the ab-plane and 0.31 nm along the c-axis; the former value seems to be longer than the one in the hole-doped T phase, $(La_{0.925}Sr_{0.075})_2CuO_4$, which is about 3 nm. The coherence length along the c-axis, however, appears to be very short as well.

5. Summary

An essential feature of the electronic structure of the layer-type cuprate is represented by the strong coulomb correlation, especially on Cu sites. This leads to the opening of the correlation gap when the conduction band, which is supposed to be the 2D $Cu_{3d}O_{2p}$ anti-bonding band, is half-filled. Either holes or electrons can be doped into the starting insulating composition in which Cu is divalent. The doped holes possess mainly O_{2p} character, while it is not clear if the doped electrons exhibit Cu_{4s} character or not. In any case, the system becomes more metallic as the dopant concentration increases and HTSC sets in. Upon further doping, the metallic nature is intensified but the superconductivity decreases in its T_c and is finally lost. In the mean time, the Hall coefficient changes its sign from p to n for a hole-doped system and from n to p for an electron-doped system. The behavior is quite symmetric upon doping for the two types of superconducting cuprate systems. It is suggestive that we once encountered a similar change in the electronic structure in the other HTSC superconducting system, bismuthates.

The important fact to be considered for the HTSC mechanism is that HTSC occurs not in the typical metallic region but rather in the region where the correlation gap (CDW gap for bismuthates) is just destroyed by the doping of carriers.

The origin of the Fermi state has not yet been understood in detail. Some of the properties can be better understood based on the model of the Fermi liquid state but the others are quite unlike what is expected from this model.

References

1. K. Kitazawa, H. Takagi, K. Kishio, T. Hasagawa, S. Uchida, S. Tajima, S. Tanaka, K. Fueki: Physica C **153–155**, 9 (1988)
2. K. Kitazawa: IMB J. Res. Dev. **33**, 201 (1989)
3. J.G. Bednorz, M. Takashige, K.A. Müller: Europhys. Lett. **3**, 379 (1987)
4. S. Uchida, H. Takagi, K. Kitazawa, S. Tanaka: Jpn. J. Appl. Phys. **26**, L1 (1987)
5. A.W. Sleight, J.L. Gillson, P.E. Bierstedt: Solid State Commun. **17**, 27 (1975)
6. L.F. Mattheiss, E.M. Gyorgy, D.W. Johnson, Jr.: Phys. Rev. B **37**, 3745 (1988)
7. R.J. Cava, B. Batlogg, J.J. Krajewski, R. Farrow, L.W. Rupp, Jr., A.E. White, K. Short, W.F. Peck, T. Kometani: Nature **332**, 814 (1988)
8. S. Uchida, K. Kitazawa, S. Tanaka: Phase Transitions **8**, 95 (1987)
9. H. Sato, S. Tajima, H. Takagi, S. Uchida: Nature **338**, 241 (1989)
10. L.F. Mattheiss, D.R. Hamann: Phys. Rev. B **28**, 4227 (1983)
11. D.E. Cox, A.W. Sleight: Acta Crystallogr. B **35**, 1 (1979)
12. S. Uchida, S. Tajima, A. Masaki, S. Sugai, K. Kitazawa, S. Tanaka: J. Phys. Soc. Jpn. **54**, 4395 (1985)
13. S. Sugai, S. Uchida, K. Kitazawa, S. Tanaka, A. Katsui: Phys. Rev. Lett. **55**, 426 (1985)
14. H. Takagi, S. Uchida, S. Tajima, K. Kitazawa, S. Tanaka: Proc. Int. Conference on Physics of Semiconductors, Stockholm (1986) p. 1851
15. S. Tajima, S. Uchida, A. Masaki, H. Takagi, K. Kitazawa, S. Tanaka, A. Katsui: Phys. Rev. B **32**, 6302 (1985)
16. K. Kitazawa, S. Uchida, S. Tanaka: Physica **135B**, 505 (1985)
17. B. Batlogg, A.P. Ramirez, R.J. Cava, R.B. van Dover, E.A. Rietman: Phys. Rev. B **35**, 5340 (1987)
18. D.G. Hinks, D.R. Richards, B. Dabrowski, D.T. Marx, A.W. Mitchell: Nature **335**, 419 (1988)
19. B. Batlogg, J.P. Remeika, R.C. Dynes, H. Barz, A.S. Cooper, J.P. Garno: In *Superconductivity in d- and f-band Metals, 1982*, ed. by W. Buckel, W. Weber (Kernforschungszentrum Karlsruhe 1982)
20. J.F. Zasadzinski, N. Tralshawala, J. Timpf, D.G. Hinks, B. Dabrowski, A.W. Mitchell, D.R. Richards: Physica C **162–164**, 1053 (1989)
21. Y. Iye: In *Mechanisms of High Temperature Superconductivity*, ed. by H. Kamimura, A. Oshiyama, Springer Ser. Mater. Sci., Vol. 11 (Springer, Berlin, Heidelberg 1989) p.263
22. S. Kambe, K. Kitazawa, M. Naito, A. Fukuoka, I. Tanaka, H. Kojima: Physica C **160**, 35 (1989)
23. S. Kambe, K. Kishio, K. Kitazawa, K. Fueki, H. Takagi, S. Tanaka: Chem. Lett. 547 (1987)
24. K.K. Singh, P. Ganguly, J.B. Goodenough: J. Solid State Chem. **52**, 254 (1984)
25. P.P. Freitas, T.S. Plaskett: Phys. Rev. B **36**, 5723 (1987)
26. L.F. Mattheiss: Phys. Rev. Lett. **58**, 1028 (1987)
27. K. Takegahara, H. Harima, A. Yanase: Jpn. J. Appl. Phys. **26**, 352 (1987)
28. N.P. Ong, Z.Z. Wang, J. Clayhold, J.M. Tarascon, L.H. Greene, W.R. McKinnon: Phys. Rev. B **35**, 8807 (1987)
29. N.F. Mott: Philos. Mag. **6**, 287 (1961)
30. A. Fujimori, E. Takayama-Muromachi, Y. Uchida, B. Okai: Phys. Rev. B **35**, 8814 (1987)
31. R.L. Greene, H. Maletta, T.S. Plaskett, J.G. Bednorz, K.A. Müller: Solid State Commun. **63**, 379 (1987)
32. K. Kumagai, I. Watanabe, H. Aoki, Y. Nakamura, T. Kimura, Y. Nakamichi, H. Nakajima: Physica **148B**, 480 (1987)
33. Y. Kitaoka, S. Hiramatsu, K. Ishida, K. Asayama, H. Takagi, H. Iwabuchi, S. Uchida, S. Tanaka: J. Phys. Soc. Jpn. **57**, 737 (1988)
34. N. Nishida, H. Miyatake, S. Okuma, T. Tamegai, Y. Iye, R. Yoshizaki, K: Nishiyama, K. Nagamine: Physica C **156**, 625 (1988)
35. D. Vaknin, S.K. Sinha, D.E. Moncton, D.C. Johnston, J. Newsam, C.R. Satinya, H.E. King, Jr.: Phys. Rev. Lett. **58**, 2802 (1987)

36 S. Mitsuda, G. Shirane, S.K. Sinha, D.C. Johnston, M.S. Alvarez, D. Vaknin, D.E. Moncton: Phys. Rev. B **36**, 822 (1987)
37 H. Fukuyama, Y. Hasagawa: Physica **148B**, 204 (1987)
38 P.W. Anderson: Science **235**, 1196 (1987)
39 R.J. Birgeneau, D.R. Gabbe, H.P. Jenssen, M.A. Kastner, P.J. Picone, T.R. Thurston, G. Shirane, Y. Endoh, M. Sato, K. Yamada, Y. Hidaka, M. Oda, Y. Enomoto, M. Suzuki, T. Murakami: Phys. Rev. B **38**, 6614 (1988)
40 T. Takahashi, H. Matsuyama, H. Katayama-Yoshida, Y. Okabe, S. Hosoya, K. Seki, H. Fujimoto, M. Sato, H. Inokuchi: Nature **334**, 691 (1988)
41 N. Nucker, J. Fink, J.C. Fuggle, P.J. Durham, W.L. Temmerman: Phys. Rev. B **37**, 5158 (1988)
42 R. Claessen, R. Manzke, H. Carstensen, B. Burandt, T. Buslaps, M. Skibowski, J. Fink: Phys. Rev. B **39**, 7316 (1989)
43 M. Matsumoto, M. Sasaki, M. Tachiki: Solid State Commun., in press
44 H. Matsuyama, T. Takahashi, H. Katayama-Yoshida, T. Kashiwakura, Y. Okabe, S. Sato, N. Kosugi, A. Yagishita, K. Tanaka, H. Fujimoto, H. Inokuchi: Physica C **160**, 567 (1989)
45 M. Suzuki: Phys. Rev. B **39**, 2312 (1989)
46 A.J. Arko: Presented at Am. Phys. Soc. HTSC Symp., St. Louis (March 1987)
47 M. Gurvith, A.T. Fiory: Phys. Rev. Lett. **59**, 1337 (1987)
48 C. Uher, A.B. Kaiser: Phys. Rev. B **36**, 5680 (1987)
49 R.S. Kowk, S.E. Brown, J.D. Thompson, Z. Fisk, G. Gruner: Physica **148B**, 346 (1987)
50 J.R. Cooper, B. Alabi, L.-W. Zou, W. Beyermann, G. Gruner: Phys. Rev. B **35**, 8794 (1987)
51 J. Clayhold, N.P. Ong, P.H. Hor, C.W. Chu: Phys. Rev. B **38**, 7016 (1988)
52 J.B. Torrance, Y. Tokura, A.I. Nazzal, A. Bezinge, T.C. Hung, S.S.P. Parkin: Phys. Rev. Lett. **61**, 1127 (1988)
53 N. Tanahashi, Y. Iye, T. Tamegai, C. Murayama, N. Mori, S. Yomo, N. Okazaki, K. Kitazawa: Jpn. J. Appl. Phys. **28**, 762 (1989)
54 Y. Tokura, H. Takagi, S. Uchida: Nature **337**, 345 (1989)
55 H. Takagi, S. Uchida, Y. Tokura: Phys. Rev. Lett. **62**, 1197 (1987)
56 J.T. Markert, E.A. Early, T. Bjornholm, S. Ghamaty, B.W. Lee, J.J. Neumeier, R.D. Price, C.L. Seaman, M.B. Maple: Physica C **158**, 178 (1989)
57 A.C.W.P. James, S.M. Zahurak, D.W. Murphy: Nature **338**, 240 (1989)
58 J. Akimitsu, S. Suzuki, M. Watanabe, H. Sawa: Jpn. J. Appl. Phys. **27**, 1859 (1988)
59 E. Takayama-Muromachi, Y. Matsui, Y. Uchida, F. Izumi, M. Onoda, K. Kato: Jpn. J. Appl. Phys. **27**, 2283 (1988)
60 H. Takagi, Y. Tokura, S. Uchida: Proc. MRS Spring Meeting, San Diego, Calif. (April 1989) Vol. 156, p. 389

Oxygen Nonstoichiometry and Valence States in Superconductive Cuprates

B. Raveau, C. Michel, M. Hervieu, J. Provost, and F. Studer

Laboratoire CRISMAT, ISMRA, Bd. du Maréchal Juin,
F-14032 Caen Cedex, France

Abstract. The problems of oxygen nonstoichiometry are presented and discussed for the high temperature superconductors in connection with the mixed valence of the metallic atoms and in particular with that of copper which is known to play a prominent role for the superconducting properties of these materials. For "123"-type and La_2CuO_4-type superconductors, the lone mixed valence ion is copper which may allow large deviations from stoichiometry, according to the structure. In the case of bismuth, thallium and lead cuprates, the possibility of these elements taking different valencies in these compounds, as it appears from recent investigations by X-ray absorption spectroscopy, shows that the rock salt-type layers play probably also a role for the superconducting properties of these oxides.

1. Introduction

Three years after the discovery of superconductivity close to 40K by Bednorz and Müller [1] in copper oxides the issue of oxygen nonstoichiometry in those materials remains up to date and is the subject of controversy. This problem is of capital importance since it governs the valency of the metallic elements and consequently is the key for controlling the superconducting properties of the copper oxides. The structural principles of those oxides which all belong to the same family $(ACuO_{3-x})_m(A'O)_n$ [2] have previously been described and will not be recalled here (see ref. 3-4). Just to summarize, it can be said that their structure consists of multiple oxygen deficient perovskite slabs built up themselves of m copper layers intergrown with multiple rock salt-type slabs formed of n AO layers. Thus, roughly they can be represented by the symbol [m, n]. In fact the situation is more complicated than described in a first approach, owing to the possible oxygen nonstoichiometry in the perovskite layers (x ranging from 0 to 1), and also in the rock salt-type layers. Table 1 summarizes the different layered cuprates with their idealized formulations which have been synthesized up to now. In spite of their great similarity these different oxides do not have the same behavior versus an oxidizing

Table I - The different layered copper oxides $(AO)_n(A'CuO_{3-y})_m$ symbolized by [m,n]. The oxides labelled N.S. are not superconductors.

m \ n	1	2	3
1	**[1,1]** $La_{2-x}A_xCuO_4$ La_2CuO_4 (A=Ca,Sr,Ba) $Tc \approx 20K-40K$	**[1,2]** $TlBa_2CuO_{5-\delta}$ N.S. $TlSr_2CuO_{5-\delta}$ Traces $Tl_{1-x}Pr_xSr_{2-y}Pr_yCuO_{5-\delta}$ $Tc \approx 40K$ (x=0.2, y=0.4) $Tl_{0.5}Pb_{0.5}Sr_2CuO_{5-\delta}$ N.S. $TlBa_{1+x}La_{1-x}CuO_{5-\delta}$ $T_c \approx 50K$ (x=0.2)	**[1,3]** $Tl_2Ba_2CuO_6$ $Tc \approx 30K$ $Bi_2Sr_2CuO_6$ $Tc \approx 10-22K$
1.5	**[1.5,1]** $Pb_2Sr_2Y_{1-x}Ca_xCu_3O_8$ $T_c = 50K$ (x=0.5) $Pb_{2-x}Bi_xSr_2Y_{1-y}Ca_yCu_3O_8$ $T_c = 79K$ (x=0.6, y=0.5, 1)		
2	**[2,1]** $La_{2-x}A_{1+x}Cu_2O_6$ A=Ca,Sr N.S.	**[2,2]** $TlBa_2CaCu_2O_7$ $Tc \approx 60K$ $TlSr_2CaCu_2O_7$ $Tc \approx 50K$ $Tl_{0.5}Pb_{0.5}Sr_2CaCu_2O_7$ $Tc \approx 85K$ $TlBa_2LnCu_2O_7$ Ln=Pr,Y,Nd N.S. $Pb_{0.5}Sr_{2.5}Y_{0.5}Ca_{0.5}Cu_2O_{7-\delta}$ $T_c = 59K$	**[2,3]** $Tl_2Ba_2CaCu_2O_8$ $Tc \approx 105K$ $Bi_2Sr_2CaCu_2O_8$ $Tc \approx 85K$ $Bi_{2-x}Pb_xSr_2Ca_{1-x}Y_xCu_2O_8$ $Tc \approx 85K$ to N.S. $Tl_{3-4x/3}Ba_{1+x}LnCu_2O_8$ Ln=Pr,Nd,Sm (x=0.25) N.S.
3	**[3,1]** $PbBaYSrCu_3O_8$ N.S.	**[3,2]** $TlBa_2Ca_2Cu_3O_9$ $Tc \approx 120K$ $Tl_{0.5}Pb_{0.5}Sr_2Ca_2Cu_3O_9$ $Tc \approx 120K$	**[3,3]** $Tl_2Ba_2Ca_2Cu_3O_{10}$ $Tc \approx 125K$ $Bi_{2-x}Pb_xSr_2Ca_2Cu_3O_{10}$ $Tc \approx 110K$
4	**[4,1]**	**[4,2]** $TlBa_2Ca_3Cu_4O_{11}$ $Tc \approx 108K$	**[4,3]** $Tl_2Ba_2Ca_3Cu_4O_{12}$ $Tc \approx 115K$

atmosphere. It has indeed been observed that the cuprates containing only rare earth and alkaline earth elements have to be annealed under an oxygen flow in order to optimize their superconducting properties, whereas those containing bismuth must be prepared in air, and the lead cuprates at lower oxygen pressure. These differences come from the fact that in the first case, rare earth cations exhibit generally only one oxidation state, whereas for bismuth and lead a mixed valent state is always possible i.e. Bi(V)/Bi(III) or Pb(IV)/Pb(II) in competition with the mixed valence state Cu(II)/Cu(III). The case of thallium cuprates is even more complex since besides the two oxidation states Tl(I)/Tl(III) everybody knows the great volatility of Tl_2O leading to a possible thallium deficiency. Thus we will examine here these families seperately and we will try to understand the problems of oxygen nonstoichiometry in connection with the valent state of the metallic atoms and of course with the superconducting properties.

2. Oxygen nonstoichiometry and valency of copper in alkaline/rare earth cuprates

These two families of superconductors, $La_{2-x}A_xCuO_4$ (A = Ca, Sr, Ba) [5-11] and $YBa_2Cu_3O_7$-type [12-19], which are known as the 40K and 92K-superconductors, respectively, have as common feature the fact that La or Y exhibit only the trivalent state whereas the alkaline earths are bivalent so that the mixed valency is only possible for copper. In both families of oxides the superconducting properties were very early attributed to the mixed valency of copper, Cu(II)-Cu(III), associated with the bidimensional character of the structure [20, 21]. However the problems of oxygen non stoichiometry are different in the two types of oxides. It is this issue that we would like to discuss here.

2.1. The oxides $YBa_2Cu_3O_{7-\delta}$

The oxygen nonstoichiometry in this system can be understood by considering two limit structures, the orthorhombic 92K-superconductor $YBa_2Cu_3O_7$ ($\delta = 0$) and the tetragonal insulating phase $YBa_2Cu_3O_6$ ($\delta = 1$).

The first one, $YBa_2Cu_3O_7$, consists of layers of corner-sharing CuO_5 pyramids linked through rows of corner-sharing CuO_4 square planar groups, forming triple copper layers interleaved with yttrium planes (Fig. 1a). $YBa_2Cu_3O_6$ derives from $YBa_2Cu_3O_7$ by removing the oxygen atoms of the CuO_4 groups located at the same level as copper so that the pyramidal layers are linked through CuO_2 sticks (Fig. 1b). It is worth pointing out that $YBa_2Cu_3O_7$ is characterized by a mixed valency Cu(II)-Cu(III) according to the formulation $YBa_2Cu^{II,III}_3O_7$ with certainly a complete delocalisation of the holes

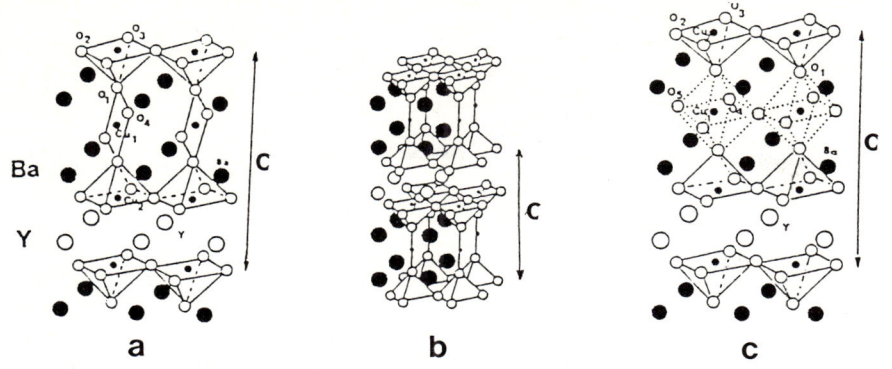

FIG. 1. Crystal structure of $YBa_2Cu_3O_7$ (a), $YBa_2Cu_3O_6$ (b), $YBa_2Cu_3O_{6.2}$ (c)

over the copper-oxygen framework, whereas in $YBa_2Cu_3O_6$, one observes a mixed valency Cu(II)-Cu(I) which corresponds in fact to a localisation of the two sorts of cations, with divalent copper in pyramidal configuration and univalent copper in two-fold coordination according to the formulation $YBa_2Cu^{II}_2Cu^{I}O_6$

The deviation from "O_7" stoichiometry leading to the general composition $YBa_2Cu_3O_{7-\delta}$ strongly affects the superconducting properties of this material. From the work done by several groups [22-25], it has been shown that T_c decreases as δ increases from 0 to 1. Nevertheless, it can be seen from Fig. 2, that this evolution of T_c is different from one author to the other. This type of behaviour is not the result of chance but corresponds to the fact that the arrangement of the oxygen vacancies in the structure depends upon the experimental method of synthesis. Clearly, order-disorder phenomena play an important role in the distribution of the oxygen vacancies. In order to understand this type of evolution we have to determine the exact distribution of the oxygen and vacancies in the layer intermediate between the two pyramidal copper layers. X-ray diffraction and neutron diffraction studies can bring information about the structure of these phases but can only give an "average structure" owing to the fact that the oxygen atoms can be distributed in the form of microdomains even in single crystal. This is for instance the case of the tetragonal non superconductor phase $YBa_2Cu_3O_{6.2}$ obtained by quenching the superconductor $YBa_2Cu_3O_7$ from 900°C to room temperature in air. All the authors [26-29] agree from X-ray and N.D. studies on a statistical distribution of oxygens and vacancies in the basal plane of the intermediate CuO_6 octahedra located between the pyramidal layers (Fig. 1c). Such a structure does not correspond to the reality since it supposes that a non negligible number of copper atoms would have a

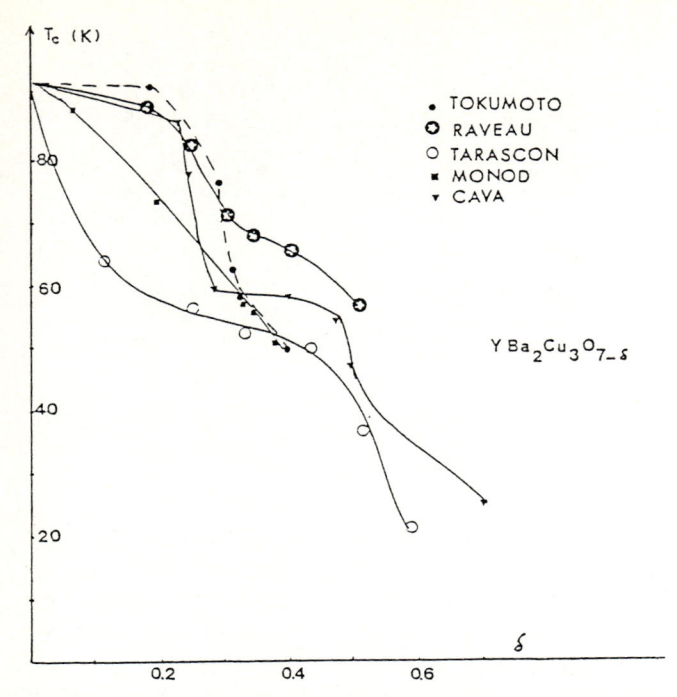

FIG. 2. Oxygen stoichiometry dependence of T_c from different authors: Cava [22], Monod [23], Tarascon [18], Tokumoto [24], Raveau [25].

three-fold coordination which has never been observed in copper oxides. The electron microscopy study of this phase confirms this point of view [28-30] it shows indeed that the crystal chemistry is more complex than deduced from powder data. Numerous crystals can be strongly disturbed and coated with an amorphous layer. The variations in the image contrast suggests significant variations of the oxygen content from one part of the crystal to the other.

Coming back to the oxygen nonstoichiometry in the orthorhombic superconducting phase YBa$_2$Cu$_3$O$_{7-\delta}$, one has to distinguish the low deviations from stoichiometry from the significant ones ($\delta \geq 0.05$).

For low values ($-0.05 < \delta < 0.05$), the chemical analysis as well as the neutron diffraction refinement of the occupancy factors of the anionic sites cannot be considered as significant. Nevertheless the electron microscopy investigation suggests that even the "stoichiometric" phase YBa$_2$Cu$_3$O$_7$ exhibits in fact an oxygen deficiency or excess. Concerning the possibility of an oxygen excess one must keep in mind the systematic microtwinning of this material [30-37] which may lead to the formation of an oxygen excess at the

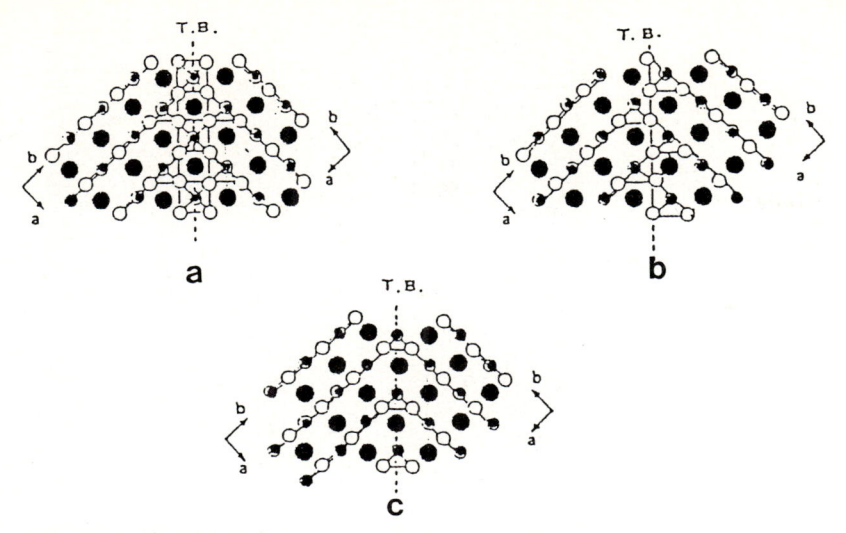

FIG. 3. Idealized models for the twinning boundary through (a) CuO$_6$ octahedra, (b) CuO$_5$ pyramids, (c) CuO$_4$ tetraedra.

twinning boundary with CuO$_6$ octahedra (Fig. 3a), or CuO$_5$ pyramids (Fig. 3b) replacing the CuO$_4$ square planar groups. However it must be pointed out that such a twinning may also take place without any change of the oxygen stoichiometry, CuO$_4$ tetrahedra replacing CuO$_4$ square planar groups at the junction (Fig. 3c) ; nevertheless this latter scheme is less likely if one observes that the formation of tetrahedra would involve a very large distortion at the junction between the domains and would lead to a higher energy for the system. Another interesting feature concerns the HREM observation along [001] of domains which exhibit a strong variation of the contrast with respect to the normal contrast of the crystal (Fig. 4) and whose size varies from 30 Å^2 to 100 Å^2. A careful simulation of the images allows such a variation to be interpreted as resulting from the presence of additional oxygen. Enlargement of this image (Fig. 5a) indeed shows that the contrast corresponding to big and small white spots, correlated to the positions of "2 oxygens + 1 vacancy" and "3 oxygens" respectively has disappeared and is replaced by equivalent spots, in agreement with the simulated images (Fig. 5b). Such a contrast can only be observed on the thin edges of the crystals in order to allow a valuable interpretation and cannot in any case lead to a determination of the occupancy factor of the sites by additional oxygen, so that the domains can be characterized by the formula YBa$_2$Cu$_3$O$_{7-\delta}$. This results in the formation of CuO$_5$ pyramids or CuO$_6$ octahedra between the pyramidal layers (Fig. 6). The observations of the crystals along [100] or [010] confirm the existence of these domains (Fig. 7), whose contrast has

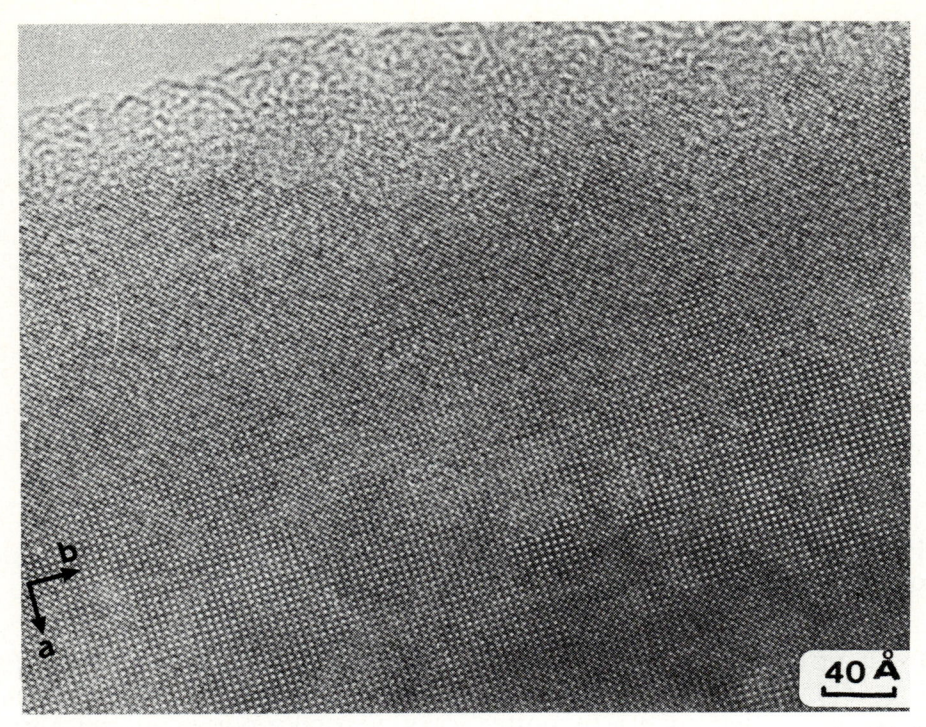

FIG. 4. [001] HREM image showing variations of the contrast which can be correlated with the presence of additional oxygen.

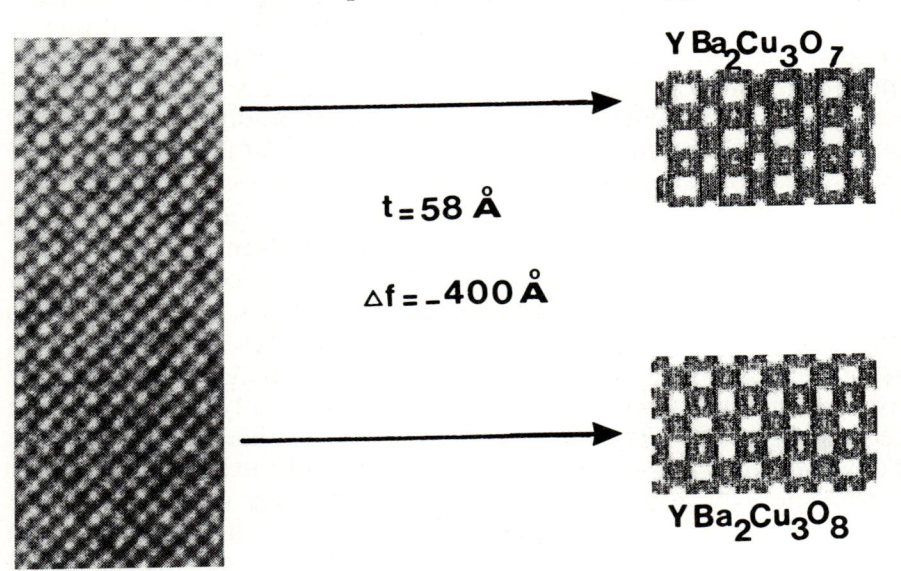

FIG. 5. Enlarged (a) and simulated (b) image corresponding to these local variations of oxygen content.

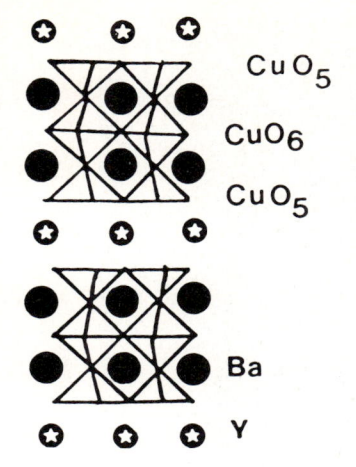

FIG. 6. Idealized model of YBa$_2$Cu$_3$O$_{8-\delta}$.

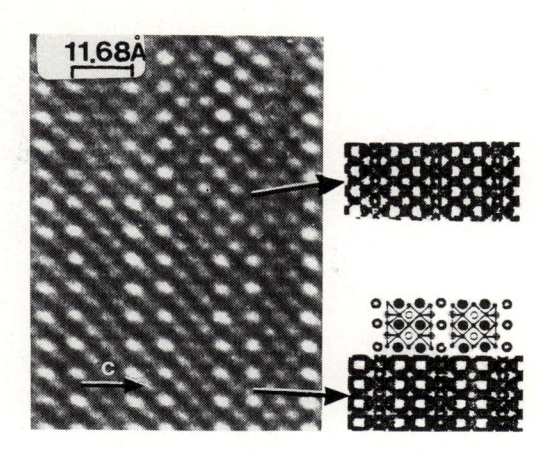

FIG. 7. [010] image of such oxygen rich domains.

been also simulated [38]. For the "stoichiometric YBa$_2$Cu$_3$O$_7$" phase one also observes defects corresponding to a decrease of the molar ratio (Y+Ba)/Cu, and leading consequently to an oxygen deficiency with respect to YBa$_2$Cu$_3$O$_7$. Such extended defects were observed for the first time as soon as 1987 [39] and were shown (Fig. 8a) to correspond to the replacement of a single row of corner-sharing CuO$_4$ square planar groups by double rows of edge-sharing CuO$_4$ square planar groups (Fig. 8b). The YBa$_2$Cu$_4$O$_8$ superconductor synthesized as a pure phase for the first time by Kaldis et al. [40] exhibits the structure of this defect.

For significant deviations from stoichiometry, it appears clearly from HREM observations that the distribution of oxygen in the crystal is most of the time inhomogeneous and strongly depends on the thermal treatment. An interesting example is that of the 60K-superconductors obtained for δ = 0.37-0.45. For those compositions local superstructures were observed [41] setting up along [10̄0], [2̄10] and [3̄10] directions.

Local "2a" superstructures are often observed (Fig. 9) without any change of the c parameter. A systematic modulation, corresponding to a mean "2a" periodicity, whose direction varies with the "a" parameter from one twinning domain to the other can also extend throughout the whole crystal (Fig. 10). In the same way the doubling of the "a" parameter is sometimes associated with a doubling of the "c" parameter. Such "2a x 2c" superstructures are shown in Fig. 11a where it can be seen that the contrast consists of one bright dot out of two along $\vec{a}$ in each triple copper layer, this contrast being translated from one layer to the adjacent one. Local "3a" superstructures are also often

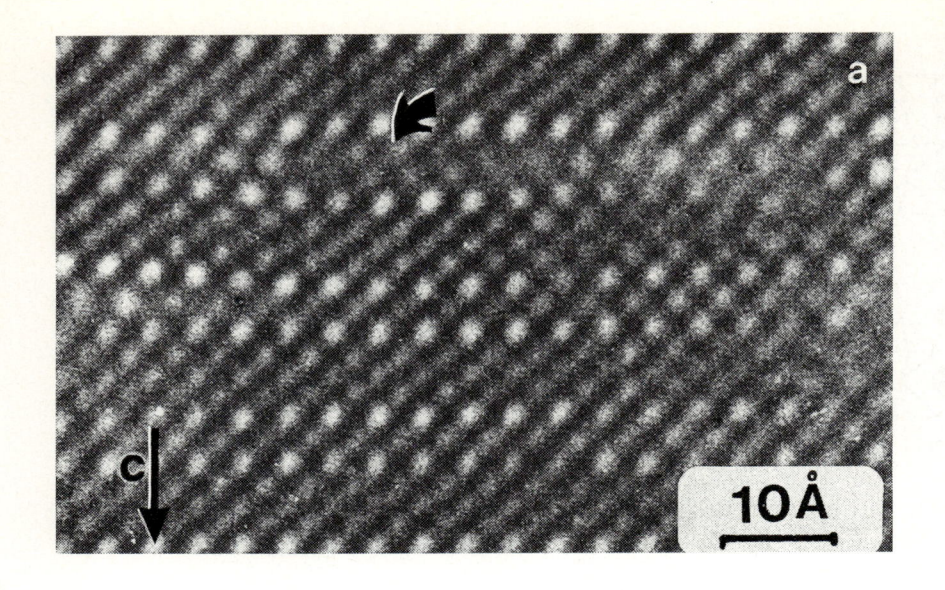

FIG. 8. (a) HREM image of a defect interpreted by the existence of a double row of edge-sharing CuO$_4$ square planar groups replacing a single row of CuO$_4$ square planar groups; (b) idealized model of the defect.

observed (Fig. 12a). Such phenomena which correspond to local ordering of oxygen vacancies were also observed from electron diffraction studies by several authors [42-47]. Nevertheless they were interpreted according to our opinion in an erroneous way since they were based either on a wrong composition [45] or on an unrealistic threefold coordination of copper [42-45]. Our models [41] proposed to explain these oxygen and vacancy orderings are based on the following points :
(i) Copper should have a usual coordination, i.e. two-fold coordination for Cu(I) and 4, 5 or 6 fold coordinations for Cu(II)-Cu(III).

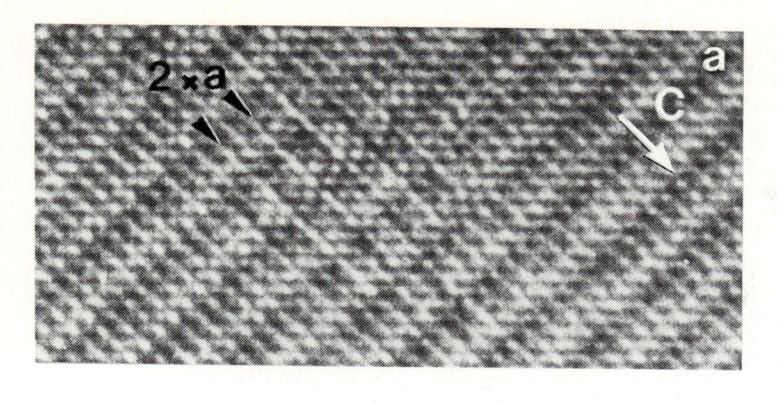

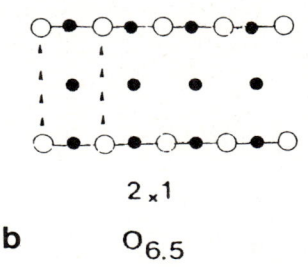

FIG. 9. Local 2a superstructure in the 60K superconductor : HREM image (a), model (b).

(ii) The observed superstructures should have a composition as close as possible to the nominal investigated composition ($\delta \sim 0.37$-0.45).
(iii) Copper should exhibit a disproportionation, according to the equation $2Cu(II) \rightarrow Cu(I) + Cu(III)$, in order to explain its coordination. Such a hypothesis [48] is founded on X-ray absorption measurements [49-51] which clearly show the existence of Cu(I) in those oxides. Moreover this latter study shows that the Cu(I) content increases as δ increases and that the holes are more distributed on the oxygen than on the copper leading for copper to a configuration Cu $3d^9$ $\underline{L}$ ($\underline{L}$: ligand hole) which excludes the existence of Cu^{3+} ions which would be $3d^8$.

In this model the multiplicity of the $\vec{a}$ axis, n x a, implies that (n-1) $[Cu^{II,III}O_2]_\infty$ rows of corner-sharing CuO_4 parallel to $\vec{b}$, alternate with one $[Cu^IO]_\infty$ row of Cu^IO_2 sticks in a periodic way. Thus the "2a" superstructure consists of one single $]Cu^IO]_\infty$ row alternating with one $[CuO_2]_\infty$ (Fig. 9b) according to the formulation $Y_2Ba_4(Cu^{II,III}{}_4O_{10})^{py}(Cu^{II,III}O_2)^{sq}(Cu^IO)^{st}$ ($\delta = 0.50$), whereas the "2a x 2c" superstructure which has the same formulation differs only by the fact that two successive layers along c are shifted with respect

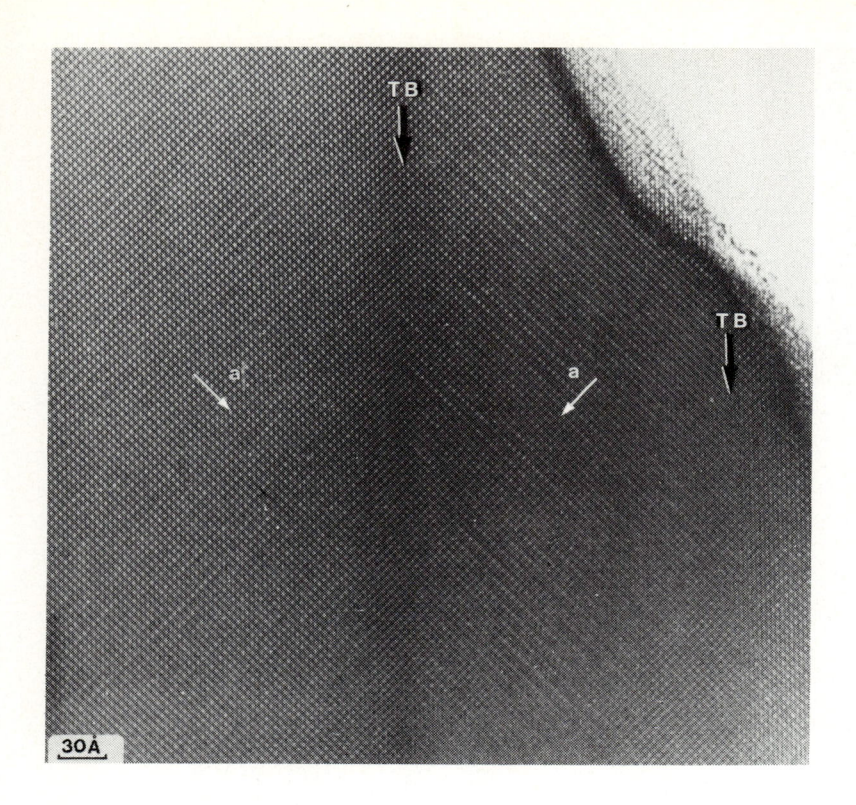

FIG. 10. Systematic modulation corresponding to a mean 2a periodicity whose direction varies from one twinning domain to the other.

one to the other (Fig. 11b). In the same way the "3a" superstructure corresponds either to the periodic sequence of one single $[Cu^IO]_\infty$ with two $[CuO_2]_\infty$ rows parallel to b (Fig. 12b) according to the formulation $Y_3Ba_6(Cu^{II,III}_6O_{15})^{py}(Cu^{II,III}O_2)_2^{sq}(Cu^IO)^{st}$ ($\delta = 0.66$) or to the sequence of one single $[CuO_2]_\infty$ row alternating with one $[Cu^IO]_\infty$ row which corresponds to $\delta = 0.33$ i.e. to the formulation $Y_3Ba_6(Cu^{II,III}_6O_{15})^{py}(Cu^{II,III}O_2)^{sq}(Cu^IO)_2^{st}$ (see Fig. 12c). Recently Amelinckx et al. [52] have shown the possibility of isolating single crystals with such superstructures ; their simulation of the structure confirms the above models.

Besides these simple micrographs, one also observes for $\delta = 0.37-0.45$ [41] other more complex local superstructures such as those running along [110] (Fig. 13) which exhibit the sequence "4a √2 - 3a √2 - 4a √2 - 3a √2 - 3a√2" or along [210] and [310] (Fig. 14). Models can easily be proposed for these supercells which respect the usual coordination of copper as shown for instance for the 2a √2 x 2a √2 or

76

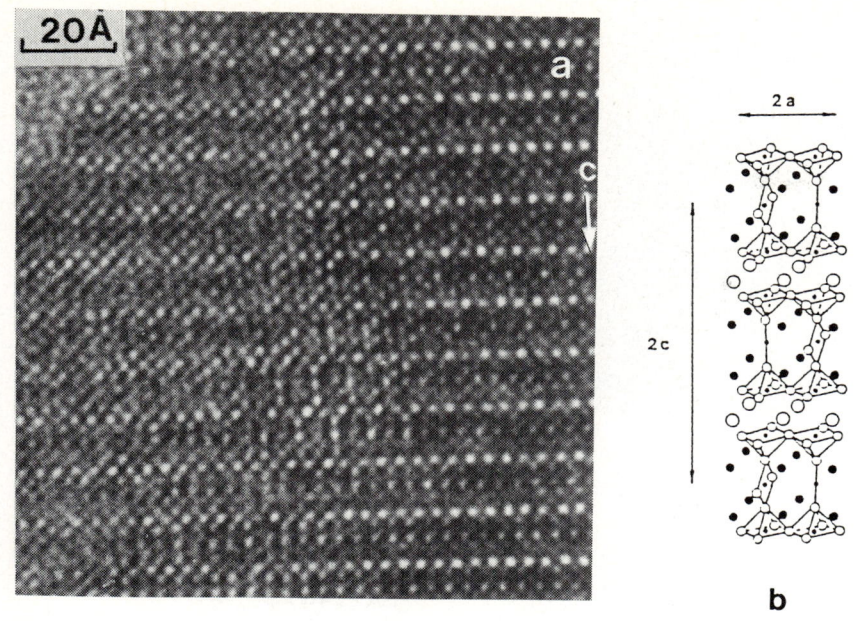

FIG. 11. (a) 2a x 2c superstructure, (b) idealized model corresponding to a formulation $YBa_2Cu_3O_{6.5}$.

the "2a x a $\sqrt{10}$" superstructures (Fig. 14a) in which CuO_5 pyramids are lined up along a, with alternated positions of the vertex which ensure the connection of segments parallel to b, formed of two CuO_4 square planar groups. This latter model can be generalized to various lengths of the segments n x CuO_4, the first one corresponding to a supercell "2a x a$\sqrt{5}$" (along [210]), with n = 1 (Fig. 14b). Thus the n value can range from n = 1 (O_7) to n = ∞ ($O_{6.5}$) and the a and b parameters of the corresponding supercell are 2a x a $\sqrt{1+(n+1)^2}$ and the oxygen content $O_{6.5}$ + (1/n+1). Moreover it must be pointed out that such an ordering of the oxygen vacancies may also be accompanied by a slight displacement of the atoms with respect to the nominal positions of the "stoichiometric" $YBa_2Cu_3O_7$ oxide, leading to a distortion of the cell.

As a conclusion the nonstoichiometry in the oxides $YBa_2Cu_3O_{7-\delta}$ can easily be understood by the coexistence in the same crystal of ordered microdomains with variable oxygen compositions, in which the usual coordination of copper is respected according to the formula $(YBa_2Cu^{II,III}_3O_7)_{1-\delta}(YBa_2Cu^{I,II}_3O_6)_\delta$. The homogeneity of the crystals and their oxygen content will then depend on the size of such domains. Clearly, the achievement of a perfect order of the oxygen vacancies for intermediate δ values will only be obtained for long annealing times and

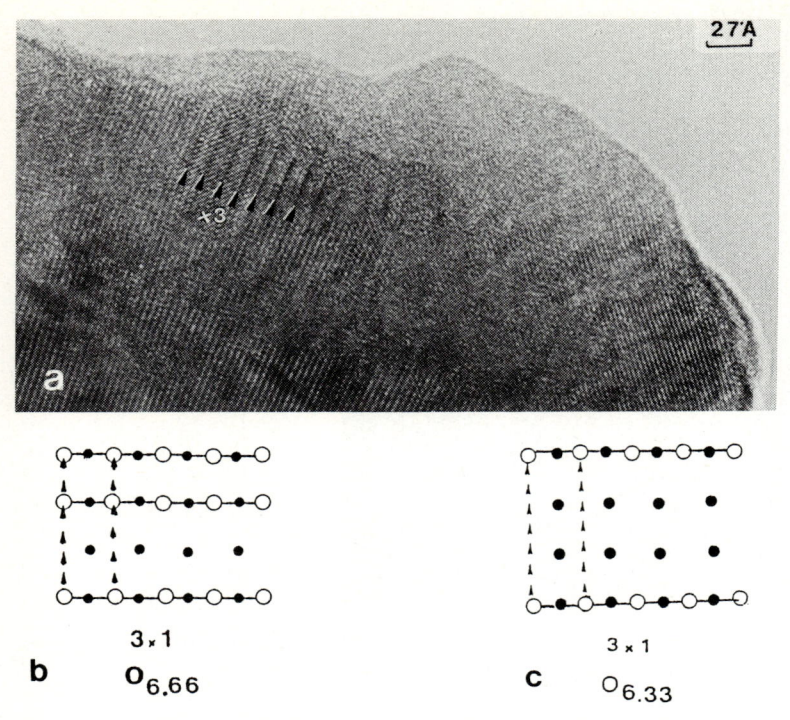

FIG. 12. (a) local 3a superstructure, (b) idealized model corresponding to general formulation $YBa_2Cu_3O_{6.66}$, (c) model corresponding to a general formulation $YBa_2Cu_3O_{6.33}$.

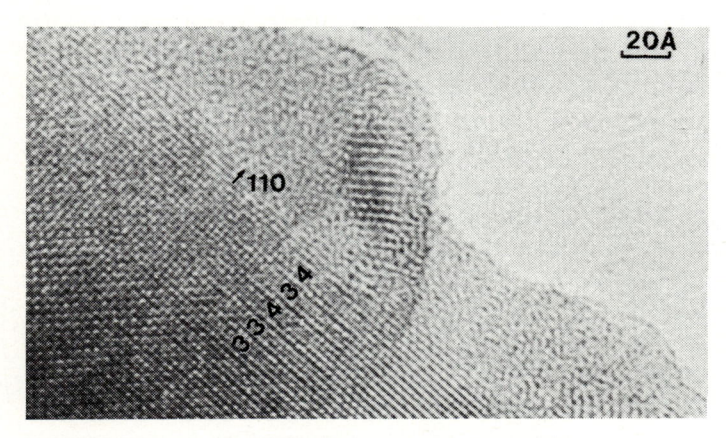

FIG. 13. Local superstructure along [110].

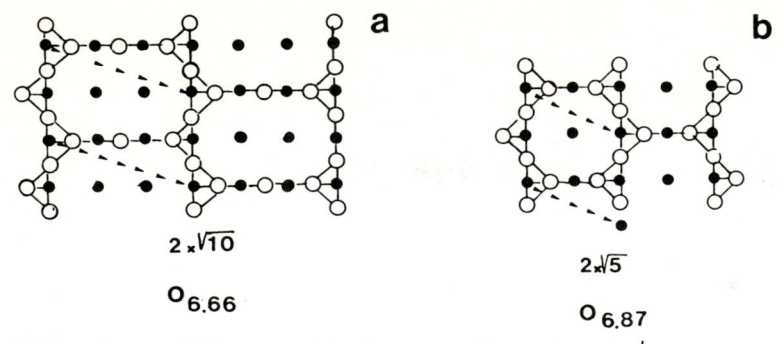

FIG. 14. Idealized model of supercell (a) $2a \times a\sqrt{10}$, (b) $2a \times a\sqrt{5}$.

by controlling the oxygen pressure in order to extend only one type of ordering to a whole crystal. On the other hand, most of the time, the diffusion of oxygen through the crystal being difficult, one will observe this coexistence of various microdomains. Consequently, it is very likely that the plateau observed in T_c near 60K corresponds to a particular predominant order, whereas the absence of this plateau for the same composition would be due to the coexistence of microdomains of very small size and having different oxygen contents.

2.2. The La_2CuO_4-type oxides

The oxides $La_{2-x}A_xCuO_4$ with A = Ca, Sr, Ba which were synthesized several years ago [5, 6] have their ideal structure formed of single rock salt layers [(A, Sr)O]$_\infty$ intergrown with single perovskite layers built up of corner-sharing CuO_6 octahedra (Fig. 15). From these previous studies [5] it was shown that as x increases the tendency to form anionic vacancies in the basal plane of the octahedra increases, leading to the formulation $La_{2-x}A_xCuO_{4-\delta}$. It results that copper which has mainly the octahedral coordination, can also present for $\delta = 0$, from time to time CuO_5 pyramids or CuO_4 square planar groups.

An interesting compound is La_2CuO_4, which was found by many physicists not to be superconducting in a first investigation. In fact it was shown very early at the beginning of 1987 [53-54] that this phase can be prepared as a massive superconductor below 37K. The oxygen nonstoichiometry in this phase is at the present time still an enigma. Pure La_2CuO_4 should be not be a superconductor since it would contain only divalent copper [55] in agreement with the fact that a non-superconducting phase can be prepared in air. Recently, Jorgensen et al. [56] have shown the existence of two forms of "La_2CuO_4" by neutron diffraction. However the deviation from oxygen stoichiometry is too low to allow a formula to be established without any ambiguity. Two models can be considered to introduce the oxygen excess, i.e. the

FIG. 15. Structure of La_2CuO_4.

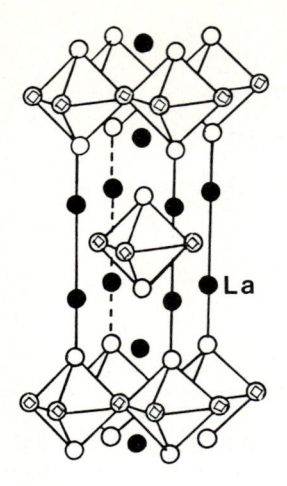

mixed valency Cu(II)-Cu(III) necessary for superconductivity. The first one can be represented by the formula $La_2CuO_{4+\delta}$; it corresponds to the presence of an oxygen excess in the rock salt-type layer. This model is supported by the fact that the isostructural $La_2NiO_{4+\delta}$ presents such a structure for $\delta = 0.18$ [57] with reasonable interatomic distances. However the stoichiometry deviation, $\delta = 0.03$, observed for the copper oxide is much lower, and the refinement of the structure performed from neutron diffraction data [58] is not really convincing especially if one observes that the O-O distances obtained by the authors are abnormally short involving the existence of "peroxide ions". The stability of such peroxides at rather low oxygen pressures (near to 1 bar) and at high temperature (T > 500°C) would be rather exceptional ! The second model, corresponding to the formula $La_{2-2\varepsilon}Cu_{1-\varepsilon}O_4$ previously proposed [59], implies for the La/Cu ratio equal to 2, the simultaneous existence of vacancies on the La and Cu sites. Such a model corresponds for $\delta = 0.032$ to a very low ε value, $\varepsilon = 0.007$ and thus cannot really be checked by N.D. measurements. Moreover in this latter hypothesis one can also propose the possibility of coexistence of superconductive microdomains in the same crystal corresponding to the formula $La_2Cu_{1-\varepsilon}O_4$ and $La_{2-4\varepsilon}CuO_4$ in which lanthanum and copper sites would be fully occupied respectively. The structural data obtained by Jorgensen et al. [57], for $La_2NiO_{2+\delta}$ are compatible with this second model. They indeed show that besides the "$\delta = 0.18$"-phase, another phase exists for $\delta \sim 0.02$ which incorporates excess oxygen by forming a different defect. Thus the issue of oxygen nonstoichiometry in La_2CuO_4 remains still open. Nevertheless it is clear that this phase has a very different behaviour from the "123" phase in that it does not involve any disporportionation of copper leading to the presence of univalent copper. This point of view is confirmed by the X-ray absorption studies of the La_2CuO_4-type oxides [60] for which Cu(I) has never been observed; whereas the species Cu $3d^9$ and $Cud^9\underline{L}$, confirm the existence of the mixed valency Cu(II)-Cu(III).

3. The thallium, lead and bismuth cuprates

The study of the oxygen nonstoichiometry in these oxides is at its very beginning, so that it is very difficult to draw definitive conclusions about their behaviour. Nevertheless some general features can be outlined.

The first striking feature deals with the thallium cuprates $TlA_2Ca_{m-1}Cu_mO_{2m+3}$ (A = Sr, Ba) [61-70] and $Tl_2Ba_2Ca_{m-1}Cu_mO_{2m+4}$ [71-78] for which no deviation from the ideal oxygen stoichiometry "O_{2m+3}" or "O_{2m+4}" has been observed up to now from the various structural studies. Indeed in these oxides it seems that the nonstoichiometry is mainly governed by the thallium deficiency in the rock salt-type layers which may then, for instance, lead to formulations such as $Tl_{2-x}Ba_2Ca_{m-1}Cu_mO_{2m+4}$, allowing the mixed valence of copper in these oxides to be understood. But this latter point which deals with the existence of many possible extended defects has previously been reviewed [79] and will not be discussed here. The important point in those materials concerns the valency of the different species. Recent Xanes studies [80, 81] have shown that those phases exhibit for copper at K edge a pattern characteristic of $3d^9$ and $3d^{10}\underline{L}$ ($\underline{L}$: ligand hole) configurations with a shift of the edge due to the presence of extra-oxygen p holes introduced by doping. Again, like for La_2CuO_4-type oxides, no univalent copper was observed for those phases. The patterns of thallium at L_{III} (Fig. 16a) and at L_I (Fig. 16b) are very similar for both oxides $TlBa_2Ca_2Cu_3O_9$ and $Tl_2Ba_2Ca_2Cu_3O_{10}$ whose structures consist of triple copper layers intergrown with double rock salt-type (Fig. 17a) and triple rock salt type layers (Fig. 17b) respectively. From the comparison with the patterns of $Tl^{III}_2O_3$ and $Tl^I_2Ta_2O_6$ it appears clearly that these spectra are similar to that of Tl_2O_3 (peaks A' and B) and are located at the same energy. The main difference is observed at L_{III} edge with the presence of a peak located at the same energy as the A peak of $Tl_2Ta_2O_6$; moreover one can notice that the intensity of the peak corresponding to the prepeak B in Tl_2O_3 has increased. The A' peak corresponds to the two short Tl-O distances (2.2 Å) of the TlO_6 octahedron parallel to c like in $Tl^{III}_2O_3$, whereas the A peak is characteristic of the long Tl-O distances (~ 2.75 Å) of the basal plane of the octahedra similar to those of $Tl^I_2Ta_2O_6$, in agreement with the

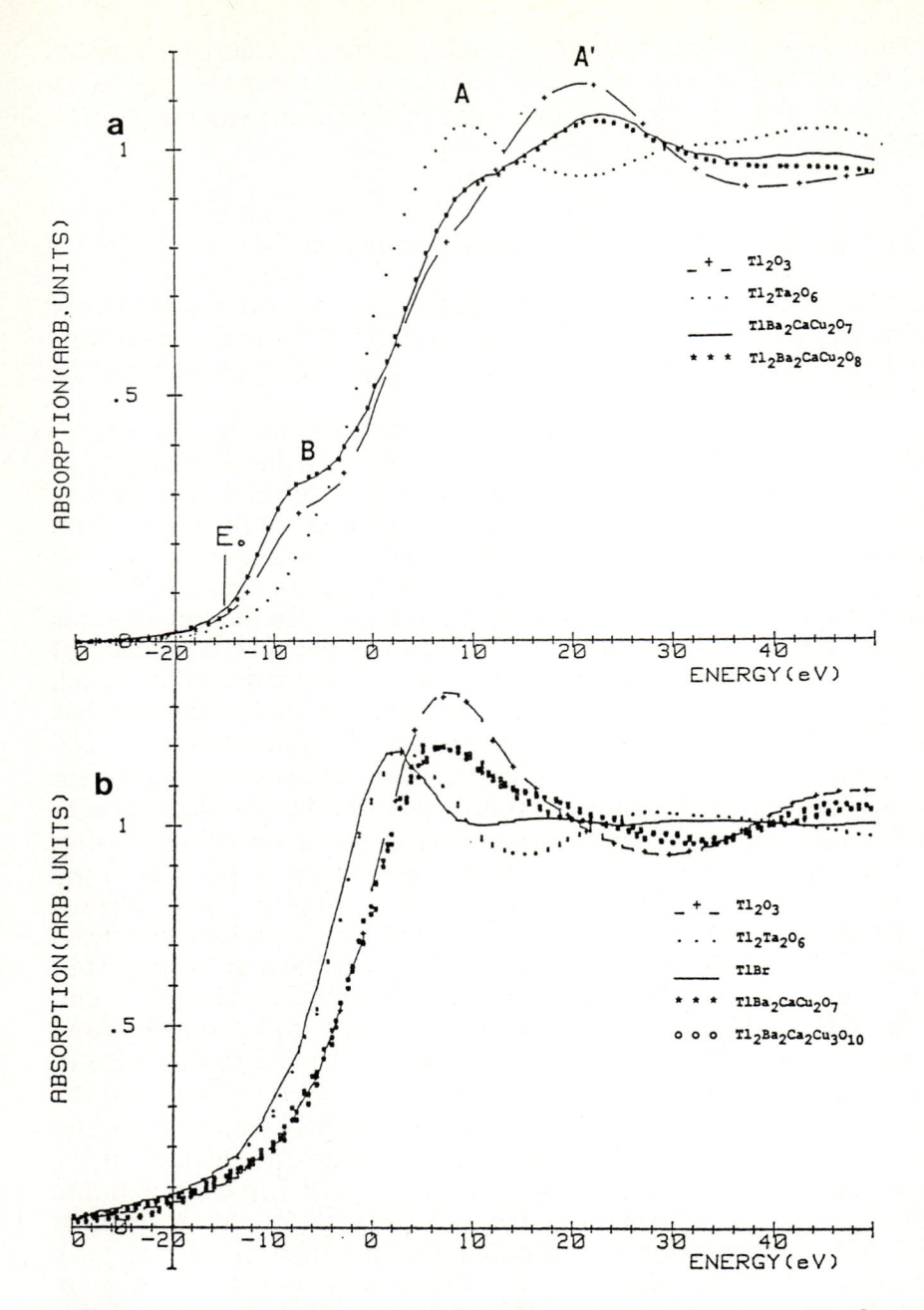

FIG. 16. a) Tl L$_{III}$-edge spectra for some reference, Tl$_2$O$_3$ and Tl$_2$Ta$_2$O$_6$ and superconductive, 1212 and 2212 compounds.
b) TL L$_I$-edge spectra for the same compounds and TlBr as a standard for Tl(I) valence state.

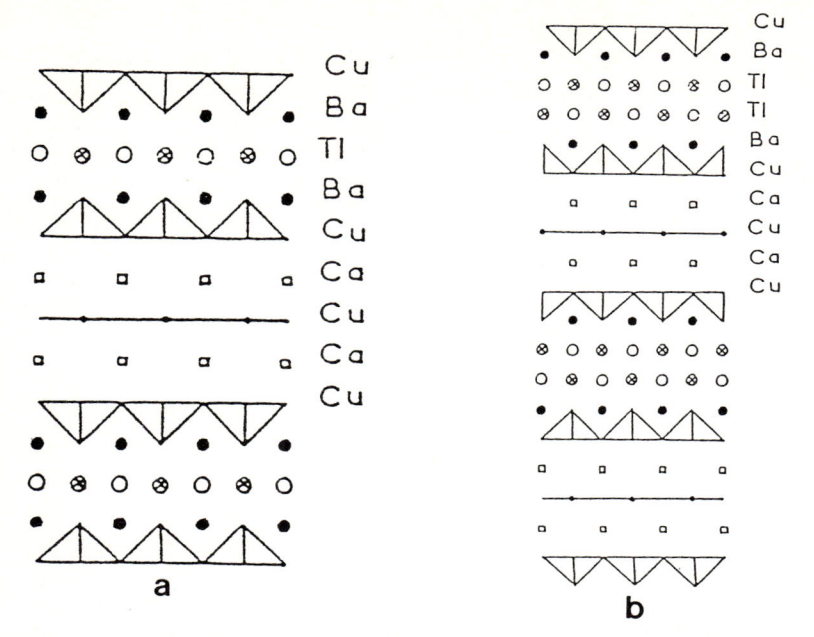

FIG. 17. Schematic drawing of the structure of TlBa$_2$Ca$_2$Cu$_3$O$_9$ (a) and Tl$_2$Ba$_2$Ca$_2$Cu$_3$O$_{10}$ (b).

X-ray diffraction studies. Thus from this point of view the patterns of the superconductive thallium cuprates appear as intermediate between those of Tl(I) and Tl(III) oxides. Nevertheless the edge shifts observed for those phases as well as the increase of prepeak B, show the absence of electrons in the 6s level and allow the presence of univalent thallium to be ruled out within the limit of detection (± 10 %). Thus it is not possible to decide whether the thallium oxygen layers play the role of reservoir of holes or not for the copper oxygen layers. The issue of the competition between the systems Tl(I)-Tl(III) and Cu(II)-Cu(III) remains still open.

The oxygen nonstoichiometry in superconductive bismuth cuprates [82-98] is so far not completely understood because of the existence of incommensurate satellites [84, 88, 96, 99-105] which makes difficult the accurate determination of the structure and especially of the oxygen network in the bismuth bilayers. Nevertheless after the synthesis of isostructural iron oxides [106, 107], the single crystal study [108, 109] has shown that the "waving" rock salt-type bismuth oxygen layers are not only characterized by modulated displacements of the bismuth and oxygen ions but that they contain an excess oxygen as shown for instance for Bi$_2$Sr$_2$Ca$_2$Fe$_3$O$_{12}$ (Fig. 18). This ability to present an excess oxygen with respect to the ideal

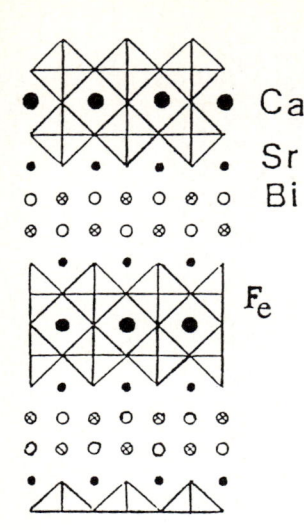

FIG. 18. Schematic drawing of the structure of the non-superconducting oxide $B_2Sr_2CaFe_2O_9$.

formula $Bi_2Sr_2Ca_{m-1}Cu_mO_{2m+4}$, is confirmed by the recent neutron diffraction study [110] of the m = 2-superconductor $Bi_2Sr_2CaCu_2O_8$. If there is no doubt that, the Cu-O layers are always characterized by the mixed valence Cu(II)-Cu(III), involving holes in the oxygen band according to the configurations Cu3d^{9}L observed for all other superconductors, the issue of the valence states of bismuth in these oxides is so far not clear. Nevertheless a preliminary X-ray absorption investigation of these phases [111] sheds some light on this problem. The patterns at L(III) edges (Fig. 19) of the three superconductors (m = 1, 2 and 3) can usefully be compared with those of Bi_2O_3, $Bi_2SrNb_2O_9$ in which bismuth is only trivalent, with that of $NaBiO_3$ in which bismuth is pentavalent, and with that of the isostructural phase $Bi_2Sr_2CaF_2O_9$ which does not superconduct. Considering the reference compounds (Fig. 19a) it appears clearly that the Bi_2O_3 and $Bi_2SrNb_2O_9$ (Aurivillius phase built up from pyramidal Bi_2O_2 layers) exhibit absolutely identical patterns, characteristic of Bi(III). From the comparison of the spectra of the superconductive cuprates (Fig. 19b) and of $Bi_2Sr_2CaFe_2O_{12}$ (Fig. 19c) with that of $NaBiO_3$ it can also be seen that both oxides exhibit small amounts of Bi(V) identified by the prepeak corresponding to empty 6s levels. Nevertheless the amount of Bi(V) in the bismuth cuprates is smaller than that observed for the isostructural iron compound. But the most striking feature concerns the shifting at the Bi L(III) edge (Fig. 19b) of the curves of the superconductive cuprates towards lower energy with respect to other Bi(III) oxides. Such a feature is indeed not observed in either the layered Aurivillius phase $Bi_2SrNb_2O_9$ (Fig. 19a) or in the isostructural iron oxide $Bi_2Sr_2CaFe_2O_9$ (Fig. 19c) which seems to exhibit similar Bi-O distances due to similar incommensurate structure.

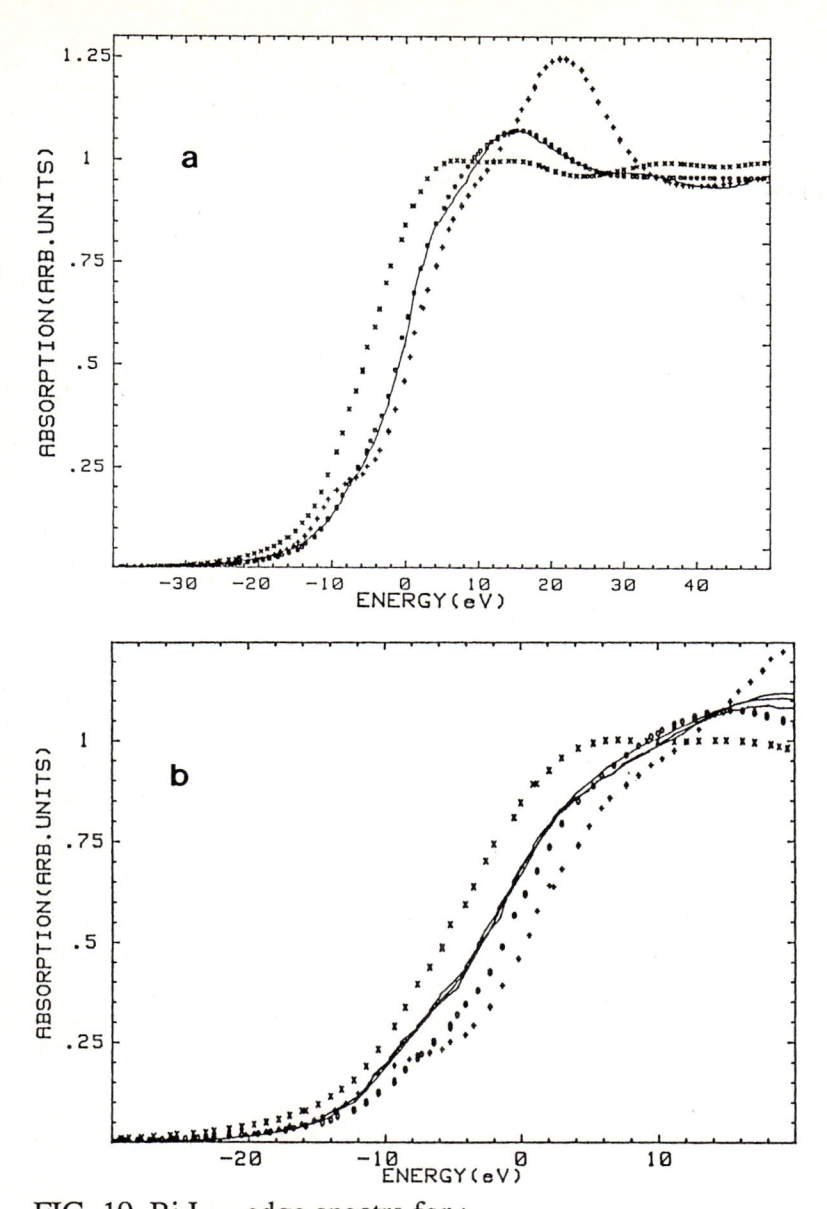

FIG. 19. Bi L$_{III}$-edge spectra for :
a) xxx Bi metal ; ooo Bi$_2$O$_3$; +++ NaBiO$_3$; ___ Bi$_2$SrNb$_2$O$_9$
b) xxx Bi metal ; ooo Bi$_2$O$_3$; +++ NaBiO$_3$; ___ the three superconductors 2201, 2212 and 2223,

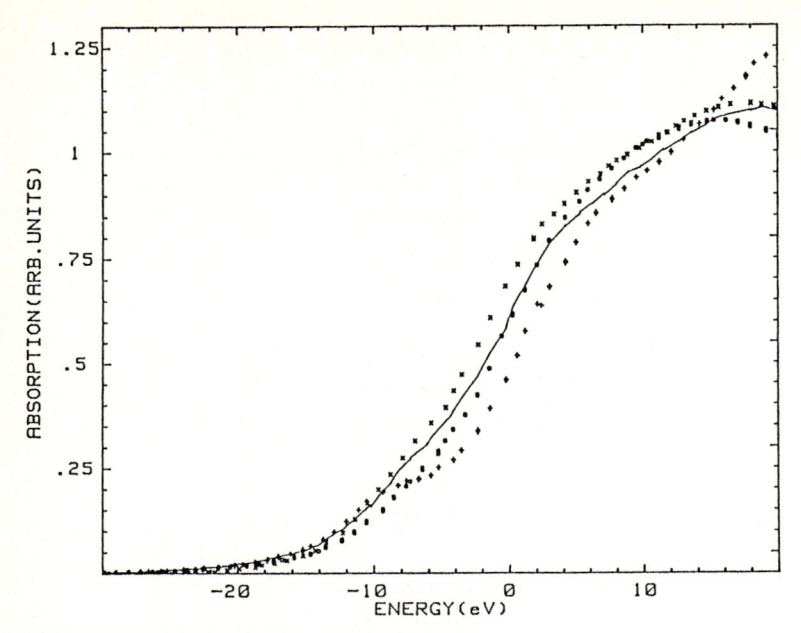

FIG. 19c. Bi L_{III}-edge spectra for :
xxx Bi metal ; ooo Bi_2O_3 ; +++ $NaBiO_3$; ___ $Bi_2Sr_2CaFe_2O_9$

Such a feature suggests for bismuth a mean oxidation state smaller than three which has never been observed to our knowledge. An alternative explanation would be a hybridization of the 6s-6p of bismuth allowing a narrow band to be built up by overlapping with the 2p orbitals of oxygens.In this hypothesis electrons would be transferred from the copper-oxygen layers towards the bismuth layers; consequently n-type semi-metallic properties should be observed for the bismuth oxygen layers, whereas p-type superconductivity would exist in the copper oxygen layers. The presence of Bi(V) would only be a secondary phenomenon resulting from the excess oxygen introduced in the bismuth-oxygen layers.

Very little is known about the oxygen nonstoichiometry in the lead cuprates whose study is just starting. However, it is worth pointing out the large oxygen deficiency in the oxides $Pb_2Sr_2Y_{0.5}Ca_{0.5}Cu_3O_8$ [112] and $Pb_{2-x}Bi_xSr_2Y_{1-y}Ca_yCu_3O_8$ [113]. The structure of these phases (Fig. 20a) is indeed built up from double copper layers intergrown with single rock salt-type layers [(Pb, Sr)O]$_\infty$ and layers of CuO_2 sticks. These latter slabs are very similar to those observed in $YBa_2Cu_3O_6$ (Fig. 1b). In fact this structure can be deduced from the hypothetical phase "$Pb_2Sr_2YCu_3O_{10}$" (Fig. 20b) which corresponds to a double intergrowth of [2,1] structure with a [1,1] structure by eliminating the oxygen atoms of the basal plane of the

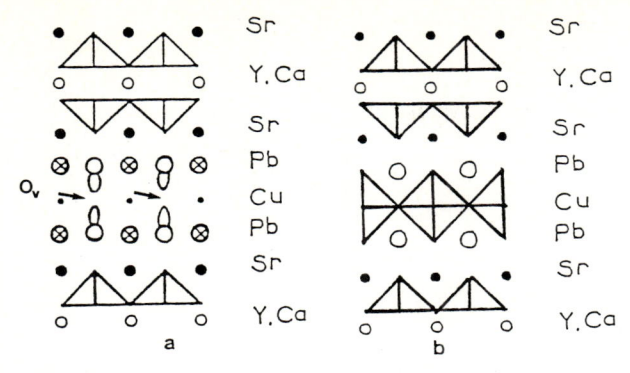

FIG. 20. Schematic drawing of the structure of $Pb_2Sr_2Y_xCa_{1-x}Cu_3O_8$-type oxides (a) and of the hypothetical oxide $Pb_2Sr_2YCu_3O_{10}$ (b).

CuO_6 octahedra of the [1,1] layer (O_V sites). Thus in that sense, this structural type corresponds to a large deviation from the ideal oxygen stoichiometry "O_{10}". This oxygen nonstoichiometry is confirmed by the fact that this phase can be easily oxidized by heating either in air or in an oxygen flow, without any destruction of the structure, the additional oxygens occupying the O_V sites (Fig. 20b). The X-ray absorptions study of $Pb_2Sr_2Y_{0.5}Ca_{0.5}Cu_3O_8$ [81] confirms for this phase the model of copper disproportionation proposed for $YBa_2Cu_3O_{7-\delta}$. One indeed observes for this oxide a K-edge pattern characteristic of univalent copper, in agreement with the X-ray diffraction study showing that a part of copper is in two-fold coordination (Cu^IO_2 sticks), whereas the superconductivity should appear in the pyramidal layers which are characterized by the mixed valency Cu(II)-Cu(III). Thus the formula of this phase can be written $Pb_2Sr_2Y_{0.5}Ca_{0.5}(Cu^{II,III}_2)_{py}(Cu^I)_{st}O_8$. The problem of valency of the different elements after oxidization has not been studied so far. Nevertheless, the disappearance of superconductivity for oxidized compounds suggests that holes are mainly transferred from copper layers to lead layers, i.e. that Pb(II) has been oxidized into Pb(IV). The recently discovered 100K-superconductor $Pb_{0.5}Sr_{2.5}Y_{0.5}Ca_{0.5}Cu_2O_{7-\delta}$ [114] shows also important problems of oxygen nonstoichiometry. The structure of this phase (Fig. 21) belongs to the [2,2] type already described for $TlBa_2CaCu_2O_7$ [64]; it consists of double pyramidal copper layers (m = 2) intergrown with double rock salt layers (n = 2). The rock salt-type slabs are in this phase formed of mixed $[Sr_{0.5}Pb_{0.5}O]_\infty$ layers sandwiched by pure $[SrO]_\infty$ layers. Nevertheless the actual structure is more complicated than this schematized model; the preliminary X-ray study of this phase suggests a significant deviation from oxygen stoichiometry in the rock salt layers ($\delta = 0.5$);

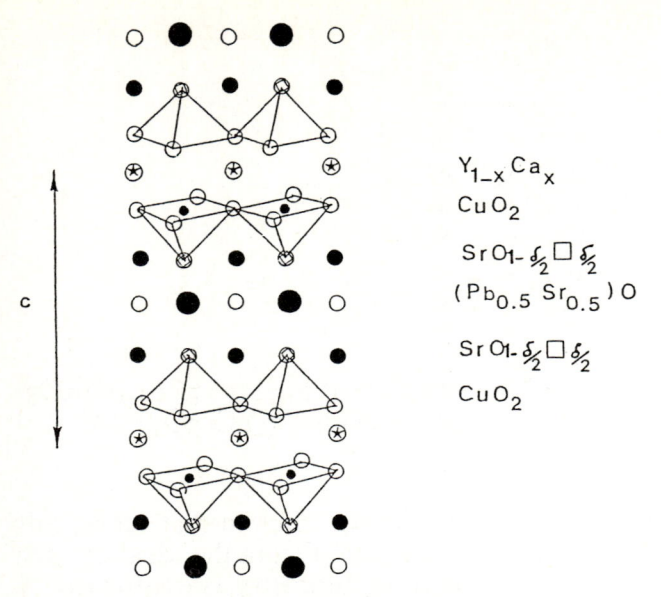

FIG. 21. Schematic drawing of the structure of the superconductor $Pb_{0.5}Sr_{2.5}Y_{0.5}Ca_{0.5}Cu_2O_{7-\delta}$

this seems to be correlated with the presence of the $6s^2$ lone pair of Pb(II) which tends to extend towards the anionic vacancies. The fact that the resistive transition is broad and changes with the experimental method and especially with the oxygen pressure and annealing (Fig. 22) shows again that the oxygen distribution in the structure should play a capital role in the superconducting properties of this phase.

Concluding remarks

The oxygen nonstoichiometry is the key to the understanding of superconductivity in high T_c superconducting oxides since : it governs the valence states of the metallic elements in these materials. It is now well established that the "123" phase is characterized by a particular behaviour involving a static disproportionation of copper into Cu(I) and the formal mixed valency Cu(II)-Cu(III) ; this phenomenon is also observed for the superconductor $Pb_2Sr_2Y_{0.5}Cu_{0.5}Cu_3O_8$. On the other hand the structure of the other superconductors is not favourable to such a disproportionation in spite of their great similarity. The issue of valence states in bismuth, thallium and lead cuprates is so far not understood owing to the multiple valencies presented by these elements able to influence the valence states of copper, but there is no doubt that for those oxides the oxygen stoichiometry plays also a capital role in the superconducting properties. A systematic study of the processing of

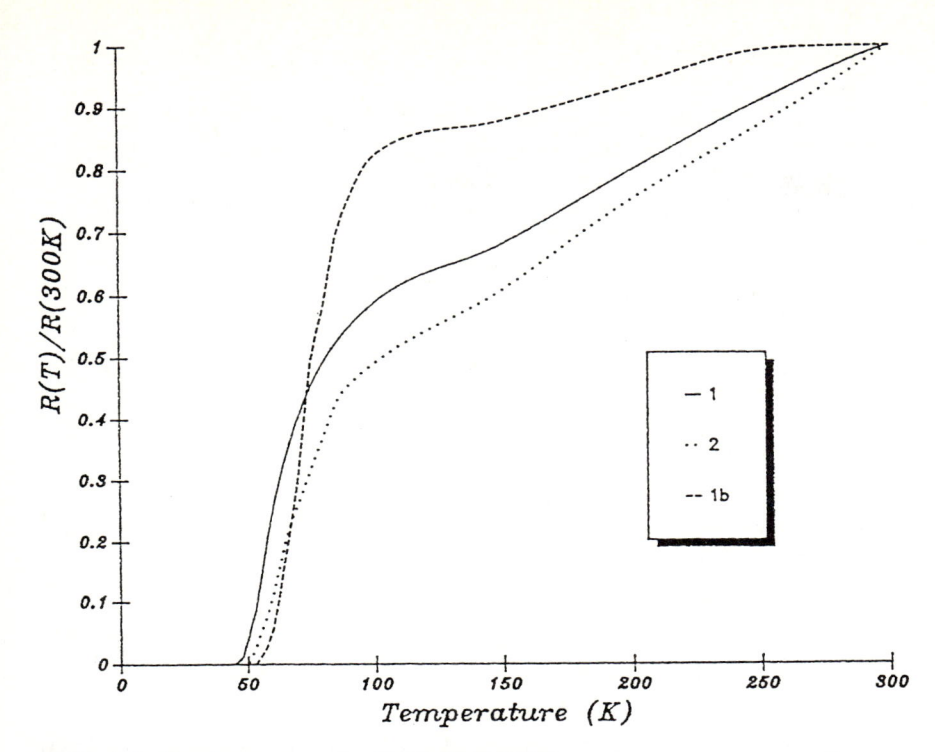

FIG. 22. Resistance ratio R(T)/R(300K) versus temperature for a sample of nominal composition $Pb_{0.5}Sr_{2.5}Y_{0.5}Ca_{0.5}Cu_2O_7$ heated at 875°C in evacuated ampoule (1) then annealed under an oxygen flow at 400°C for 12 hours (1b) and for a sample of nominal composition $Pb_{0.5}Sr_{2.5}Y_{0.5}Ca_{0.5}Cu_2O_{7.25}$ treated at 875°C in evacuated ampoule (2)

these materials needs to be carried out, in order to control the different parameters - temperature, oxygen pressure, time, etc. - which govern the kinetics of the oxygen diffusion in the bulk as well as the thermodynamics of those phases.

REFERENCES

1. Bednorz J.G., Müller K.A., Z. Phys. B64 (1986) 189.
2. Raveau B., Michel C., Hervieu M., Proc. of the 11th Int. Symp. Princeton, June 1988, Solid State Ionics, 32-33 (1989) 1035.
3. Raveau B., Michel C., Hervieu M., Groult D., Provost J., Proc. of the M.R.S. Meeting, San Diego, Apr. 1989.
4. Raveau B., Michel C., Hervieu M., Groult D., Chemistry of Superconducting materials, Ed. Vanderah T.A., Noyes Publication, 1989.

5. Nguyen N., Studer F., Hervieu M., Raveau B., J. Solid State Chem., 9 (1981) 120.
6. Michel C., Raveau B., Rev. Chim. Miner., 21 (1984) 407.
7. Bednorz J.G., Takashige M., Müller K.A., Europhys. Lett., 3 (1987) 379.
8. Cava R.J., Van Dover R.J., Batlogg B., Rietman E.A., Phys. Rev. Lett., 58 (1987) 408.
9. Tarascon J.M., Green L.H., McKinnon W.R., Hull G.W., Geballe T.H., Science, 235 (1987) 1373.
10. Rao C.N.R., Ganguly P., Curr. Sci., 56 (1987) 47.
11. Beille J., Cabanel, R., Chevalier B., Chaillout C., Demazeau G., Deslandes F., Etourneau J.,Lejay P., Michel C., Provost J., Sulpice A., Tholence J.L., Tournier R., C.R. Acad. Sci., 304 II (1987) 1097.
12. Wu M.K., Ashburn J.R., Torng C.J., Hor P.H., Meng R.L., Gao L., Huang Z.J., Wang Y.Z., Chu C.W., Phys. Rev. Lett., 58 (1987) 908.
13. Cava R.J., Batlogg B., Van Dover R.B., Murphy D.N., Sunshine S., Siegrist T., Remeika J.P., Rietman E.A., Zahurak S., Espinosa G.P., Phys. Rev. Lett., 58 (1987) 1676.
14. Michel C., Deslandes F., Provost J., Lejay P., Tournier R., Hervieu M, Raveau B., C.R. Acad.Sci., 304 II (1987) 1059.
15. Lepage Y., McKinnon W.R., Tarascon J.M., Greene L.H., Hull G.W., Hwang D.M., Phys. Rev. B, 35 (1987) 7245.
16. Beno M.A., Soderholm L., Capone D.W., Hinks D., Jorgensen J.D., Schuller I.K., Segre, C.V. Zhang K., Grace J.D., Appl. Phys. Lett., 51 (1987) 57.
17. Capponi J.J., Chaillout C., Hewat A.W., Lejay P., Marezio M., Nguyen N., Raveau B., Soubeyroux J.L., Tholence J.L., Tournier R., Europhys. Lett., 12 (1987) 1301.
18. Tarascon J.M, Greene L.H., Bagley B.G., McKinnon W.R., Barboux P., Hull G.W., Novel Superconductivity, Ed. Wolf A., Kresin V., Plenum, New York, (1987) 705.
19. Michel C., Deslandes F., Provost J., Lejay P., Tournier R., Hervieu M., Raveau B., C.R. Acad.Sci., 304 II (1987) 1169.
20. Friedel J., J. Phys., 48 (1987) 1787.
21. Labbé J., Bok J., Europhys. Lett., 3 (1987) 1225.
22. Cava R.J., Batlogg B., Chen. C.H., Rietman E.A., Zahurak S.M., Werder D., Nature 329 (1987) 423.
23. Monod P., Ribault M., D'Yvoire F., Jegoudez J., Collin G., Revcolevschi A., J. Phys., 48 (1987) 1369.
24. Tokumoto M., Ihara H., Matsubara T., Hirabayashi M., Terada N., Oyanagi H., Murata K., Kimura Y., Jpn. J. Appl. Phys., 26 (1987) L1566.

25. Raveau B., Deslandes F., Michel C., Hervieu M., Proc. of the Intern. Meeting on High T_c superconductors, Schloss Mauterndorf, Feb. 1988, High T_c superconductors, Ed. Weber H.W., Plenum N.Y., (1988) 3.
26. Izumi F., Asano H., Ishigaki T., Takayama-Maromachi E., Uchida Y., Watanabe N., Nishikawa T., Jpn. J. Appl. Phys., 26 (1987) L649.
27. Jorgensen J.D., Beno M., Hinks D.G., Soderholm L., Volkin K.J., Kitterman R.L., Grace J.D., Schuller I.K., Segre C.V., Zhang K., Kleefisch M.S., Phys. Rev. B, 36 (1987) 3608.
28. Domengès B., Hervieu M., Caignaert V., Raveau B., Tholence J.L., Tournier R., J. Microsc.Spectr. Electr., 13 (1988) 75.
29. Renault A., McIntyre G.J., Collin G., Pouget J.P., Comes R., J. Phys. 48 (1987) 1407.
30. Hervieu M., Domengès B., Raveau B., Tarascon J.MI., Post M., McKinnon W.R., Mat. Chem. Phys. 21 (1989) 181.
31. Roth G., Ewert D., Heger G., Hervieu M., Michel C., Raveau B., D'Yvoire F., Revcolevschi A.,Z. Phys. B, 69 (1987) 21.
32. Domengès B., Michel C., Heger G., Provost J., Raveau B., Phys. Rev. B, 36 (1987) 3922.
33. Hervieu M., Domengès B., Michel C., Provost J., Raveau B., J. Solid State Chem., 71 (1987) 263
34. Bordet P., Capponi J.J., Chaillout C., Hodeau J.L., Marezio M., Progress in High T_c Superconductivity, Ed. Rao C.N.R., World Scientific, (1988) 76.
35. Van Tendeloo G., Zandbergen H.W., Amelincks S., Solid State Comm., 63 (1987) 603.
36. Iijima S., Ichihashi T., Kubo Y., Tabuchi J., Jpn J. Appl. Phys., 26 (1988) 1478.
37. Hewat E.A., Dupuy M., Bourret A., Capponi J.J., Marezio M., Solid State Commun., 63 (1987) 389.
38. Hervieu M., Domengès B., Michel C., Raveau B., Europhys. Lett., 4 (1987) 205.
39. Domengès B., Hervieu M., Michel C., Raveau B., Europhys. Lett., 4 (1987) 211.
40. Bordet P., Chaillout C., Chenevas J., Hodeau J.L., Marezio M., Karpinski J., Kaldis E., Nature, 334 (1988) 596.
41. Hervieu M., Domengès B., Raveau B., Post M., McKinnon W.R., Tarascon J.M., Mat. Lett., 8 (1988) 73.
42. Chaillout C., Alario-Franco M.A., Capponi J.J., Chenavas J., Hodeau J.L., Marezio M., Phys. Rev.B, 36 (1987) 7118.
43. Alario-Franco M.A., Chaillout C., J. Solid State Chem., (1989) in press.
44. Chaillout C., Alario-Franco M.A., Capponi J.J., Chevanas J., Strobel P., Marezio M., Physica C, 156 (1988) 455.

45. Zandbergen H.W., Van Tendeloo G., Obake T., Amelincks S., Phys Status Solidi (a), 103 (1987) 45.
46. Van Tendeloo G., Zanbergen H.W., Amelincks S.A., Solid State Comm., 63 (1987) 389.
47. Ichihashi T., Iijima S., Kubo Y., Tabuchi J., Jpn. J. Appl. Phys., (1988)
48. Raveau B., Michel C., Hervieu M., Provost J., Physica C, 153 (1988) 5.
49. Oyanagi H., Ihara H., Matsubara T., Tokumoto M., Matsushita T., Hirabayashi M., Murata K., Terada N., Yao T., Iwasaki H., Kimura Y., Jpn. J. Appl. Phys., 26 (1987) L 1561.
50. Baudelet F., Collin G., Dartyge E., Fontaine A., Kappler J.P., Küll G., Itie J.P., Jegoudez J., Maurer M., Monod P., Revcolevschi A., Tolentino H., Tourillon G., Verdaguer M., Z. Phys. B, 69 (1988) 141.
51. Bianconi A., De Santis M., Di Cocco A., Flank A.M., Fontaine A. Lagarde P., Katayama-Yoshida, H., Kotani A., Marcelli A., Phys. Rev. B, 38 (1988) 7196.
52. Reyes Gasga, Krekels T., Van Tendeloo G., Van Landuyt J., Amelinckx S., Bruggink W.M.M., Verwey M., Physica C, submitted.
53. Beille J., Cabanel R., Chevalier B., Cahillout C., Demazeau G., Deslandes F., Etourneau J., Lejay P., Michel C., Provost J., Sulpice A., Tholence J.L., Tournier R., C.R. Acad. Sci., 304 II (1987) 1097.
54. Beille J., Chevalier B. Demazeau G., Deslandes F., Etourneau J., Laborde O., Michel C., Lejay P. Provost J., Raveau B., Sulpice A., Tholence J.L., Tournier R., Physica B, 146 (1987) 307.
55. Singh K.K., Ganguly P., Goodenough J.B., J. Solid State Chem., 52 (1984) 254.
56. Jorgensen J.D., Dabrowki B., Pei S., Hinks D.G. Soderholm L., Morosin B., Shirber J.E., Venturini E.L., Ginley D.S., Phys. Rev. B, 38 (1988) 11337.
57. Jorgensen J.D., Dabrowski B., Pei S., Richards D.R., Kinks D.G., Phys. Rev. B., 1989, in the press.
58 Chaillout C., Cheong S.W., Fisk Z., Lehmann M.S., Marezio M., Morosin B., Schirber J.E., Physica C, 158 (1989) 183.
59. Raveau B., Michel C., Hervieu M., Chemistry of high temperature superconductors, Ed. Nelson D.L., Whittingham M.S., George T.F., A.C.S. Washington (1987) 122.
60. Bianconi A., Budnick J., Flank A.M., Fontaine A., Lagarde P., Marcelli A., Tolentino H., Chamberland B., Demazeau G., Michel C., Raveau B., Phys. Lett. A, 127 (1988) 285.

61. Beyers R., Parkin S.S.P, Lee V.Y., Nazzal A.I., Savoy R., Gorman G. Huang T.C., La Placa S., Applied Phys. Lett., 53 (1988) 432.
62. Martin C., Bourgault D., Michel C., Provost J., Hervieu M., Raveau B.,Europ. J. Inorg. Solid State Chem., (1989), in press.
63. Bourgault D., Martin C., Michel C., Hervieu M., Provost J., Raveau B., J. Solid State Chem., 78 (1989) 26.
64. Hervieu M., Maignan A., Martin C., Provost J., Raveau B., J. Solid State Chem., 75 (1988) 212.
65. Martin C., Bourgault D., Michel, Hervieu M., Raveau B. Modern Phys. Lett. B, 3 (1989) 93.
66. Martin C., Provost J., Bourgault D., Domengès B., Michel C., Hervieu M., Raveau B., Physica C, 157 (1989) 469.
67. Martin C., Michel C., Maignan A., Hervieu M., Raveau B., C.R. Acad. Sci., 307 (II) (1988) 27.
68. Parkin S.S.P., Lee V.Y., Nazzal A.I., Savoy R., Beyers R., La Placa S. Phys. Rev. Lett., 61 (1988) 750.
69. Ihara H., Sugise R., Hirabayashi M., Terada N., Jo M., Hayashi K., Negishi A., Tokumoto M., Kimura Y., Shimomura T., Nature, 334 (1988) 511.
70. Sugise R., Hirabayashi M., Terada N., Jo M., Shimomura T., Ihara H., Physica C, 157 (1989) 131.
71. Torardi C.C., Subramanian M.A., Calabrese J.C., Gopalakrishnan J.,McCarron E.M., Morrissey K.J., Askew T.R., Flippen R.B., Chowdhry U., Sleight A.W., Phys. Rev. B, 38 (1988) 225.
72. Hazen R.M., Finger D.W., Angel R.J., Prewitt C.T., Ross N.L., Adidiacos C.G., Heaney P.J., Veblen D.R., Sheng Z.Z., El Ali A., Hermann A.M., Phys. Rev. Lett. 60 (1988) 1657.
73. Parkin S.S. P., Lee V.Y., Engler E.M., Nazzal A.I., Huang T.C., Gorman M.G., Savoy R., Beyers R., Phys. Rev. Lett., 60 (1988) 1539.
74. Politis C., Luo H., Modern Phys. Lett. B, 2 (1988) 793.
75. Maignan A., Michel C., Hervieu M., Martin C., Groult D., Raveau B.,Modern Phys. Lett.B, 2 (1988) 681.
76. Subramanian M.A., Calabrese J.C., Torardi C.C., Gopalakrishnan J., Askew T.R., Flippen R.B., Morrissey K.J., Chowdhry U., Sleight A.W., Nature, 332 (1988) 420
77. Torardi C.C., Subramanian M.A., Calabrese J.C., Gopalakrishnan J., Morrissey K.J., Askew T.R., Flippen R.B., Chowdhry U., Sleight A.W., Science, 240 (1988) 631.
78. Hervieu M., Maignan A., Martin C. Michel C., Provost J., Raveau B.,Modern Phys. Lett. B, 2 (1988) 1103.

79. Raveau B., Martin C., Hervieu M., Bourgault D., Michel C., Provost J. Proc. Intern Symp. Solid State Chemistry, Pardubice, June 1989.
80. Studer F., Retoux R., Martin C., Michel C., Raveau B., Dartyge E., Fontaine A., Tourillon G., Modern Phys. Lett. B, (1989) in press.
81. Studer F., Bourgault D., Martin C., Retoux R., Michel C., Raveau B., Dartyge E., Fontaine A., Physica C., 159 (1989) 609.
82. Michel C., Hervieu M., Borel M.M., Grandin A., Deslandes F., Provost J., Raveau B., Z. Phys. B, 68 (1987) 421.
83. Maeda M., Tanaka Y., Fukutomi M., Asano T., Jpn. J. Appl. Phys. Lett., 27 (1988) L209 and L 548.
84. Hervieu M., Michel C., Domengès B., Laligant Y., Lebail A., Ferey G., Raveau B., Modern Phys. Lett. B, 2 (1988) 491.
85. Torardi C.C. Subramanian M.A., Calabrese J.C., Gopalakrishnan J., McCarron E.M.C., Morrissey K.J., Askew T.R., Flippen R.B., Chowdhry U., Sleight A.W., Phys. Rev. B., 38 (1988) 225.
86. Tarascon J.M., Le Page Y., Barboux P., Bagley B.G., Greene L.H., McKinnon W.R., Hull G.W., Giroud M., Huang D.M., Phys. Rev. B, 37 (1988) 9382
87. Subramanian M.A., Torardi C.C., Calabrese J.C., Gopalakrishnan J., Morrissey K.J., Askew T.R., Flippen R.B., Chowdhry U., Sleight A.W, Science 239 (1988) 1015.
88. Hervieu M., Domengès B., Michel C., Raveau B., Modern Phys. Lett., 2 (1988) 835.
89. Bordet P., Capponi J.J., Chaillout C., Chenavas J., Hewat A.W. Hewart E.A., Hodeau J.L., Marezio M., Tholence J.L., Tranqui D., Physica C, 153 (1988) 623.
90. Politis C., Appl. Phys., A45 (1988) 261.
91. Von Schnering H.G., Walz L., Schwartz M., Beker W., Hartweg M., Popp T., Hettich B., Müller P., Kampt G., Angew Chem. 27 (1988) 574.
92. Kasitani T., Kusaba K., Kikuchi M., Kobayashi N., Syono Y., Williams T.B., Hirabayashi M., Jpn. J. Appl. Phys., 27 (1988) L587.
93. Sunshine S.A., Siegrist T., Schneemeyer L.F., Murphy D.W., Cava R.J., Batlogg B., Van Dover R.B., Fleming R.M., Glarum S.H., Nakahara S., Farrow R., Krajewski J.J., Zahurak S.M., Waszcak J.V., Marshall J.H., Marsh P., Rupp L;W., Peck W.F., Phys. Rev. B, 38 (1988) 893.
94. Tarascon J.M., McKinnon W.R., Barboux P., Hwang D.M., Bagley B.G., Greene L.H., Hull G.W., Le Page Y., Stoffel N., Giroud M., Phys. Rev. B, 38 (1988) 8885.

95. Kijima N., Endo H., Tsuchiya J., Sumiyama A., Mizumo M., Oguri Y., Jpn. J. Appl. Phys., 27 (1988) L821.
96. Zandbergen H.W., Huang Y.K., Menken M.J.V., Li J.N., Kadouaki K., Menovsky A.A., Van Tendeloo G., Amelinckx S., Nature, 332 (1988) 620.
97. Gao Y., Lee P., Coppens P., Subramanian M.A., Sleight A.W., Science, 241 (1988) 954.
98. Endo U., Koyamaand T., Kawai T., Jpn. Appl. Phys., 27 (1988) L1476.
99. Hervieu M., Michel C., Raveau B., J. Less Common Metal 150 (1989) 59.
100. Van Tendeloo G., Zandbergen K.W., Van Landuyt J., Amelinckx S., Appl. Phys. A, 46 (1988) 233.
101. Zandbergen H.W., Grou P., Van Tendeloo G., Van Landuyt J., Amelinckx, Solid State Comm., 66 (1988) 397.
102. Hewat E.A., Dupuy M., Bordet P., Chaillout C., Hodeau J.L., Marezio M., Nature, 333 (1988) 53.
103. Gai P.L., Day P., Physica C, 152 (1988) 335.
104. Kakayamo-Muromachi F., Uchioto Y., Ouo A, Izumi F., Ouoda M., Jpn. J. Appl. Phys., 27 (1988) L365.
105. Kijima T., Tanaka J., Bando Y., Onoda M., Izumi F., Jpn. J. Appl. Phys., 27 (1988) L369.
106. Hervieu M., Michel C., Nguyen N., Retoux R., Raveau B., Europ. J. Inorg. Solid State Chem., 23 (1988) 375.
107. Retoux R., Michel C., Hervieu M., Nguyen N., Raveau B., Solid State Comm., 69 (1989) 599.
108. Le Page Y., McKinnon W.R., Tarascon J.M., Barboux P., unpublished.
109. Tarascon J.M., Miceli P.F., Barboux P., Hwang D.M., Hull G.W., Giroud M., Greene L.H., Le Page Y., McKinnon W.R., Tselepis E., Pleizier G., Eibschutz M., Neumann D.A., Rhyne J.J., Phys. Rev. B, 39 (1989) 11587.
110. Jirak Z., private communication.
111. Retoux R., Studer F., Michel C., Raveau B., Fontaine A., Dartyge E., Phys. Rev., B, submitted.
112. Cava R.J., Batlogg B., Krajewski J.J., Rupp L.W., Schneemeyer L.F., Siegrist T., Van Dover R.B., Marsh P., Peck W.F., Gallagher P.K., Glarum S.H., Marshall J.H., Farrow R.C., Waszczak J.V., Hull R., Trevor P., Nature, 336 (1988) 211.
113. Retoux R., Michel C., Hervieu M., Raveau B., Modern Phys. Lett. B, 3 (1989) 591.
114. Rouillon T., Provost J., Hervieu M., Groult D., Michel C., Raveau B., Physica C, 159 (1989) 201.

Superconductivity and Magnetism in Chevrel Phases

Ø. Fischer

Département de Physique de la Matière Condensée, Université de Genève,
24, Quai E. Ansermet, CH-1211 Genève 4, Switzerland

The class of ternary molybdenum chalcogenides of the type $M_x\text{Mo}_6 X_8$, where M is a metal and $0 < x < 4$ (Chevrel phases), contains a series of superconducting compounds. Many of these compounds have remarkable properties. Some have very high upper critical fields, making them important candidates for high magnetic field applications. Others contain a regular lattice of magnetic ions, and display very unusual properties, like re-entrant superconductivity, coexistence of antiferromagnetic order and superconductivity, and magnetic field induced superconductivity. The essential structural element in these compounds is a Mo_6 octahedron (cluster), and most of the unusual properties observed can be understood as a result of the particular cluster structure. In addition to the original Chevrel phases, a series of new compounds, having a parent structure but with condensed clusters Mo_9, Mo_{12}, etc., have been synthesized. The final compound is a pseudo-one-dimensional compound like $\text{Tl}_2\text{Mo}_6\text{Se}_6$ showing markedly anisotropic properties.

1. Introduction

The ternary molybdenum chalcogenides, first reported by *Chevrel* et al. [1], have turned out to be part of a large class of compounds with a variety of interesting properties. These compounds have the composition $M_x\text{Mo}_6\text{S}_8$, $M_x\text{Mo}_6\text{Se}_8$ or $M_x\text{Mo}_6\text{Te}_8$. M may be one of several metals, e.g. Li, Na, Ca, Sr, Sn, Pb. The value of x is 1 or close to 1 if M is a large cation (example: Pb) but may be as large as 4 if M is a small cation (example: Cu). The fact that many of these compounds are superconducting was first reported by *Matthias* et al. [2]. They found that the critical temperature T_c could be as high as 15 K for the compound PbMo_6S_8, which at the time was quite a respectable value for T_c. A property which turned out to be remarkable in many of these compounds is the upper critical field B_{c2} [3–6]. The value of 60 T at zero temperature found for PbMo_6S_8 constituted at the time by far the highest value ever observed. It is remarkable that even after the spectacular discovery of the high temperature superconductors [7], B_{c2} of the Chevrel phases remains of the same order of magnitude as the highest B_{c2} observed.

Another central property of these materials is that, with the $RE\text{Mo}_6\text{S}_8$ compounds (RE: Rare Earth) [8], they were the first to allow coexistence of long range magnetic order and superconductivity. This observation triggered a broader search for other ternary compounds with similar properties. One result was the discovery of the $RE\text{Rh}_4\text{B}_4$ series [9] which behave in many ways analogously to the Chevrel phases. Many other classes of ternary compounds were found as well. In this lecture I shall attempt to give an overview of the remarkable properties of the Chevrel phases. These notes constitute only a brief summary, and for more complete presentations and references I refer to other reviews on these materials [10, 11].

2. The Structure and the Compounds

The key to understanding the very special properties of these compounds is to be found in their particular crystal structure [12, 13]. A representation of the basic structure of these compounds is shown in Fig. 1.

This structure can be visualized as a slightly rhombohedrally distorted simple cubic lattice of Mo_6X_8 units. These units are themselves distorted cubes, with the X atoms situated at the corners and the Mo atoms placed approximately at the center of the faces. The edges of these cubes are tilted with respect to the axes of the overall structure by turning the Mo_6X_8 units roughly 25° about the (111) axis so that the overall structure is a rhombohedral one with the rhombohedral angle close to 90°. This is the structure of the binary compounds Mo_6S_8, Mo_6Se_8 and Mo_6Te_8. The other compounds, which result when an element M is added to this lattice, have basically the same structure: The M element occupies the space between the Mo_6X_8 units. We

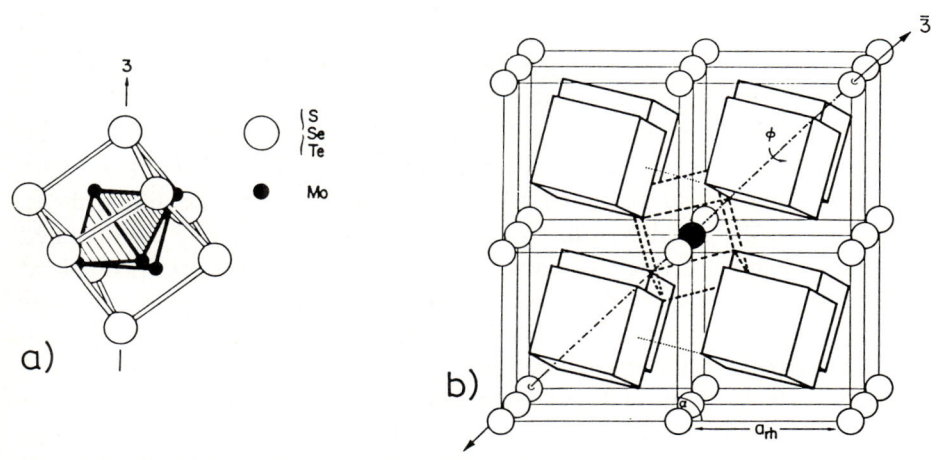

Fig. 1a,b. The crystal structure of $M\text{Mo}_6X_8$ ($X = \text{S}$, Se) compounds. (a) One Mo_6X_8 unit. (b) Stacking of the Mo_6X_8 units and the M atoms [13]

here have to distinguish between the case where M is a large cation and the one where it is a small cation. In the former case the M cation is surrounded by 8 Mo_6X_8 units and the structure is a slightly distorted CsCl structure with M replacing Cs and the Mo_6X_8 unit replacing Cl so that the formula is MMo_6X_8. This is the case in particular for $PbMo_6S_8$. In the case where M is a small cation there are two crystallographically different sites. Since these sites are delocalized with each 6 equivalent sites around the ternary (111) axis, it is possible to introduce up to 4 M atoms per unit cell. This is for instance the case for $Cu_xMo_6S_8$ where a homogeneity domain exists at room temperature between $x = 1$ and $x = 4$ [14,15]. These compounds tend to undergo a structural transition to a triclinic structure at low temperature where the M cations order.

There exist more than 100 different compounds in this class of materials. Most of them are metallic and a large number are superconducting. In Table 1 a selection of compounds are listed together with their values of T_c, B_{c2} and ξ, the coherence length.

Table 1. Typical values of critical temperature, critical field and coherence length for some Chevrel phases [16]

Compounds	T_c [K]	$(dB_{c2}/dT)_{T_c}$ [T/K]	$B_{c2}(0)$ [T]	ξ [Å]
Mo_6Se_8	6.3	1.7-3.6	9-16	45-60
$LaMo_6Se_8$	11.3	7	45	27
$LaMo_6S_8$	5.8	1.5	54	78
$Cu_{1.8}Mo_6S_8$	10.5	1.9	15	47
$SnMo_6S_8$	13.4	3.7	34	31
$PbMo_6S_8$	14.6	5.6	59	24

The overall trends of the electronic properties in this class can be understood in terms of a simple valence electron picture. For the S compounds one may to first approximation consider the sulphur atoms as having a valence -2 and the conduction electrons to be mainly of Mo-$4d$ character. The electronic levels of the M atoms are situated far from the Fermi level so that their main role is to deliver electrons to the $4d$ levels, thereby more or less filling these, depending on the valence of the M atom. This crude picture is corroborated by detailed band structure calculations. In Fig. 2 we show the results obtained by *Jarlborg* and *Freeman* [17], for $SnMo_6S_8$. The partial densities of states do indeed show that the electrons at the Fermi level have mainly Mo-$4d$ character. Neither the Sn s and p electrons nor the S p electrons contribute very much to the bands near the Fermi level. There is, however, a significant charge transfer from Mo and Sn to S in the sense of a chemical valence formula as discussed above. According to this picture, the compounds with divalent M atoms, like $PbMo_6S_8$, $SnMo_6X_8$, and $YbMo_6S_8$, are all electronically similar,

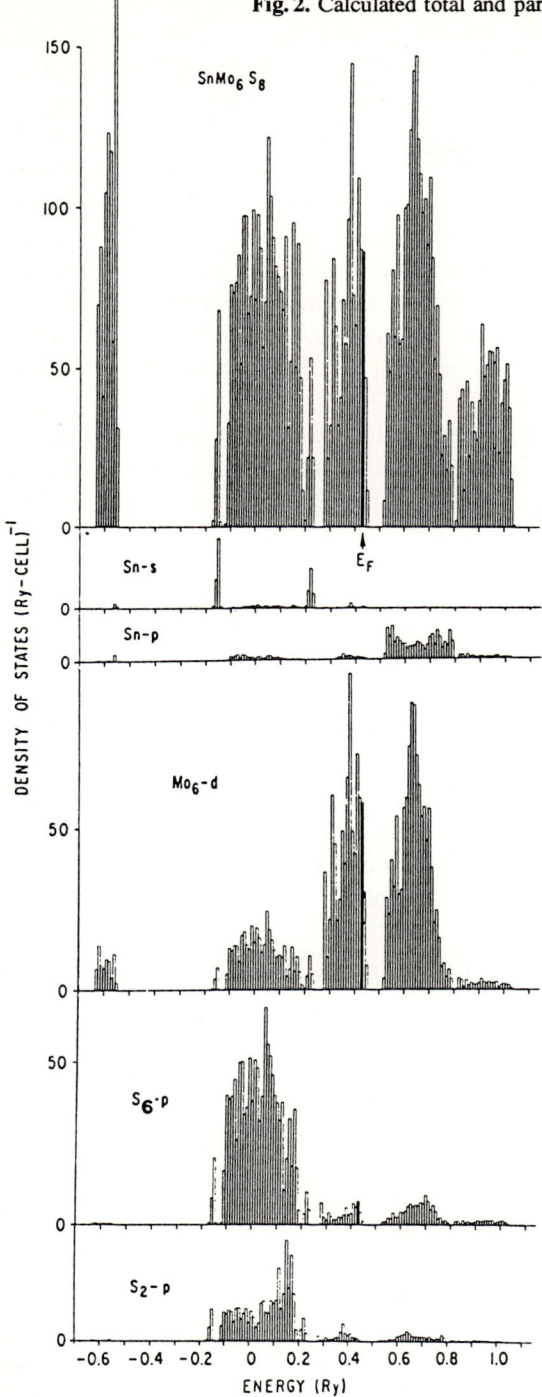

Fig. 2. Calculated total and partial densities of states for SnMo$_6$S$_8$ [17]

and indeed they all have both high T_c and B_{c2} values. A striking example of this rule is the compound $Mo_6S_6I_2$ [18]. By substituting two iodine atoms for sulphur atoms in Mo_6S_8 one formally increases the number of valence electrons on the Mo cluster to 22, exactly as in the other compounds with divalent cations. As a result the critical temperature rises from 1.6 K to 14 K! A counter example to this rule might at first sight seem to come from the compounds $CaMo_6S_8$, $SrMo_6S_8$, $BaMo_6S_8$ and $EuMo_6S_8$, which are not normally found to be superconducting. The reason for this absence of superconductivity is a low temperature phase transition to a triclinic structure. In the case of $EuMo_6S_8$ this transition can be suppressed by applying a pressure of about 12 kbar. In the rhombohedral phase this compound turns out to be superconducting with a T_c of 12 K, very close to the one expected [19]. The compound $CaMo_6S_8$ has recently been synthesized with a T_c of 8 K [20].

Another property, which at the time was unexpected, is the existence of superconductivity in the $RE Mo_6S_8$ compounds. Here RE can be any of the rare earths, and superconductivity is found in nearly all of these compounds. Superconductivity and magnetism have been known since the late 1950s to be antagonistic phenomena and before the discovery of the Chevrel phases no long range magnetic order had been found to coexist with superconductivity. The observation of superconducting Chevrel phases with a regular lattice of magnetic atoms suggested that interesting phenomena could happen at lower temperatures when the RE lattice orders magnetically. And indeed, many new phenomena, to be discussed below, have been seen as a result of the interplay of magnetism and superconductivity. The $RE Mo_6S_8$ compounds have mostly low critical temperatures, as shown in Table 2. This is not so much a result of magnetic scattering as of the trivalent state of the RE atoms. The much higher T_c of the Eu compound (under pressure) and the Yb compound is a result of the divalent state of these two elements.

Table 2. Superconducting and magnetic transition temperatures for $RE Mo_6S_8$ compounds

Compounds	T_{c1} [K]	T_{c2} [K]	T_M [K]	Magnetic order	References
$CeMo_6S_8$	–	–	2.3	AF(?)	[21]
$PrMo_6S_8$	4.0				
$NdMo_6S_8$	3.5		0.3	AF	[22]
$SmMo_6S_8$	2.9				
$EuMo_6S_8$	(12 K at 12 kbar)	–	0.3	AF complex	[23]
$GdMo_6S_8$	1.6		0.84	AF	[24]
$TbMo_6S_8$	2.1		1.05	AF	[25]
$DyMo_6S_8$	2.1		0.4	AF	[26]
$HoMo_6S_8$	2.2	0.79	0.75	F (oscillatory)	[27] [28]
$ErMo_6S_8$	2.2		0.2	AF	[25]
$TmMo_6S_8$	2				
$YbMo_6S_8$	9		(2.7?)		[29]
$LuMo_6S_8$					

One prediction of the band structure calculation is that there is a gap in the density of states at a filling of 24 electrons at the Mo_6 cluster. This can be understood in terms of a saturation of bonds between the Mo and its neighbours [30]. By replacing two Mo atoms by Ru atoms or four Mo atoms by Re atoms in Mo_6S_8 or Mo_6Se_8 this situation is indeed found. The compounds $Mo_4Ru_2Se_8$ and $Mo_2Re_4S_8$ and $Mo_2Re_4Se_8$ are semiconductors [31].

The selenides are similar to the sulphides, but with the difference that Se is less ionic than S and thus the trends in superconducting properties are different. The compounds with divalent M atoms have lower T_c values and the ones with trivalent M atoms higher T_c values than in the corresponding sulphides. Thus in particular, the $RE Mo_6 Se_8$ compounds have, relatively speaking, high T_c values [32]. Among the tellurides only a few superconducting compounds have been found. They do not form with large cations.

In the search for the compound with $M = $ In, a new structure type emerged. The compound $In_3Mo_{15}Se_{19}$ has a structure which resembles that of the Chevrel phases, but in addition to the Mo_6Se_8 unit there also appears a Mo_9Se_{11} unit [33]. The latter can be understood as a condensation of two Mo_6Se_8 units as shown in Fig. 3.

This was the first compound of a whole new class of compounds containing condensed clusters $Mo_{3n}Se_{3n+2}$ where n values of 2,3,4,5,6,8 and 10 have been found in different compounds [34]. Many of these compounds are also

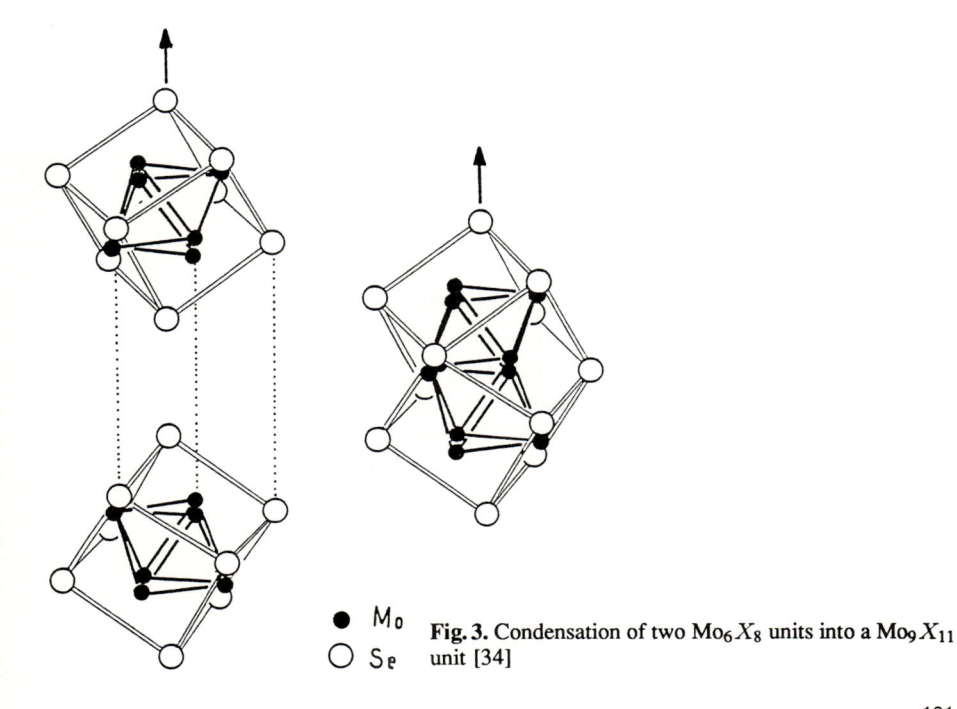

Fig. 3. Condensation of two Mo_6X_8 units into a Mo_9X_{11} unit [34]

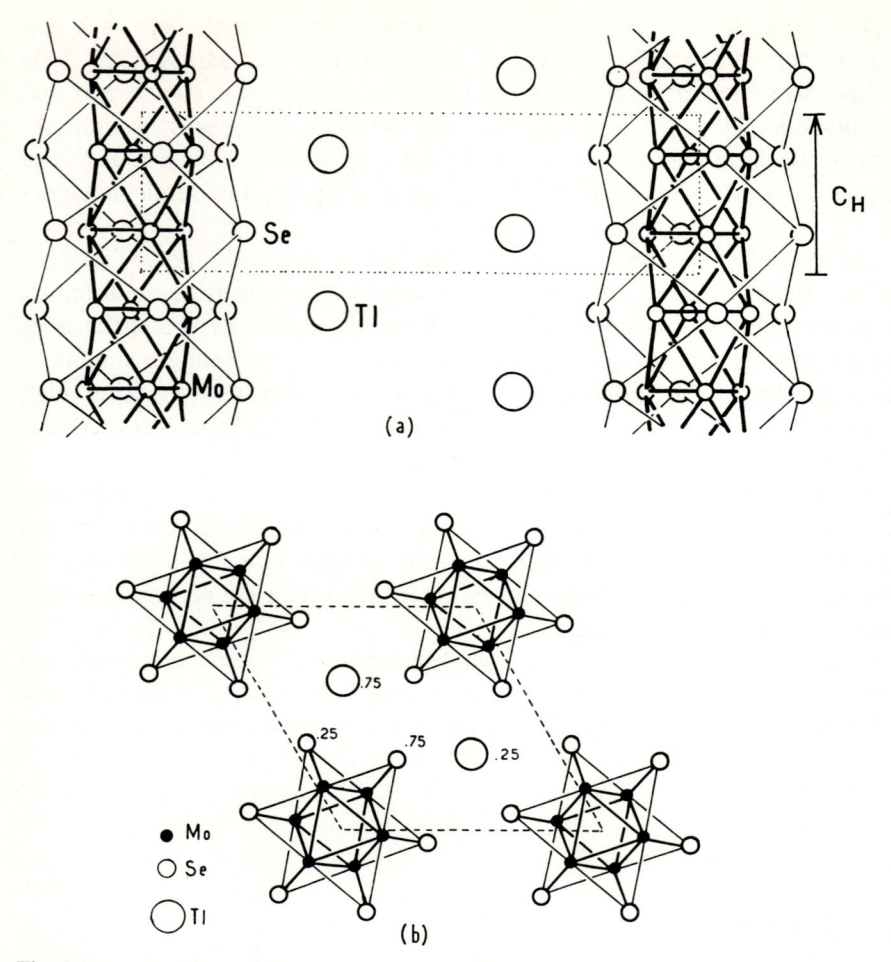

Fig. 4. Representation of the structure of $Tl_2Mo_6Se_6$ [34]

superconducting. The end member of this series is an infinite chain $[Mo_6Se_6]$. This gives rise to pseudo-one-dimensional compounds, such as $Na_2Mo_6Se_6$ and $AgMo_4Te_6$. One of these compounds is superconducting, $Ti_2Mo_6Te_6$, with a T_c of 6.5 K [35, 36]. The structure of this compound is illustrated in Fig. 4. As discussed below, the superconducting properties of this compound are very anisotropic and many questions remain to be answered still concerning the exact nature of its superconducting state.

3. Critical Fields and Critical Currents

As mentioned above, one of the striking characteristics of these materials is the very high critical magnetic field observed in some of them. The origin of this property can again be traced back to the structure. The relatively weak coupling between the Mo_6 clusters leads to narrow bands and a low Fermi velocity. This again leads to a small coherence length (25 Å for $PbMo_6S_8$) and to a high B_{c2}. Typical values for ξ and B_{c2} for some compounds are given in Table 1. The exact numbers depend somewhat on the purity of the samples. A discussion of these aspects can be found in [16]. See also [19] for a detailed treatment of $EuMo_6S_8$.

A difficulty that arises in this context is that it has turned out to be difficult to produce very homogeneous samples, especially of the high B_{c2} compounds. On the other hand, the superconducting transitions in a magnetic field are usually relatively large when measured resistively or inductively. This might arouse the suspicion that only a small part of the material really has a high critical field. To check this point we have recently carried out specific heat measurements in a magnetic field [37]. This was done on very homogeneous samples made by a high temperature synthesis. The specific heat anomaly and its field dependence are shown in Fig. 5.

The transition remains narrow in a field and the anomaly essentially shifts to lower temperatures with increasing field. For this sample, which had a T_c of 11.5 K, the slope of the critical field was found to be as high as 6.8 T/K and the extrapolated B_{c2} at $T = 0$ was 54 T. This demonstrated beyond any doubt that the whole volume of the sample has this high B_{c2}. Homogeneous samples with a T_c of 14.6 K are now routinely produced and show the same

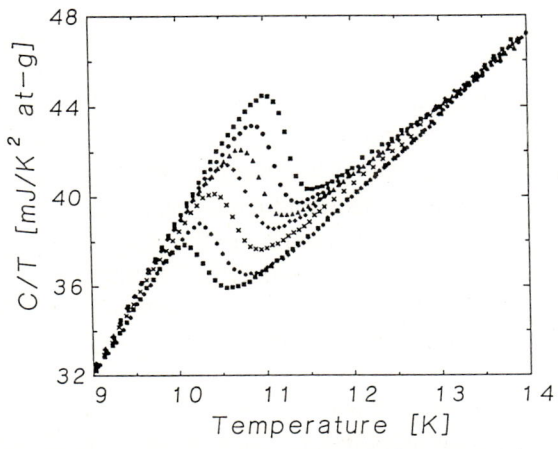

Fig. 5. Specific heat anomaly in $PbMo_6S_8$ measured in magnetic fields of 0, 1, 2, 3, 4, 5, and 6 T [37]

behaviour. The slope is usually somewhat less, typically 5.3 – 5.6 T/K and the extrapolated critical field ranges between 54 and 59 T at $T = 0$ [38].

Characteristic for these samples is that they have a smaller slope dB_{c2}/dT when the critical field is measured inductively. This difference between the two measurements of B_{c2} is understood as a result of the macroscopic screening currents present in the inductive measurements. Thus we find a behaviour which is reminiscent of the granular behaviour found in the high temperature superconducting oxides.

To what extent this granularity limits the value of the critical current in polycrystalline material, as it is believed to do in the oxides, is not yet clarified. A first test in this direction has been carried out using flux penetration measurements [39]. In these measurements one obtains a plot of the penetration of the magnetic field as a function of the distance into the sample. In the case of an ideal sample, the resulting plot is a straight line with a slope proportional to the critical current. In a granular system like the oxides the plot is composed of two linear portions [40]. One reflects the low critical current between the grains and the other the high intra-grain critical current. A similar measurement on $PbMo_6S_8$ gives only one straight line [39]. Thus the ratio between the two critical currents is not as large as in the oxides. However, this does not mean that the two are identical. In order to get an estimate of the former, a measurement was carried out on a weakly sintered powder of $PbMo_6S_8$. In this case a behaviour similar to that of the oxides was found, revealing an intra-grain critical current about an order of magnitude higher than the inter-grain critical current in the dense $PbMo_6S_8$ sample [39]. Thus a certain problem related to granularity is present in this material. It appears, however, to be much less severe than in the oxides. The highest critical currents obtained in wires are greater than 10^8 A/m^2 at 20 T [41, 42].

One reason why these materials show a less strong tendency towards granular behaviour than the oxides is probably to be found in the fact that they have mainly isotropic properties. The anisotropy of the critical field B_{c2} has been investigated in several Chevrel phases. Anisotropies of typically 20 % in H_{c2} with respect to the ternary axis have been observed [16, 43]. Thus alignment of the grains to obtain high critical currents is not expected to be crucial. However, it is nevertheless probable that the superconducting properties may be disturbed along the grain boundaries, leading to pinning in the case of weak perturbations but to decoupling between the grains and glassy behaviour in the case of more severe disturbances. On the other hand the interface energy obtained when two grains of different orientations connect may also play some role in pinning [44].

The anisotropy is, as one might expect, much higher in the condensed cluster compounds. The extreme case is represented by $Tl_2Mo_6Se_6$. The ratio of the resistivity perpendicular and parallel to the c-axis (chain-axis) has been

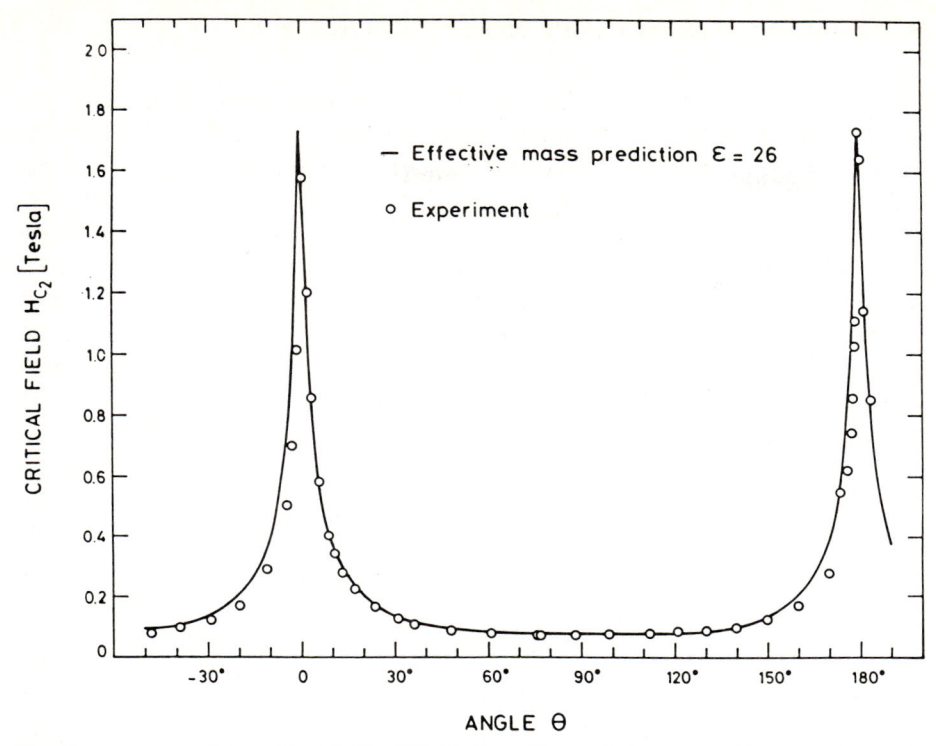

Fig. 6. Anisotropy of the critical field of $Tl_2Mo_6Se_6$. The angle is measured between the c-axis and the magnetic field [35]

found to be of the order of 10^3 and the anisotropy of the critical field is among the largest observed in a superconductor $B_{c2\|}/B_{c2\perp} = 26$ [35, 36]. Figure 6 shows the anisotropy of this material.

The other remarkable fact is that this material appears to be a low electronic density, low density of states material. *Brusetti* et al. [45] have investigated the lower critical field recently and conclude that the material has a very low B_{c1} and correspondingly a very high Ginzburg–Landau parameter, which would imply a very large penetration depth. To what extent the superconducting state is weakened by fluctuation effects is not yet elucidated. It is striking though that as soon as the material is doped with small amounts of impurities superconductivity disappears and the material becomes semiconducting. Furthermore, nearly all other compounds of this type are insulating. Whether this is the result of a Peierls transition is still an open question.

4. The Interplay of Magnetism and Superconductivity

The properties of the Chevrel phases that have achieved most attention are the ones related to the interaction of superconductivity and magnetism. In the binary and pseudo-binary compounds, studied before the discovery of the $RE\text{Mo}_6\text{S}_8$ compounds, superconductivity was destroyed by small amounts of magnetic impurities due to the exchange scattering of the conduction electrons off the magnetic moments. The result was that all efforts to produce a material with a large enough concentration of magnetic moments to obtain a long range magnetic order in the superconducting state failed. In the Chevrel phases the exchange interaction between the Mo-$4d$ conduction and the RE-$4f$ electrons is weak. This is a direct result of the cluster-type crystal structure, where the Mo atoms are situated far away from the RE atoms, providing a small overlap between the electrons on the two sites. A more extreme case in this sense has since been found in the $RE\text{Ba}_2\text{Cu}_3\text{O}_7$ compounds, where there seems to be practically no interaction at all between the superconducting electrons and the magnetic moments. The situation in the Chevrel phases and some other ternary compounds, like $RE\text{Rh}_4\text{B}_4$, is that the exchange interaction is sufficiently weak that the phenomena of long range magnetic order and superconductivity can be directly confronted with each other, but also sufficiently strong that the two phenomena interfere with each other so that new and unusual properties occur. In what follows a brief summary is given of some of the more striking phenomena. For a more detailed discussion we refer to more extensive reviews [10, 11, 46].

4.1 Ferromagnetism and Superconductivity One of the RE Chevrel phases, HoMo_6S_8, orders ferromagnetically at a temperature T_m below the superconducting critical temperature T_{c1}. In this compound superconductivity is destroyed at a temperature T_{c2} just below T_m [27]. In Fig. 7 we show the first observation of this re-entrant superconductivity in HoMo_6S_8.

There are three interesting temperature regions: i) In the paramagnetic and superconducting region ($T_m < \text{T} < T_{c1}$) superconductivity is very little disturbed by the presence of the magnetic moments since the exchange scattering is very weak. When one applies a magnetic field, superconductivity will be influenced both because there is an extra field given by the magnetization and because there will be a polarization effect due to the exchange interaction. Even if the latter is weak, the polarization is a first-order effect that may nevertheless be very large. This is demonstrated by the magnetic field induced superconductivity to be discussed in Sect. 5. However, for HoMo_6S_8, experiments show that the electromagnetic effects dominate [47, 48].

ii) In the narrow region between T_{c2} and T_m a new phenomenon occurs: the magnetic order is not simply ferromagnetic, but has an oscillatory nature.

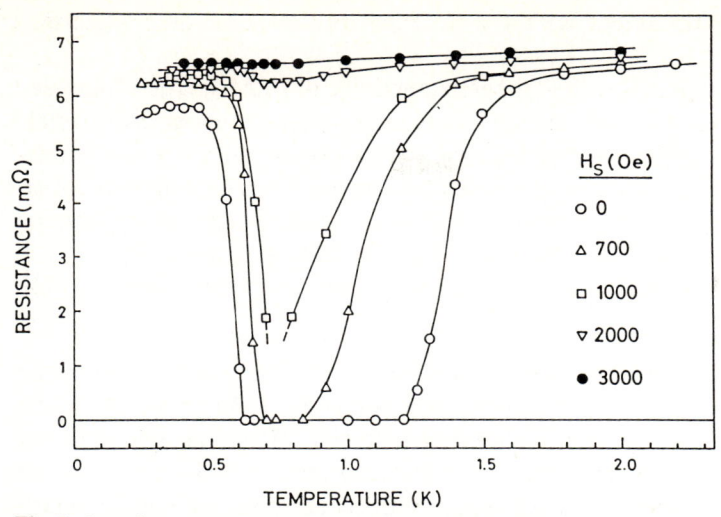

Fig. 7. dc resistance versus temperature in magnetic fields of 0, 70, 100, 200, 300 mT [27]

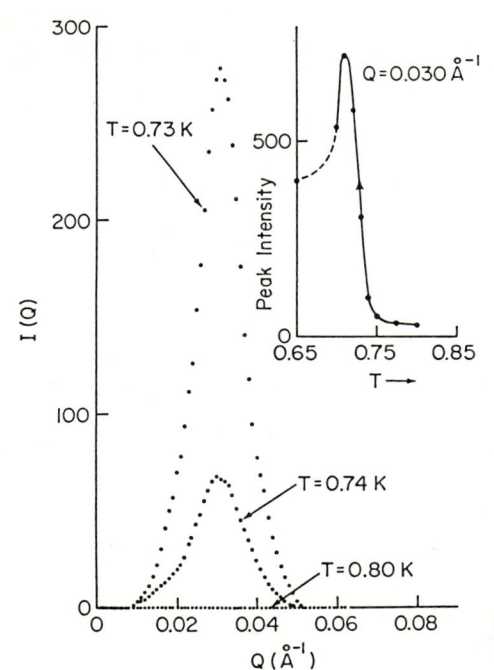

Fig. 8. Net scattering intensity, after subtraction of the scattering at 2 K, versus scattering vector for HoMo$_6$S$_8$ [49]

Figure 8 displays neutron scattering data showing the occurrence of a low angle diffraction peak in this temperature interval [49]. This phenomenon occurs because the superconductor tends to suppress any uniform magnetic field or polarization. Thus the interplay between the two collective phenomena leads to this oscillatory behaviour with a wavelength of a few hundred angstroms [50].

iii) In the domain below T_{c2}, superconductivity in HoMo$_6$S$_8$ is suppressed because the saturation magnetization M_0 is larger than the upper critical field B_{c2}. However, the difference between the two is very small and shape effects and hysteresis effects may lead to a situation where $M < B_{c2}$, and thus superconductivity reappears under certain conditions. In particular the application of a weak magnetic field may lead to this superconducting state [51].

Very similar behaviour has also been seen in the compound ErRh$_4$B$_4$ [52].

4.2 Antiferromagnetic Superconductors Nearly all REMo$_6$X$_8$ compounds order antiferromagnetically at a temperature below T_c. [53]. In all cases one observes a coexistence between the two phenomena. This is at first sight not surprising since no average polarization effects are present. However, in many cases the superconducting state is strongly influenced when the antiferromagnetic order sets in. In fact, this was how the coexistence was first observed in DyMo$_6$S$_8$ [54]. In Fig. 9 we show the upper critical field, displaying a strong anomaly at T_m, together with the intensity of the $(1/2, 0, 0)$ peak in the neutron diffraction, reflecting the sublattice magnetization [26]. There is

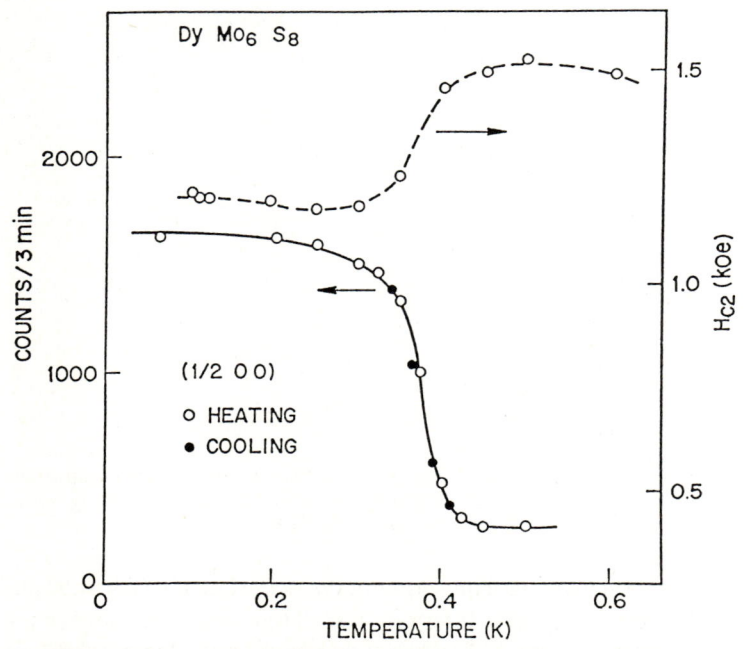

Fig. 9. The temperature dependence of the $(1/2, 0, 0)$ antiferromagnetic peak in DyMo$_6$S$_8$ shown together with the critical field anomaly [26]

an obvious correlation between the two phenomena. The origin of this effect is to be found in the modifications of the electronic structure and the electron phonon interaction at the antiferromagnetic transition [50].

The case of $HoMo_6S_8$ is particularly interesting. This compound, which one might suspect to become ferromagnetic like its sulphide counterpart, in fact shows antiferromagnetic ordering. However, this is of an oscillatory, long wavelength (100 Å) nature, very similar to that observed in $HoMo_6S_8$ in the narrow temperature interval [55]. It is therefore plausible to assume that $HoMo_6S_8$ is in reality a ferromagnet which is forced into the oscillatory state as a result of the presence of superconductivity. The occurrence of this state in the whole temperature range below T_m is certainly related to the fact that the selenide has a much higher T_c and B_{c2} than the sulphide.

Many other antiferromagnetic superconductors have been found in other ternary superconductors such as $RERh_4B_4$ and in the $REBa_2Cu_3O_7$ series.

5. Magnetic Field Induced Superconductivity: The Jaccarino–Peter Effect

In 1962 *Jaccarino* and *Peter* [56] proposed that under certain conditions it would be possible to make a ferromagnet superconducting by applying a strong magnetic field. The basic idea here is that superconductivity in a ferromagnet is primarily made impossible because of the strong spin polarization effects. These can be described by an effective exchange field, B_J, proportional to the magnetization, but much stronger than the latter. A typical value for the exchange field at saturation is of the order of 100 T. In certain substances one finds that the exchange field is in the opposite direction to the magnetization and thus to an external field. The latter will counteract the exchange field and, for values of the external field roughly equal to $-B_J$, polarization effects will disappear and superconductivity will appear.

The observation of this effect in a non-superconducting ferromagnet is made difficult because various conditions must be fulfilled. Since the material is not superconducting to start with, it is difficult to control and adjust the material parameters. However, the effect can also be observed in a paramagnet since the high external field will in any case align the magnetic moments at low temperature. In this case one expects to observe superconductivity at low fields also. As soon as the magnetic moments align due to the external field, superconductivity will be destroyed. Then at much higher fields superconductivity should come back and then finally disappear again at very high fields. The Chevrel phases appeared to be ideal materials in which to look for this effect, since one condition for its occurrence is that B_{c2} in the absence of magnetic ions be high. $EuMo_6S_8$ in its rhombohedral phase has properties which come very close to the ones necessary. In order to tune the parameters it

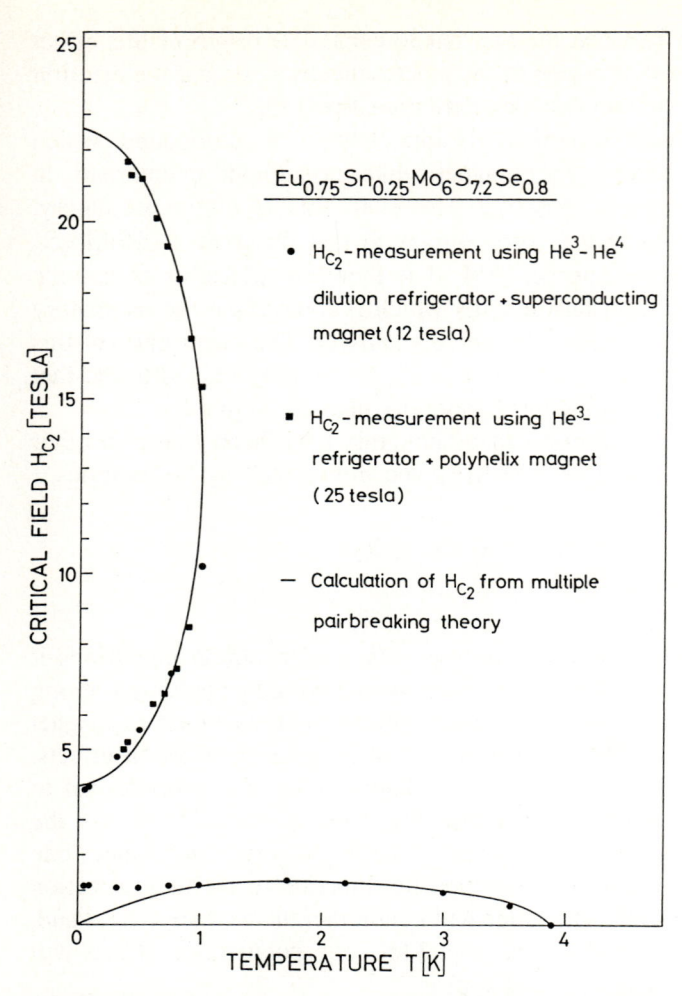

Fig. 10. Upper critical field versus temperature for the compound $Eu_{0.75}Sn_{0.25}Mo_6S_{7.2}Se_{0.8}$ [57]

was necessary to make an alloy. Figure 10 shows the phase diagram that was observed for a sample with the composition $Eu_{0.75}Sn_{0.25}Mo_6S_{7.2}Se_{0.8}$ [57]. The full line gives the theoretically calculated B_{c2} curve using reasonable, but adjusted parameters. Further work has confirmed the explanation of this phenomena. Note that due to this effect this sample has an upper critical field of 20 T although its critical temperature is less than 4 K!

Unexpected magnetic properties of this field induced state have been predicted [58]. Contrary to all other superconductors this state will have a paramagnetic superconducting magnetization. The field will be enhanced in the superconducting state and the flux lines will be anti-vortices instead of vortices. For certain fields and temperatures it is also predicted that the field modulation due to the flux lines will disappear altogether. There will be a homogeneous and complete field penetration in the superconducting state. A first indication of this effect has been observed by magnetization measurements, but further local measurements are necessary to confirm experimentally this unusual behaviour.

References

1   R. Chevrel, M. Sergent, J. Prigent: J. Solid State Chem. **3**, 515 (1971)
2   B.T. Matthias, M. Marezio, E. Corenzwit, A.S. Cooper, H.E. Barz: Science **175**, 1465 (1972)
3   R. Odermatt, Ø. Fischer, H. Jones, G. Bongi: J. Phys. C **7**, L13 (1973)
4   Ø. Fischer: Colloques Int. CNRS No. 242, Grenoble (1974)
5   S. Foner, E.J. McNiff, E.J. Alexander: Phys. Lett. **49A**, 269 (1974)
6   Ø. Fischer, H. Jones, G. Bongi, M. Sergent, R. Chevrel: J. Phys. C **7**, L450 (1974)
7   J.G. Bednorz, K.A. Müller: Z. Phys. B **64**, 189 (1986)
8   Ø. Fischer, A. Treyvaud, R. Chevrel, M. Sergent: Solid State Commun. **17**, 721 (1975)
9   B.T. Matthias, E. Corenzwit, J. Vandenberg, H. Barz: Proc. Natl. Acad. Sci. USA **74**, 1334 (1977)
10 Ø. Fischer, M.B. Maple (eds.): *Superconductivity in Ternary Compounds I, II*, Topics Curr. Phys., Vols. 32 and 34 (Springer, Berlin, Heidelberg 1982)
11 Ø. Fischer: In *Ferromagnetic Materials*, Vol. 5, ed. by E.P. Wohlfarth, K.H.J. Buschow (North-Holland, Amsterdam 1990)
12 R. Chevrel, M. Sergent: In [10], Vol. I, p. 25
13 K. Yvon: In [10], Vol. I, p. 87
14 R. Flükiger, A. Junod, R. Baillif, P. Spitzli, A. Treyvaud, A. Paoli, H. Devantay, J. Muller: Solid State Commun. **23**, 699 (1977)
15 R. Flükiger, R. Baillif: In [10], Vol. I, p. 113
16 M. Decroux, Ø. Fischer: In [10], Vol. 2, p. 57
17 T. Jarlborg, A.J. Freeman: Phys. Rev. Lett. **44**, 178 (1980)
18 M. Sergent, Ø. Fischer, M. Decroux, C. Perrin, R. Chevrel: J. Solid State Chem. **22**, 87 (1977)
19 See, e.g., M. Decroux, S.E. Lambert, M.B. Maple, R.P. Guertin: J. Low Temp. Phys. **73**, 283 (1988)
20 C. Geantet, J. Padiou, O. Pena, M. Sergent, R. Horyn: Solid State Commun. **64**, 1363 (1987)
21 M. Pelizzone, A. Treyvaud, P. Spitzli, Ø. Fischer: J. Low. Temp Phys. **29**, 453 (1977)
22 N.E. Alexeevski, G. Wolf, V.N. Narozhnyi, A.S. Rudenko, H. Hohlfeld: Sov. Phys. – JETP **62**, 617 (1985)
23 S. Quezel, P. Burlet, E. Roudaut, J. Rossat-Mignot, A. Benoit, J. Flouquet, O. Pena, R. Horyn, R. Chevrel, M. Sergent: Proc. Journées Materiaux Supraconducteurs, Rennes. Ann. Chim. Fr. **1984 9**, 1057 (1984)
24 C.F. Majkrzak, G. Shirane, W. Thomlinson, M. Ishikawa, Ø. Fischer, D.E. Moncton: Solid State Commun. **31**, 773 (1979)
25 W. Thomlinson, G. Shirane, D.E. Moncton, M. Ishikawa, Ø. Fischer: Phys. Rev. B **23**, 4455 (1981)
26 D.E. Moncton, G. Shirane, W. Thomlinson, M. Ishikawa, Ø. Fischer: Phys. Rev. Lett. **41**, 1133 (1978)
27 M. Ishikawa, Ø. Fischer: Solid State Commun. **23**, 37 (1977)
28 J.W. Lynn, D.E. Moncton, W. Thomlinson, G. Shirane, R.N. Shelton: Solid State Commun. **26**, 493 (1978)
29 P. Bonville, J.A. Hodges, P. Imbert, G. Jehanno, R. Chevrel, M. Sergent: Rev. Phys. Appl. **15**, 1139 (1980)

30   K. Yvon: In *Current Topics in Materials Science*, Vol. 3, ed. by E. Kaldis (North-Holland, Amsterdam 1979)
31   A. Perrin, R. Chevrel, M. Sergent, Ø. Fischer: J. Solid State Chem. **33**, 43 (1980)
32   R.N. Shelton, R.W. McCallum, H. Adrian: Phys. Lett. **56A**, 213 (1976)
33   B. Seeber, M. Decroux, Ø. Fischer, R. Chevrel, M. Sergent, A. Gruttner: Solid State Commun. **29**, 419 (1979)
34   M. Potel, R. Chevrel, M. Sergent: Acta Crystallogr. B **36**, 1319, 1545 (1980)
     M. Potel: Thèse de Doctorat d'Etat, Université de Rennes (1981)
     M. Gougeon: Thèse de troisième cycle, Université de Rennes (1984)
35   J.C. Armici, M. Decroux, Ø. Fischer, M. Potel, R. Chevrel, M. Sergent: Solid State Commun. **33**, 607 (1980)
36   R. Lepit, R. Monceau, M. Potel, P. Gougeon, M. Sergent: J. Low. Temp. Phys. **56**, 219 (1984)
37   D. Cattani, J. Cors, M. Decroux, B. Seeber, Ø. Fischer: Physica C **153–155**, 461 (1988)
38   J. Cors: Thesis, University of Geneva (1990), to be published
39   D. Cattani: Thesis, University of Geneva (1990), to be published
40   H. Küpfer, I. Apfelstedt, R. Flükiger, C. Keller, R.I. Meier-Hirmer, B. Runtsch, A. Turowski, U. Wiech,
     T. Wolf: Cryogenics **28**, 650 (1988)
41   B. Seeber, M. Decroux, Ø. Fischer: Physica B **155**, 129 (1989)
42   Efforts to improve the critical currents in Chevrel phases are being carried out in a European collaboration within the EUREKA 96 project
43   M. Decroux, Ø. Fischer, R. Flükiger, B. Seeber, R. Deleclefs, M. Sergent: Solid State Commun. **25**, 393 (1978)
44   C. Rossel: Thesis, University of Geneva (1981)
45   R. Brusetti, P. Monceau, M. Potel, P. Gougeon, M. Sergent: Solid State Commun. **66**, 181 (1988)
46   L.N. Bulaevskii, A.I. Buzdin, M.L. Kulic, S.V. Panjukov: Adv. Phys. **34**, 175 (1985)
47   P. Burlet, A. Dinia, S. Quezel, W.A.C. Erkelens, J. Rossat-Mignod, R. Horyn, O. Pena, C. Geantet, M. Sergent, J.L. Genicon: Physica **148B**, 99 (1987)
48   M. Giroud, J.L. Genicon, R. Tournier, C. Geantet, O. Pena, R. Horyn, M. Sergent: Physica **148B**, 113 (1987)
49   J.W. Lynn, A.Raggazoni, R. Pynn, J. Joffrin: J. de Phys. Lett. **42**, L-45 (1981)
50   For a discussion see P. Fulde, J. Keller: In [10], Vol. II, p. 249
51   M. Giroud, J.L. Genicon, R. Tournier, C. Geantet, O. Pena, R. Horyn, M. Sergent: J. Low Temp. Phys. **69**, 419 (1987)
52   M.B. Maple, H.C. Hamaker, L.D. Woolf: In [10], Vol. II, p. 99
53   M. Ishikawa, Ø. Fischer, J. Muller: In [10], Vol. II, p. 143
54   M. Ishikawa, Ø. Fischer: Solid State Commun. **24**, 747 (1977)
55   J.W. Lynn, J.A. Gotaas, R.W. Erwin, R.A. Ferrell, J.K. Bhattacharjee, R.N. Shelton, P. Klavins: Phys. Rev. Lett. **52**, 133 (1984)
56   V. Jaccarino, M. Peter: Phys. Rev. Lett. **9**, 290 (1962)
57   H.W. Meul, C. Rossel, M. Decroux, Ø. Fischer, G. Remenyi, A. Briggs: Phys. Rev. Lett. **53**, 497 (1984)
58   Ø. Fischer, H.W. Meul, M.G. Karkut, G. Remenyi, U. Welp, J.C. Piccoche, K. Maki: Phys. Rev. Lett. **55**, 2972 (1985)

Organic Superconductivity: The Role of Low Dimensionality and Magnetism

D. Jérome

Laboratoire de Physique des Solides (associé au CNRS)
Université Paris Sud, F-91405 Orsay, France

Abstract.

The search for organic conductors in the early seventies challenged by the possibility of superconductivity at high temperature has enabled metallic-like electrical conduction to be extended to molecular crystals. Superconductivity has been discovered in 1980 in the quasi one dimensional cation radical salt (TMTSF)$_2$PF$_6$ with $T_c \approx 1.2K$. Over the last decade new materials with enhanced T_c up to 10 K or so have been discovered. Besides superconductivity a wealth of new and remarkable phenoma has been observed; for example, the field induced spin density wave states, the quantization of the Hall effect in a three dimensional material and giant magnetoresistance oscillations related to the 2-D character of the Fermi surface of the (ET)$_2$X family. We shall also discuss the role of Coulombic interactions through an overview of the NMR data.

I. Introduction.

Unlike the important discovery of superconductivity above 30 K which occurred somewhat suddenly in 1986 in a series of cuprate compounds belonging to a broader family of oxide compounds [1](LiTiO, BaPbBiO, etc...) in which no particularly high transition temperatures had been found before, superconductivity in organic solids took some time to be obtained. About 15 years of intense research activity have proved to be necessary [2] after Little [3] suggested a new mechanism for pair formation in low dimensional organic conductors leading possibly to high temperature superconductivity. Little's proposal has certainly boosted the research in the field of organic conductors which started in the early seventies with the discovery of the first organic conductor TTF-TCNQ showing a high and metal-like conductivity in a wide temperature domain below room temperature [4] figure1a.

Organic superconductivity was initially discovered in the (TMTSF)$_2$X series [5] with T_c in the one Kelvin range [2]. However, T_c has been rising very quickly over the last eight years, especially with the discovery of organic salts of the cation radical BEDT-TTF leading to conductors with an enhanced two-dimensional character and T_c above 8K [6,7]. T_c is now approaching 10K at ambient pressure in the material κ-(BEDT-TTF)$_2$Cu(SCN)$_2$ [8] figure1a. From now on the molecule BEDT-TTF will be labelled ET.

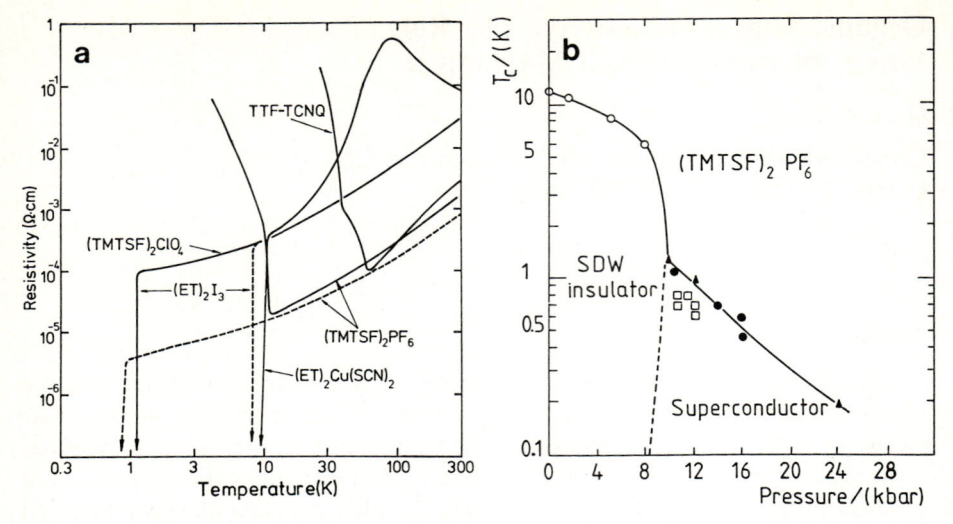

Fig.1: (a) Temperature dependence of several organic conductors. A Peierls ground state is achieved in TTF-TCNQ below 54 K. An antiferromagnetic SDW state establishes in $(TMTSF)_2PF_6$ at 12 K. Superconductivity can be stabilized under pressure in $(TMTSF)_2PF_6$ (P>9 kbar) and even at atmospheric pressure in $(TMTSF)_2ClO_4$ or in several $(ET)_2X$ salts. (b) Phase diagram (P-T) of $(TMTSF)_2PF_6$ showing the common border between the SDW insulating state and the superconducting state. Squares, triangles and circles are data from different laboratories.

The words "organic conductors" are justified by the numerous peculiarities of these new conducting materials:
i) metallic-like conduction arises from the overlap of π-molecular orbitals between neighbouring entities.
ii) the inorganic ingredients which are necessary for the formation of organic superconducting solids do not contribute to the electron delocalization.
iii) the existence of a novel mechanism for the pair binding, non-phonon mediated, seems to be likely for at least some of the organic superconductors ; those displaying a common phase border between superconductivity and magnetism, figure 1b.

According to figure 2, the time dependence of T_c in organics is remarkably fast. It would even be more striking if the time dependence were normalized by the number of researchers involved in the search for new organic superconductors, definitely much smaller than in the domain of high T_c cuprates. It would thus be unjustified to conclude that T_c should saturate at 10K in organic conductors. Furthermore, the many facets of organic conductors, besides superconductivity, must be studied and understood carefully before a satisfying level of understanding of superconductivity in organic conductors can be reached.

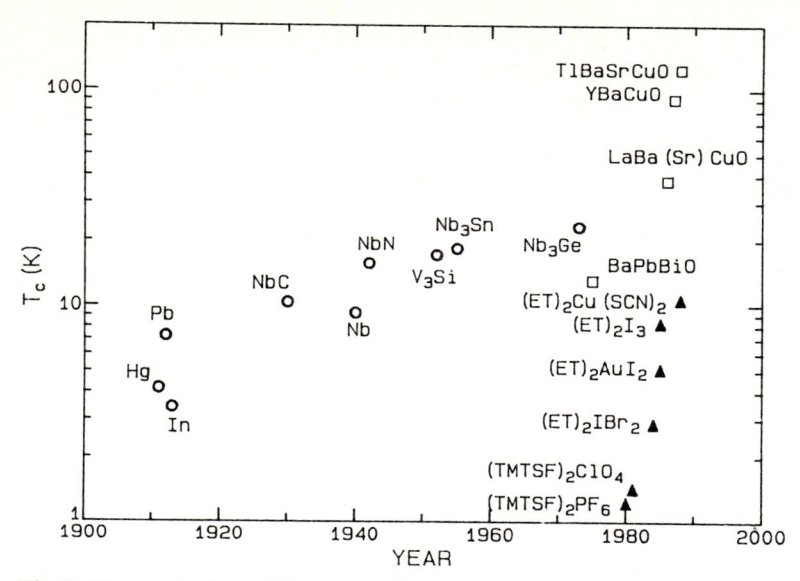

Fig.2: The evolution of T_c versus time in various inorganic and organic conductors.

These lectures attempt to provide a brief overview of this rapidly expanding field with special emphasis on the two important concepts of organic conductors which are: dimensionality and interactions between electrons (magnetism).

We shall first recall the basic theoretical behaviour of 1-D and quasi 1-D electron systems. Secondly, we provide several experimental evidences showing that the band picture is a good starting point for the description of these materials with Q-1-D and 2-D bands for $(TMTSF)_2X$ and $(ET)_2X$ series respectively. Then we show that the intermediate Coulomb coupling picture is a natural consequence of the interpretation of various magnetic measurements. Finally we emphasize that the pairing mechanism in the Q-1-D series $(TMTSF)_2X$ may involve the interchain exchange of antiferromagnetic spin fluctuations. However, the pairing mechanism may not necessarily be the same in the 2-D series $(ET)_2X$.

II. The One Dimensional Problem.

An array of 1-D conducting chains is characterized by a pair of planar (open) Fermi surfaces (FS) intersecting the $k_{//}$ axis at $\pm k_F$ where the Fermi wave vector is related to the density of carriers (ρ) per unit cell of a size by the equation $2k_F = \rho\pi/a$, figure 3. However, real organic conductors are only approximately 1-D conductors [9]. There exists a non-negligible interchain coupling resulting from the overlap of molecular orbitals giving rise to warped Fermi surfaces. Tubular and open F.S. are also present for 2-D conductors but so far no completely closed surfaces (see fig.3) have yet been encountered in organic conductors.

The peculiarity of the 1-D electron gas is already present in the system of non-interacting particles. The response $\Delta E = -\chi_0(q)V_q^2$ of the non-interacting 1-D

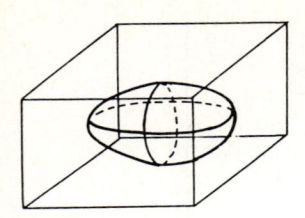

Fig.3: Various possibilities of Fermi surfaces, anisotropic 3-D (top), quasi-1-D open and warped surfaces (middle) and 1-D surfaces (bottom).

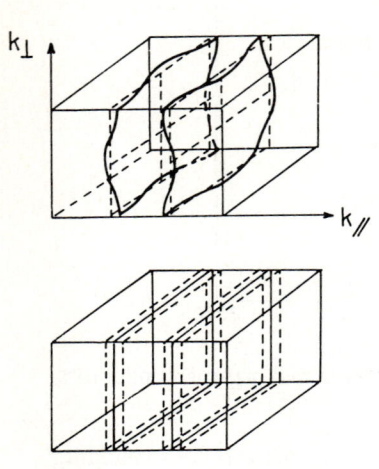

electron gas to the perturbing potential V_q is maximum for the wave vector q_o which minimizes the energy denominator $E(k+q)-E(k)$ in the calculation of the free electron response:

$$\chi_0(q) = \frac{2}{V} \sum_k \frac{f(E_k) - f(E_{k+q})}{E_{k+q} - E_k} .$$

For a 1-D system, $q_o = 2k_F$ and at $T = 0$ K $\chi_0(q) \approx \log|q-2k_F|$.

Similarly, the response at the wave vector $2k_F$ is logarithmically divergent, $\chi_0(2k_F) \sim \log T/E_F$. Switching on the interactions in the low dimensional electron gas the response (within the random phase approximation) becomes $\chi(q)=\chi_0(q)/1-I\chi_0(q)$ where the interaction I is approximated by a δ-like behaviour. Accordingly, the system develops an instability at a critical temperature given by the equation

$$1 - I \chi_0(Q) = 0 \qquad (1)$$

where Q is the vector which optimizes the nesting Q-1-D Fermi surface.

For an orthorhombic structure and near neighbour interactions the tight-binding energy dispersion reads

$$E(k) = -2t_a\cos k_x a - 2t_b\cos k_y b - 2t_c\cos k_z c . \qquad (2)$$

116

However, it is a great simplification to linearize the dispersion (2) along k_x when dealing with F.S properties as long as kT, t_b, $t_c \ll t_a$ we derive

$$E(k) = E_0 + v_F(|k_x| - k_F) + C(|k_x| - k_F)^2 - 2t_b \cos k_b y \, . \quad (3)$$

In the equation (3) a 2-D model has been used, forgetting about the c direction. Consequently, the FS is determined by the equation

$$k_x = k_F + 2t_b/v_F \cos k_y b + O(t_b^2 \cos^2 k_y b) + \ldots \quad (4)$$

Equ.4 gives a sinusoidal shape for the F.S in the k_x-k_y plane in first order in t_b. Instead of the dispersion relation (3) it is much easier to use the approximate relation

$$E(k) = E_0 + v_F(|k_x| - k_F) - 2t_b \cos k_y b - 2t_b' \cos 2k_y b \, . \quad (5)$$

Eqs (3) and (5) give the same F.S. when $t_b'(\approx t_b^2/t_a)$ is $\ll t_b$. For real materials of the (TMTSF)$_2$X series $t_b' \approx$ 10K and $t_b \approx$ 100K. As long as t_b' remains small compared to the characteristic energy of the Q-1-D electron gas instability T_0 the nesting of the F.S. may be considered as perfect and long range order (spin density wave for example) establishes according to equ.(1). The modulation wave-vector is thus the transverse vector $Q_t=(2k_F,\pi/b)$ which provides a perfect nesting of the Q-1-D F.S., fig.4, as long as the t_b' term in eq 5 is neglected. This modulation accompanies the onset of phase transition at T_0. If t_b' is large ($t_b'>T_0$) the nesting of the F.S. is frustrated and no long range order can develop at low temperature. Instead, the susceptibility shows a non-divergent maximum at a wave vector Q_0 close to $Q_0=(2k_F+q_{//}, q_\perp)$ where $q_{//}$ is a small vector and $q_\perp$ is different from π/b. Q_0 is the vector which optimizes the nesting of the non-sinusoidal F.S. given by equ 5 (see figure 4). We shall see in the following that the description of the F.S. of Q-1-D conductors in terms of a slight deviation from perfectly nested surfaces has proved to be particularly useful for the instabilities of the electron gas under high magnetic fields [10].

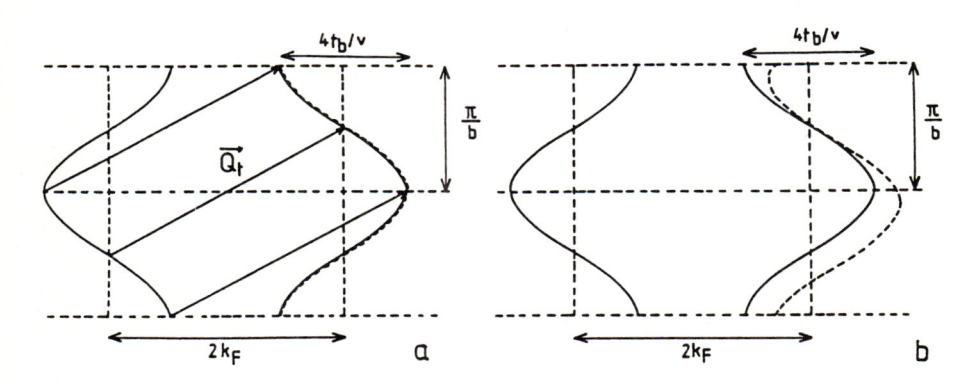

Fig.4: Nesting of the Fermi surface (a) Perfect nesting Q_t for negligible t_b' in eq (5). (b) Best nesting for non-negligible t_b'.

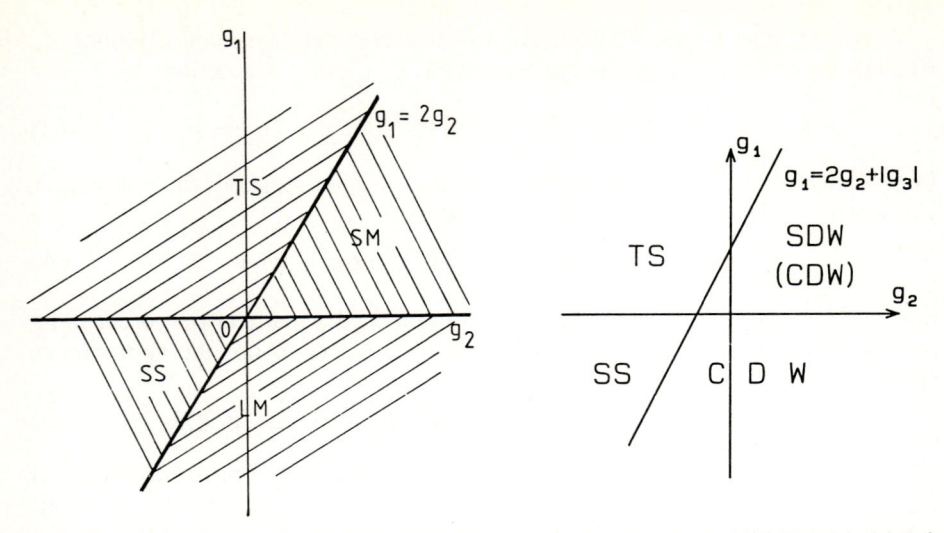

Fig.5: Diagram of most divergent 1-D correlations without (left) and with (right) the Umklapp scattering contribution. Triplet superconductivity (TS), Singlet superconductivity (SS), Spin modulation (SM or SDW), Lattice modulation (LM or CDW).

A well-established characteristic feature of 1-D electrons is the mixture between Peierls and Cooper diverging channels at low temperature [11]. The diagrammatic expansion of the Cooper pair response of the 1-D interacting electron gas contains diverging contributions coming from density fluctuations (Peierls or SDW) and vice versa for the Peierls channel [12]. Since both electron-electron and electron-hole channels diverge with the same logarithmic rate at low temperature the coupling between them cannot be ignored. When a mean field theory is used a finite transition temperature is found, the value of which depends on the Fourier components of the electron-electron (e-e) interaction, g_1 and g_2 for the $2k_F$ and $q=0$ parts respectively, figure 5.

This solution is not satisfactory for a 1-D electron gas since the basic theorem about the non-existence of long-range order in 1-D systems with finite range interactions is violated [13].

A significant improvement has been provided by the introduction of renormalization techniques [14]. Hence, correlation functions follow power law divergences at low temperature with exponents depending explicitly on the values of coupling constants (non-universal)[14]. For spin independent interactions, in lowest order, pair (Cooper) and density (CDW or SDW) correlations are most diverging when $g_1 > 2g_2 + g_3$ and $g_1 < 2g_2 + g_3$ respectively. Here g_3 is an Umklapp scattering contribution which is present only when each unit cell contains one carrier (half-filled band). The renormalization procedure does not give instabilities at finite temperature for a 1-D electron gas. The "phase diagram" thus derived for the dominant divergences is shown in figure 5.

An important consequence of the renormalization techniques is to make the effective interactions temperature dependent. For example, starting with bare interactions g_1 and $g_2>0$ at infinite temperature the renormalized interactions become [15]

$$g_1(T) = \frac{g_1}{1 - \frac{g_1}{\pi v_F} \ln \frac{T}{E_F}} \qquad (6a)$$

$$g_2(T) = g_2 - \frac{1}{2}(g_1 - g_1(T)) \ . \qquad (6b)$$

Following eqs (6), $g_1(T)$ renormalizes to zero (weak coupling) and $g_2(T)$ becomes attractive (<0) if the condition $g_1>2g_2$ is fulfilled at high temperature. A direct consequence of eq (6) is a temperature dependence of the uniform spin susceptibility:

$$\chi_s(T) = \chi_p \frac{1 + \mathbf{g_1}(T)}{1 - \mathbf{g_1}} \qquad (7)$$

where $\mathbf{g_1} = g_1/\pi v_F$. Channels mixture and renormalization effects become important below $T_g \sim E_F \exp - 1/\mathbf{g_i}$, [14]. With $\mathbf{g_i}$ in the range of 0.5 to 1 for organic conductors and $E_F \approx 1000$ K T_g lies above room temperature and all problems related to 1-D physics cannot be ignored (at least in a certain domain of temperatures).

1-D physics is limited towards low temperatures by the size of the interchain coupling (called $t_\perp$ from now on). At $T<t_\perp$ in a mean-field treatment the log divergence of individual correlation functions stops and below the cross-over temperature $T_{x1}^0 \sim t_\perp/\pi$ the Fermi liquid picture for an anisotropic 3 D system recovers its entire meaning [16] with a concomitant decoupling of pairing and density correlations. It is important in a given 1-D conductor to be able to derive the cross-over temperature from experiments. For a non-interacting electron gas $T_{x1} = T_{x1}^0 \sim t_\perp/\pi$ and thus ≈ 100K in the $(TMTSF)_2X$ series. Therefore, no power law divergences should be observed below 100K for those materials. The observation of power laws down to 10K suggests that T_{x1} is indeed renormalized below T_{x1}^0 by intrachain electron-electron interactions [17]. This topic will be discussed in section (VI).

III. Materials.

The vast majority of molecular crystals are insulators or at best semiconductors. Even if the electrons are well delocalized over a molecule there exists a finite energy gap which they have to overcome to give rise to macroscopic conductivity. Intermolecular conduction has been achieved only in peculiar molecular assemblies such as charge-transfer complexes $D^{+\gamma} A^{-\gamma}$, organic salts $D_x^{+1} X^{-1}$ or transition metal coordination complexes $D^{+\gamma} M_x^{-\gamma}$ where D, A, X and M are respectively a donor or acceptor molecule, an inorganic ion or a metal ions coordination complex M = Ni, Pd, Pt, Au...

Charge transfer complexes, TTF-TCNQ or so, have provided the initial momentum for the research in organic conductors. However, organic salts and transition metal complexes are the only organic materials which have given rise so far to superconductors.

An adequate packing of the molecules in the crystal allowing a strong intersite overlap of the molecular orbitals between highest occupied molecular orbitals (HOMO) for cation radical salts is also a prerequisite for high conductivity, figure6.

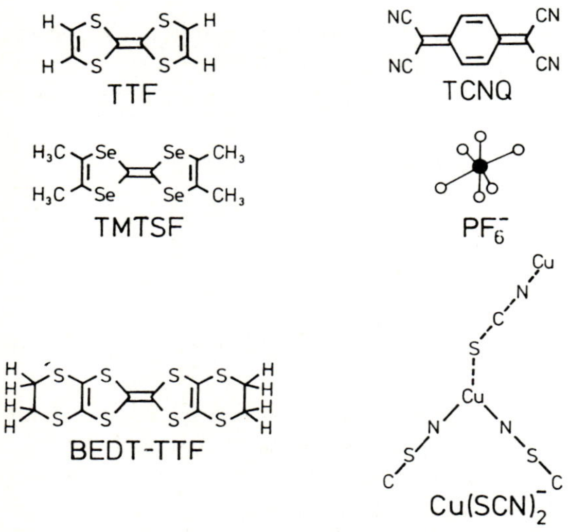

Fig.6: Some examples of donor molecules (left) and acceptor or anions (right).

Segregated chains of donor molecules give rise to 1-D or Q-1-D conductors, for ex. (TMTCF)$_2$X C=S, Se, and X=ClO$_4$,PF$_6$,ReO$_4$.... On the other hand, the packing of segregated layers separated by anion sheets leads to more 2-D materials, (ET)$_2$X. An interesting crystal structure is the one observed for the κ-phase of ET compounds, for example κ-(ET)$_2$I$_3$ or κ-(ET)$_2$ Cu(SCN)$_2$. It is in the latter compound that the highest T$_c$ has been achieved, see figure 7.

As far as the D$_2^+$X$^-$ family is concerned the periodicity of the lattice is determined by the anion spacing. Although all molecular sites experience the same potential from the 3-D anion lattice a slight dimerization of the intermolecular bond lengths exists. Its amplitude varies between different salts. It is large, ≈2.8% in the sulfur family and much smaller ≈0.8% in the selenium series [18]. A direct consequence of the bond length dimerization is the existence of a small (dimerization) gap Δ at k = ± π/a which makes the energy band half-filled (2k$_F$= π/a) instead of quarter-filled as it can be inferred from the stoichiometry [19].

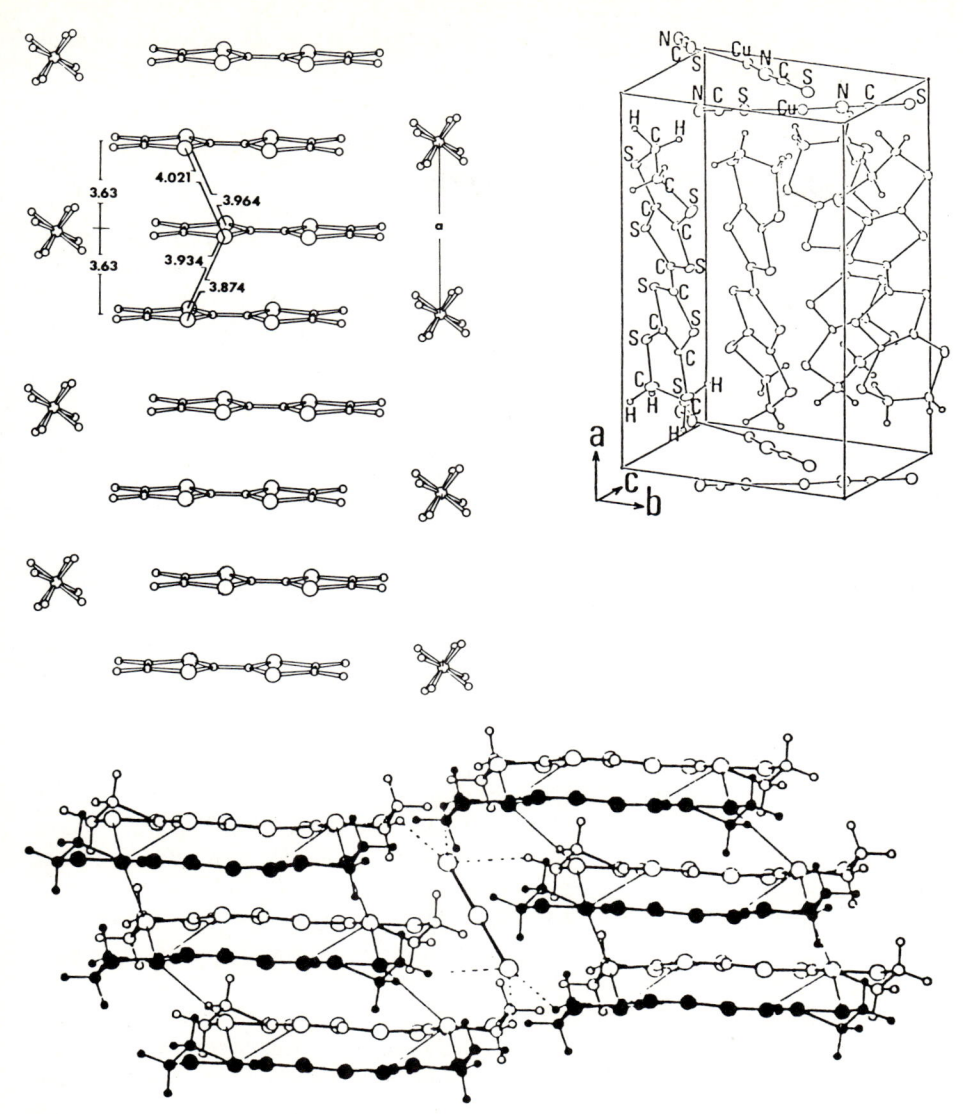

Fig.7: Crystal structure of (TMTSF)$_2$X (upper left) κ–(ET)$_2$ Cu(SCN)$_2$ (upper right) and β-(ET)$_2$I$_3$ (bottom)

Consequently, umklapp scattering characterized by the coupling constant $g_3 \sim g_1$ Δ/E_F must be taken into account at T less than Δ [20].

Band parameters are typically 0.25 eV, 25 meV and 1 meV and 0.15 eV, 15 meV and 0.5 meV for $t_a: t_b: t_c$ in (TMTSF)$_2$X and (TMTTF)$_2$X series respectively.

Additional requirements for electron delocalization are first the minimization of the intrasite Coulomb repulsion and secondly the absence of other periodic distortions opening a gap at the Fermi level (Peierls, SDW or anion ordered insulating phases).

IV. Experimental determination of Fermi surfaces.

This section will show how the existence and the shape of Fermi surfaces of organic superconductors can be investigated via high magnetic field studies. Two experiments are briefly surveyed: the first one refers to the existence of giant magnetoresistance oscillations in 2-D superconductors in the $(ET)_2X$ series, the second one has shown how the application of a high magnetic field along the low conductivity axis can drastically modify the shape of a Q-1-D Fermi surface in the $(TMTSF)_2X$ series and lead to new physical phenomena.

In the β-phase of $(ET)_2I_3$ exhibiting highly conducting planes perpendicular to the c direction two superconducting phases can be stabilized at low temperature depending on the cooling procedure. When the sample is cooled under atmospheric pressure superconductivity occurs at $T_c=1.5K$ in the $β_L$-phase [21]. This phase is characterized by an incommensurate lattice modulation involving the terminal CH_2 groups of ET molecules which establishes at 175 K [22]. The formation of an incommensurate modulation at low temperature can be avoided provided a moderate hydrostatic pressure (>350 bar) is applied while cooling between 300 and 100K. At the latter temperature the pressure can be released and superconductivity is observed at 8.1 K [27] in a metastable phase called $β_H$ which is expected to be free from superlattice structure.

The behaviour of the magnetoresistance of the $β_H$ phase at T=0.38K is presented up to 12 T and for the limited field range 9-12T in figure 8, [24]. The sharp onset of resistance around 2T is the signature of the critical field $H^⊥_{c2}$ for the magnetic field perpendicular to the conducting planes. The fast oscillations which become visible above 9T are periodic in 1/H with a fundamental field H_0 of 3730 T and show extremely large amplitudes at 12T. In addition, a low frequency (beating) oscillation is clearly observed with the frequency $H_1=36.8T$.

The fundamental field of 3730 T corresponds to a cross-sectional area which amounts to ≈ 50% the area of the first Brillouin zone. Furthermore, the large oscillation amplitudes can be understood when the conductor is highly anisotropic with different effective masses m_c along the field and $m_a(<<m_c)$ in the plane perpendicular to the field (where an isotropic motion is assumed). In the usual description of the Shubnikov-de Haas oscillations for an isotropic FS the amplitude of the pth harmonic is of the order of $(H/pH_0)^{1/2}$ [25]. However, there exists an enhancement factor $(m_c/m_a)^{1/2}$ in case of a tube-like FS. It is indeed already inferred from superconducting critical fields and conductivity anisotropies that the ratio of the transfer integral t_a/t_c is at least of the order of 40 in this conductor [26]. All experimental data point towards a pronouced 2D shape for the F.S. The beating frequency H_1 measures directly the amplitude of the warping since $H_0/H_1 = (k_Fa)^2 t_a/t_c$ assuming a dispersion law such as $E = h^2k^2/2m_a + 2t_c\cos k_z c$.

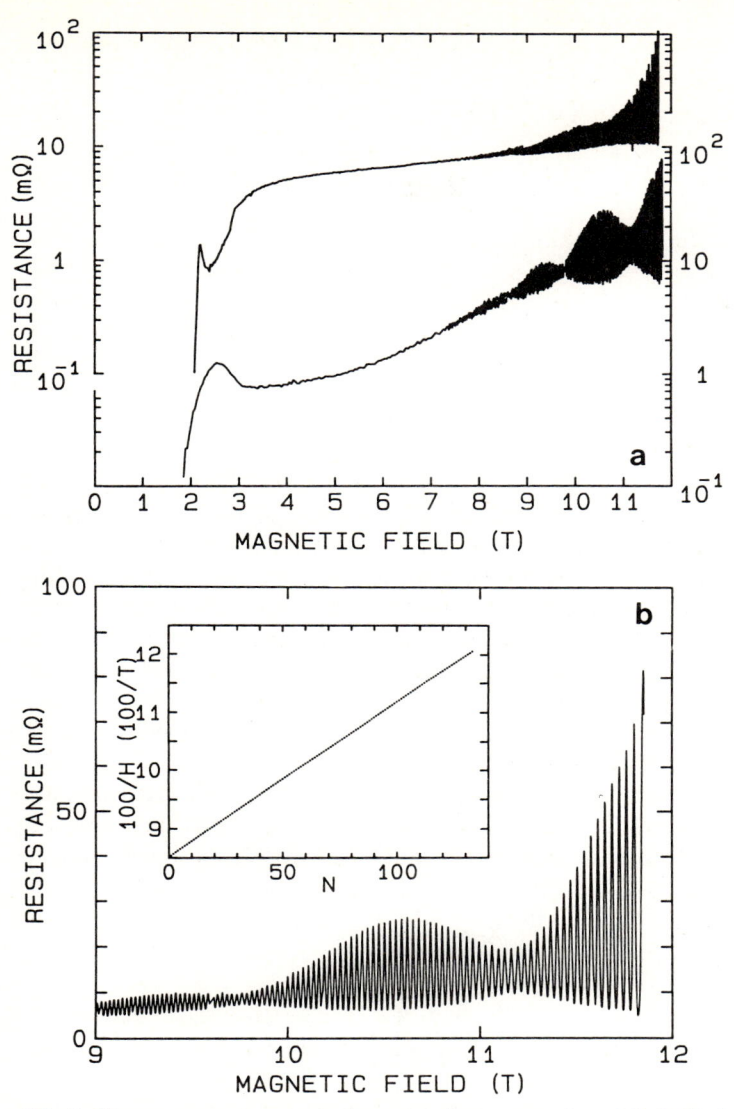

Fig.8: Transverse magnetoresistance of two samples of β-(ET)$_2$I$_3$ up to 12 T at T = 0.38K (a). Detailed behaviour of the magnetoresistance between 9 and 12 T (b). The insert shows that the field positions of the extrema are periodic in 1/H.

With H$_1$ = 36.8T and a = 3.5 Å, t$_a$/t$_c$ = 140 is obtained. A direct estimate of t$_c$ = 0.5meV is also derived from H$_1$/H= 2t$_c$/ħω$_c$ and the measured cyclotron mass of 4.

The above experimental results emphasize the 2D nature of the FS in β-(ET)$_2$I$_3$ conductors. Similar but somewhat weaker features have also been observed in κ-(ET)$_2$ Cu(SCN)$_2$ [27].

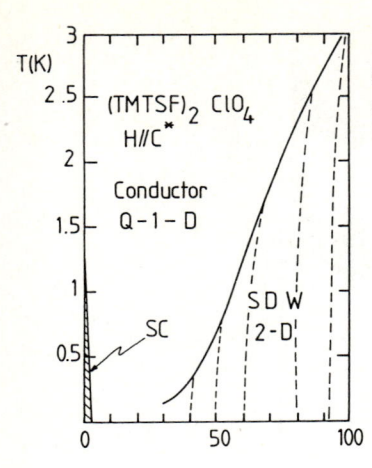

Fig.9: Field-induced spin density wave states (FISDW) in the R(relaxed) state of $(TMTSF)_2ClO_4$.

The open and planar character of the F.S. in Q-1-D $(TMTSF)_2X$ superconductors is best illustrated by an indirect but remarkable effect of the magnetic field, namely, the field-induced spin density wave states, FISDW, [10], figure 9. Superconductivity of $(TMTSF)_2X$ compounds is obtained either under pressure with the anions PF_6, TaF_6, ReO_4 [12] or at ambient pressure with ClO_4 [28] provided the cooling procedure of the latter compound is slow enough to allow the uniform ordering of ClO_4 ions below 24 K [29].

Following the theoretical model presented in Sec II superconductivity is stabilized when the frustrated nesting does not allow the onset of the SDW ground state ($t'_b > T_0$). Thus, the electron susceptibility $\chi_0(q)$ no longer presents any logarithmic divergence at Q_t but only a relative (non-divergent) maximum at Q_0 (close to Q_t), see figure 10. However, a comb of logarithmic divergences is

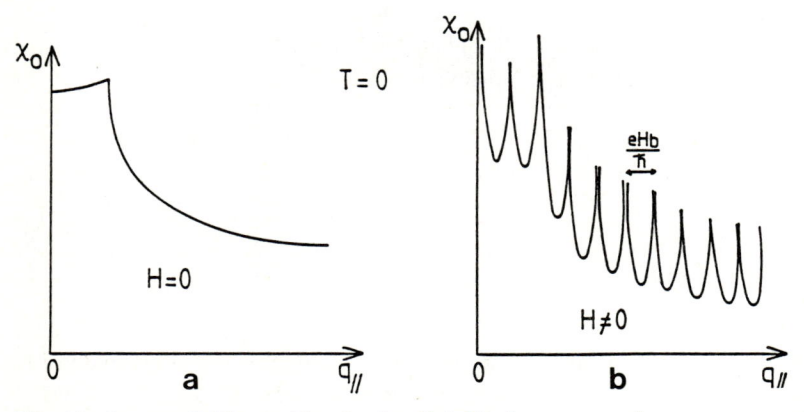

Fig.10: Susceptibility at T = 0 of a Q-1-D electron gas in zero magnetic field (a) displaying a non-divergent maximum at a wave vector ($q_{//}\neq 0$) near the transverse nesting wave vector ($q_{//}=0$) and in a finite magnetic field (b) which restores a comb of logarithmic divergences (after Montambaux thesis).

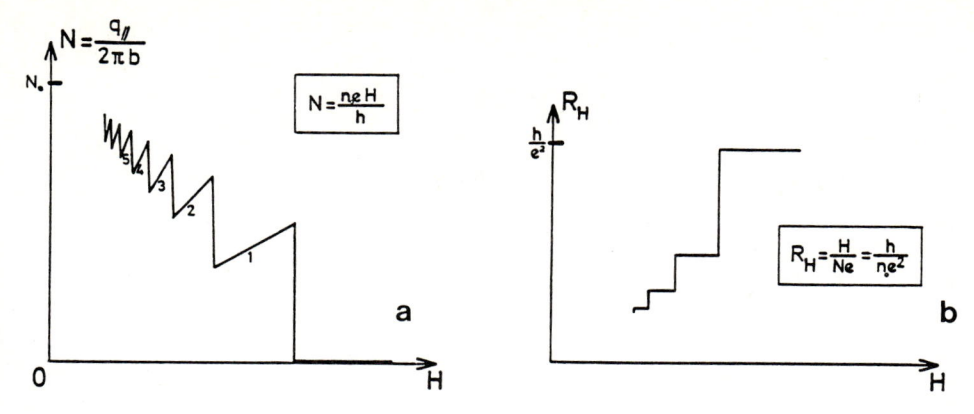

Fig.11: Linear dependence of the density of carriers versus magnetic field within each sub-phase characterized by the quantum number n = 1,2,... (a) and behaviour of the Hall resistance (b)

restored by the application of a magnetic field along the low conductivity c* axis making the electronic motion (under magnetic field) more one dimensional than under zero field [10], fig 9. A sequence of phases is observed as the magnetic field is varied above a certain threshold field. Each phase is semimetallic with a very low density of carriers ($\approx 10^{-2}$/unit cell) corresponding to the remaining "unnested" carriers. Furthermore, the Q vector of the SDW modulation related to the strongest log divergence in fig 10 becomes field dependent in order to keep the number N of Landau levels associated with the "unnested" carriers completely filled over the extended field range corresponding to the stability domain of the SDW phase labelled by the quantum number N. This procedure minimizes the diamagnetic energy at the expense of a slight deviation from the optimum nesting condition in zero field. If, however, the Q vector becomes too far from the optimum nesting a discontinuous jump of Q occurs with a concomitant change of the quantum number by one unit (first order phase transition). The model predicts a sequence of first order transitions between FISDW labelled by successive integers N and a field dependent density of carriers n = NeH/h, figure 11. The quantization of the Hall voltage is a direct consequence of the model since $R_H = H/ne = h/Ne^2$, fig 11. It is the finding of a field-independent Hall voltage in (TMTSF)$_2$ClO$_4$ [30] which has stimulated the development of the theory of the Q-1-D electron gas under high fields. But one complication pertaining to (TMTSF)$_2$ClO$_4$ is the anion ordering at 24K [31] which modifies the Fermi surface, opening small gaps at ($\pm k_F$, $\pm \pi/2b$) on the FS.

In principle, measurements on the PF$_6$ salts in which anion ordering effects are absent should be easier to interpret in terms of the model. However, the price to pay is the use of high pressures to stabilize the conducting state at low temperature.

Hall data for (TMTSF)$_2$PF$_6$ (P=8Kbar) [32] are shown in figure 12. Below 18T well defined Hall plateaus whose heights are in the ratio 1:1/2:1/3:1/4 are observed. The absolute values for the largest plateaus are 11.6 and 10.2 kΩ/molecular layer. Hence, not only are their ratios given by successive integers but the magnitude of

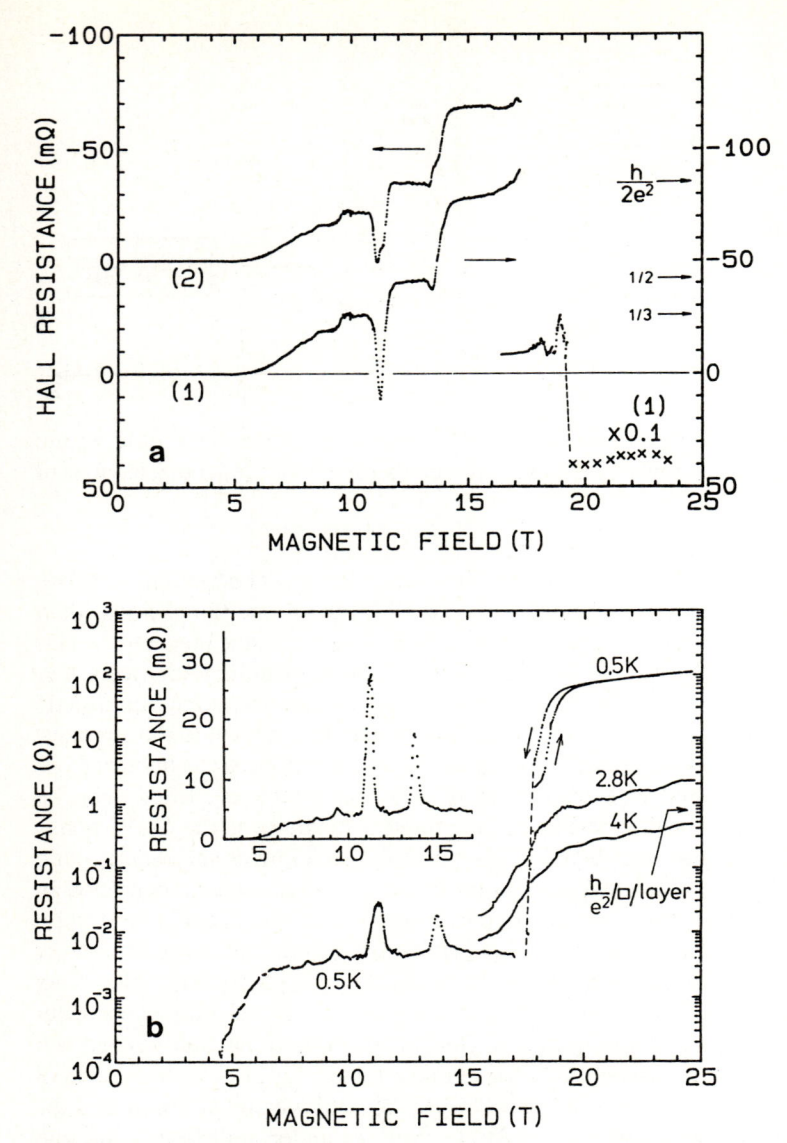

Fig.12: a) Hall resistance versus magnetic field //c* for two samples of (TMTSF)$_2$PF$_6$ under pressure. The quantized values h/2ne^2 (12.9/n kΩ) per molecular layer are marked on the right for sample # 1 (T = 0.5 K).
b) Field dependence of the longitudinal resistance showing peaks of magnetoresistance between Hall plateaux and the transition towards the n = 0 state at 18 T (after reference [32]). The insert shows the behaviour of the magnetoresistance.

the largest plateau corresponds rather well to $h/2e^2$=12.9 kΩ/layer, the value expected for the quantum Hall effect [32] in the presence of spin degeneracy. Magnetoresistance data, fig 12, show well defined peaks at the fields where the Hall voltage jumps to the next plateau. Above 18T, the magnetoresistance increases rapidly by a factor 10^4 reaching a value which is much larger than h/e^2 expected in case of fractional Hall effect [33]. Thus, the H>18T ground state is probably the N=0 state in which all carriers are removed from the Fermi level. The N=0 state would thus be similar to the H=0 SDW ground state observed at ambient pressure. A different high field behaviour has been obtained in the ClO_4 compound where a reentrance of the phase exhibiting no long range order has been reported by various techniques [34]. The differences between PF_6 and ClO_4 behaviours at high field are not yet understood. It may be caused by a subtle consequence of the anion ordering or the different energy scale for the onset of a SDW state in the two compounds.

Summarizing, the existence and the shape of low dimensional Fermi surfaces is firmly established experimentally in the 1-D and 2-D classes of organic superconductors. In addition, Q-1-D conductors have enabled the first observation of the Hall effect quantization in a 3-dimensional material.

V. Interacting low dimensional electrons.

Although a Fermi liquid picture with 1-D or 2-D FS is probably a good first order assumption according to the data presented in the previous section, the variety of different ground states observed in the $(TMTCF)_2X$ series (Spin-Peierls, SDW, SC) suggests that electron-electron interactions cannot be forgotten. It is the matter of this section to present the role of the interplay between interactions and dimensionality. Electron-electron interactions are of two kinds, either direct (Coulomb repulsion g_1, g_2, g_3) or mediated by the electron-phonon (el-ph) coupling [35] ($g_1^{ph} = -\lambda^2/\kappa$, where λ and κ are the el-ph coupling and elastic constant respectively).

The importance of el-ph interactions is well established for some classes of 1-D conductors where Peierls ground states have been clearly identified by diffraction techniques i.e. linear platinum chain compounds, transition metal tri-chalcogenides, etc. [37].

As far as organic conductors are concerned, the many phase transitions of TTF-TCNQ below 60K have been attributed to the Peierls mechanism in the presence of Coulomb repulsions [37]. For one-chain conductors, Peierls transitions are usually not observed except for the very 1-D series of compounds such as $(Arene)_2PF_6$ and related compounds in which $2k_F$ 1-D lattice fluctuations are observed below room temperature and a 3-D condensation of a superlattice occurs around 150K or so [38].

In the $(TMTCF)_2X$ series el-ph and Coulombic interactions are both important since these conductors have narrower bands than arene salts (<1eV instead of 2 eV and $\lambda \sim t_{//}$ whereas $g_i \approx t_{//}^{-1}$ in a tight-binding model). The phase diagram of the $(TMTTF-TMTSF)_2X$ series provides a remarkable illustration of the existence of several competing interactions, fig 13. A spin-Peierls ground state (total spin S=0)

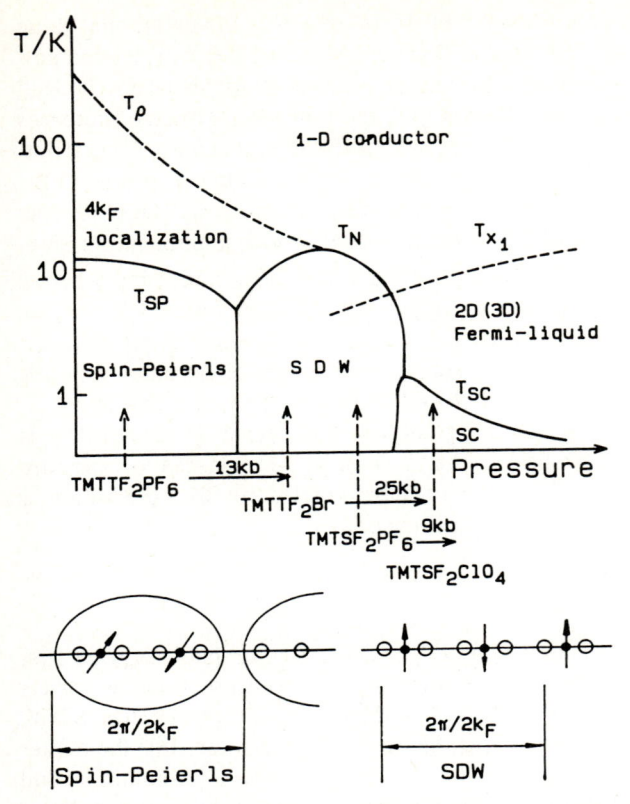

Fig.13: Generalized phase diagram for the (TMTTF-TMTSF)$_2$X series.

together with a lattice distortion is the ground state of TMTTF$_2$PF$_6$ [39]. Above 10 kbar, the same material becomes antiferromagnetic as detected by NMR investigations [40]. A similar state is observed as the ground state for (TMTTF)$_2$Br [41] and the selenium compound (TMTSF)$_2$PF$_6$ [42] which becomes a superconductor under pressure. With the smaller anion ClO$_4^-$, superconductivity can be obtained under ambient pressure in (TMTSF)$_2$ClO$_4$. It is also interesting to look at the behaviour of the various sulfur based materials at high temperature. There is a loss in the charge degrees of freedom below the temperature T_ρ (compound and pressure dependent) which has been understood in terms of a Coulomb driven 4k$_F$ charge localization [43]. Moving from left to right in the diagram of fig 13, the bond lengths dimerization diminishes with concomitant decrease of the g$_3$ contribution and of T_ρ. When the situation for a superconducting ground state is achieved the condition $g_3 \ll g_1, g_2$ should prevail. However g_1 and g_2 do influence strongly the conduction and magnetic properties of materials undergoing a superconducting transition at low temperature. It is therefore a major task to measure the amplitude of these Coulomb repulsions via appropriate experimental investigation such as NMR.

VI. Power laws and Coulomb interactions in 1-D conductors.

The electron degrees of freedom modulate the hyperfine interaction in a conductor and can in turn be the dominant source of spin-lattice relaxation [44]:

$$(T_1T)^{-1} = 2\gamma_n^2 |A|^2 \int dq^d \chi''_\perp(q, \omega_n)/\omega_n \qquad (8)$$

where $\chi''_\perp$ is the imaginary part of the transverse spin susceptibility of the interacting electron system. T_1^{-1} in eq(8) involves a summation of $\chi''_\perp(q, \omega_n)$ over all q-vectors which simplifies for a 1-D conductor since only $q = 0$ and $q = 2k_F$ processes are allowed. Hence T_1^{-1} for a 1-D conductor contains only the $q = 0$ and $q = 2k_F$ contributions probing the uniform and antiferromagnetic correlations of the 1-D electron gas respectively [45]. The $2k_F$ contribution to $(T_1T)^{-1}$ is proportional to $\chi(2k_F,T)$ which in turn follows a power law temperature dependence $\chi(2k_F,T) \alpha (T/E_F^*)^{-\gamma_{1D}}$ in the 1-D regime ($T_{x1} < T < E_F^*$) where γ_{1D} is the one dimensional interaction dependent power law exponent [46] and E_F^* is the renormalized Fermi energy ($<E_F$). In the extreme situation of strongly repulsive interactions and half-filling of the band (1-D quantum antiferromagnet) $\gamma_{1-D} = 1$. Thus $T_1^{-1}(2k_F)$ becomes temperature independent. The small q excitations of the 1-D gas are not singular at low temperature. They are responsible for the temperature dependence of the uniform spin susceptibility $\chi_s(T)$ and their contributions to the nuclear relaxation read [47]

$$(T_1T)^{-1}_{q=0} \propto \chi_s^2(T) . \qquad (9)$$

Summarizing, the nuclear relaxation of a 1-D electron system becomes

$$T_1^{-1} \approx C_0 T \chi_s^2 + C_1 \qquad (10)$$

where C_1 is either constant or weakly temperature dependent for localized ($\gamma_{1D}=1$) and delocalized carriers ($\gamma_{1D} \sim 0.7$-0.9) respectively. Therefore, the interpretation of the NMR relaxation provides a unique possibility to probe both ($q = 0$) and antiferromagnetic ($q \approx 2k_F$) correlations. Some of the NMR data obtained in the $(TMTCF)_2X$ series [48] are summarized in figure 14a and b. The very good agreement between the data of $(TMTTF)_2PF_6$ and the relation (9) taking $C_1 = $ const between 40 and 200 K indicates that the electron system behaves in this T-domain as a purely 1-D quantum antiferromagnet with $2k_F$ spin fluctuations extending up to 200 K (contributing to C_1) and a temperature dependence of $\chi_s(T)$ governed by 1-D paramagnon effects ($q = 0$). Let us recall that the charge degrees of freedom are frozen in below $T \approx 200K$ in $(TMTTF)_2PF_6$.

The influence of $2k_F$ correlations is not as strong in selenium (superconducting) systems, fig 14a. However, antiferromagnetic correlations can clearly be seen in $(TMTSF)_2PF_6$ below 100 K and in $(TMTSF)_2ClO_4$ below 30 K when the temperature dependence of $(T_1T)^{-1}$ does not follow any longer the temperature

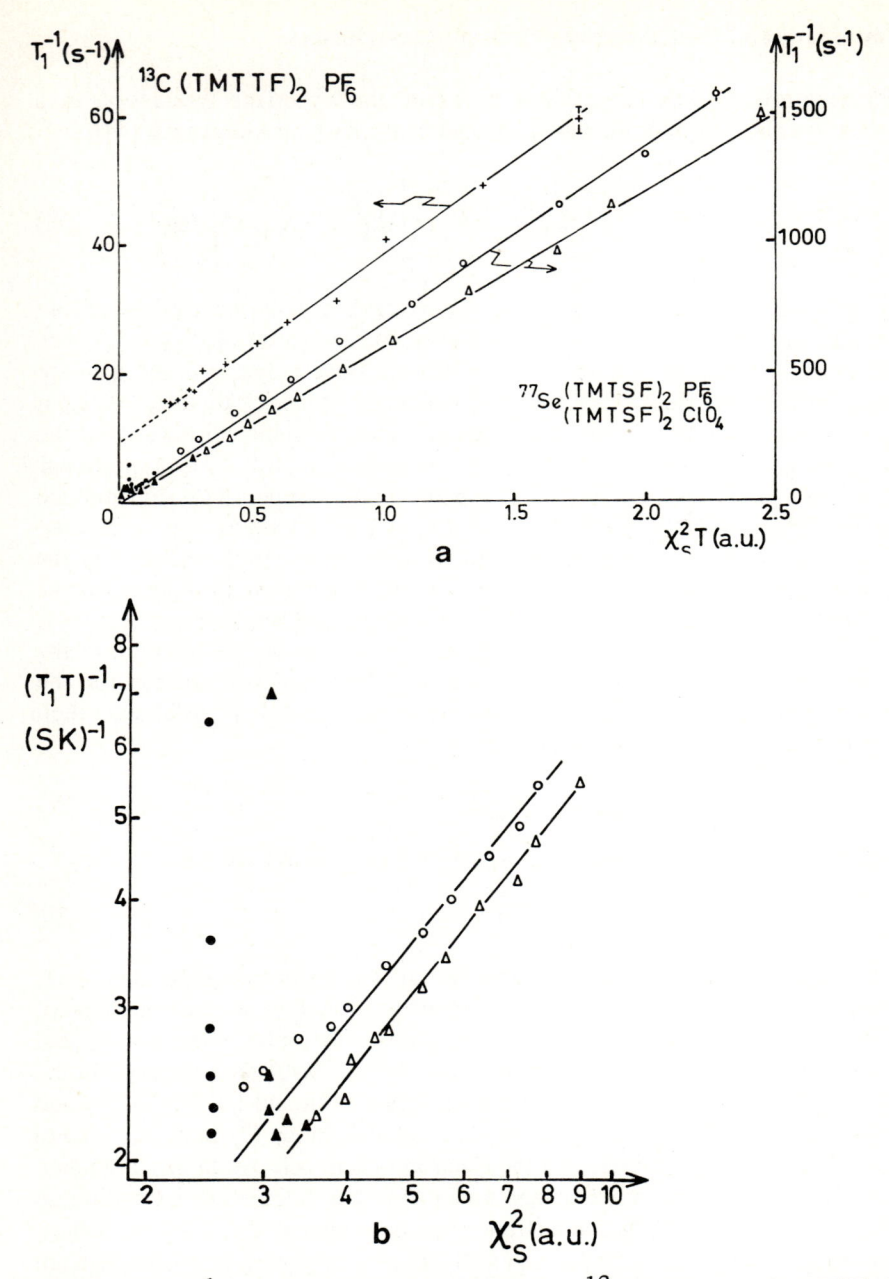

Fig.14: (a) T_1^{-1} versus $\chi^2_s T$ for (TMTTF)$_2$PF, ^{13}C-NMR and (TMTSF)$_2$PF$_6$, ClO$_4$, ^{77}Se-NMR.
(b) $(T_1 T)^{-1}$ enhancement in (TMTSF)$_2$ClO$_4$ (triangles) below 30 K and in (TMTSF)$_2$PF$_6$ (circles) below 100 K. Open and closed symbols are data from two different experimentalists.

dependence of χ^2_s, fig 14b. The T_1^{-1} versus T plot of figure 14b shows that the divergence of $\chi(2k_F, T)$ (which is responsible for the plateau of T_1^{-1}) levels off below 8K. A classical Fermi liquid behaviour with a renormalized Fermi energy is thus recovered below 8K in the vicinity of the superconducting transition. It is tempting to consider the region around 8 K as the cross-over region where the temperature dependence of the correlations changes from 1-D at high temperature to 2D(3D) below $T_{x1} \sim 8K$.

A somewhat quantitative analysis of the amplitude of Coulomb interactions can be extracted from the temperature dependence of T_1^{-1} and χ_s, figures 15 and 16 leading to $g_i/\pi V_F \approx 0.8$ for an organic superconductor such as $(TMTSF)_2ClO_4$. The major result of these studies is the experimental fact that superconductivity of

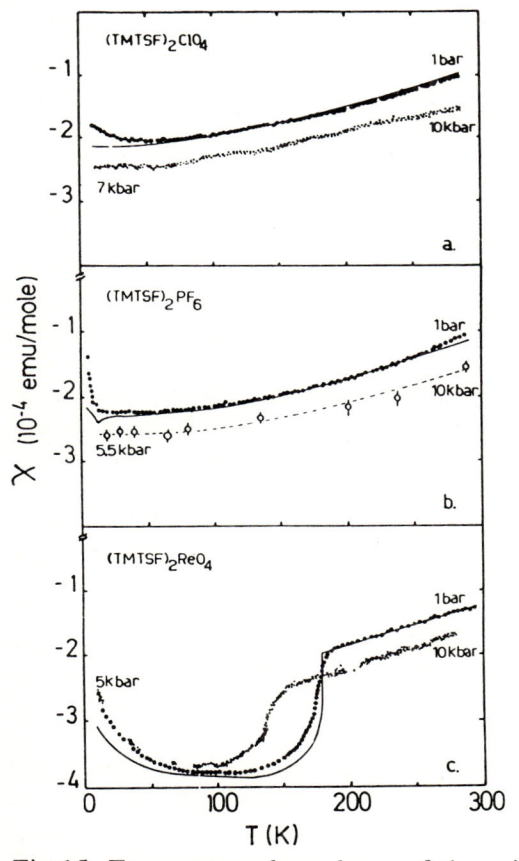

Fig.15: Temperature dependence of the spin susceptibility of various organic conductors. The spin contribution amounts to $\approx 3 \times 10^{-4}$ emu/mole at room temperature. The onset of a SDW state and of an anion ordered state are visible in $(TMTSF)_2PF_6$ and $(TMTSF)_2ReO_4$ respectively (after L.Forro et al Synth.Metals 19, 339 (1987).

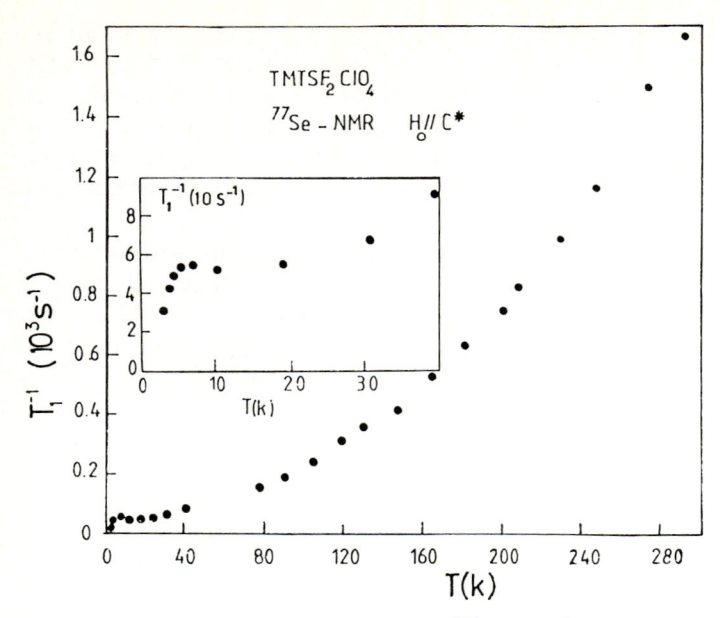

Fig.16: Temperature dependence of ^{77}Se T_1^{-1} in (TMTSF)$_2$ClO$_4$. The inset emphasizes the departure from the Korringa law at low temperature.

Q-1-D materials arises in a Fermi liquid within a strong background of itinerant antiferromagnetic fluctuations.

The low value of $T_{x1} \approx 8K$ as inferred from NMR data, figure 16, can be considered as a reduction of the bare cross-over at $T^o_{x1} \sim 100K$ due to intrachain electron-hole (e-h) correlations [49]. Intrachain correlations tend to bind an electron and a hole together on a given chain making thus the hopping of single particles more difficult than in the absence of these 1-D correlations. Consequently, $t_\perp$ is renormalized and a ratio $T_{x1}/T^o_{x1} \approx 0.1$ is indeed compatible with $t_\perp/E_F \sim 0.1$ and g_i values in the vicinity of unity as shown by the analysis of NMR data.

The onset of long range SDW at $\approx$ 19K in TMTTF$_2$Br and similar compounds (i.e in the 1-D localized regime $T>T_{x1}$ since T_{x1} is likely to be even smaller than 8K in sulfur compounds) raises a serious theoretical problem. What is the interchain exchange coupling $J_\perp$ which can be efficient for the establishment of AF order in the absence of Fermi surface?

Bourbonnais and Caron [50] have proposed that such an interchain exchange for a correlated 1-D conductor can be generated by the renormalization of 1-D interactions and of the single particle tunneling. The interchain exchange $J_\perp$ couples AF spin fluctuations of neighbouring chains at $T>T_{x1}$. For itinerant electrons it reads

$$J_\perp(T) \sim \pi v_F \left(\frac{g_i}{\pi v_F}\right)^2 \left(\frac{t_\perp}{\pi T}\right)^2 .$$

In the limit of carriers localized by a strong Coulomb repulsion the usual AF exchange $2t^2_\perp/U$ between stacks is recovered.

VII. Possibility for unconventional pairing in (TMTSF)$_2$X.

The model of interchain exchange for AF long range ordering can have important consequences for superconductivity [50a]. As noticed by Bourbonnais and Caron [50b] 1-D AF correlations persist at low temperature even if the deviation from perfect nesting Q_t no longer allows SDW long range order. Thus, the Cooper channel remains the only diverging channel at low temperature (in the absence of pair breaking mechanisms, impurities, etc...). When looking at the behaviour of the Cooper channel the same interchain exchange mechanism provides attractive forces between carriers on neighbouring chains. This is equivalent to the exchange of a $2k_F$ spin fluctuation between two carriers of opposite spin and momentum located on neighbouring chains, fig 17a. The generated attractive e-e interaction enters the 3-D mean-field gap equation for singlet superconductivity at $T<T_{x1}$ (and $t'_b>t'_b{}^*$) and can possibly overcome the intrachain repulsive Coulomb interaction. A finite transition temperature follows from this situation.

In this model, pairing two electrons on different chains takes advantage of the attractive interchain coupling and reduces the local repulsion since a minimum distance between carriers exists. Furthermore, this minimum distance b should give rise to an anisotropic superconducting gap

$$\Delta(k) \approx \Delta_0 \cos k.b \qquad (11)$$

exhibiting lines of zeros on the FS at $k_\perp b = \pi/2$. Several experimental results could support the possibility for an anisotropic gap in (TMTSF)$_2$X superconductors.

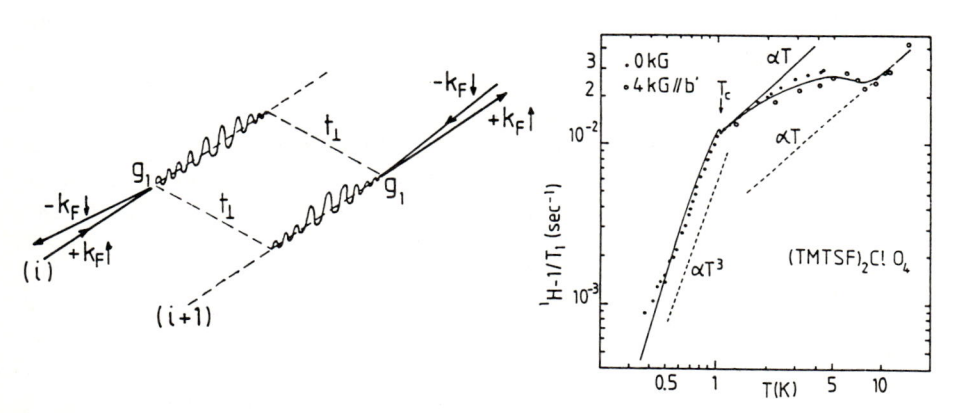

Fig.17: Schematic representation of the interchain exchange of $2k_F$ spin fluctuations between carriers on different chains (left). Temperature dependence of T_1^{-1} in (TMTSF)$_2$ClO$_4$. Below T_c data are from Takigawa et al.[51] in zero field. Above T_c data includes 4kG results obtained at Orsay (right).

i) The nuclear relaxation profile does not show [51] the typical enhancement which is often observed slightly below T_c in regular isotropic superconductors, figure 17b.

ii) T_1^{-1} has no activated behaviour below T_c but instead exhibits a power law temperature dependence, αT^3 [51], in agreement with the theory for the cosine law [52].

iii) Non-magnetic defects or impurities are known to be very strong pair breakers in $(TMTSF)_2X$ [53] in agreement with the theory of anisotropic superconductors [54].

The mediation of attractive pairing via the exchange of spin fluctuations between chains following Bourbonnais and Caron could also allow the coexistence in the high temperature domain of AF and superconducting correlations usually forbidden in a 1-D situation. The observation of a zero frequency collective mode in the far infrared properties of these conductors above T_c [55] could support experimentally this model.

VIII. Higher T'$_c$s in organic conductors.

Can one reach values of Tc's higher than the so far observed 1-2K in the $(TMTSF)_2X$ series? This is a reasonable question since T_c is known to be a rapidly decreasing function of pressure and higher values of T_c could be expected in a situation of larger cell volumes or "negative" pressure. This was the motivation for the pressure study of $(TMTSF)_2ReO_4$ [56]. At ambient pressure the non-centrosymmetric ReO_4 anions order at 180K doubling the periodicity in all three directions (1/2, 1/2, 1/2) [57]. Consequently, a gap opens at the Fermi level of the Q-1-D electron gas and the system undergoes an anion ordered metal-insulator transition at 180K [57]. However, anion ordering is modified by the application of a high pressure [58]. Above 10 kbar the anion ordering (0, 1/2, 1/2) which becomes stable does not affect significantly the stability of the conducting state at low temperatures: superconductivity is achieved at 1-2 K [59].

As T_c is also very much pressure dependent in $(TMTSF)_2ReO_4$ there was a hope to raise it up to 10K or so at ambient pressure as long as the formation of the anion-ordered insulating state can be prevented. The stabilization of a conducting state at ambient pressure below 30K with anions ordered in the (0, 1/2, 1/2) configuration is feasible following the P-T procedure shown in figure 18. Unfortunately, instead of a superconducting state, the intrinsic SDW instability of the Q-1-D electron gas is recovered below 8 kbar. The salient conclusions of this experiment are first, the anions can be trapped in a metastable configuration which does not open a significant gap at the Fermi level but one cannot avoid the intrinsic competition between superconductivity and antiferromagnetism and secondly T_c higher than 1-2 K should be looked for in materials where superconductivity is no longer competing with antiferromagnetism.

One other possibility is to look for superconductivity in organic conductors which are still less one dimensional than those pertaining to the $(TMTSF)_2X$ family. Higher values for T_c have indeed been obtained with salts of the ET radical

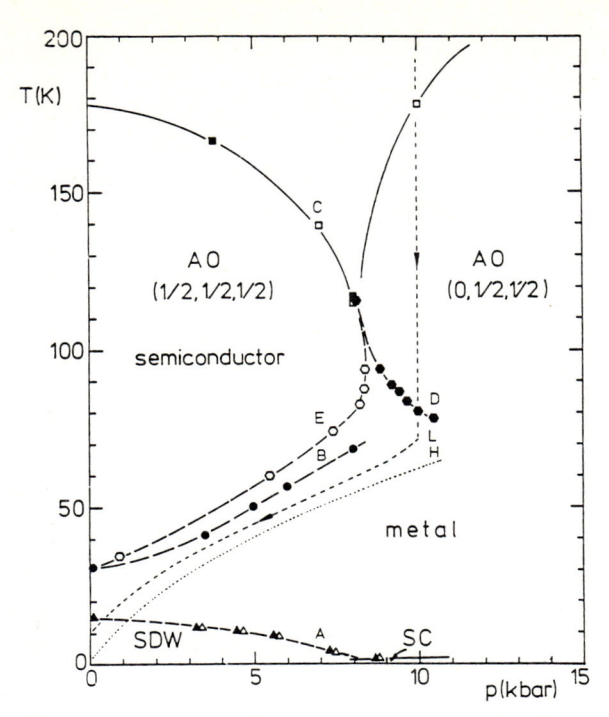

Fig.18: Intrinsic phase diagram of (TMTSF)$_2$ReO$_4$ obtained with the "Orsay procedure" going around the anion-ordered (AO) state. The cooling path is indicated by the dashed line with the arrows (after reference [56]).

cation which display an enhanced 2-D character. T_c=2.7, 5 and 8.1 K are observed with β-(ET)$_2$X for X = IBr$_2$, Au I$_2$ and I$_3$ respectively. The highest transition temperature is observed in the 2-D conductor κ-(ET)$_2$ Cu(SCN)$_2$ [8], (T_c~9-10K), figure 19. For this latter compound the anisotropy of conductivity between the conducting plane direction and the perpendicular direction is about 10^3 at room temperature and there is practically no anisotropy in the conducting plane.

The parameters of the superconducting phase are: T_c≈10K and $H_{c1}^{//(\perp)}$ = 1G(45G), $H_{c2}^{//(\perp)}$ = 20 T (2.5 T), $\xi_{//}^{b(c)}$~174 Å(118Å) and $\xi_\perp$~7.8 Å at 4.2 K [60].

The transverse coherence length, derived from a Ginzburg-Landau analysis of the critical fields is less than the corresponding lattice parameter 16.4 Å. This fact together with the small value of the interplanar coupling $t_\perp$~6K raises the problem of the applicability of the G-L description to these layered superconductors. In other words can β-(ET)$_2$X superconductors be considered as 3-D anisotropic superconductors or as 2-D superconductors [61] in which 3-D order at T_c^{3D} results from a Josephson coupling between layers with T_c^{3D}<T_c^{2D} (mean-field)? The answer is still not clear since in this series of materials the ratio $(2t_\perp/T_c)^2$ lies in the region 1-2 where both interpretations meet.

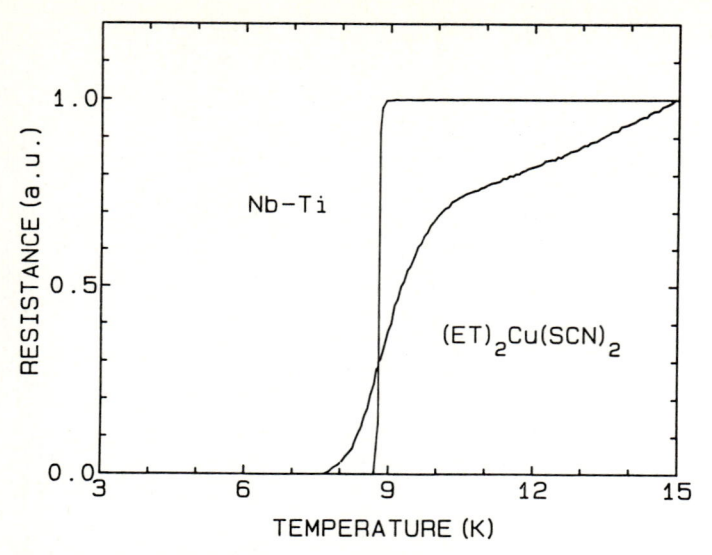

Fig.19: The superconducting transition of (ET)$_2$Cu(SCN)$_2$ compared to a Nb-Ti reference sample. The width of the transition in the organic sample is probably due to the existence of impurities or lattice defects, it may vary from sample to sample.

IX. Conclusions.

At the end of the present survey it should become clear to the reader that new families of synthetic conductors in which superconductivity has been discovered have also given rise to a wealth of new basic physical effects. We have emphasized that a Fermi liquid description is the proper picture at low temperature for all classes of conducting materials with an adequate renormalization of the interactions and that various materials can be classified according to the shape of their Fermi surface. Those pertaining to the (TMTSF)$_2$X series exhibit quasi planar surfaces and are very sensitive to the application of a magnetic field which favours the stability of spin density wave ground states. This remarkable effect has been discovered in (TMTSF)$_2$ClO$_4$ but recent studies in (TMTSF)$_2$PF$_6$ have provided high quality data showing that the Hall effect can be quantized in a 3D material very much as it is in a 2D electron gas. However, everything is not clear yet in this phenomenon of field induced spin density wave states, in particular the modification of the Fermi surface due to the ordering of non-centrosymmetric anions such as ClO$_4$. Quasi tubular (2D) Fermi surfaces have been observed through magnetoresistance oscillations in the (ET)$_2$X series. Spectacular oscillations have been obtained in β-(ET)$_2$I$_3$ with a frequency corresponding to a cross-section of 1/2 the first Brillouin zone. Smaller cross-sections are obtained with the κ structure X = Cu(SCN)$_2$[62].

Electron-electron interactions do play a crucial role in the physics of organic superconductors. Their existence is responsible for the generalized phase diagram of the (TMTTF-TMTSF)$_2$X series with the possibility of spin-Peierls, spin density wave and superconducting ground states [63]. NMR studies have shown that superconducting long range order of the (TMTSF)$_2$X series establishes in a background of well developed antiferromagnetic correlations. This has led theoreticians to propose a pairing mechanism based on the exchange of antiferromagnetic fluctuations between neighbouring chains. Some experimental data support the anisotropy of the gap which should result from the proposed interchain pairing. However, more experiments are needed. Very little is known yet regarding the mechanism of superconductivity in 2-D superconductors. The role of magnetism in the latter series is not as evident as for the 1-D series. Isotope effect experiments should clarify in the future the influence of various intramolecular modes on superconductivity. A third family of organic superconductors exists and has not been reviewed in the present article. They are the two or one chain charge transfer compounds based on the molecule M(dmit)$_2$, M = Ni, Pd. For these interesting materials the ground states in competition are superconducting and charge density wave states [64].

The effect of impurities (non-magnetic) on organic superconductivity has also not been reviewed in this article. Experimentally speaking the pair-breaking effect of non-magnetic impurities is fairly well established in both 1-D and 2-D series of superconductors. This unusual pair-breaking effect could be related to the anisotropic character of the superconducting gap. Finally, there was no mention in this review of the collective motion of SDW which has been observed recently in the antiferromagnetic ground state of (TMTSF)$_2$NO$_3$ [65] and (TMTSF)$_2$PF$_6$ [66] when the electric field overcomes a certain threshold field.

In conclusion, the discovery of superconductivity in organic conductors has been the seed for the development of new families of conductors displaying remarkable physical properties. It is our goal, together with our colleagues in Synthetic and Structural Chemistry to improve the level of understanding of these molecular materials at a still more microscopic scale and to engineer new compounds displaying well controlled properties (structural disorder, dimensionality, nature of ground state, value of T_c, etc...).

I wish to acknowledge the constant cooperation of my colleagues in the various disciplines, K.Bechgaard, P.Batail and their groups for the chemistry, C.Bourbonnais and G.Montambaux for their contributions to the theory. Experiments have been performed at Orsay and Grenoble (SNCI) together with P.Auban, J.R.Cooper, F.Creuzet,S.Tomic, W.Kang and P.Wzietek. We have largely benefited from the technical help of J.C.Ameline. The activity on organic conductors is partly supported by the ESPRIT contract BRA 3121.

REFERENCES.

[1] J.G.Bednorz and K.A.Müller, Z.Phys. B64, 189 (1986)
[2] D.Jérome, A.Mazaud, M.Ribault and K.Bechgaard, J.Phys.Lett.Paris, 41, L-95 (1980)
[3] W.A.Little, Phys.Rev.134A, 1416 (1964)
[4] J.Ferraris, D.O.Cowan, V.Walatka and J.H.Perlstein ,J.Am.Chem.Soc.95, 948 (1973) and L.B.Coleman M.J.Cohen, D.J.Sandman, F.G.Yamagishi, A.F.Garito and A.J.Heeger, Solid State Comm, 12, 1125 (1973)
[5] K.Bechgaard, C.S.Jacobsen, K.Mortensen, H.J.Pedersen and N.Thorup, Solid State Comm. 33, 1119 (1980)
[6] V.N.Laukhin, E.E.Kostyuchenko, Y.V.Susko, I.F.Schegolev and E.B.Yagubskii, JETP Lett 41, 81 (1985) and K.Murata, M.Tokumoto, H.Anzai, H.Bando, G.Saito, K.Kajimura and T.Ishiguro, J.Phys.Soc.Japan 54, 1236 (1985)
[7] F.Creuzet, G.Creuzet, D.Jérome, D.Schweitzer and H.J.Keller, J.Physique Lett 46, L-1079 (1985)
[8] H.Urayama, H.Yamochi, G.Saito, K.Nozawa, M.Kinoshita, S.Sato, K.Oshima, A.Kawamoto and J.Tanaka, Chem.Lett 55 (1988)
[9] J.Friedel and D.Jérome,Contemp.Phys.23, 583 (1982)
[10] G.Montambaux in Low Dimensional Conductors and Superconductors, p.233, edited by D.Jérome and L.G.Caron, NATO ASI Series, Plenum 1987
[11] Y.Byschkov, L.P.Gorkov and I.E.Dzyaloshinskii, Sov.Phys.JETP 23, 489 (1966)
[12] D.Jérome and H.J.Schulz, Adv.in Physics 31, 299 (1982)
[13] L.D.Landau and E.M.Lifshitz, Statistical Physics p.482. Pergamon, London (1959)
[14] J.Solyom, Adv.in Physics 28, 201 (1979)
[15] S.Barisic, Mol.Cryst.Liq.Cryst.119,413 (1985)
[16] V.J.Emery, J.Physique C-3, 44, 977 (1983)
[17] C.Bourbonnais in Low Dimensional Conductors and Superconductors, p.155, Loc cit.
[18] B.Gallois, thesis, Univ.of Bordeaux (1987)
[19] J.P.Pouget in Low Dimensional Conductors and Superconductors, p.17, Loc cit.
[20] S.Barisic and S.Brazovskii, in Recent Developments in Condensed Matter Physics, edited by J.T.Devreese Plenum (1981)
[21] E.B.Yagubskii, I.F.Schegolev, V.N.Laukhin, P.A.Kononovitch, M.V.Kartsnovik, A.V.Zvarykina and L.J.Burarov, JETP Lett 39, 12 (1984) and G.W.Crabtree, K.D.Carlson, L.N.Hall,P.T.Copps, H.H.Wang, T.J.Emge, M.A.Beno and J.M.Williams,Phys.Rev.B30, 2958 (1984)
[22] P.C.W.Leung, T.J.Emge, M.A.Beno, H.H.Wang, J.M.Williams,V.Petricek and P.Coppens, J.Am.Chem.Soc.107, 6184 (1985)
[23] W.Kang, G.Creuzet, D.Jérome and C.Lenoir, J.Physique 48, 1035 (1987)

[24] W.Kang, G.Montambaux, J.R.Cooper, D.Jérome, P.Batail and C.Lenoir, Phys.Rev.Lett.62,2559 (1989)
[25] D.Shoenberg, Magnetic Oscillations in Metals Cambridge Univ.Press, Cambridge (1984)
[26] K.Murata, M.Tokumoto, H.Anzai, H.Bando, K.Kajimura, T.Ishiguro and G.Saito, Synth.Metals 19, 151 (1987)
[27] K.Oshima, T.Mori, H.Inokuchi, H.Urayama, H.Yamochi and G.Saito, Phys.Rev.B37
[28] K.Bechgaard, K.Carneiro, M.Olsen and F.B.Rasmussen, Phys.Rev.Lett.46, 852 (1981)
[29] S.Tomic, D.Jérome, P.Monod and K.Bechgaard, J.Physique C-3, 44, 1083 (1983)
[30] M.Ribault, D.Jérome, J.Tuchendler, C.Weyl and K.Bechgaard, J.Physique 44, 953 (1983)
[31] J.P.Pouget, R.Moret, R.Comès, K.Bechgaard, J.M.Fabre and L.Giral,Mol.Cryst.Liq.Cryst.79, 129 (1982)
[32] J.R.Cooper, W.Kang, P.Auban, G.Montambaux, D.Jérome and K.Bechgaard, Phys.Rev. Lett. 63,1984 (1989)
[33] K.von Klitzing, Rev.Mod.Physics, 58, 519 (1986) D.C.Tsui, H.L.Stormer and A.C.Gossard, Phys.Rev.Lett.48, 1559 (1982)
[34] M.J.Naughton, R.V.Chamberlin, X-San, S.Y.Hsu, L.Y.Chiang, M.Ya.Azbel and P.M.Chaikin, Phys.Rev.Lett 61, 621 (1988)
[35] C.Bourbonnais and L.G.Caron, Mol.Cryst.Liq.Cryst 119, 287 (1985)
[36] R.Moret and J.P.Pouget in Crystal Chemistry and Properties of Materials with Quasi-One-Dimensional Structures p.87, J.Rouxel editor, D.Reidel publisher (1986)
[37] S.Barisic in Low Dimensional Conductors and Superconductors ,Loc cit.
[38] P.Penven, D.Jérome, S.Ravy, P.A.Albouy and P.Batail, Synth.Metals 27, B405 (1988)
[39] J.P.Pouget, R.Moret, R.Comès, K.Bechgaard, J.M.Fabre and L.Giral, Mol.Cryst.Liq.Cryst 79, 129 (1982)
[40] F.Creuzet, C.Bourbonnais, L.G.Caron, D.Jérome and K.Bechgaard, Synth. Metals 19, 289 (1987)
[41] F.Creuzet, T.Takahashi, D.Jérome and J.M.Fabre, J.Physique Lett.43,L-255 (1982)
[42] T.Takahashi, Y.Maniwa, H.Kawamura and G.Saito, J.Phys.Soc.Japan 55, 1364 (1986)
[43] V.Emery, R.Bruinsma, S.Barisic, Phys.Rev.Lett, 48, 1039 (1982)
[44] T.Moriya, J.Phys.Soc.Japan, 18, 516 (1963)
[45] G.Soda, D.Jérome, M.Weger, J.Alizon, J.Gallice, H.Robert, J.M.Fabre and L.Giral, J.Physique 38, 931 (1977)
[46] C.Bourbonnais, Synth.Metals 19,57 (1987)
[47] C.Bourbonnais, P.Wzietek, D.Jérome, F.Creuzet, L.Valade and P.Cassoux, Europhys.Lett.6,177 (1988)

[48] C.Bourbonnais, P.Wzietek, F.Creuzet, D.Jérome, P.Batail and K.Bechgaard, Phys.Rev.27, 1532 (1989)
[49] C.Bourbonnais, Mol.Cryst.Liq.Cryst.119, 11 (1985)
[50a] V.J.Emery, J.Physique 44, C-3, 977 (1983)
[50b] C.Bourbonnais and L.G.Caron, Synth.Metals 27, A515 (1988)
[51] M.Takigawa, H.Yasuoka and G.Saito, J.Phys.Soc.Japan 56, 873 (1987)
[52] Y.Hasegawa and H.Fukuyama, J.Phys.Soc.Japan 56, 877 (1987)
[53] L.Zuppiroli in Low Dimensional Conductors and Superconductors p.307, loc cit.
[54] Y.Suzumura and H.J.Schulz, Phys.Rev 39, 11398 (1989)
[55] T.Timusk in Low Dimensional Conductors and Superconductors p.275 loc.cit.
[56] S.Tomic and D.Jérome, J.Phys.Condens.Matter 1, 4451 (1989)
[57] R.Moret, J.P.Pouget, R.Comès and K.Bechgaard, Phys.Rev.Lett.49, 1008 (1982)
[58] R.Moret, S.Ravy, J.P.Pouget, R.Comès and K.Bechgaard, Phys.Rev.Lett 57, 1915 (1986)
[59] S.S.P.Parkin, D.Jérome and K.Bechgaard, Mol.Cryst.Liq.Cryst.79, 213 (1981)
[60] T.Sugano et al, Synth Metals 27, A325 (1988)
[61] L.N.Bulaevskii, Uspekhi Fiz.Nauk 116, 449 (1975) and Adv. in Physics 37, 443 (1988)
[62] K.Oshima, T.Mori, H.Inokuchi, H.Urayama, H.Yamochi and G.Saito, Phys.Rev.B, 38, 939 (1988)
[63] P.Auban, D.Jérome, K.Lerstrup, I.Johansen, M.Jorgensen and K.Bechgaard, J.Physique 50, 2727 (1989)
[64] S.Ravy, J.P.Pouget, L.Valade and J.P.Legros, Europhysics Lett.9, 391 (1989)
[65] S.Tomic, J.R.Cooper and K.Bechgaard Phys.Rev.Lett.62, 462 (1989)
[66] W.Kang, S.Tomic, J.R.Cooper and D.Jérome, Phys.Rev.B(Rapid Comm.) (1990)

Flux Quantisation and Quantum Coherence in Conventional and HTC Superconductors and Their Application to SQUID Magnetometry

C.E. Gough

Superconductivity Research Group, University of Birmingham,
Birmingham, B15 2TT, UK

Abstract. A brief introduction to the theoretical foundations and experimental confirmation of flux quantisation and quantum coherence in conventional LTC and HTC superconductors is given including a description of recent experiments involving composite HTC/LTC superconducting circuits. The physics of the rf and dc SQUID (Superconducting QUantum Interference Device) is outlined including a discussion of the potential and current limitations on the practical application of liquid nitrogen cooled SQUIDs.

1. Introduction.

Superconductivity provides us with a macroscopic manifestation of quantum mechanical processes originating at the microscopic level – the Meissner effect and zero conductivity arising from the formation of Cooper pairs brought about by the interaction of electrons with virtual phonons. Even more dramatically, superconductivity also allows quantum mechanical processes themselves to be demonstrated on a macroscopic scale – such as the interference of quantum mechanical waves in the SQUID magnetometer. Indeed, at yet another level, superconductors can be used to test the boundaries between classical physics and quantum mechanics, in circuits in which physical parameters such as charge and flux must be treated as quantum mechanical rather than classical variables – as in Macroscopic Quantum Tunneling (MQT) of a superconducting ring from one quantum state to another by quantum mechanical tunneling of flux across an energy barrier.

This paper is concerned with those properties of superconducting systems that reveal quantum mechanics operating at a macroscopic level. In particular, we describe the classic manifestations of quantum coherence in superconductors, including the quantisation of flux in units of h/2e and the periodic field dependence of quantum interference around circuits containing weak links.

The first part of the paper is devoted to the theoretical ideas and early experiments which demonstrated quantum coherence and interference in conventional LTC, HTC and, more recently, composite HTC/LTC superconducting systems. The second part of the paper is concerned with the application of some of these properties to the realisation of very sensitive SQUID magnetometers including HTC devices which operate at liquid nitrogen temperatures. Their likely limitations are briefly described. The emphasis throughout is on physical principles rather than operational details, for which the reader is referred to more detailed accounts by Barone [1] and van Duzer et al. [2].

2. Theoretical background

2.1. The macroscopic wave function

Flux quantisation is a direct consequence of a quantum mechanical description of the macroscopic superconducting state. London [3], in his classic monograph on *Superfluids*, proposed that the superconducting state could be described by a wave function of the form $\Psi(r)$, where $|\Psi(r)|^2$ represents the local supercurrent electron density, n_s. Using such a description and assuming superconductivity was associated with single electrons, London obtained a value for the flux quantum equal to h/e, whereas it is now known that the pairing of electrons leads to quantisation in units of h/2e. The measurement of the flux quantum therefore provides a rather direct method of determining the effective charge of the electronic carriers in the superconducting state.

At around the same time as the publication of London's book, Ginsburg and Landau [4], with considerable intuitive insight, postulated an expansion for the difference in energy between the superconducting and normal states of the form

$$f_s = f_n + \alpha|\Psi|^2 + \frac{\beta}{2}|\Psi|^4 + \frac{1}{2m^*}|(\frac{\hbar}{i}\nabla - e^*A)\Psi|^2$$
$$+ \text{ magnetic energy terms,}$$

where $\Psi(r)$ is now treated as a spatially-varying complex order parameter and α and β are temperature dependent variables. Although originally intended to describe superconductors in the immediate vicinity of phase transitions, under certain conditions this expansion can also be used to describe superconductors throughout the superconducting phase diagram (see Tinkham[5] and Tilley and Tilley [6] for an account of the application of the Ginsburg–Landau theory). Gorkov [7] subsequently justified such an expansion from conventional BCS theory $m^*=2m$ and $e^*=2e$, corresponding to the pairing of electrons in the superconducting state.

Without an established microscopic theory it is impossible to justify rigorously a similar expansion for HTC superconductors, though the direct observation of a flux lattice [8] with a detailed structure similar to that expected from the Abrikosov model of the superconducting mixed state [9] strongly suggests the validity of a Ginsburg–Landau (GL) description.

In some theoretical models of HTC superconductors, such as Anderson's RVB model [10], the superconducting ground state may well involve the d-state pairing of electrons [11]. This would require a modified GL-expansion with an increased number of degrees of freedom of the order parameter, which would have a different symmetry from that of conventional BCS superconductors with s-state symmetry.

It is therefore important to consider whether experiments could be devised which might provide information on the underlying symmetry of the HTC wave function. Such an experiment might, for example, involve the coupling of the HTC and conventional LTC wave functions across an interface, which provided the motivation for the experiments on quantum coherence and flux quantisation in composite HTC/LTC rings reported in this paper.

Within the framework of the Ginsburg–Landau model and by application of Maxwell's equations, it is straightforward using a variational approach [5,6] to derive an expression for the local supercurrent density J(r). The resulting expression

$$J(r) = -\frac{ie\hbar}{2m}(\Psi^* \nabla \Psi - \Psi \nabla \Psi^*) - \frac{2e^2}{m}\Psi^*\Psi A,$$

is identical to the standard quantum mechanical expression expected for a current of particles of mass 2m and charge 2e, where A is the vector potential defining the magnetic field B = curl A. If $\Psi(r)$ is expressed as $|\Psi|\exp(i\theta(r))$, the above equation can be recast as

$$J(r) = \frac{e}{m}(\hbar\nabla\theta - 2eA)|\Psi|^2.$$

2.2 Fluxoid quantisation

Consider a superconductor enclosing a hole containing magnetic flux. Surface diamagnetic screening currents will flow within the magnetic penetration depth, λ, to exclude field from the interior of the superconductor. Sufficiently far from the surface ($\gg \lambda$), B(r) and J(r) will become negligibly small. If a line integral of the above equation is taken around a path encircling the hole but well inside the bulk, the requirement that $\Psi(r)$ is single valued requires that the phase difference on completely encircling the hole must be an integral multiple of 2Π. For such an integration path, this implies an enclosed magnetic flux, Φ, quantised in units of $\Phi_0 = h/2e = 2.07 \times 10^{-15}$ T.m². Note that Φ_0 is a fundamental constant and is unchanged by band structure effects in real metals.

If the integration path is taken close to the surface or if the flux is contained within a thin-walled cylinder (thickness $\lesssim \lambda$), the contribution to the line integral from the current can no longer be neglected and it is the fluxoid,

$$[\Phi + \frac{\mu_0 \lambda^2}{2} \int j \cdot dl],$$

$$\text{where } \lambda^2 = \frac{m}{\mu_0 e^2 n_s},$$

that is quantised [3], rather than simply the flux enclosed.

2.3 Experimental evidence for flux quantisation

The first demonstrations of flux quantisation [12,13] were performed on thin-walled superconducting cylinders formed by evaporating a superconductor onto the surface of a glass fibre. The thickness of the superconducting layers was much larger than the magnetic penetration depth to ensure that it was the total flux enclosed that was quantised rather than the fluxoid. Deaver and Fairbank [12] verified flux quantisation by observing the flux expelled from such a cylinder as it was cooled through its critical temperature in an external field, whereas Doll and Nabauer [13] measured the magnetic moment associated with the trapped flux directly, by vibrating their sample and observing the induced voltage in a coil nearby.

Flux quantisation, confirming the pairing of electrons in a HTC superconductor, was first demonstrated by the Birmingham Group [14] by direct measurement of the flux trapped inside a ceramic ring of multi-phase YBCO using a conventional liquid helium cooled SQUID magnetometer as the detector. Unlike a solid ring of conventional superconductor, the HTC ring was only weakly superconducting, so that it was relatively easy to excite the ring between its metastable quantum states corresponding to different integral numbers of the flux quantum trapped within the ring. Transitions between such states, excited by periodic bursts of rf radiation, are shown in Fig.1a.

Such an experiment would not have been possible with a conventional LTC superconductor as the energy barrier for flux transitions through the bulk superconducor would have been far too high. The Birmingham experiment was possible because the multi-phase YBCO ring used was only weakly superconducting and allowed single quanta of flux to move cleanly in and out of the ring, presumably across a single weak point on a percolating superconducting path between grains.

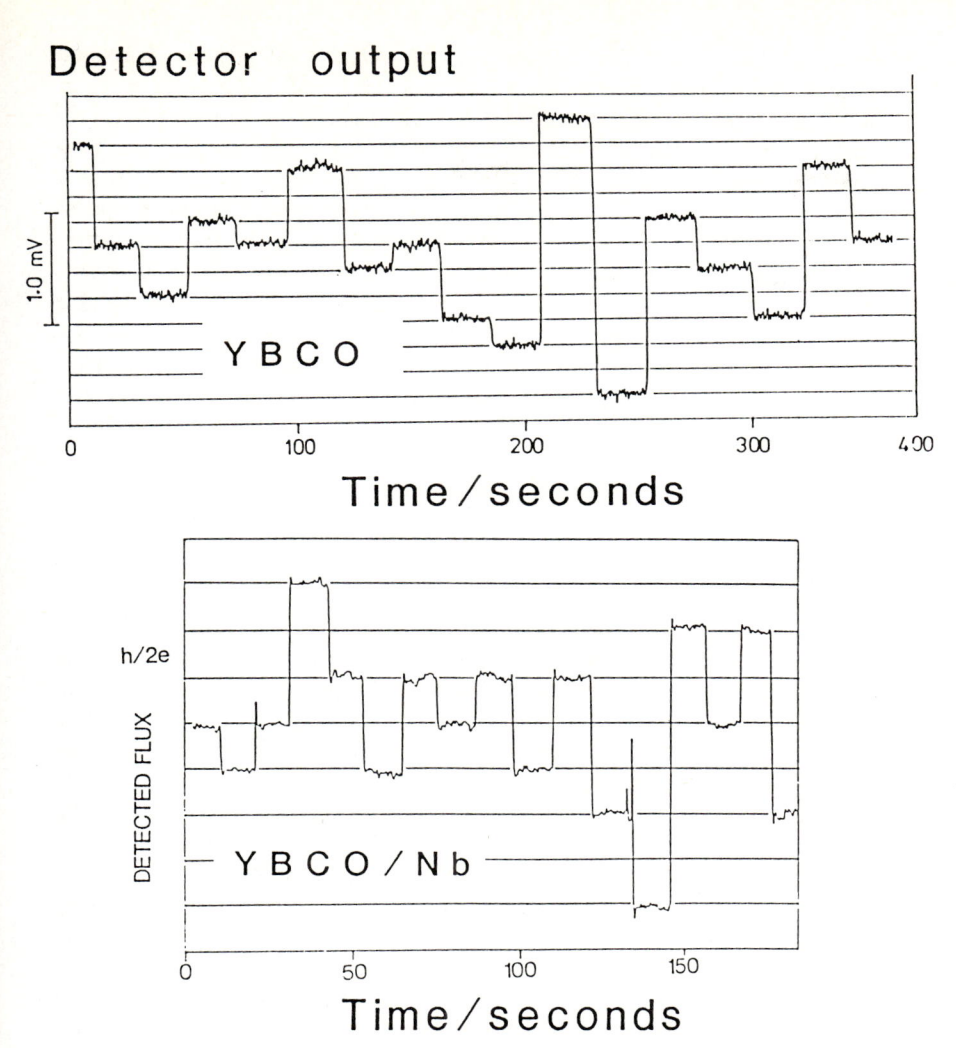

Fig. 1. *Flux quantisation in (a) an YBCO ring, (b) a composite YBCO/Nb ring.*

In addition to demonstrating flux quantisation and therefore the pairing of electrons, this experiment verified the existence of long-range phase coherence of the superconducting wave function around superconducting paths involving many thousands of individual grains, which were only weakly connected at their points of contact.

The original demonstration of flux quantisation in a HTC superconductor was restricted to liquid helium temperatures; however, subsequent absolute measurements of the periodicity of an rf SQUID [15], fabricated from single-phase 123-YBCO material,

145

confirmed the expected flux quantum periodicity at liquid nitrogen temperatures and above.

In a recent variant of our initial flux quantisation experiment, we have been able to verify the quantisation of flux in a composite ring containing sections of HTC and conventional LTC superconductor [16]. The composite ring was formed by pressing a niobium point contact into a slot cut across a superconducting ring of high quality, single phase YBCO. The observation of flux quantisation in units of h/2e in such a ring, Fig 1b, clearly demonstrates the phase coherence around the ring of the wave functions describing the conventional LTC and HTC superconductors. In particular, it verifies the coherent coupling of such wave functions across an interface between a conventional LTC and HTC superconductor. Unfortunately, this cannot prove that HTC and conventional LTC superconductors have the same underlying symmetry, as electron reflection at an interface causes symmetry breaking.

2.4 Flux quantisation in the mixed state

In the Abrikosov mixed state [9], flux penetrates in the form of flux-lines each associated with a quantum of flux. At low fields penetration is by isolated flux lines, where the superconducting order parameter $\Psi(r)$ is reduced to zero along the axis of the flux line where the magnetic field is largest. Diamagnetic circulating screening currents exclude flux from the bulk material. On any integration path round the flux line the fluxoid must be quantised. Well away from the core ($\gg \lambda$), the total flux enclosed is thus quantised in units of Φ_0. The density of flux lines is therefore B/Φ_0.

At larger fields the flux lines are more closely spaced and interact strongly. In a perfect superconductor free of microstructural defects, the flux lines arrange themseves on a regular lattice. The Abrikosov model predicts a triangular lattice though interactions between the superconducting state and the underlying crystal lattice can lead to other flux lattice symmetries [17]. For a regular lattice, an integration path clearly exists around each lattice site where the diamagnetic shielding current density is zero, so that the total flux associated with each flux line is again Φ_0. Multiply quantised flux lines do not occur as they are energetically unstable.

The flux line lattice can be directly visualised by magnetic decoration techniques [18,19], in which a fog of finely divided ferromagnetic particles is allowed to fall on the surface of a superconductor in which flux has been trapped. The magnetic particles tend to accumulate on the flux line cores, where they remain on warming to room temperature allowing subsequent examination under an electron microscope. In a series of beautiful experiments, Dolan and coworkers [19] have recently verified the existence of near perfect flux line lattices in regions of YBCO relatively free of gross defects such as twinning planes. By examining the microscopic distribution of flux lines in the presence of crystalline defects a wealth of important information on flux-line/defect interactions can be obtained.

2.5 Flux quantisation in weak-link rings

Another experimental situation in which quantum coherence effects are important is the superconducting ring containing a weak-link, as illustrated in Fig. 2. Weak links can be formed as thin insulating layers between two superconducting surfaces (an SIS junction), across a thin normal region between two superconductors (an SNS or proximity junction), across a narrow

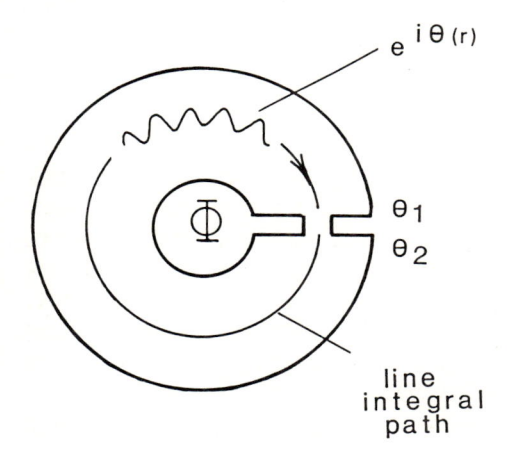

Fig. 2. *Superconducting ring containing a weak link.*

constriction on a superconducting path (a microbridge) or across a junction made by pressing a superconducting point onto another superconductor (a point-contact junction).

The current through such a junction is often well-described by the Josephson current-phase relationship, $I = I_c \sin \theta$, where I_c is the maximum supercurrent that the weak-link can support and $\theta = (\theta_1 - \theta_2)$ is the phase difference of the superconducting wave function across the weak-link (see Feynman [20] for a simple derivation of the above expression). If a current path is taken well inside the bulk of the superconductor on either side of the junction (Fig.2), continuity of the phase of the superconducting wave function requires that $2n\Pi = -\theta + 2\Pi\Phi/\Phi_0$, where Φ is the magnetic flux within the ring.

The circulating supercurrent can therefore be expressed as $I = I_c \sin 2\Pi\Phi/\Phi_0$. In the absence of an applied field, the flux within

147

the ring is determined by the circulating current, $\Phi = LI$, (where L is the ring inductance) so that $\Phi = LI_c \sin 2\Pi\Phi/\Phi_o$. The magnetic behaviour of such a ring is determined by the parameter $\beta = 2\Pi LI_c/\Phi_o$. If $\beta < 1$ there is only one solution, $\Phi = 0$, and the ring is unable to trap flux. However, for $\beta > 1$, more than one solution is possible corresponding to the trapping of flux. For $\beta >> 1$, a large number of flux trapping states exist, in which the flux enclosed is a near multiple of h/2e. Strictly speaking, the Birmingham HTC flux quantisation experiment [14] should more correctly have been interpreted in such terms, as it now recognised that the ring was essentially a weak-link ring. However, the verification of flux quantisation in units of h/2e is unaffected, since $\beta >> 1$ for the ring used.

A weak-link ring can be biased by an externally applied flux Φ_{ext}, as produced by a current carrying coil within the ring. The diamagnetic induced shielding current is then given by $(\Phi - \Phi_{ext})/L$. The flux within the ring is then given by

$$\Phi_{ext} = \Phi + LI_c \sin (2\Pi\Phi/\Phi_o) \ .$$

The magnetic behaviour of such a ring depends critically on the value $\beta = 2\Pi LI_c/\Phi_o$, as illustrated by the representative magnetisation curves drawn in Fig. 3. For $\beta < 1$ the magnetic behaviour is reversible, whereas for $\beta > 1$ the magnetic behaviour is hysteretic, as illustrated by the magnetisation cycle ABCDEFG. It will be noted that in both cases the magnetic properties, whether reversible or hysteretic, are periodic in the *externally* applied flux with periodicity φ_o. The hysteretic magnetic behaviour of the weak-link ring provides the basis of operation of the rf SQUID to be described in a later section.

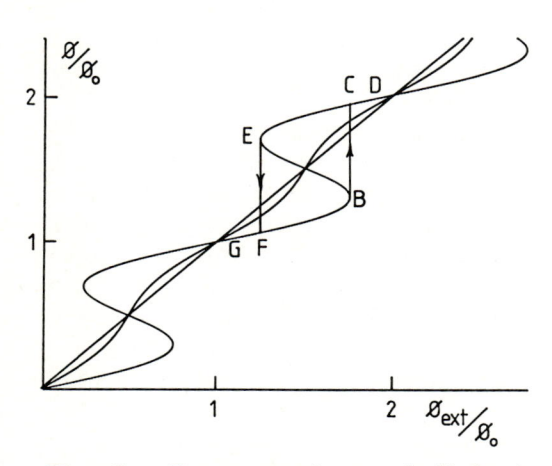

Fig. 3. *Representative weak-link ring magnetisation curves.*

2.6. Influence of thermal fluctuations and macroscopic quantum tunnelling

The above description neglects complications from thermal fluctuations. Their influence can most easily be considered in terms of the energy of the weak-link ring, which can be expressed as

$$E = (\Phi - \Phi_{ext})^2/2L - E_J \cos 2\Pi\Phi/\Phi_o ,$$

where the first term represents the magnetic energy and the second is the energy associated with Josephson tunneling across the weak link, $E_J = LI_c\Phi_o/2\Pi$. Typical energy curves are shown in Fig. 4 for zero bias and for an external bias of one flux quantum. Metastable flux trapping states occur at the local minima.

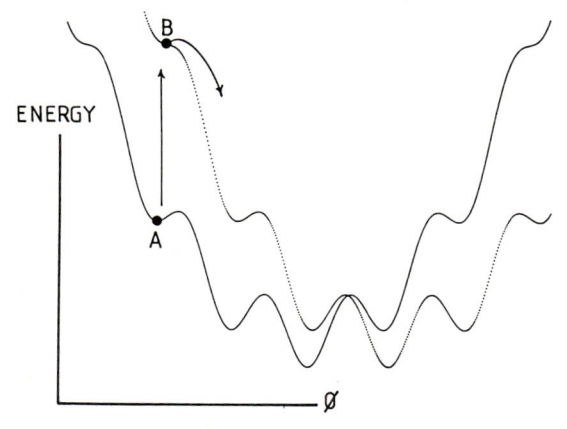

Fig. 4. *Energy of weak-link ring as a function of enclosed flux. Solid curve $\Phi_{ext} = 0$, dotted line $\Phi_{ext} = \Phi_o$.*

As the external bias is changed, the energy of a particular trapping state changes (e.g. from A to B in Fig. 4) until it becomes unstable allowing a transition to another metastable state of lower energy, as illustrated. The subsequent dynamics of the resulting flux transition depends on a number of factors including the capacitance C of the junction and any parallel damping resistance R. For a heavily damped junction with McCumber parameter [21] $\beta_c = L/CR^2 \gg 1$, transitions occur to the next lowest state; however, for a lightly damped junction with $\beta_c < 1$ multiple flux quantum transitions occur.

At finite temperatures, thermal fluctuations can cause premature transitions between the metastable flux states as illustrated in Fig.5a. These processes can become important even at quite low temperatures, $kT \simeq E_J/40$ [23]. This results in a stochastic

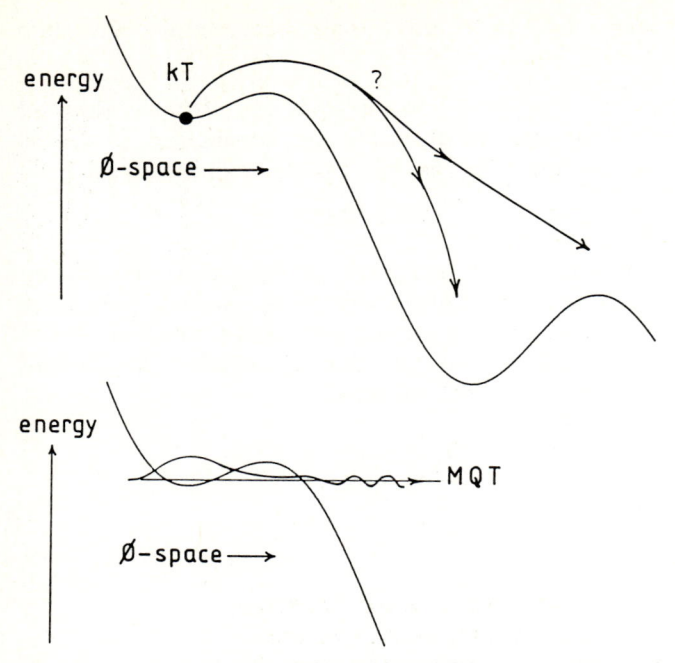

Fig. 5. (a) Thermally activated flux transitions. (b) Quantum mechanical tunnelling of flux.

variation in the value of external flux at which transitions between trapped flux states occur leading to noise in rf SQUIDs and to a depression of the effective critical current that can be supported by a weak-link ring. Measured critical currents in rings (and across weak-links in general) will therefore always be less than the intrinsic critical current, $I_c(T)$. The temperature dependence of critical currents deduced from flux trapping or voltage/current measurements will therefore not necessarily correspond to the underlying temperature dependence of $I_c(T)$, so that some care must be taken in interpreting such measurements.

Premature transition can also occur by the quantum mechanical tunneling of flux in Φ-space, Fig. 5b, just like the quantum mechanical tunneling of a particle under a potential barrier. This is known as Macroscopic Quantum Tunneling (MQT) and was first predicted by Widom and Clark [23] and Lhikarev [24]. It has been confirmed in a number of elegant experiments by the Berkeley group [25] and others [26,27]. Note that MQT involves a second level of quantisation overlaying the quantum description of the superconducting state by a quantum mechanical wave function.

MQT is generally only observed at low temperatures (<<1K), when competing thermal excitation processes can be ignored, and then only for very low capacitance junctions ~10^{-15} F or less, which inevitably has to include the reactive loading of the outside world [28]. MQT between the small grains of a sintered ceramic superconductor may become important at sufficiently low temperatures.

The superconducting weak-link ring enables a number of *gedanken* experiments to be devised to test quantum mechanics as a description of the physical world on a macroscopic scale [29]. In particular, it enables questions to be posed, which are in principle capable of experimental test, about the apparent conflict between the certainties of physical measurement and the uncertainties of quantum mechanics (e.g. the Schrodinger Cat Paradox). Legget [30] has written extensively on these intriguing problems.

3. Quantum coherence

3.1. The two-junction interferometer

We now consider the critical current through a superconducting ring containing two weak-links in parallel, as illustrated in Fig. 6.

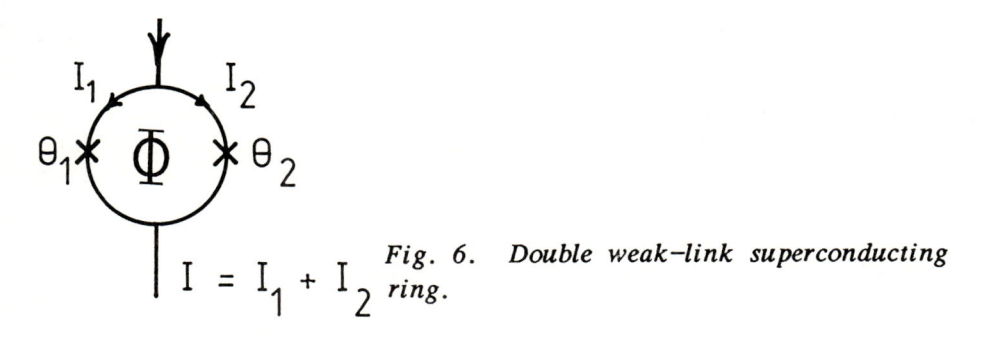

Fig. 6. *Double weak-link superconducting ring.*

The supercurrent divides along the two paths with the current in each path determined by the phase differences, θ_1 and θ_2, across the weak-links. The total current assuming identical strength junctions is then given by $I_c[\sin\theta_1 + \sin\theta_2]$. Using the now familiar arguments of phase coherence around a superconducting circuit, we can also write $\theta_1 - \theta_2 = 2\Pi\Phi/\Phi_0$, where Φ is the sum of any flux applied externally plus the flux produced by any supercurrents encircling the ring. The total curent through the ring is $2I_c\cos(\Pi\Phi/\Phi_0)\sin(\theta_1+\theta_2)/2$. Therefore, the maximum supercurrent that can be passed through the ring is $2I_c |\cos \Pi\Phi/\Phi_0|$, which is periodic in Φ_0.

This field dependence is easily understood in terms of the coherent interference of the currents passing through the two weak-links and there is a very close formal similarity with the familiar double-slit interference pattern in optics. In practice, just as in optical diffraction experiments, the difraction pattern is often modulated by the difraction pattern from quantum interference in the individual junctions themselves, caused by field penetration across the junction, as for example observed by Jaklevic [31].

The above dependence of critical curent on enclosed flux provides the basis of operation of the DC SQUID magnetometer (see refs 1 and 2 for a detailed description of SQUIDs and their applications). DC SQUIDs fabricated from conventional superconductor thin films operating at liquid helium temperatures are by far the most sensitive magnetometers available [32]. Like any device, their sensitivity depends on the bandwidth over which measurements are made. Sensitivities of a few times $10^{-5} \Phi_0/\sqrt{Hz}$ have been achieved corresponding to energy sensitivites $(\Delta\Phi)^2/2L$ approaching the quantum limit, h/Hz. The sensitivity is sufficient to monitor the miniscule magnetic fields produced by neurological signals inside the brain, which is one of the most exciting potential applications of such devices outside physical instrumentation, where SQUIDs are very widely used.

3.2 Quantum interference SQUID behaviour in a composite HTC/BCS ring

In recent experiments on composite niobium/YBCO structures, we have, in addition to demonstrating flux quantisation, observed double junction quantum interference (the dc SQUID) and rf SQUID behaviour.

The measurements have been performed using what is known as a 2-hole SQUID structure, which behaves essentially identically to a single ring but is only sensitive to external field gradients and is thus less sensitive to the influence of external magnetic noise. The structure is illustrated in Fig. 7. Two holes were drilled through a

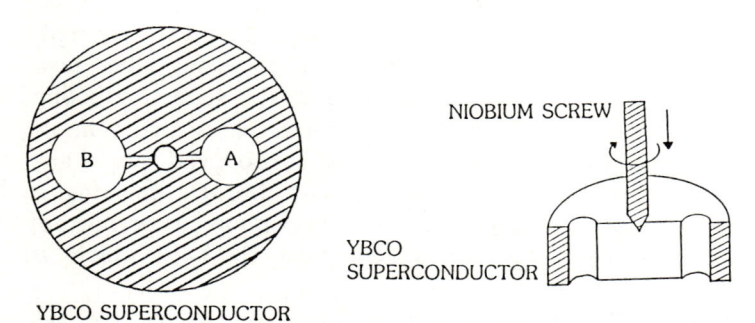

Fig. 7. Two-hole composite YBCO/Nb SQUID.

solid disc of high quality YBCO ceramic with a narrow slot cut between them. An externally adjustable niobium point contact was screwed down into the slot to create a central bridging region with two weak-link junctions at the points of contact.

The strength of the weak links could be adjusted by a screw mechanism operating from room temperature. All experiments were performed at liquid helium or up to Tc of the niobium by warming the assembly in helium gas temperatures.

Fig. 8 is a composite figure showing several operating

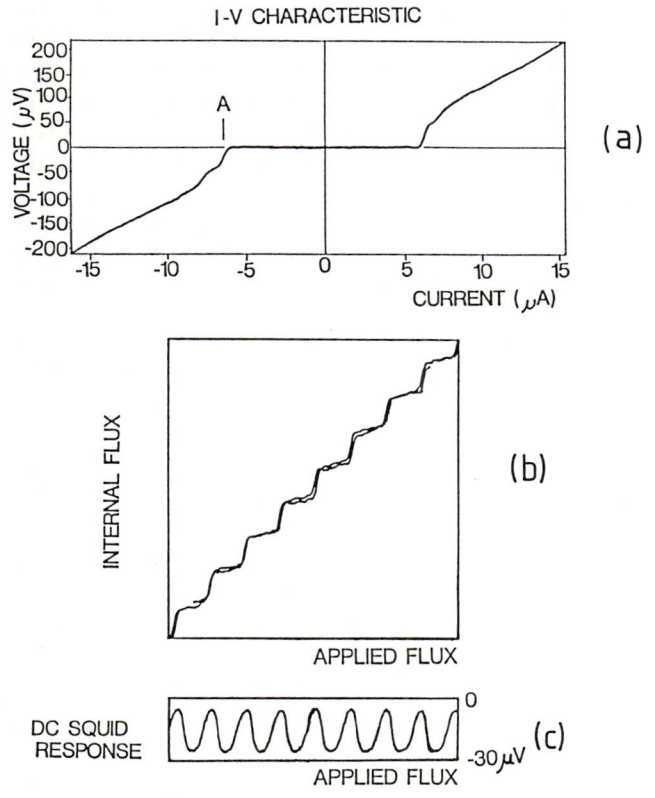

Fig. 8. Operating characteristics of composite YBCO/Nb SQUID.

characteristics of the arrangement. Fig. 8a shows some typical V/I characteristics with a well defined critical current but rounded by the thermal fluctuation effects discussed earlier. The additional structure is almost certainly associated with self-induced microwave resonances of the system excited via the inverse ac Josephson effect [1,2].

For the same point settings, Fig. 8b, shows the flux in hole B as the flux in hole A is linearly ramped using a toroidal coil to precisely define the flux sensitivity. The resulting magnetisation curves are essentially identical to those predicted in section 2.4 for a weak link ring with $\beta = 2\pi L I_c/\Phi_0 \gtrsim 1$, with quasi-reversible magnetic properties on the experimental time scale because of thermal fluctuations. These measurements provide another demonstration of flux quantisation around a composite HTC/LTC superconductor loop.

In Fig. 8c, the voltage across the junction for a current near I_c is plotted as a function of the externally swept flux on the same horizontal axis as the magnetisation curves. Quantum interference of the current passing across the two weak links gives rise to a periodic modulation in the critical current resulting in the observed periodicity in voltage across the device. The depth of modulation of the critical current is relatively small, partly because the strengths of the two junctions were unlikely to have been the same and partly because of the influence of the self-flux produced by circulating currents in the ring [1,2].

Nevertheless, despite the relative crudity of the device, a flux noise sensitivity of 2×10^{-4} $\Phi_0/\sqrt{Hz}$ has been measured at 4.2K, with a 1/f-knee in the noise (see later section) setting in at around 5Hz.

On lowering the temperature, $I_c(T)$ and hence $\beta(T)$ increases, so that the magnetic properties of the composite YBCO/Nb becomes increasingly hysteretic, as illustrated in Fig. 9. This hysteretic

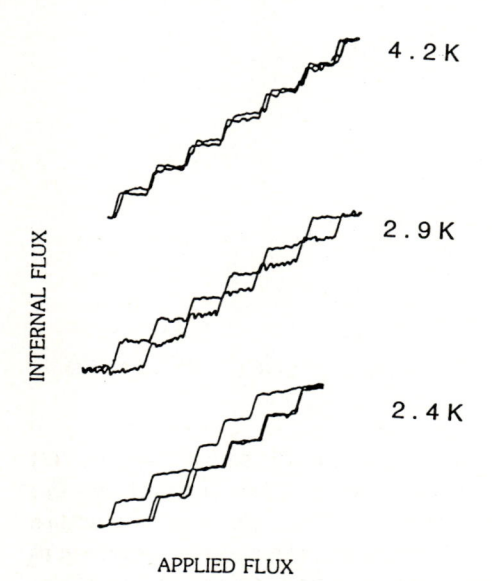

Fig. 9. Magnetisation of composite SQUID as a function of temperature.

magnetic behaviour enabled the device to be used as an rf squid, as described in the next section, with a flux noise sensitivity at 4.2K of $3 \times 10^{-4} \Phi_0 / \sqrt{Hz}$.

3.3 rf SQUID behaviour.

As the theory of operation of an rf SQUID is rather complicated [1,2], we will confine our discussion to the underlying principles. An rf SQUID operates by measuring the hysteretic magnetic energy losses induced when a resonant LC-tank circuit is magnetically coupled to a superconducting ring containing a weak link. The hysteretic properties of such a ring were derived in section 2.4 and are very similar to those illustrated in Fig. 9.

Energy is fed into the resonating tank-circuit through a large resistance, so that the oscillating magnetic field within the rf coil

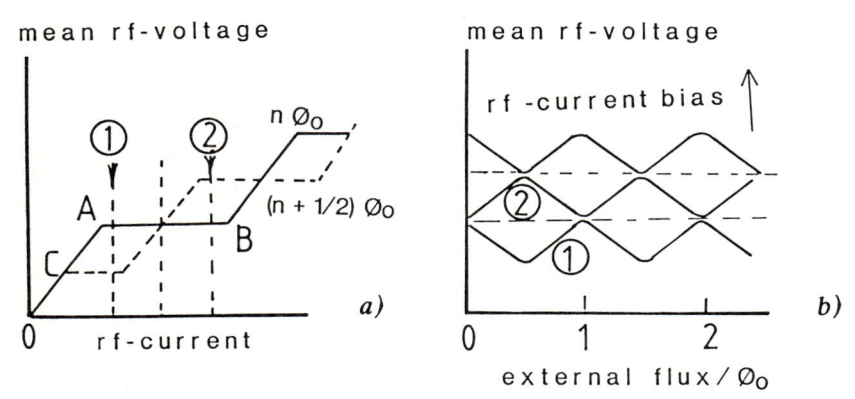

Fig. 10. Schematic operating characteristics of rf SQUID.

builds up in time. As soon as the magnetic field generated is sufficient to induce a hysteretic transition, as indicated in Fig. 3, energy is lost and the stored rf energy is decreased to below that necessary to give hysteresis. The process repeats itself indefinitely so that for a given rf charging current, the rf voltage across the LC circuit assumes an averaged steady value.

Fig. 10 illustrates the peak rf voltage across the tank circuit as a function of rf bias current. The first knee in the characteristics occurs when the bias current is just sufficient to take the loop round a single hysteresis loop. On increasing the charging current, the voltage remains nearly constant but the charging rate between hysteresis cycles is decreased. Eventually, one hysteresis loop is traversed every rf cycle, so that further increase of the charging current simply increases the stored energy until the amplitude is sufficient to include 2 transitions/cycle, whereupon another plateau in the voltage output is reached.

The above analysis assumes zero static applied flux. When a static flux is applied, a correspondingly smaller rf amplitude is required to reach the first hysteretic magnetic transition. The onset of the plateau region therefore depends on externally applied magnetic flux. Because the magnetic properties are periodic in Φ_{ext}, the maximum reduction in the plateau voltage occurs for $\Phi_{ext} = (n+1/2)\Phi_0$ as, indicated in Fig. 10. If the rf SQUID is biased at a constant rf current (e.g. at position 1 in fig. 10a) the voltage across the tank circuit will vary periodically with Φ_0, with a characteristic "triangle pattern" as shown in fig. 10b. The triangle pattern can be interpreted simply as the periodic change in Q of the tank-circuit expected from the periodic hysteretic properties of the magnetically coupled superconducting weak-link ring.

The first successful realisation of rf SQUID behaviour was performed with point-contact structures [33]. Nb point-contact rf SQUIDs have continued to be widely used to the present day because they are relatively easy to construct and have a surprisingly good noise performance. More recently, new designs have evolved using thin films to fabricate the weak-link loop. These are often refered to as hybrid rf-SQUIDs.

4. HTC SQUIDs

4.1 RF and DC SQUIDs

The first observation of rf SQUID behaviour in HTC superconductors was made by Colclough et al. [34] using bulk ceramic material, where the weak-link rings were the naturally occuring weak-link loops betweeen the grains of the sintered material. The distribution in size of the intergranular supercurrent loops resulted in a quasi-periodic field dependence of the SQUID output with contributions from a number of triangular dependences of different periodicities but with an average field "periodicity" given by Φ_0/A, where A is an effective average area of the intergranular current loops. Similar quasi-periodic field dependences were subsequently observed in the V/I characteristics of a thinned down section of YBCO [35], which can be understood in terms of an assembly of intergranular dc SQUIDs. Because the effective area is relatively small, the field sensitivity is rather poor; it is also impossible to couple flux efficiently into such devices.

SQUID magnetometry has been identified as one of the most likely early applications of HTC superconductors. Koch and co-workers at IBM had some limited early success in fabricating a thin-film dc SQUID from YBCO with a geometrically defined superconducting loop [36]. However, the SQUID only operated up to 68K and the observed field dependence was only quasi-periodic, with a field dependence rather reminiscent of current loops between the grains of thin film. Other groups had little initial success in improving on the peformance of such devices. More recently, the IBM group [37] have succeeded in fabricating a very promising dc SQUID using a Tl-based HTC film. The noise performance of one such SQUID at liquid nitrogen temperatures was shown to exceed the performance of commercially available hybrid rf SQUIDs. However, it remains difficult to control the strength of weak-links in thin film devices.

Progress on HTC rf SQUIDs was rather more immediate. Zimmermann et al. [38] was the first to realise an rf SQUID fabricated out of bulk YBCO, which operated very successfully at liquid nitrogen temperature. In this device the weak-link was formed by what is known as a break-junction. A radial slot was cut in a ceramic disc towards a central 1mm hole. The slot was cracked open to the central hole under liquid nitrogen or helium by driving a wedge into the slot. The ceramic material was then allowed to relax back into contact to form the weak-link.

In an attempt to circumvent the lack of mechanical and thermal stability of such a structure, the Birmingham group [39] devised a 2-hole structure with a very small connecting region as shown in the insert of Fig. 11. The interconnecting region was mechanically filed

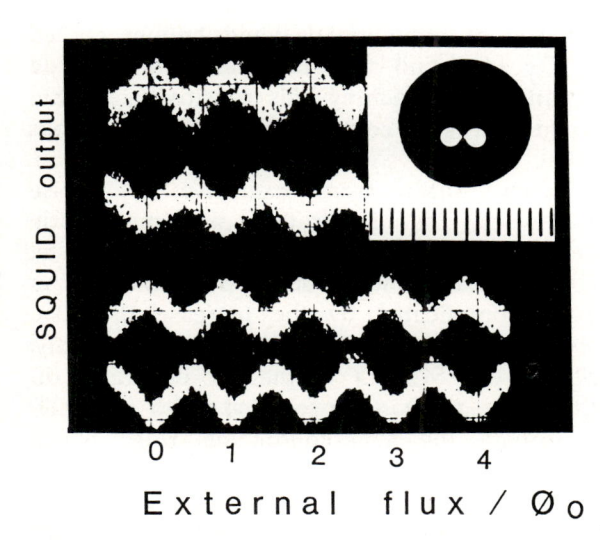

Fig 11. Liquid nitrogen temperature rf SQUID characteristics for a number of rf bias currents. The insert shows a typical structure with larger holes than used in practice.

down to provide a weak-link of the appropriate strength for currents circulating the two holes. Oscilloscope traces of rf SQUID output from such a device at 77K are shown in Fig. 11. They exhibit the expected triangular field dependence with periodicity Φ_0. This was proved by applying a known flux with a toroidal coil passing through one of the holes. These traces also demonstrate the expected phase reversal of the triangles associated with field modulation on increasing the rf bias current. Almost identical traces have recently been obtained with the rf-SQUID fabricated using the niobium point-contact, slotted, YBCO ring geometry referred to in section 3.2.

4.2 The performance of rf and dc HTC SQUIDs at liquid nitrogen temperatures

It is obvious that the sensitivity of a liquid nitrogen cooled SQUID system will always be inferior to a liquid helium operated device, because of thermally excited flux noise $(\Delta\Phi)^2/2L \sim kT$. To minimise such noise L must clearly be as small as possible. But small L implies small size, so that it becomes increasingly difficult to couple the external field to be measured into the device. Furthermore, for successful operation of both rf and dc SQUIDs $\beta = 2\pi L I_c/\Phi_0$ is normally required to be around 3-6. This provides constraints on the value of I_c for a given L. In an early analysis of the likely performance of dc SQUIDs at liquid nitrogen temperatures, Pegrum [40] predicted a likely noise figure of order $10^{-4}\Phi_0/\sqrt{Hz}$, which is comparable with the performance of commercial liquid helium cooled rf SQUIDs. HTC rf SQUIDs [41] and the Thallium thin-film dc SQUID already have performances approaching this sensitivity, so the potential application for liquid nitrogen cooled HTC SQUIDs looks highly promising.

However, all HTC SQUIDs suffer from excessive 1/f noise contributions at low frequencies, as illustrated by the noise measurements on a YBCO and BISCCO Birmingham SQUIDs, at 77K and 4.2K, Fig. 12. Noise measurements on rf-SQUIDs over a large range of temperatures is complicated by the increase in $\beta(T)$. This leads to additional noise from stochastic multiple flux transitions and from "telegraph noise" on the SQUID output as illustrated in Fig.13 It seems likely that this noise is associated with partially trapped flux lines passing through the superconducting ring, which are thermally excited between two metastable states. From the temperature dependence of the transition rates a pinning energy of around 2000K is deduced, which is very similar to the pinning energy deduced from flux creep in the same material [42].

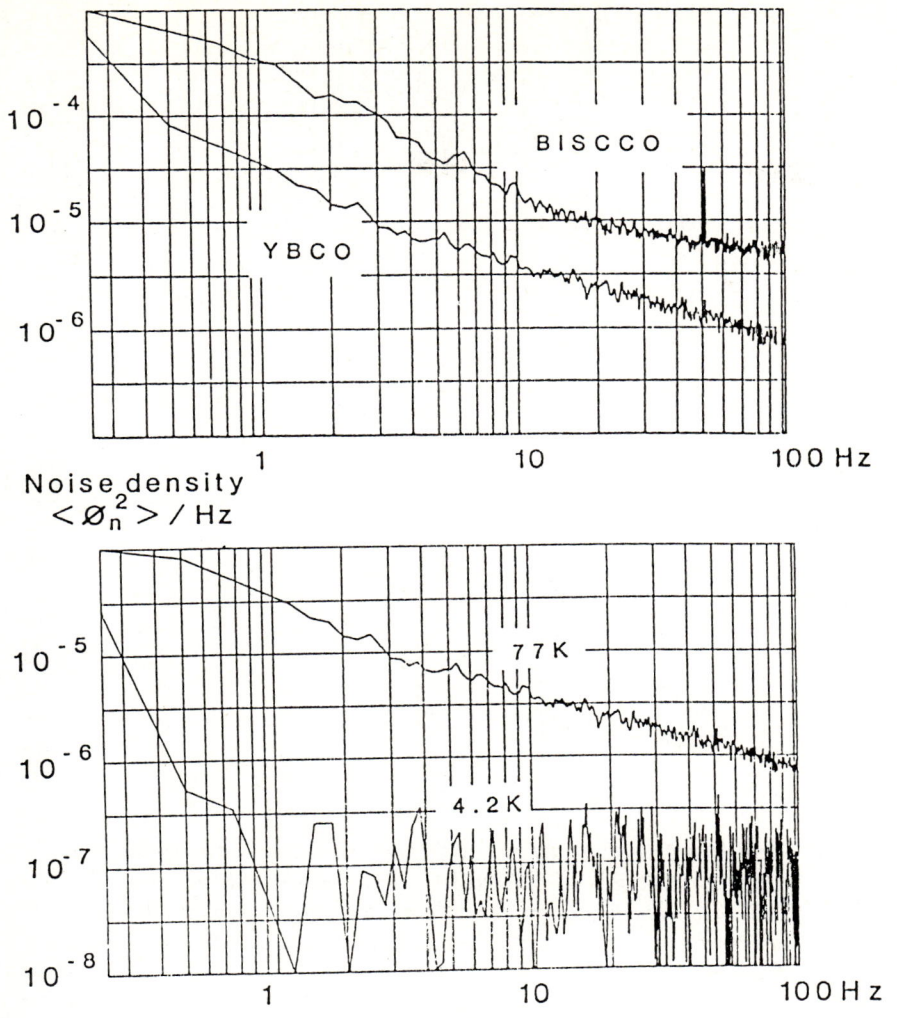

Fig. 12. Noise measurements on HTC rf SQUIDs. Upper curve compares YBCO and BISCCO SQUIDs at 77K, lower curve compares YBCO flux noise at 77K and 4.2K.

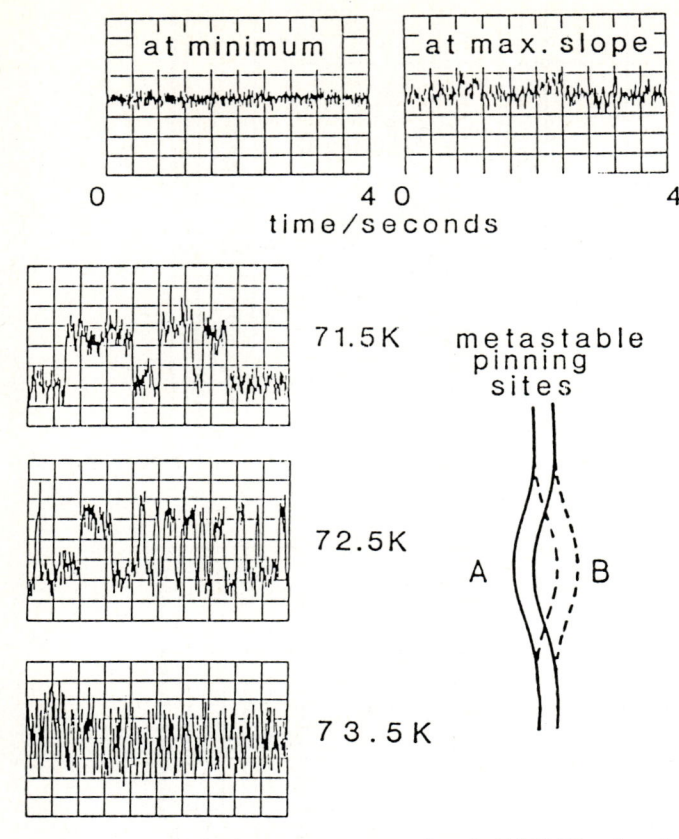

Fig.13. Real-time output of rf SQUID showing additional flux noise on slopes of triangles. The lower traces indicate temperature dependent telegraph noise which originates from thermal unpinning of a trapped flux-line.

Acknowledgements.

I am particularly grateful to members of the Birmingham Superconductivity Research Group for many valuable discussions and for their permission to quote unpublished measurements in this paper. The Birmingham research is supported by the SERC with valuable additional contributions from Lucas and ICI.

References.

1. A. Barone and G. Paterno, *The Physics and Applications of the Josephson Effect*, John Wiley, New York, 1982.
2. T. van Duzer and C.W. Turner, *Principles of Superconducting Devices and Circuits*, Elsevier, New York, 1981.
3. F. London, *Superfluids*, Dover Publications, New York, 1950
4. V.L. Ginsburg and L.D. Landau, Zh.Eksperim. i Teor. Fiz. **20** (1950) 1064.
5. M.Tinkham, Introduction to Superconductivity, McGraw Hill, New York, 1975.
6. D.R. Tilley and J. Tilley, *Superfluidity and Superconductivity (2nd edition)*, Adam Hilger, Bristol, 1985.
7. L.P.Gorkov, Soviet Physics-JETP **9** (1959) 1364.
8. G.J. Dolan, G.V. Chandrashekhar, T.R. Dinger, C. Feild and F. Holtzberg, Phys. Rev. Lett. **62** (1989) 827.
9. A.A. Abrikosov, Soviet Physics-JETP **5** (1957) 1174.
10. P.W. Anderson, Science **235** (1987) 1196.
11. C. Gros, D. Poilblanc, T.M. Rice and F.C. Zhang, Physica **C153-155** (1988) 543.
12. B.S. Deaver and W.M. Fairbank, Phys. Rev. Lett. **7** (1961) 43.
13. R. Doll and M. Nabauer, Phys. Rev. Lett. **7** (1961) 51.
14. C.E. Gough, M.S. Colclough, E.M. Forgan, R.G. Jordan, M.Keene, C.M. Muirhead, A.I.M. Rae, N. Thomas, J.S. Abell, S. Sutton, Nature **326** (1987) 855.
15. S. Harrop, C.M. Muirhead, M.S. Colclough, C.E. Gough, Physica **153-155** (1988) 1411.
16. M.N. Keene, T.J. Jackson and C.E. Gough, to be published in Nature, **340** (1989).
17. B. Obst, Phys. Stat. Sol. (b) **45** (1971) 467.
18. U. Essmann and H. Trauble, Phys. Letts. **A24** (1967) 526.
19. G.J. Dolan, F. Holtzberg, C. Feild and T.R. Dinger, Phys. Rev. Lett. **62** (1989) 2184.
20. R.P. Feynman *Lectures on Physics vol 3 ch 21* (1965), Addison-Wesley, New York.
21. D.E. McCumber, J.Appl.Phys. **39** (1968) 2503.
22. J. Kurkijarvi, Phys. Rev. **B6** (1972) 832.
23. A. Widom and T.D. Clark, Nuovo Cimento Lett. **28** (1980) 186
24. K.K. Likharev, Rev. Mod. Phys. **51** (1979) 101.
25. M.H. Devoret, J.M. Martinis and J. Clarke, Phys.Rev.Lett **55** (1985) 1908.
26. D.B. Schwartz, B. Sen, C.N. Archie and J.E. Lukens, Phys. Rev. Lett. **55** (1985) 1547.
27. S. Washburn, R.A. Webb, R.F. Voss and S.M. Farris, Phys. Rev.Lett **54** (1985) 2712.

28. A.I.M. Rae and C.E.Gough, Jnl. of Low Temp. Phys. **65** (1986) 399.
29. A.I.M. Rae, *Quantum Mechanics: Illusion or Reality*, Cambridge University Press (1986).
30. A.J. Leggett, Contemp. Phys. **25** (1984) 583.
31. R.C. Jakelvic, J. Lambe, J.E. Mercereau and A.H. Silver, Phys.Rev.**140** (1965) A1628.
32. M. Nisenoff, Cryogenics **28** (1988) 47.
33. A.H. Silver and J.E. Zimmermann, Phys. Rev. Lett. **15** (1965) 888.
34. M.S. Colclough, C.E. Gough, M. Keene, C.M. Muirhead, N. Thomas, J.S. Abell, S. Sutton, Nature **328** (1987) 47.
35. D. Robbes, Y. Monfort, M.L.C. Sing, D. Bloyet, J. Provost, B. Raveau, M. Doisy, and R. Stephan, Nature **331** (1988) 51.
36. R.H. Koch, C.P. Umbah, G.J. Clark, P. Chaudhari, and R.B. Laibowitz, Appl.Phys.Lett. **51 (1987)** 200.
37. R.H. Koch et al, Jnl. Less Common Metals **151** (1989).
38. J.E. Zimmerman, J.A. Beall, M.W. Cromar and R.H. Ono, Appl.Phys.Lett. **51** (1987) 617.
39. S. Harrop, C.M. Muirhead, M.S. Colclough, C.E. Gough, Physica **C153-5** (1988) 1411.
40. C.M. Pegrum and G.B. Donaldson, *EEC Workshop on HTSC and Potential Applications* (1987) 125.
41. S.P. Harrop, C.E. Gough, M.N. Keene, C.M. Muirhead, Supercond. Sci. and Technol. **1** (1988) 68-70.
42. C. Mee, A.I.M. Rae, W.F. Vinen and C.E. Gough, to be presented at Stanford HTSC conference (1989).

Similarities and Differences Between Conventional and High-T$_c$ Superconductors*

A. Barone

Dipartimento Scienze Fisiche, Università di Napoli,
P. le Tecchio, Napoli, Italy, and
Istituto di Cibernetica CNR, Via Toiano 6, Arco Felice, Napoli, Italy

Abstract. A comparative discussion is given between high-Tc superconductors (HTS) and conventional ones (LTS) on the basis of investigations performed in superconducting junction structures. To provide an introduction to the field the lecture covers basic elements of both superconductive tunneling and weak superconductivity. Information which can be inferred by tunneling spectroscopy is discussed on a simple ground. The state of the art and possible projections of HTS in this area are considered. Far from being mature the subject of the new superconductors confirms its paramount interest and stimulating perspectives.

1. Introduction

With the purpose of giving a lecture which is tutorial in character it is necessary not only to avoid erudite temptations but also to confine attention to those concepts which can be regarded as most essential and to those notions which can be expected to be of long lasting value.

The difficulty in satisfying such prescriptions lies in the complexity of the matter and in the "fluidity" of the whole picture emerging from the phenomenology of the new HTS.

Since the discovery of HTS by Bednorz and Müller [1] a significant part of the vast literature on the subject, produced at an impressive rate all over the world, has been dedicated to investigations of junctions of various types. The reason is twofold: on one side, superconductive tunneling spectroscopy represents a powerful tool for the investigations of the fundamental properties of the superconductor employed (e.g. the energy gap), on the other side these structures, in particular the Josephson junctions [2], represent the heart of the superconductive electronics and therefore of the whole sector of "small scale applications".

To date the situation is not yet clear. The experiments contain a large degree of uncertainty in their interpretation.

---

* Work supported by the Consiglio Nazionale delle Ricerche, under the Progetto Finalizzato "Tecnologie Superconduttive e Criogeniche"

In spite of the large number of results and the wide variety of approaches, a number of ambiguities emerge which are closely related to the mechanism of superconductivity itself in the new HTS. Indeed in this situation reliable junction devices do not yet exist for good spectroscopy nor enough reliable spectroscopic data to make good devices. There are however, in my opinion, very stimulating indications. There are already various well developed technologies to prepare high quality films (even up to 125K) and encouraging results on SNS and bridge type junctions.

The scope of the present lecture is to provide a basic knowledge of the phenomenology and some of the theoretical concepts concerning superconductive junctions in the framework of conventional superconductors and, afterwards, to discuss, also in the light of these notions, the problems, the state of the art and the perspectives of this topic in the context of HTS.

The reason for starting with low Tc materials reflects both the need of these notions for the necessary comparison with HTS and the attempt to have a self-contained lecture as properly required by this course.

2. Superconductive Tunneling and Josephson Junctions

A tunneling junction in the most archetypal configuration consists of two metal films separated by a thin (say 30-50 Å) dielectric barrier. The current-voltage characteristics (I-V) of such a device can provide useful information on the physics which governs such a structure. The I-V curve in the case of normal metal electrodes (i.e. N-I-N structure) at low voltage (a few mV) is ohmic. This is a direct consequence of the quantum mechanics; the passage of electrons across the insulating barrier occurs due to the tunneling effect. The original experiments in this context were performed by Giaever (for the whole Section and related references see [2]) A simple model based on the Fermi "Golden rules" can be outlined as follows. The tunneling current $I_{L \to R}$ from the left to the right metal is assumed to be proportional to the density of filled states on the left metal, to the density of empty states on the right metal and to the tunneling probability. That is, integrating over all energies

$$I_{L \to R} = 2\pi/h \int_{-\infty}^{+\infty} |M|^2 \, n_L F_L \, n_R (1-F_R) \, dE \qquad 1)$$

where F_L (F_R) and n_L (n_R) are the Fermi factors and the density of states on the left (right) metal respectively. M is the matrix element connecting states of equal energy between the two metals and can be assumed to be energy independent since the "particle" energy is of the order of ~1 meV, whereas the barrier height is of the order of 1 eV. Moreover, around E_F, the density of states can be considered energy independent as well.

Writing the analogous expression for the tunneling current from the right metal to the left one it is straightforward to show that the net current $I_{NN} = I_L - L_R$ reduces to

$$I_{NN} = V/R_N$$

where the constant R_N can be interpreted as the tunneling resistance of the "normal" junctions.

If we consider now for L and R two superconductors (T<Tc) such a behavior is drastically modified. For the superconductive junctions we make the further assumptions that "quasiparticles" can be regarded as normal electrons and that the matrix element M remains unchanged. The densities of states are now referred to two superconductors so that, using BCS theory, we shall write

$$n_s = n_n |E|/|E^2 - \Delta^2| \quad \text{for } |E| \geq \Delta \quad \text{and} \quad n_s = 0 \text{ for } |E| < \Delta$$

where Δ is the energy gap. Equation 1) is therefore modified and the superconductive tunneling current $I_{SS} = I_L - I_R$ is now

$$I_{SS} = \text{const.} \int_{-\infty}^{+\infty} \frac{|E|}{\sqrt{E^2 - \Delta_L^2}} \frac{|E+eV|}{\sqrt{(E+eV)^2 - \Delta_R^2}} |F(E) - F(E+eV)| \, dE \, .$$

This integral can be calculated numerically giving a logarithmic singularity for the current at $V_1 = |\Delta_L - \Delta_R|/e$ and a finite discontinuity at $V_2 = \pm |\Delta_L + \Delta_R|/e$. The resulting I-V characteristics are sketched in Fig. 1. If the two superconductors are equal ($\Delta_L = \Delta_R$) the I-V curve reduces to that of Fig. 1b.

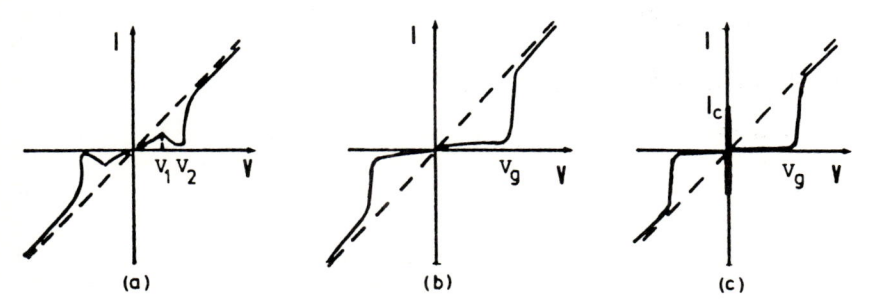

Fig. 1 - Current voltage characteristics of a superconductive tunneling structure. a) single particle tunneling with $\Delta_L \neq \Delta_R$; b) $\Delta_L = \Delta_R$; c) Josephson effect.

The phenomenological model just described, although largely oversimplified, provides the essential qualitative features. This has been proven "a posteriori" on the basis of a rigorous theory.

In the case of the HTS however the validity of the expression of I_{SS} given above is not obvious. Indeed the assumption of energy independent tunneling probability cannot be safely made since the condition of gap and voltages, very small compared to E_F, does not hold. Indeed the BCS itself is for HTS an assumption and to some extent even the "gap" value is a controversial argument.

Let us observe that the case $\Delta_L \neq \Delta_R$ is of interest when dealing with experiments employing HTS since in many cases junction structures (e.g. YBCO-Nb) are considered; in that case however $\Delta_{HTS} \gg \Delta_{LTS}$. Let us now discuss what happens when the tunneling barrier thickness is further reduced. In this condition Josephson effect can occur.

Under the name "Josephson Effect" fall a number of phenomena, closely related to superconductivity, whose general physical insight is of paramount interest. We shall give here just a brief outline of the basic ideas in a rather oversimplified version.

As is well known, a superconductor can be described as a whole by a macroscopic wave function $\Psi = \rho^{1/2} e^{\alpha}$ where $\rho = |\Psi|^2$ is the density of Cooper pairs and α the phase. In a single isolated superconductor the number N of pairs is fixed, therefore the uncertainty relation $\Delta N \Delta \alpha \cong 2\pi$ tells us that the phase α is undefined though if we fix its value at a given point it is automatically fixed at all points ("long range order"). When two superconductors S_L and S_R are placed very close to each other (say at a distance of ~10Å) then their macroscopic wave functions can overlap leading to a "weak coupling". If the energy involved in the coupling is greater than the thermal fluctuation energy, there is a "phase correlation" between the two superconductors and pairs of electrons can tunnel from one another In this way, a supercurrent, which is a function of the relative phase, $\phi = \alpha_{S1} - \alpha_{S2}$, between the two superconductors, can flow leading to a situation in which the two "joint" superconductors behave to some extent as a single superconductor. $\Delta N \Delta \alpha \cong 2\pi (n \ N_L - N_R)$, so that n can change and ϕ is modified accordingly.

The current-phase $I(\phi)$ and voltage-phase $V(\phi)$ relations

$$I = I_c \sin \phi \qquad (a)$$

$$\partial \phi / \partial t = 2e/\hbar \ V \qquad (b)$$

are the well known constitutive equations of the Josephson effect. I_c is the critical current. We see that V=0 implies ϕ= constant and therefore the possibility of a finite zero voltage current flowing through the structure (d.c. Josephson effect). Moreover $V \neq 0$ across the structure implies the occurrence of an oscillating current $I = I_c \sin (\phi + 2e/h \ Vt)$ with frequency $\nu = 2eV/h$ (a.c. Josephson effect). This gives 483.6 GHz/μV.

For tunneling type junctions and for equal superconductors at finite temperature

$$I_c(T) = \frac{\pi}{2} \frac{\Delta(T)}{eR_N} \tanh \frac{\Delta(T)}{2k_B T}$$

and

$$I_c(0) = \frac{\pi}{4} \frac{V_g}{R_N}$$

where $V_g = 2\Delta/e$ is the gap voltage. This relation provides a ready estimation of the expected critical current from the current voltage curve. For different gaps $I_C(T)$ can be obtained numerically.

The weak coupling between the two superconductors can be realized not only via quantum mechanical tunneling of Cooper pairs through a suitable barrier (tunnel junction) but also providing a narrow, with respect to the coherence length, path for the pairs between the two superconductors (Dayem bridge, variable thickness bridge, etc.), or by pressing a sharp superconducting point over a flat superconductor (point contact) which can be regarded as a rather ambiguous combination of the previous two kinds of link. Over the last decades a large variety of Josephson structures has been investigated using dielectric, semiconductor or metal barrier (proximity effect), as well as different kinds of superconducting bridges and a number of more or less exotic configurations.

As we shall see in the case of HTS junctions, instead of the archetype tunnel junctions we are very likely faced with some of these structures. As we have seen (eq. b) the phase ϕ is modulated in time by voltage. There is also a spatial modulation of ϕ by magnetic field. It is

$$\nabla_{x,y} \phi = (2e/h \ d) \underline{H} \times \underline{n}$$

where $d = (t + \lambda_{L1} + \lambda_{L2})$; t is the barrier thickness and λ_L the London penetration depth. $\underline{n}$ is a unit vector normal to the junction phane (x,y).

The dependence of the critical current vs applied magnetic field reveals dramatically the wave-like nature of superconducting electrons. Indeed for a rectangular barrier junction we have the well known diffraction Fraunhofer-like pattern

$$I_c(\emptyset) = I_c(0) \ \frac{\sin \pi \emptyset}{\pi \emptyset}$$

where $\emptyset = \Phi/\Phi_0$ with $F = H_y L d$ the magnetic flux threading the junction and Φ_0 the flux quantum.

Interference phenomena can occur as well. In particular if we consider two Josephson junctions connected in parallel by a superconductive path we get an interference pattern (see the lecture of Gough in this volume).

Josephson relations can be combined giving the sine-Gordon equation, in normalized units,

$$\frac{\partial^2 \phi}{\partial x^2} + \frac{\partial^2 \phi}{\partial y^2} - \frac{\partial^2 \phi}{\partial t^2} = \frac{1}{\lambda^2 \phi} \sin \phi \ .$$

$\lambda_j = (hc^2/8\pi \ eJ_c \ d)^{1/2}$ is the Josephson penetration length through which we can classify "small" ($L,W<\lambda_j$) and "large" ($L,W>\lambda_j$) junctions, L and W being the transverse junction dimensions.

3. High-Tc Superconductors

Since the beginning of the new era of high-Tc superconductors, great attention was dedicated to investigations concerning superconductive tunneling spectroscopy. Due to the difficulty of realizing good films, early experiments were carried out with bulk samples. The junctions considered were obtained just by pressing two pieces of superconductor or by using a superconducting ceramic pellet with a counter electrode consisting of a point contact either of ceramic or of conventional superconductor (e.g. Nb). A variety of structures configurations were tested, including break junctions, point-crystal contacts and so on [3,4].

The interest in trying to perform such experiments was twofold: on one side, to infer fundamental information on the nature of the high-Tc superconductivity, and on the other, to establish the optimal procedure to realize a "junction device" which, as stated at the beginning of the lecture, would represent the main goal for the development of small scale applications (namely superconductive electronics).

Unfortunately, although Josephson effects have been observed in a large variety of experiments on HTS a reliable S-I-S tunneling film type junction made by HTS has not yet been obtained.

Before going into the details of "why" such junctions have not been yet realized or "how" it would be possible to reach such a goal, we shall discuss about point contact junctions formed with a superconducting point either on a pellet or on a crystal.

The importance of such structures, though far from actual devices, lies not only in their simplicity. Indeed the possibility of varying the point pressure on the pellet counter electrode can be very helpful. Such "coupling adjustability" of two superconductors allows one, in fact, to investigate a variety of situations and, provided a careful "reading" of the experiment results, can give useful information. Of couse the limits of reliability of such structures are rather severe in that, a point contact (e.g. Nb point on YBCO pellet) can explore only a specific region of the sample and is usually quite difficult to perform a sort of scanning to probe the entire area. A junction can also be realized when a pellet of HTC is substituted by a good crystal (point-on-crystal junctions). In this case however, in order to get enough current, the point is usually pushed on the crystal with a significant pressure. As a result it can likely occur that the contact area of the crystal deteriorates so that the crystal structure is actually lost.

A large amount of experimental results using point contacts and the other above mentioned junction structures has been concerned with the determination of the energy gap, (if any), of the HTS. The investigations are based on the study of current-voltage (I-V) and dI/dV characteristics. While in the case of conventional superconductors the well defined structure of the I-V curve is sufficient to clearly identify the gap structure, in the case of HTS no evident signature of the gap is present in the I-V characteristics. It is therefore necessary to resort to dI/dV (or dV/dI) measurements. The zoology of the results obtained in this way is rather wide and the degree of ambiguity suggests some caution [5]. There are indeed several phenomena which can lead to finite voltage structures in the dV/dI characteristics (a partial list is reported in the above quoted references).

It is instructive to compare results obtained by tunneling measurements with point contact junctions using HTS pellets with those obtained by analogous LTS junctions. Some structures in the dV/dI curves appear which can be ascribed to the morphology of the sample (granularity) rather than to more intrinsic effects. This has been also shown [6] by comparing directly a Nb point on a sintered YBCO pellet with a Nb point on a sintered Nb pellet. However a feature which is usually present in HTS junctions is the linear dI/dV vs V characteristics such as that reported in Fig. 2 [see 6 and references reported therein].

Just during the preparation of the manuscript I was informed from a preprint by a group in Argonne (Qiang Huang, J.F. Zasadzinski, K.E. Gray, J.Z. Liu and H. Claus) that measurements on Bi compound reproducible point contacts exhibit a decreasing conductance at high voltages.

There is also a large number of experiments performed on junctions realized by a sintered pellet or crystal overlayed by a conventional superconductor film. For such a planar junction configuration the interface problems become essential. Various HTS materials have been

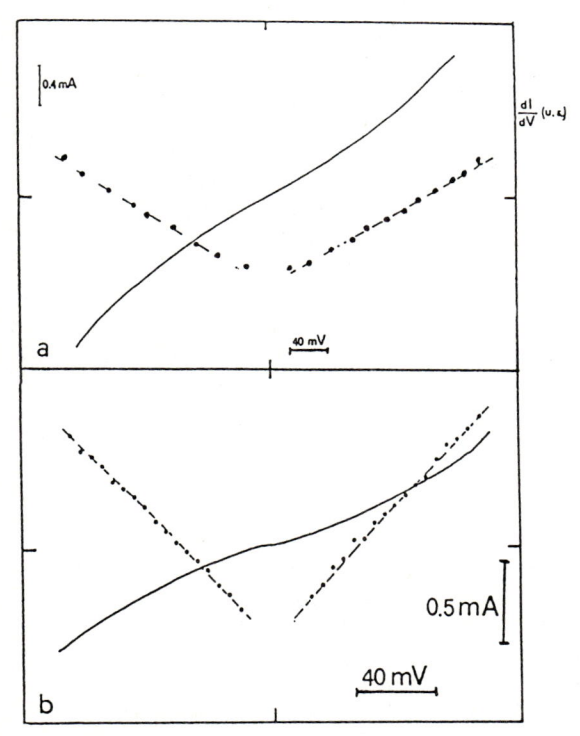

Fig.2 - Behavior of the dI/dV for two different YBCO-YBCO point contact junctions.

employed in this context. Very recently a tunnel junction employing the non-cuprate new superconductor $Ba_{1-x} K_x BiO_3$ (x=0.375; $T_c \sim$ 30 K) has been realized [7]. Incidentally we observe that tunneling data on these samples show evidence of phonon mediated coupling.

As mentioned above, the lack of HTS film (S-I-S) junctions is determined by interface problems which, in turn, are originated by the very low coherence length ξ of these materials. In this respect reliable results were obtained in S-N-S type junctions namely YBCO-Au-Nb [8] in which the proximity effect realized by the use of Au played an essential role. Quite recently heteroepitaxial YBCO-PBCO-YBCO Josephson junctions using a multi-target laser deposition technique have been proposed [9]. In both cases good results of SNS junctions type were obtained.

Another important aspect of HTS materials is their layered structure. This can lead to a significant anisotropy. Several consequences can arise from this circumstance such as a modification of the density of states [10]. The sine-Gordon equation describing the electrodynamics of a Josephson junction (see Section 2) has to be modified in a tensorial form [11] with quite significant differences in the current-magnetic field dependence etc.

In conclusion, looking at the differences between LTS and HTS materials within the framework of superconductive tunneling, some conclusions can be drawn. There is no evidence so far of an S-I-S junction made by HTS thin films which would show in a I-V characteristic both a Josephson effect and a clear gap structure simultaneously. The reason for this is not fully understood though interface effects in connection with the very short coherence length, can to a large extent explain the problem. The layered anisotropic structure plays an essential role also in connection with the dimensionality of the system [10]. (See also the lecture by Dr. T. Schneider in this volume). Of great importance will be therefore a further detailed experimental investigations of the tunneling parallel and perpendicular to, the planes.

In connection with the problem of anisotropic layered structure, it is also of interest to distinguish between the occurrence of Abrikosov and Josephson type vortices corresponding to the two cases $\xi_c(T) \gg d$ and $\xi_c(T) \ll d$ where d is the layer spacing [12].

REFERENCES

[1] J. G. Bednorz and K. A. Müller, Z. Phys. 64, 189 (1986)
[2] A. Barone and G. Paterno, "Physics and Applications of the Josephson Effect", John Wiley Publish. New York (1982)
[3] Proc. Novel Superconductivity - June 22-26 Berkeley, Ca. S.A. Wolf and V.Z. Kresin Eds., Plenum Press (1987)
[4] Proc. HTSC-M2 Interlaken (1988) J. Miller and J.C. Olsen Eds., North-Holland (1988)
[5] A. Barone, above quoted reference pag. 1712

[6] A. Barone, A. Di Chiara, F. Fontana, G. Paterno, L. Maritato, G. Peluso G. Pepe, U. Scotti di Uccio - Applied Superconductivity Conference, San Francisco, August 1988
[7] J. F. Zasadzinski, N. Tralshawala, D.G. Hinks, b. Dabrowski, A. W. Mitchell, D. R. Richards, Physica C (1989)
[8] H. Akoh, F. Shinoki, M. Takahashi and S. Takada, Jap. J. Appl. Phys. 27, L519 (1988)
[9] C. T. Rogers, A. Inam, B. Dutta, X. D. Wu, T. Venkatesan (Preprint)
[10] T. Schneider, H. De Raedt, M. Frick - Z. Phys.(to be published)
[11] R. G. Mints, Modern Phys. Lett. B3, 51 (1989)
[12] A. Barone, A.I. Larkin and Yu. Ovchinnikov (to be published)

Part II

Coherence-Length-Related Properties

Short Coherence Length and Granular Effects in Conventional and High T_c Superconductors

G. Deutscher

Raymond and Beverly Sackler Faculty of Exact Sciences, School of Physics and Astronomy, Tel Aviv University, Ramat Aviv, Tel Aviv 69978, Israel

Abstract. We first review the properties of conventional granular superconductors: critical temperature enhancement, effective coherence length and penetration depth, critical fields and critical currents, fluctuation effects. This review serves as the basis for the discussion of granular effects in the high T_c oxides. We emphasize the role played by the short coherence length, which is an important common characteristic of the two types of materials.

1. Introduction

The granular character of the high T_c oxides has been noticed since their discovery by Bednorz and Müller [1]. Granular characteristics include a poor conductivity in the normal state and in the general vicinity of the Metal-Insulator (M-I) transition, low critical currents and first vortex penetration below the bulk thermodynamic critical field H_{c1}.

The main purpose of this article will be to try and clarify the similarities and the differences between the new oxides and the conventional granular superconductors. One obvious motivation is that a granular behavior has profound implications for practical applications of the new superconductors - hence it is important to review the main features of (conventional) granular superconductivity as a guide towards understanding the properties and possible limitations of the new oxides. This will be the main purpose of Section 2. On a more fundamental level, one can also ask oneself whether the critical temperature enhancement often observed in conventional granular superconductors - such as granular aluminum [2] - has anything to do with high T_c superconductivity. A few remarks on this are also included in the Section 2.

Section 3 deals with granular effects - or as we shall say more accurately, short coherence lengths effects - in the high T_c oxides. A conventional granular superconductor is a mixture of a large coherence length superconductor with an insulator, generally surrounding the superconducting grains. As a result of a reduced macroscopic coefficient of diffusion, this mixture can have a very short effective coherence length. In contrast, we shall emphasize that a short coherence is a fundamental property of the pure high T_c oxides. This short coherence length, rather than the presence of a large amount of insulating second phase material, leads to their granular behavior. It is on this basis that we shall compare the two kinds of granular superconductors.

2. Conventional Granular Superconductors

Granular superconductors have been extensively researched during the last twenty years [2]. They present a number of distinctive features that have rendered them attractive as a test ground for the unfolding theories on percolation, electron localization and the M-I transition [3]. But the interest in granular superconductivity stemmed orginally from the enhanced critical temperature observed in granular $A\ell$ films. We briefly discuss this topic before reviewing other properties of granular superconductors.

2.1 T_c Enhancement in Granular Superconductors

The archetype granular superconductor is granular $A\ell$. Granular $A\ell$ films are easily prepared by evaporating $A\ell$ in a vacuum system in the presence of a reduced pressure of oxygen [2]. One thus obtains films composed of metallic $A\ell$ grains surrounded by $A\ell$ oxide. By controlling the oxygen partial pressure, the $A\ell$ evaporation rate and the temperature of the substrate, one can obtain films with a well defined and controlled granularity [4]. The grain size can be varied from several hundred down to about 20Å, and the normal state resistivity from that of pure metallic $A\ell$ into the insulating regime.

The enhancement of the critical temperature can be substantial, by more than a factor of 3. It was shown that this enhancement is a function of the surface to volume ratio, in specimens for which the grain size distributions were established in detail [5] (Fig. 1). This suggests that the $A\ell$-$A\ell$ oxide interface is favorable for superconductivity. A remarkable structural property of the granular metals is that the thickness of the insulating barriers is of atomic size and apparently fairly uniform [6]. A granular material is not a random mixture, but rather some sort of structured compound. Random metal-insulator mixtures also exist [7], but do not in general show T_c enhancement.

In spite of many efforts, the origin of the T_c enhancement in the granular superconductors was never completely clarified. The possibility of an excitonic mechanism at the metal-insulator interfaces was originally considered [8], but eventually the prevailing view became that the enhancement was rather due to a softening of the phonon spectrum of the superconductor, presumably because of increased

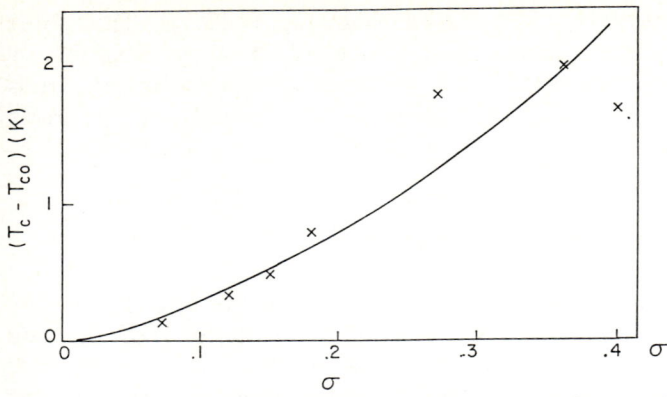

Fig. 1. Critical temperature enhancement of granular Aℓ as a function of the grains' surface to volume ratio τ. (After M. Gershenson, Ref. 11).

crystalline disorder at small grain sizes [9]. As such, it did not open a promising route to high T_c superconductivity, since according to the prevalent theories the electron phonon interaction could not in any case lead to a T_c higher than about 30K. This pessimistic view however was not universally accepted. A more complex picture emerged from tunneling measurements and also from detailed T_c versus grain size determinations carried out by our group at Tel Aviv in the mid-seventies. Electron tunneling measurements and Eliashberg gap equation inversion carried out by M.Dayan on regular and crystalline Aℓ films gave very interesting results [10]. M. Dayan concluded a reduction of the electron screening, resulting in an increase of both the repulsive and the attractive interactions. This was proposed as the main mechanism for T_c enhancement, rather than phonon softening. We shall go back to the point of reduced electron screening when we discuss the high T_c oxides. In analyzing his T_c versus grain size data, M. Gershenson used a proximity effect model assuming that the grains are composed of a core with bulk Aℓ properties, and a surface having a different excitation (phonon or other) energy and coupling constant. He concluded that the surface had a higher excitation energy, a possible indication for an excitonic mechanism [11]. Dayan's and Gershenson's results left the door open for a non-trivial explanation of the T_c enhancement.

The role played by the atomic size oxide layer is not the only puzzling resemblance between granular Aℓ and the high T_c oxides. Another one is the fact that in both cases the highest T_c is achieved close to the M-I transition. Starting from the insulating phase and increasing the Aℓ volume fraction, the insulator-to-metal transition occurs simultaneously with the insulator-to-superconductor transition, the onset of the superconducting transition quickly reaching its

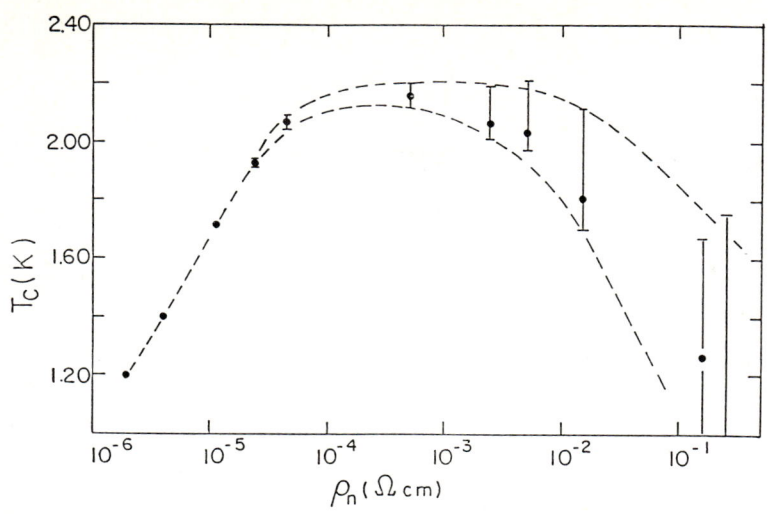

Fig. 2. Critical temperature enhancement of granular Aℓ as a function of the normal state resistivity ρ_n. (After M. Gershenson, Ref. 11)

maximum value [4]. (Fig.2) A further increase in the Aℓ content - resulting in an increased carrier concentration and normal state conductivity - is accompanied by a decrease of T_c. This behavior is strikingly similar to that observed in $Ba_{1-x}K_xBiO_3$ (maximum T_c 30K) and in $BaBi_{1-x}Pb_xO_3$ (maximum T_c 11K) [12]. The highest T_c reached with granular Aℓ is about 8K [13].

Another puzzling resemblance between granular Aℓ and the high T_c oxides is the critical field anisotropy. Critical field measurements on granular Aℓ films showed a rather large anistropy, the parallel critical field being many times larger than the perpendicular one [14] (Fig. 3). Because they were obtained on very thick films d > (ξ,λ), these results could not be explained by the well known dimensionality effects for thin films. Instead, the existence of a layered structure $Aℓ/Aℓ_2O_3/Aℓ$ had to be assumed. The very existence of the granular structure shows that $Aℓ/Aℓ_2O_3/Aℓ$ interfaces are favorable for the stabilization of the compound; therefore the formation of a layered structure during film growth is not too surprising. Recent structural studies have given some direct evidence for the existence of layers in granular films [15].

We thus have a few reasons to believe that there might well indeed exist a link between the T_c enhancement seen in granular Aℓ, and high temperature superconductivity in the perovskites discovered by Bednorz and Müller: the role played by the metal/oxide interface in enhancing T_c, the proximity of the M-I transition, the tendency to from a layered structure and the reduced electron screening (a point we shall come back to later). Maybe this link can give us some clue as to the origin of the high T_c.

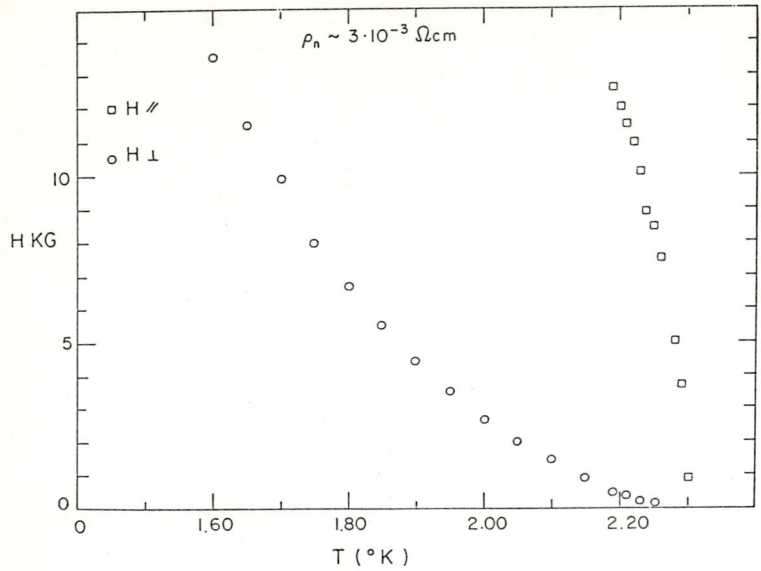

Fig. 3. Critical field anisotropy of granular Aℓ films (After Ref. 14).

In the remainder of this lecture we shall review the effects of granularity on the other superconducting properties: critical fields, penetration depth, critical currents and fluctuation effects.

2.2 Effective Coherence Length and Upper Critical Field

Leaving aside the question of the origin of the enhanced T_c, we shall now assume that the superconducting grains have well defined thermodynamic properties (T_c, H_c, etc...), which include the interface effects. This is a valid assumption in the limit where the coherence length ξ_0 of the superconductor that constitutes the grains is much larger than the grain size d. One can then treat the grain's properties in the Cooper limit [11]. Another important feature of the large ξ_0/d limit is that one can assume that the order parameter Δ is constant in each grain. This follows from the standard Landau Ginzburg boundary condition $(\Delta\nabla)_n=0$, where n designates the normal to the boundary. This condition is fully justified when ξ_0 is much larger than the interatomic distance a. A conventional granular superconductor is thus characterized by the inequalities

$$\xi_0 \gg d \gg a .$$

A Landau-Ginzburg free energy expression can then be built, replacing the usual $|\nabla\Delta|^2$ term by a sum of terms of the form $C|\Delta_i-\Delta_j|^2$, where i and j designate neighboring grains. The intergrain boundaries are treated as Josephson junctions having a normal state resistance R_n:

$$F_s = F_n + \sum_i V_i \left[A|\Delta_i|^2 + B\frac{1}{2}|\Delta_i|^4 \right] + \frac{1}{2}\sum_{ij} C|\Delta_i - \Delta_j|^2 , \quad (1)$$

where $C = (\pi\hbar/16\, R_n k_B T_c)$.

With this free energy expression, one can then calculate an effective coherence length $\bar{\xi}(T)$ for the granular medium [16]:

$$\bar{\xi}(T)^2 = -\left(\frac{C}{2dA}\right) . \quad (2)$$

Assuming that the grain size and the junction coefficients of transmission are uniform through the sample, the normal state resistivity is given by $\rho_n = R_n d$. One can then express $\bar{\xi}(T)$ in terms of measurable parameters:

$$\bar{\xi}(T)^2 = \frac{\pi\hbar}{32 e^2 N(0)\rho_n k_B T_c} \cdot \left[\frac{T_c}{T_c - T}\right] , \quad (3)$$

where we have used $A = N(0)((T_c-T)/T_c)$, $N(0)$ being the normal state density of states at the Fermi level. The dependence $\xi^{-2} \propto \rho_n$ is in fact quite general. We note that by using Einstein's relation $\sigma = 2e\, N(0)D$, where D is the coefficient of diffusion, Eq. 3 can be rewritten as:

$$\bar{\xi}(T)^2 = \frac{\pi\hbar D}{16 k_B T_c} \cdot \left[\frac{T_c}{T_c - T}\right] ,$$

As is well known, this relation applies to homogeneous alloys in the dirty limit [17]. It also applies to many realizations of disordered superconductors, for instance to the case of a percolating mixture of a superconductor with an insulator [18]. The condition of validity of Eq. 3, is that $\bar{\xi}(T)$ be larger than the inhomogeneity scale. For the granular case, this scale is the grain size d. For the percolating case it is the percolation correlation length ξ_p [19]. As long as $\bar{\xi}(T)$ is larger than the relevant inhomogeneity scale, it is determined by the superconductor's properties and by the macroscopic normal state transport properties (long range coefficient of diffusion). In this limit, the detailed structure of the disordered medium is unimportant. In particular, if $\bar{\xi}$ is shorter than an effective London penetration depth $\bar{\lambda}$ (discussed below), one can show that the upper critical field is given in the homogeneous limit by the usual expression:

$$H_{c2} = \frac{\phi_0}{2\pi\bar{\xi}(T)^2} .$$

If we add more insulating material to the mixture, ρ_n becomes larger and $\bar{\xi}$ shorter, and one eventually reaches the inhomogeneous limit $\bar{\xi} \sim d$ or $\bar{\xi} \sim \xi_p$. H_{c2} is then independent of ρ_n but becomes structure sensitive [18,19]. For instance, in the granular case and in the assumed limit $\xi_0 \gg d$, it is not possible to accommodate a vortex line inside the grains. The closest the vortex can get is the grain size. For $\bar{\xi} \sim d$, the upper critical field of the granular material becomes equal to that of the grains themselves [20]. For the percolating case, H_{c2} in the limit $\bar{\xi} < \xi_p$ depends on the fractal structure of the infinite cluster at scales smaller than ξ_p, also independent of ρ_n [18].

This description of the upper critical field of granular and other disordered superconductors is well verified experimentally. The linear dependence of H_{c2} on ρ_n was first seen in granular Aℓ by Cohen et al. [21]. Its saturation at large values of ρ_n was later reported by Deutscher and Dodds [14]. As first noticed by Deutscher et al. [20], the crossover at $\bar{\xi}(T) \sim d$ can also be produced by varying the temperature at a fixed value of ρ_n. The homogeneous regime is always reached near T_c, and if ρ_n is sufficiently large that $\bar{\xi}(0) < d$, the crossover condition $\bar{\xi}(T) \sim d$ is met at some finite temperature. The crossover temperature T^* is easily identified since at $T > T^*$, H_{c2} has the bulk linear L.G. temperature dependence $H_{c2} \propto (T_c-T)$, while at $T < T^*$ it has the temperature dependence of an isolated grain smaller than the penetration depth, $H_{c2} \propto (T_c-T)^{1/2}$ (Fig.4).

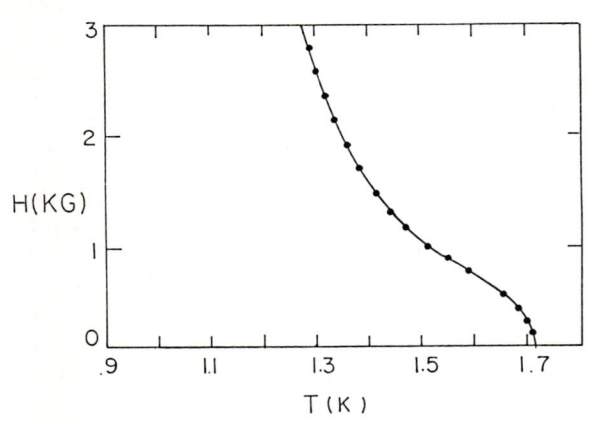

Fig. 4. Transition from 3D behavior (near T_c) to 0D behavior (lower temperatures $T < T^*$ see text) in granular Aℓ-Ge films (After Ref. 20).

The homogeneous and inhomogeneous regimes have also been studied in detail in the percolating superconductor In-Ge, showing in particular the saturation and anomalous temperature dependence (determined by the fractal dimensionality of the infinite cluster) in the inhomogeneous limit [18] (Fig. 5).

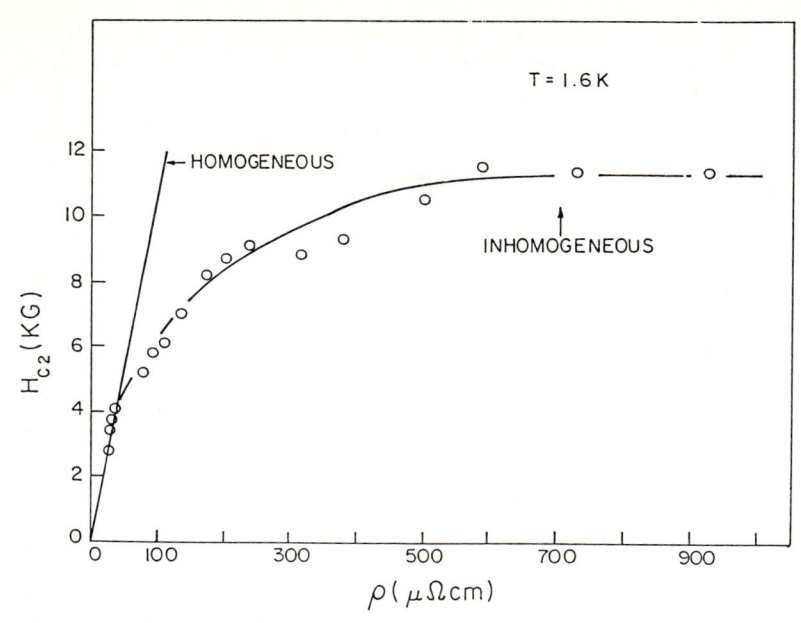

Fig. 5. Upper critical field of In-Ge films as a function of their normal state resistivity ρ_n: transition from the homogeneous regime (small ρ_n, $H_{c2} \propto \rho_n$) to the inhomogeneous one (large ρ_n, H_{c2} = const.). (After Ref. 18).

The essential results that we must retain from this discussion of the coherence length in conventional granular and other disordered superconductors are that: 1) in the homogeneous limit H_{c2} is given by the standard expression for a dirty superconductor in terms of the macroscopic coefficient of diffusion; and 2) in the inhomogeneous limit H_{c2} is independent of the normal state resistivity but becomes structure sensitive. Note that in this limit the order parameter is in general not destroyed uniformly everywhere in the sample at a well defined critical field. In the granular case for instance, H_{c2} in the inhomogeneous limit is that of the individual grains. Since the latter is size dependent, a distribution of grain sizes will automatically lead to a distribution of critical fields. Generally speaking, we expect that in the inhomogeneous limit the field transitions will be much broader than in the homogeneous limit: the diluted nature of inhomogeneous superconductivity will lead to increased flux creep etc... Broad transitions have been clearly observed in percolating superconductors near the threshold [18].

2.3 Effective Penetration Depth and Lower Critical Field

As is well known, the penetration of a magnetic field in a Josephson junction is much larger than that in a bulk superconductor [17]. One thus expects the field penetration in the granular structure, composed of grains coupled through junctions, to be larger than that in the superconductor that constitutes the grains.

The penetration depth λ_J in a single junction is given by

$$\lambda_J = \left(\frac{\hbar c^2}{16\pi e J_c \lambda_L}\right)^{1/2}, \tag{4}$$

where J_c is the critical current density across the junction and λ_L the London penetration depth in the superconducting banks of the junction. J_c can be orders of magnitude smaller than the bulk critical density, λ_J is then of macroscopic size [17].

The granular superconductor can be modelled by a lamellar geometry with an inter-junction distance d, the magnetic field being applied parallel to the planes of the junctions. As first noticed by Deutscher and Entin-Wohlman [22], a new situation arises when $d \lesssim \lambda_L$ (Fig. 6). Screening in the banks then becomes weaker, λ_L in Eq. 4 must be replaced by d/2 (the effective screening in the banks). In the limit $\lambda_L \gg d$, the field penetration is now uniform and characterized by an effective penetration depth

$$\bar{\lambda} = \left(\frac{\hbar c^2}{8\pi e d J_c}\right)^{1/2}, \tag{5}$$

where J_c is given by Josephson's expression [23]

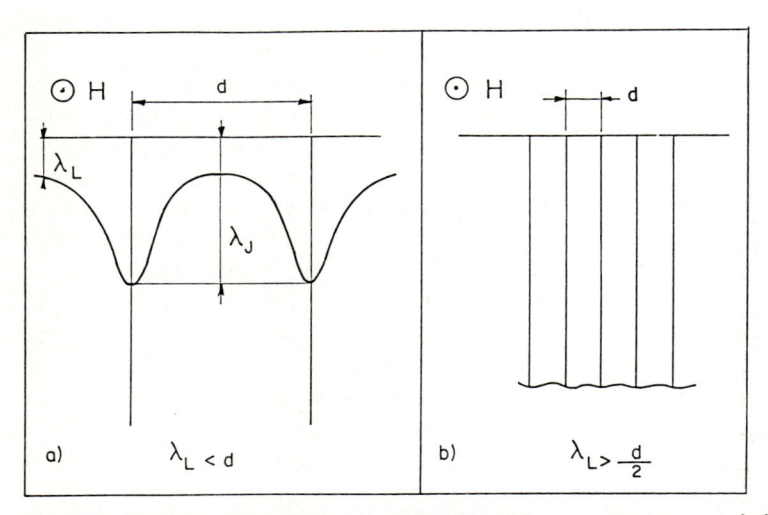

Fig. 6. Penetration of a magnetic field in a sample containing junctions separated by a distance d : a) $\lambda_L < d$, inhomogeneous regime; b) $\lambda_L > d/2$, homogeneous regime.

$$J_c = \frac{\pi \Delta(T)}{R_n d^2} \tanh\left[\frac{\Delta(T)}{2k_B T}\right] .$$

Using again $\rho_n = R_n d$, we can write Eq. 5 as

$$\bar{\lambda}^2 = \hbar c^2 \rho_n / \left[8\pi^2 e \Delta(T) \tanh\left[\frac{\Delta(T)}{2k_B T}\right]\right] . \tag{6}$$

This expression predicts $\bar{\lambda} \propto (\rho_n)^{1/2}$, as for a homogeneous dirty superconductor, and near T_c, $\bar{\lambda} \propto (T_c - T)^{-1/2}$ again as in Landau Ginzburg. Although Eq. (5,6) were derived for a lamellar structure, they should also essentially apply to a granular one. Note that the grain size does not appear in Eq. (6), much the same as it does not appear in Eq. 3 for the effective coherence length in the homogeneous limit.

Using the relation $H_{c1} = (\phi_0/4\pi \bar{\lambda}^2) \ln \bar{\kappa}$, where $\bar{\kappa} = (\bar{\lambda}/\bar{\xi})$, we obtain near T_c

$$H_{c1} = \frac{\pi^2}{2c\rho_n} \frac{\Delta^2(T)}{k_B T_c} \ln \bar{\kappa} . \tag{7}$$

The field for first vortex penetration can be very small, for high normal state resistivities. For a typical granular superconductor $\rho_n = 1 \cdot 10^{-5} \Omega m$, $\Delta(0) = 1 meV$, we obtain $H_{c1} \tilde{\sim} .1G$.
The effective L.G. parameter $\bar{\kappa}$ can be computed from Eq. 3 and Eq. 6:

$$\bar{\kappa} = 0.16 \, c \, e \, N(0)^{1/2} \rho_n , \tag{8}$$

again similar to that for a homogeneous dirty type II superconductor [17].

2.4 Critical Currents

Critical currents of granular superconductors are typically much weaker than for homogeneous ones, since they are dominated by the critical current across the individual grain boundaries. For far apart junctions (large grains) one expects the critical current density to be equal to that of the individual junctions. However, as we shall see, this is not the case for small grains: again, we find a homogeneous and an inhomogeneous limit [23].

Following Josephson's relation, a current induces a phase difference ϕ between neighboring grains. The free energy of a granular structure, per lamella, with the current flowing perpendicular to the lamellae, is given near T_c by

$$F_s = F_n + A|\Delta|^2 + \frac{B}{2}|\Delta|^4 + \frac{1}{LS} \frac{1}{R_n} \frac{\pi \hbar}{8e^2} \frac{\Delta(T)}{k_B T_c}(1-\cos\phi) ,$$

where L and S are respectively the lamellae thickness and area. For a given value of ϕ one minimizes the free energy to obtain the equilibrium value of Δ and one thus obtains a relation between the current and the reduced order parameter $f=(\Delta/\Delta_0)$, where Δ_0 is the zero current equilibrium value of Δ:

It is useful to consider the parameter

$$c^{-1} = \frac{4e^2}{\pi\hbar} R_n \, LS \, N(o) \, k_B T_c \left[\frac{T_c-T}{T_c}\right] . \tag{9}$$

In the weak coupling limit $c \gg 1$ (high resistance, large grain size) one recovers Josephson's result, $J_c(T)=J_j(T)$. In the strong coupling limit (low resistance, small grains, vicinity of T_c), one obtains

$$J_c^2(T) = \frac{8}{27} J_j(T) \, J(T) \quad,$$

where

$$J(T) = \left(\frac{2e}{\hbar}\right) LS \, (N(o)\Delta^2(o)/2) \, [(T_c-T)/T_c] .$$

We notice that in this limit $J_c \propto ((T_c-T)/T_c)^{3/2}$, the classical L.G. result, and also $J_j \propto \rho_n^{-1/2}$, again as for a homogeneous dirty superconductor (and in contrast with Josephson's result $J_j \propto \rho_n^{-1}$). As a matter of fact, a comparison between Eq. 3 and Eq. 9, shows that within a numerical coefficient, the condition $c=1$ corresponds to $\bar{\xi}(T)= d$. It is quite satisfactory that in the homogeneous limit $\bar{\xi}(T) > d$, both the upper critical field and the critical current density have the same behavior as that for a homogeneous superconductor. Conventional low T_c granular superconductors fall generally in this limit. In contrast, we shall see in Section 3 that the high T_c oxides fall always in the inhomogeneous limit (as far as the upper critical field and critical current densities are concerned).

So far we have treated the case where the material is characterized by one grain size and one intergrain junction resistance. This situation is seldom realized (except in artificial networks). In particular, it is reasonable to expect a significant distribution of intergrain resistances. A percolation description is then more appropriate. The inhomogeneous scale is then the percolation correlation length ξ_p, which can be much larger than the grain size:

$$\xi_p = d(p_c - p)^{-\nu} \quad,$$

where p is the fraction of effectively superconducting junctions, p_c the critical value at which an infinite superconducting cluster is formed, and ν a critical exponent ($\nu_{3D} = 0.88$, $\nu_{2D} = 1.3$).

In the percolation case, one falls easily in the inhomogeneous limit $\bar{\xi}(T) < \xi_p$. The critical current density is then determined by the density of independent superconducting paths [19]:

$$J_c \propto \xi_p^{-(D-1)} \quad ,$$

where D is the dimensionality. Thus

$$J_{c3D} \propto (p_c-p)^{-2\nu} \sim (p_c-p)^{-1.8} \quad ,$$

$$J_{c2D} \propto (p_c-p)^{-\nu} \sim (p_c-p)^{-1.3} \quad .$$

It is useful to compare these expressions to those of the percolating resistivity [19]:

$$\rho_{3D} \propto (p_c-p)^{2.0} \quad ,$$

$$\rho_{2D} \propto (p_c-p)^{1.3} \quad .$$

Notice that for both dimensionalities, J_c is essentially proportional to ρ_n^{-1} - i.e., in the inhomogeneous percolation limit we nearly recover the inhomogeneous granular Josephson limit.

2.5 Fluctuation Effects

The crossover from the homogeneous to the inhomogeneous regime is accompanied by a change in the effective dimensionality of the system. Below the critical temperature this produces as we have seen a change in the temperature dependence of the upper critical field. Near and above T_c, it affects the thermodynamic fluctuations of the order parameter. In the granular case, the fluctuations of the order parameter in neighboring grains become uncorrelated when $\bar{\xi}(T) \lesssim d$, and therefore zero dimensional ($\xi_0 > d$, the order parameter cannot vary inside the grain). This change in dimensionality has some important consequences for a number of physical properties, such as the excess fluctuation conductivity and the behavior of the heat capacity in the vicinity of the transition [16].

The transition to zero dimensionality affects the width of the transition [4] and the temperature dependence of the excess conductivity [14]. The effects are particularly pronounced for very small grains. In an isolated grain[7], the temperature range where the fluctuations of the order parameter are very large (critical region) is of the order of T_c itself when $d < d_c$ where

$$d_c = (N(0)k_B T_c)^{-1/3} \quad .$$

In that case, the sharp heat capacity jump characteristic of bulk superconductivity is essentially washed out [16]. If the grains are very weakly coupled, we also expect the jump to be replaced by a broad transition. This is indeed observed experimentally [24]. As the grains

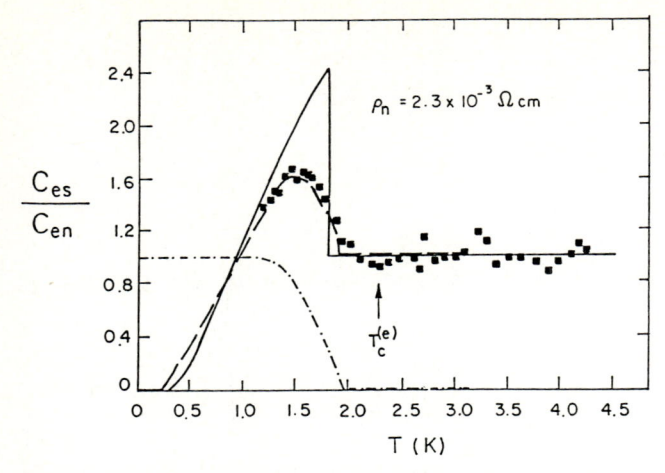

Fig. 7. Heat capacity of a high resistivity granular Aℓ film: continuous line, mean field behavior; broken line, fit to percolation model (After Ref. 24 and 25).

are being decoupled by adding more oxide to granular Aℓ, the jump is strongly broadened and its height reduced well before superconductivity is quenched altogether (there is still a resistivity transition). The detailed shape of the heat capacity transition measured in Ref. 16 for the case of very small grains (d < d_c) has been given a percolation interpretation [25] (Fig. 7). The idea is that in the presence of a sufficiently broad distribution of intergrain resistances (intergrain coupling strength), finite clusters of effectively coupled grains (coupling energy larger than $k_B T$) develop progressively as the temperature is lowered below the individual grains' critical temperature T_{cg}. Eventually, at some temperature $T_c < T_{cg}$, an infinite cluster with zero resistance is formed. The model then divides the grains into two categories: i) those that are effectively connected to the infinite cluster (Josephson coupling energy larger than $k_B T$) have a 3D behavior: the thermodynamic fluctuations are effectively quenched and the "mass" of the infinite cluster contributes to the heat capacity jump; ii) the unconnected grains, which do not contribute to the heat capacity jump because of zero D behavior. This model, although evidently oversimplified, seems to contain most of the physics involved and gives an excellent fit to the experiment.

The behavior in the large grain case (d > d_c) is just the opposite: a heat capacity jump is observed even when the grains are so weakly coupled that the sample is an insulator and does not show a superconducting transition [26].

Outside of the critical region the fluctuation conductivity is given by

$$\Delta \sigma \propto \left(\frac{T-T_c}{T_c} \right)^{2-D/2} ,$$

where D is the dimensionality. The crossover between 0D behavior far from T_c and 2D or 3D behavior near T_c has been observed in granular Aℓ [14].

2.6 Localization Effects Near The M-I Transition

We have so far treated the granular material as being a reasonably good metal in the normal state. This approximation is in general appropriate in the homogeneous limit, but often fails in the inhomogeneous case (high resistivities).

A review of localization effects on the superconducting transition is well beyond the scope of this lecture, but it is definitely an important topic for the comprehension of the high T_c oxides.

Weak localization effects in granular Aℓ have been studied in great detail by the Rutgers group, and are in general agreement with the weak localization theories [27]. In particular, the anomalous negative magnetoresistance characteristic of the weak spin-orbit interaction limit is clearly observed.

Strong localization effects near the M-I transition have also been studied in granular superconductors. One always observes a decrease of the critical temperature as the M-I transition is approached [4]. One interesting question is whether superconductivity is quenched before, at or after the M-I transition.

Abeles [2] has given a heuristic derivation of the characteristic resistivity at which the M-I transition takes place in a granular system. Assume that two grains are coupled through a junction of resistance R_n. The junction has a certain capacitance C, and accordingly a time constant $\tau = R_n C$. Now suppose that the capacitance is charged by the transfer of one electron from one grain to the other, the energy of the capacitor is now (e^2/C). However, according to the uncertainty principle, the charged state is well defined only if its energy is larger than $\hbar/\tau$. Hence, the capacitor is insulating only if $(\hbar/R C) < (e^2/C)$, or:

$$R_n > \frac{\hbar}{e^2} .$$

This criterion leads to a <u>universal</u> critical sheet resistance, $R_\square \sim \hbar/e^2$, for a sample consisting of one layer of grains, and a nonuniversal (grain size dependent) resistivity $\rho_n \sim (\hbar/e^2) d$ in a 3D granular medium. By and large, this simple prediction is fairly well verified experimentally, within numerical factors.

The superconductor-normal transition may or may not coincide with the M-I transition. For the 3D granular case, and for small grains $d<d_c$, containing on the average less than one Cooper pair per grain, the two transitions appear to coincide [28]. On the other hand, if $d>d_c$, the superconductor-normal transition may take place directly into the insulator phase (superconductor-insulator transition), as reported for larger-grain $A\ell$-Ge films [29].

Recent experiments by the Minnesota group on ultra-thin films indicate that in 2D superconductivity is quenched at a universal value of the sheat resistance, within numerical factors the same as that given by Abeles for the metal-insulator transition in 2D [30].

Superconductivity near the M-I transition remains an active field of research with many unanswered questions. Unlike the case of weak localization, there is a lack of detailed theoretical guidelines. Empiricialy it is found that details of the localization mechanism, such as the spin-orbit interaction, have a definite influence on the drop in T_c [28].

The separation of localization versus interaction effects also remains largely an open problem.

3. Granular Effects in the High T_c Oxides

In their original publication on the discovery of superconductivity in LaBaCuO, Bednorz and Müller [1] reported a slightly semiconducting behavior above T_c and a strong shift of the transition under weak currents ($\sim 10 A^2/cm$). They interpreted this behavior in terms of percolating superconductivity, and/or localization effects. Two series of experiments by the Zurich group quickly confirmed the granular character of the oxides.

In the microwave absorption experiments by Blazey et al. [31], samples are submitted to a d.c. field with a superimposed modulation, and to an orthogonal microwave field. The microwave absorption is detected synchronously with the modulation. It measures primarily vortex motion in Josephson junctions, because the absorption is inversely proportional to the vortex viscosity, which is several orders of magnitude smaller in junctions than in the bulk. Vortex motion is detected by these experiments at fields much smaller than the bulk H_{c1}, both in ceramics and in single crystals. This topic is discussed in detail in the contribution by K.W. Blazey.

Magnetization measurements were also interpreted in terms of a granular structure. Non-ergodic (irreversible) and time dependent effects [32] were interpreted in terms of a glass model [33], by modelling the ceramics as a disordered network of superconducting loops of typical size S. In the presence of a magnetic field of the order of:

$$H^* = \frac{\phi_0}{2S},$$

frustration effects become important due to the inability of the network to satisfy the flux quantization requirement in all the loops of the system. Glass behavior sets in at $H > H^*$. In particular, there exists an irreversibility line $T^*(H)$ that separates the domains of irreversible and reversible magnetization. Numerical simulations are in agreement with the experimental result:

$$(T_c - T^*(H)) \propto H^{2/3} \tag{10}$$

similar to the de Almeida-Thouless line in spin glasses. An indication of the possible existence of <u>intragrain</u> junctions was the fact that the typical loop size deduced from the experiments of Ref. 32 was somewhat smaller than the grain size. Glassy behavior in ceramics and in single crystals is discussed in detail in the contribution by I. Morgenstern.

For the following discussion, we retain that: 1) vortices penetrate in the high T_c oxides at fields much lower than the bulk; 2) this penetration occurs at sites where the viscosity is small, i.e., at junctions; 3) both the magnetization behavior and the low field microwave absorption show qualitatively similar behavior in the ceramics and in the single crystals.

The presence of junctions in samples containing no significant amount of second phase non-superconducting material is undoubtedly one of the most striking properties of the high T_c oxides. We shall show that they result from the short <u>intrinsic</u> coherence length of the oxides, combined with the presence of extended crystallographic defects <u>of atomic size</u>, such as grain boundaries in polycrystalline ceramics, stacking faults and twin boundaries in single crystals [34]. This is in contrast with low T_c, large coherence length superconductors, where atomic size defects have essentially no influence on superconductivity. Also in contrast with conventional granular superconductors, the junctions are the site of a depressed order parameter. This depression plays an important role in reducing the strength of superconductivity, particularly in polycrystalline samples and near T_c.

In this section, we first review recent estimates of the coherence length. We follow this with a detailed presentation of short ξ junctions properties, and their applications to the calculation of critical current densities across grain boundaries and in polycrystalline materials. We finally discuss fluctuations and the influence of internal boundaries on the properties of single crystals.

3.1 Determination of the Coherence Length

The determination of ξ is straightforward in a homogeneous type II superconductor, from a measurement of the upper critical field:

$$\xi^2 = \frac{\phi_0}{2\pi H_{c2}} .$$

Unfortunately, the experimental determination of H_{c2} meets with a number of difficulties that have been discussed at length elsewhere [35]. The resistive transition in the presence of an applied field broadens considerably, with the onset temperature hardly dependent on the applied field and the offset temperature T_{cof} often varying as $(T_c - T_{cof}(H)) \propto H^{2/3}$. The transition width $(T_c - T_{cof})$ is large even for single crystals. Specific heat measurements in the presence of an applied field also point to an onset temperature which is field independent within the accuracy of the measurement [36]. Also within experimental accuracy, the magnetization is field independent in high fields [37].

With these caveats, $1T/K < (dH_{c2}/dT) < 10K/T$ for the applied field along the c axis, and about 5 times larger in the orthogonal direction in YBCO. This gives $5Å < \xi_{ab} < 15Å$, $1Å < \xi_c < 5Å$. Following our discussion in Section 2, we shall therefore describe the high T_c oxides as granular superconductors in the inhomogeneous limit, $\xi < d$: i) in ceramics, d is the grain size (several micron), since the weakest junctions are at grain bondaries; ii) in a single crystal, d is not known a priori, but even the smallest identified inhomogeneity scale such as the intertwin distance in YBCO (several 100Å or more) is still much larger than ξ. A description of the high T_c oxides as inhomogeneous granular superconductors immediately explains the experimental observation that H_{c2} is independent of the normal state resistivity.

In conventional granular superconductors the intrinsic coherence length is larger than the grain size, $\xi_0 > d$, so that the effective coherence length can be either larger or smaller than d, depending on the normal state resistivity, grain size and temperature. In contrast, the high T_* oxides are always in the inhomogeneous limit since $\xi_0 < d$.

Another modification concerns the boundary condition for the order parameter. In the conventional case, $\xi_0 > d \gg a$. In the oxides, $\xi_0 \sim a \ll d$. The boundary condition must then be replaced by

$$\frac{1}{\Delta}(\nabla\Delta)_n = -\frac{1}{b}, \tag{11}$$

where $b \sim (\xi_0^2/a) \sim a$ [17]. There is a depression of the order parameter at surfaces and interfaces, and more generally at crystallographic defects of atomic size (Fig. 8).

On the phenomenological level, this is probably the most fundamental difference between the high T_c oxides and the conventional superconductors. It is well known empirically that defects such as surfaces and grain boundaries do not affect the order parameter in low T_c superconductors, and it is well understood that this is because $\xi_0 \gg a$. Accordingly, all the fundamental measurements (penetration depth, depairing critical current, tunneling density of states) can be safely carried out on polycrystalline samples, and at surfaces. In contrast, the use of single crystals is essential for the quantitative study

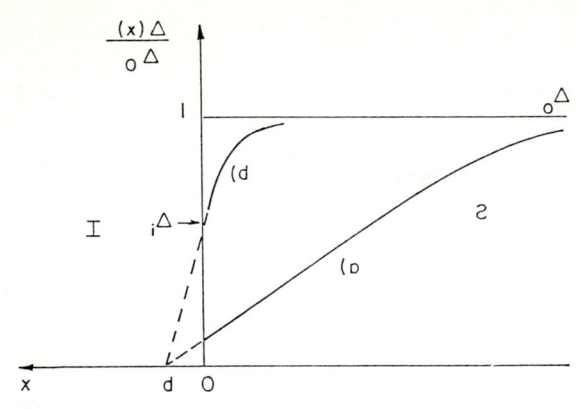

Fig. 8. Depression of the order parameter near a free boundary in a short ξ superconductor: a) near T_c; b) at low temperature. (After Ref. 34).

of high T_c superconductivity, and all surface measurements are a priori suspect of not giving the true bulk properties.

One can easily convince oneself that the short $\xi_0 \sim 10\text{Å}$ is entirely reasonable as an intrinsic value for the oxides. Based on the uncertainty principle,

$$\xi_0 = \frac{\hbar v_F}{\pi \Delta} ,$$

where v_F is the Fermi velocity and Δ the pair potential. The short v_F (due to the small carrier density) and the large Δ (due to the high T_c), give a coherence length about two orders of magnitude smaller than that of Pb ($T_c = 7K$, $\xi_0 = 800\text{Å}$).

We have seen in Section 2 that in conventional granular superconductors the effective coherence length can also be very short, $\bar{\xi} \lesssim \sim 20\text{Å}$, not much larger than ξ_0 of the oxides. But there is one essential difference: in the conventional case, $\bar{\xi}$ measures the range of the intergrain correlation of the order parameter; inside the grains, superconductivity is still governed by a large ξ_0. In contrast, the short ξ_0 is, as we have seen, an intrinsic and fundamental property of the high T_c oxides. The short $\bar{\xi}$ of low T_c granular superconductors has a structural origin; in the high T_c oxides, granularity results effectively from the short ξ_0.

3.2 Critical Currents

Having established that the high T_c oxides are granular superconductors in the inhomogeneous limit, we can now calculate their critical current in the weak coupling limit established in Section 2, Eq. 5'. In a typical ceramic, $\rho_n \approx 10^{-5}$ Ωm, d $\approx 10^{-5}$m. Taking $\Delta = 20$ meV [38], we obtain $J_c(o) \approx 3.10^5$ A/cm². This is significantly higher than the typical experimental value $\lesssim 10^4$ A/cm². At 77K, the discrepancy is even stronger, the calculated value being about 10^5 A/cm² and a typical experimental value a few 10^2 A/cm².

The origin of this discrepancy lies in the boundary depression of the order parameter. We discuss successively the behavior of a single junction and that of polycrystalline samples.

Boundary Critical Current

The critical current across a boundary is determined by the value of the order parameter on the banks and by the boundary's resistance in the normal state.

The value of the order parameter at the boundary can be calculated by solving the non linear Landau-Ginzburg equation for the order parameter [17]:

$$\xi^2(T) \left(\frac{df}{dx}\right)^2 = \frac{1}{2}(1-f^2)^2 \ ,$$

where $f = (\Delta(x)/\Delta_0)$, Δ_0 being the equilibrium value of the order parameter far from the boundary.

The general solution is:

$$f(x) = \tanh\left(\frac{x+x_0}{\sqrt{2}\xi(T)}\right)$$

where the value of x_0 is determined from the boundary condition Eq.11. The value of the order parameter at the boundary $\Delta_i(T)$ can thus be obtained as a function $(b/\xi(T))$ (Fig. 9).

It is useful to consider the normalized critical current:

$$\frac{(J_c(T))}{J_c(0)} = \frac{\Delta_i(T)}{\Delta_i(0)} \tanh\left(\frac{\Delta_i(T)}{2k_BT}\right) \ , \tag{12}$$

which can be calculated as a function of the parameter $\delta(0) = (b/\sqrt{2}\,\xi(0))$. A fit to experimental data obtained on a single grain boundary parallel to the c axis [39] leads to $\delta(0) = 0.66$, $(\Delta_i(0)/\Delta_0(0)) = 0.5$ (Fig. 10). Notice that the experimental curve is well fitted over the entire temperature range, including the tail near T_c. In this limit, $(\Delta_i/\Delta_0) \propto (b/\xi(T)) \propto (T_c-T)^{1/2}$ and $J_c(T) \propto \Delta_i^2(T) \propto \Delta_0^2(T) \cdot (T_c-T) \propto (T_c-T)^2$.

The value $(\Delta_i(0)/\Delta_0(0)) = 0.5$ obtained from the fit to the experiment of Mannhart et al. [39] is extremely useful, because the expression $b = (\xi_0^2/a)$ is only an approximation. Moreover, ξ_0 is not well known, as we have discussed above. Hence, at the present stage, the exact boundary

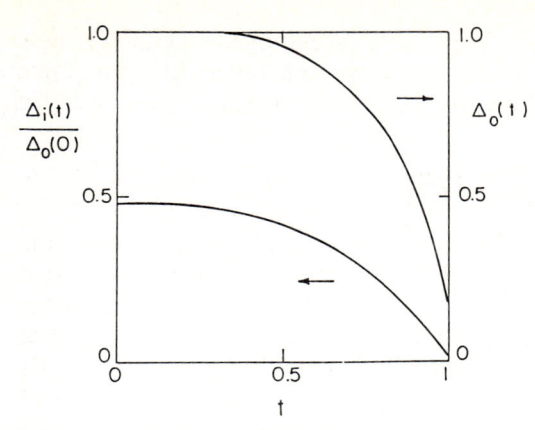

Fig. 9. Temperature dependence of the boundary order parameter $\Delta_i(t)$ compared to that of the bulk $\Delta_0(t)$, for $\left(\dfrac{b}{\sqrt{2}\xi(o)}\right) = 0.66$. Note the linear behavior of $\Delta_i(t)$ near $t = 1$. ($t = (T_c-T)/T_c$).

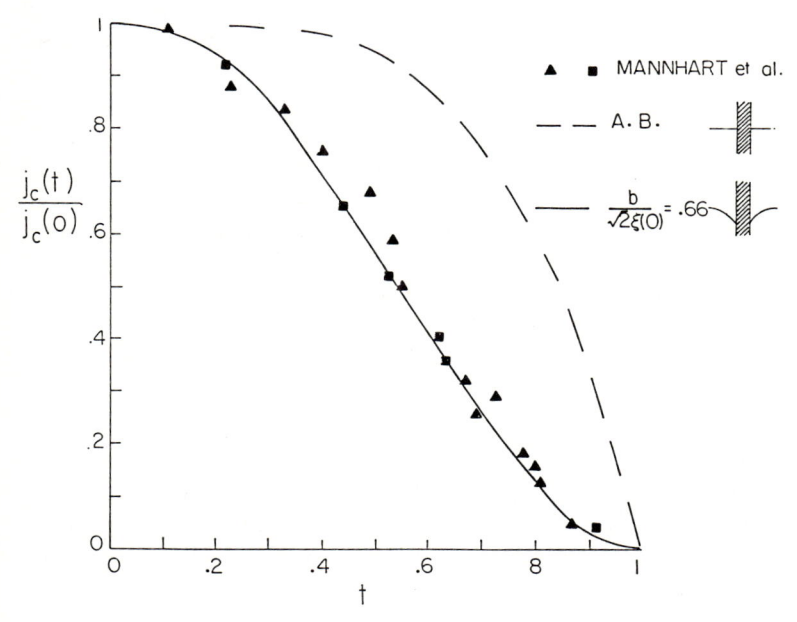

Fig. 10. Temperature dependence of the normalized critical current $j_c(t)/j_c(o)$ for $\left(\dfrac{b}{\sqrt{2}\xi(o)}\right) = 0.66$. (After Ref. 35).

condition cannot be calculated from first principles. Having now determined it by a fit to experimental data, we can use it to make some practical evaluations, for instance of the critical current in polycrystalline samples.

Critical Current of Polycrystalline Samples

The anisotropy of the coherence length plays a crucial role in lowering the critical current of ceramic samples. Consider two grains separated by a boundary perpendicular to the c axis. The relevant ξ for the determination of the boundary order parameter is now $\xi_c = 0.2\xi_{ab}$. The parameter $\delta(0)$ is then about 5 times smaller and $\Delta_i(0) = 0.1\Delta_0(0)$. Accordingly, $J_c(0)$ will be about 10 times smaller than for a junction of the same resistance without a boundary depressed order parameter.

In a polycrystalline sample we must expect a majority of poorly oriented junctions. Only a few will have the preferred orientation parallel to the c axis. Also, experiments have shown that a few degrees misorientation is sufficient to depress the critical current down to its fully misoriented value [39]. The critical current density across a grain boundary oriented perpendicular to the c axis should therefore be close to that of ceramic samples. We calculate $J_c(0) = 3.10^4$ A/cm², $J_c(77) = 1.10^3$ A/cm². As noticed by Dwir [40], these figures are indeed typical of ceramics. Hence, the granular description with a boundary depressed order parameter due to the short ξ_0 gives a quantitative description of the critical current of bulk ceramics. The technologically important and somewhat discouraging value $J_c(77) = 1.10^3$ A/cm² has its origin in fairly fundamental considerations. But it can be improved in several ways:

i) For a given boundary coefficient of transmission, $R_n \propto (1/d^2)$ and $J_c \propto d$. Larger grains can improve the critical current density, if the boundary's quality can be preserved at large grains.

ii) Texturing can considerably improve J_c

iii) Less anisotropic oxides would be more favorable. YBCO is already more favorable than the higher T_c Bi and Tℓ oxides.

Field Dependence of the Critical Current in the Ceramics

The critical current of the ceramics is strongly field dependent, showing typically a decrease of one order of magnitude in fields of a few 100G. The typical screening distance is given by

$$\delta = \frac{H_{c1}}{J_c},$$

with $H_{c1}(77) \approx 100$G, $J_c \approx 100$ A/cm², we get $\delta \approx 1$cm. Hence the field penetration in the intergrain junctions is essentially complete, even at very low applied fields. In this regime, the junctions [7] critical current is proportional to the distance L between vortices,

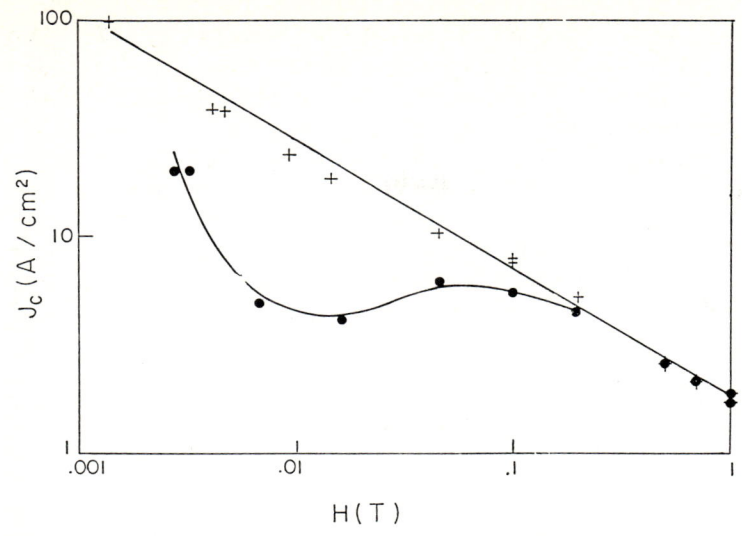

Fig. 11. Field dependence of the critical current in a ceramic sample. (After U. Dai et al., Ref. 41).

$$L = \frac{\phi_0}{2\lambda_L H},$$

times the interference factor $|\sin(\pi d/L)|$ [17]. Because of the distribution of grain sizes this factor will be averaged out, and:

$$J_c = J_c(H=0) \frac{H_0}{H} \quad (13)$$

when $H > H_0$, with $H_0 = (O_0/2\lambda_L d)$, $H_0 \sim 10G$. A typical decrease of the critical current by one order of magnitude in a field of 100G is indeed observed (Fig. 11).

In detail, the behavior of $J_c(H)$ is more complex. In increasing fields, the rapid initial decrease is often followed by a plateau and then by a further and rapid decrease at fields > 1000G. $J_c(H)$ is strongly hysterestic. In decreasing fields, the plateau is not reproduced, the data often fits a power law $J_c(H) \propto H^{-n}$, with n varying from 0.5 to 2 depending on the sample structure. A detailed understanding of this complex behavior has not yet been achieved, but it must involve an enhancement of the local field at grain boundaries due to the grains' diamagnetism in increasing fields ($H < H_{c1}$), and a reduced local field in decreasing fields due to flux trapping in the grains [41].

3.3 Fluctuations-Critical Region

Before discussing in the last section granular effects in single crystals, it will be useful to calculate for the oxides the range of temperature where large thermodynamic fluctuations of the order parameter occur. The width of this so called "critical region" is closely related to the intrinsic pinning energy, and hence to the intrinsic critical currents in the presence of a magnetic field.

We have seen in Section 2 that in the case of a small grain $d \ll \xi_0$ fluctuation effects become large when there is on the average one Cooper pair per grain or less. The equivalent condition in a 3D superconductor is that there will be about one Cooper pair in a volume ξ_0^3. In this limit the mean field approximation that so well describes conventional superconductivity breaks down. Since we have seen that in the high T_c oxides $\xi_0 \approx a$, we do expect them to have a broad critical region.

Following Ginzburg [42], the width of the critical region in a 3D superconductor is given by

$$t_g = \frac{1}{32\pi^2} \left[\frac{k_B}{\Delta C \bar{\xi}^3_{(o)}} \right]^2 , \qquad (14)$$

where ΔC is the (mean field) jump of the heat capacity at T_c, and $\bar{\xi}$ is the appropriate average of the anisotropic coherence length, $\bar{\xi}^3 = \xi_{ab}^2 \xi_c$. Deviations from the mean field heat capacity are then given by

$$\delta C = \Delta C \left[\frac{t}{t_g} \right]^{-1/2} ,$$

where $t = (T_c - T)/(T_c)$. A similar expression holds below T_c.

Using the thermodynamic relation [17]

$$H_c(o) = (\pi \Delta C T_c)^{1/2} , \qquad (15)$$

and the classical expression

$$H_c(o) = \frac{\phi_0}{2\pi\sqrt{2}\lambda\xi} ,$$

Eq. (14) can be written as

$$t_g = 2\pi^4 \left[\frac{k_b T_c \lambda_{ab}^2}{\phi_0^2 \xi_c} \right]^2 . \qquad (16)$$

The advantage of this expression is that t_g is expressed in terms of two fundamental Landau-Ginzburg lengths, both in principle measurable. This is more advantageous than to use Eq. (14) because the measurement of ΔC may be open to question, particularly if fluctuation effects are important, which is precisely the case that we wish to investigate here. The length $(\phi_0^2/k_B T_c) = (293/T_c)$cm, with T_c in Kelvin.

In a conventional superconductor such as NbTi with $T_c = 10K$, $\lambda = 1,000\text{Å}$, $\xi = 100\text{Å}$, one gets $t_g \approx 1.10^{-9}$. For YBCO ($T_c = 93K$, $\lambda_{ab} = 1,500\text{Å}$, $\xi_c = 2\text{Å}$), $t_g \approx 3.10^{-3}$. For the highest T_c Bi and Tℓ compounds ($T_c = 125K$, $\lambda_{ab} = 1,500\text{Å}$, $\xi_c = 0.5\text{Å}$), $t_g = 0.07$. For comparison, $t_g \approx 1.10^{-3}$ in superfluid HeII. Since, as is well known, there is no mean field region in superfluid He, the same is expected for the oxides.

Detailed experiments on the temperature dependence of the penetration depth [43] and on the fluctuation conductivity [44] in YBCO show that the behavior of this oxide is in fact mostly mean field; the beginning of a critical behavior of the heat capacity is seen within about 1K of T_c [45]. Hence, $t_g \lesssim 1.10^{-4}$. In order to reconcile this thermodynamic result with Eq. (16) we must assume that the currently accepted value of λ_{ab} is somewhat overestimated, by at least a factor of 1.5 (taking the upper bound $\xi_c = 5\text{Å}$). Following our discussion of granular effects on the effective value of λ in Section 2, the experimentally measured λ_{ab} may well be enhanced by the presence of internal boundaries such as twins. These twins are typically 1000Å apart, i.e., a distance of the order of λ, and it is quite possible that they enhance λ_{ab} by the required factor. A shorter intrinsic λ_{ab} has important consequences, that have been discussed elsewhere [46].

But the main conclusion to be drawn from Eq.(16) is that for the highest T_c, more strongly anisotropic Bi and Tℓ compounds, <u>no mean field region is to be expected.</u> Even with a smaller λ_{ab}, we still have for these compounds $t_g \gtrsim 1.10^{-3}$. We submit that this is the fundamental reason why no mean field heat capacity jump has been seen in any of these compounds.

At the present time, YBCO appears to be a borderline case, being the highest T_c oxide that still has by and large a 3D mean field behavior. The higher T_c oxides, which have a stronger anisotropy, are basically two dimensional, i.e., the coherence between successive stacks of CuO layers is weak and easily broken.

3.4. Granular Effects in Single Crystals

With the help of Eq. (15), Eq. (14) can also be written as:

$$t_g = \frac{1}{32\pi^2} \left[\frac{\pi k_B T_c}{H_c^2(o)\, \xi^3(o)}\right]^2 . \qquad (17)$$

Within a numerical factor, the expression inside parentheses in the r.h.s. of Eq. (17) is equal to $(k_B T_c/U(0))$, where $U(0)$ is the condensation energy per coherence volume. $U(0)$ is the basic energy scale for core pinning of the vortices, and near T_c the quantity $(U(0)/k_B T_c)$ determines the extent of flux creep (activated motion of vortices [47]). In a conventional type II superconductor, we have obtained $t_g \sim 1.10^{-9}$, $(U(0)/k_B T_c) \sim 100$ to 1000, and flux creep effects are very small. In YBCO, $t_g \lesssim 1.10^{-4}$, $(U(0)/k_B T_c) \sim 1$ to 10, flux creep effects are large. In the higher T_c Bi and $T\ell$ compounds, $t_g \approx 0.01$, $U(0)/k_B T_c) \gtrsim 1$, flux creep effects are "giant" [48]. Core pinning becomes inefficient at preventing flux motion.

When the condensation energy per coherence volume is of the order of $k_B T_c$ the superconductor may be called intrinsically granular by analogy to the case of the small grains discussed in Section 2, for which a critical size d_c was given following the same criterion (condensation energy in the grain of the order of $k_B T_c$).

But one should remember that the reason for the large value of t_g in the oxides is primarily the small value of ξ_c. Fluctuation effects mainly occur through a loss of coherence between adjacent stacks of CuO layers (similarly to the loss of coherence between neighboring grains in the granular case). Such loss of coherence is observed through a loss of rigidity of the 3D vortex lattice [49].

Granularity effects are further enhanced by the presence of planar defects, numerous in the high T_c oxides. The order parameter is depressed at these internal boundaries, as discussed in 3.2. Such depression reduces even further the local value of $U(0)$ at the boundary. In YBCO, a typical energy scale for flux creep along twin boundaries is then 20 meV [50], in good agreement with experiments [51]. Many measurements give direct indications for the presence of internal junctions in single crystals: the microwave absorption experiments already mentioned [31], powder magnetization measurements [52], and very recently the direct observation of vortices trapped inside twin boundaries [53].

4. Conclusions

The early discovery of granular effects in sintered samples of the high T_c oxides have been confirmed by a number of experiments on single grain boundaries and on single crystals - leaving no doubt that granular effects in the oxides are of a fundamental nature.

The short coherence length - and more particularly, the very short coherence length along the c axis - provides the theoretical basis for this granular behavior:

  a) in polycrystalline samples, the critical current across grain boundaries is seriously affected by the depressed order parameter, particularly for boundaries perpendicular to the c axis.

  b) in single crystals, planar defects can behave as junctions with a reduced activation energy for vortex depinning.

  c) in the highest T_c oxides (Bi and $T\ell$ compounds), ξ_c is so short that the width of the 3D critical region is comparable to that in superfluid He. Granularity effects are then truly intrinsic, and result in easy decoupling of neighboring stacks of CuO layers.

Acknowledgements

The review on granular effects in conventional superconductors owes much to a number of former and current members of the superconductivity group at Tel Aviv University: in particular, M. Dayan, M. Gershenson, Y. Shapira, A. Kapitulnik, O. Entin-Wohlman and E. Grunbaum. I am also very thankful to Alex Müller for the attention that he has paid to granular superconductivity, and for constantly pointing out to its relevance to superconductivity in the oxides. This work has been supported in part by the GIF and by the US-Israel Binational Science Foundation, and by the Oren Family Chair of Experimental Solid State Physics.

References

[1] J. G. Bednorz and K. A. Müller, Z. Physik B64, 188 (1986).
[2] B. Abeles, Adv. Phys. 24, 407 (1975); B. Abeles, Applied Solid State Science, ed. R. Wolfe (Academic Press, N.Y.) 1976, p1.
[3] G. Deutscher in Percolation, Localization and Superconductivity, eds. A. B. Goldman and S. A. Wolf (Plenum 1984), p. 95.
[4] G. Deutscher et al., J. of Low Temp. Phys. 10, 231 (1973).
[5] G. Deutscher et al., J. Vac. Sci. Technol. 10, 697 (1973).
[6] Y. Shapira and G. Deutscher, Thin Solid Films, 87, 29 (1982).
[7] G. Deutscher et al., Solid State Commun. 28, 593 (1978).
[8] M. Strongin et al., Phys. Rev. B1, 1078 (1970).
[9] C. E. Jackson et al., Physica 55, 447 (1971).
[10] M. Dayan, J. of Low Temp. Phys. 32 643 (1978).
[11] M. Gershenson, Ph.D. Thesis, Tel Aviv 1975.
[12] D. G. Hinks et al., proceedings of the 1989 Spring MRS Meeting.
[13] A. M. Lamoise et al., J. de Physique Lettres 36, L-271 (1975).
[14] G. Deutscher and D. Dodds, Phys. Rev. B16, 3936 (1977).
[15] R. S. Newrock et al., Physica A157, 220 (1989).
[16] G. Deutscher et al., Phys. Rev. B10, 4598 (1974).
[17] P. G. de Gennes, Superconductivity of Metals and Alloys, Benjamin, New York 1966.
[18] A. Gerber and G. Deutscher, Phys. Rev. B35, 3214 (1987).
[19] G. Deutscher, in Chance and Matter, Eds. J. Souletie, J. Vannimenus and R. Stora, (Elsevier Science Publishers B. V. 1987), p.1.
[20] G. Deutscher et al., Phys. Rev. B22, 4264 (1980).
[21] R. W. Cohen and B. Abeles, Phys. Rev. 168, 444 (1968).
[22] G. Deutscher and O. Entin-Wohlman, J. of Phys. C. Lett. 10, L433 (1977).
[23] G. Deutscher, Revue de Physique Apliquée 8, 127 (1983).

[24] R. L. Filler et al., Phys. Rev. B21, 5031 (1980).
[25] G. Deutscher et al., Phys. Rev. B21, 5041 (1980).
[26] Y. Shapira and G. Deutscher, Phys. Rev. B30, 166 (1984).
[27] K. C. Mui et al., Phys. Rev. B30, 2951 (1984).
[28] Y. Shapira and G. Deutscher, Phys. Rev. B27, 4463 (1983).
[29] M. Kunchur et al., Phys. Rev. Lett. 59, 1232 (1987).
[30] D. B. Haviland et al., Phys. Rev. Lett. 62, 2180 (1980).
[31] K. W. Blazey, to appear in the Proceedings of the 9th General Conf. of the Condensed Matter Division of the E.P.S., Nice France March 6-9, 1989.
[32] K. A. Müller et al., Phys. Rev. Lett. 58, 1143 (1987); A. C. Mota et al., Phys. Rev. B36, 401 (1987).
[33] I. Morgenstern et al., Z. Phys. B69, 33 (1987).
[34] G. Deutscher and K. A. Müller, Phys. Rev. Lett. 59, 1745 (1987).
[35] G. Deutscher, to appear in IBM J. of Res. and Development, May 1989.
[36] R. A. Fisher et al., J. Superconduct. 1, 231 (1988).
[37] D. K. Finnemore et al. in "Advances in Superconductivity", Eds. K. Kitzawa and T. Ishiguro, Springer-Verlag (Tokyo), 1989, p. 389.
[38] For a review, see A. Barone, Physica C153-155, 1712 (1988).
[39] J. Mannhart et al., Phys. Rev., Lett. 61, 2476 (1988).
[40] B. Dwir, preprint.
[41] U. Dai et al, preprint.
[42] V. L. Ginzburg, Physica C153-155, 1619 (1988).
[43] L. Krusin-Elbaum et al., Phys. Rev. Lett. 62, 217 (1989).
[44] M. Hikita and M. Suzuki, Phys. Rev. B39, 4756 (1988).
[45] G. Deutscher, to appear in Physica Scripta, Proceedings of the 9th General Conf. of the Condensed Matter Division of the E.P.S., Nice March 6-9, 1989.
[46] G. Deutscher, to appear in J. de Physique.
[47] P. W. Anderson, Phys. Rev. Lett. 9, 309 (1962).
[48] Y. Yeshurun and A. P. Malozemoff, Phys. Rev. Lett. 60, 2202 (1988).
[49] P. L. Gammel et al., Phys. Rev. Lett. 61, 1666 (1988).
[50] G. Deutscher, J. of Less Common Metals, 150, 1, (1989).
[51] C. W. Hagen and R. Griessen, Phys. Rev. Lett. 62, 2857 (1989).
[52] M. Daeumling et al., to be published in Cryogenics; H. Küpfer, et al., to be published in Cryogenics.
[53] G. J. Dolan et al., Phys. Rev. Lett. 62, 827 (1989).

Critical Currents in Single-Crystal and Bicrystal Films

P. Chaudhari[1], D. Dimos[1], and J. Mannhart[2]

[1]IBM Research Division, T.J. Watson Research Center,
 Yorktown Heights, NY 10598, USA
[2]Physikalisches Institut 2, Auf der Morgenstelle 14,
 D-7400 Tübingen, Fed. Rep. of Germany

Abstract. Possible mechanisms for vortex pinning and for the weak-link behavior of grain boundaries in high-T_c superconductors are discussed.

The thermodynamic critical current density for single crystals of the high temperature superconductors is estimated to be between 10^8 and 10^9 A/cm^2 at 4.2 K. These estimates are obtained by balancing the condensation energy against the kinetic energy acquired during the flow of a supercurrent. In type II superconductors, such as the high-T_c materials, the critical current density is determined not by the thermodynamic limit but by a lower limit set by the motion of flux lines or vortices. The motion of these lines is limited by dissipative processes distributed uniformly along the line and by localized spatial perturbations of the superconducting order parameter; the latter are called pinning centers. In crystalline materials, pinning centers are generally obtained by introducing defects, such as grain boundaries and nonsuperconducting phases, into the solid. However, vortices can also be pinned by periodic variations of the order parameter that may be connected with the intrinsic crystal structure of the superconductor.

For polycrystalline samples of the high-T_c superconductors, the critical current densities are substantially lower than their values in single crystals. There are two reasons for this behavior. First, the high-temperature superconductors have an anisotropic crystal structure, which leads to a substantial anisotropy in the critical current density. In particular, the critical current density parallel to the Cu-O planes is about an order of magnitude larger than in the perpendicular direction for YBa$_2$Cu$_3$O$_7$ [1]. Second, the critical current densities are strongly limited by poor superconducting coupling across grain boundaries.

In this contribution, the mechanisms which could limit the critical current density in single crystal and polycrystalline films are examined. In particular, we shall be concerned with describing the behavior of these materials at low temperatures and low magnetic fields. The papers by Deutscher and by Mannhart provide additional details on the temperature and magnetic field dependence.

1. Single Crystal Films

In single crystal films of YBa$_2$Cu$_3$O$_7$ prepared by laser ablation, critical current densities of about 6×10^7 (4.2 K), 6×10^6 (77 K), and 1.8×10^6 A/cm^2 (88 K) have been obtained [2]. In this section, we are concerned primarily with the pinning mechanism responsible for the very high J_c values obtained in these epitaxial films. Table 1 summarizes our current understanding of the role of possible pinning sources in these materials.

Table 1. Pinning sources and their role in YBa$_2$Cu$_3$O$_{7-\delta}$

Surface roughness	Minimal
Anisotropy	None
Precipitates	None
Twins	Not dominant
Dislocations	Not dominant
Point defects	Not known
Intrinsic structure	Not dominant
Self-trapping	Not known

Expitaxial films both with smooth (specular) and with rough surfaces have been measured and no connection has been observed between the surface quality and critical current density. If anything, the laser deposited films, which have specular surfaces, yield the highest critical current density. In some of the early results on epitaxial films, it was observed that the c-axis was not always perpendicular to the substrate, but in some regions lay in the plane of the film along one of two possible orthogonal directions. It was, thus, conceivable that the vortices could be trapped by this crystallographic arrangement. However, recent experiments have shown that the critical current is optimal in films which are single crystals over the entire substrate. Hence, anisotropy induced pinning cannot be playing a dominant role.

In some of the early films, second phase particles were also frequently observed. As our control of composition has improved, the concentration of second phase particles has decreased and the critical currents have steadily increased. Therefore, it appears that precipitates do not play a significant role in pinning vortices in epitaxial films.

Epitaxial films are also highly twinned. *Deutscher* and *Müller* [3] have pointed out that the short coherence length results in a depression of the order parameter near a twin boundary. We might, therefore, expect twin boundaries to act as pinning sites. Indeed decoration experiments [4] clearly show vortex arrays outlining twin boundaries. However, the same experiments also provide evidence that most vortices are pinned at sites that could not be associated with an identifiable crystalline defect, such as twin boundaries.

Dislocation lines strongly perturb the lattice in a number of ways and, hence, are expected to depress the order parameter locally. If the vortex is parallel to the dislocation line, we might expect the vortex to be strongly pinned and the critical current associated with depinning of the vortex to be high. A rough estimate of the depinning current density suggests it is also proportional to H_c/λ and about half the Ginzburg-Landau value [5]. This estimate is based on the assumption that the order parameter is zero at the core of the dislocation. If it is assumed that dislocation lines are perpendicular to the plane of the film, we can estimate the dislocation density required to give rise to the observed critical current density. For an applied field of 2000 Oe, the spacing between vortices is approximately 1100 Å. If the dislocation spacing is on the average xÅ, the effectiveness of pinning is reduced by $(1100/x)^2$ due to vortex lattice effects. Equating $(1100/x)^2 (5 \times 10^8)/2$ to a critical current density of 10^7 A/cm^2 yields $x = 5500$ Å. Although dislocations have been observed in epitaxial films, their spacing is generally much greater than the half micrometer estimated here. Also, the geometry we have assumed here is the most favorable for dislocation induced pinning. If the vortex and dislocation lines are not parallel, pinning is reduced. We therefore conclude that, at the present level of experimental observations, dislocations do not appear to be the dominant pinning source.

Point defects are generally not viewed as strong pinning points in conventional superconductors, the argument being that the coherence length is much larger than the defect so that the local change in condensation energy is small. In contrast, the high-temperature superconductors have very short coherence lengths: in YBa$_2$Cu$_3$O$_7$ the coherence lengths are estimated to be 3 Å along the c-axis and about 15 Å in the Cu-O planes. Furthermore, these materials appear to be sensitive to deviations in composition, particularly oxygen. Combining these two facts, we might expect that point defects are capable of providing pinning [6]. However, despite its appeal, there is little experimental evidence to either support or reject this idea. It has also been suggested that the layered structure of the high-temperature superconductors can provide intrinsic pinning of vortices. In this case, vortices parallel to the Cu-O planes can be trapped between these sheets because of the periodic modulation of the order parameter [7, 8]. Unfortunately, this model can only work for one unique direction of vortex motion; hence, it cannot explain the existing data. In the talks by Beyers and Raveau at this conference, it was shown that many defects can be present in these materials. In particular, we have heard that clusters of point defects can give rise to ordered rows of defects. If such defects are present at random or have a spacing comparable to or larger than the coherence length, we might expect them to serve as pinning sites. It is thus possible that a combination of the intrinsic structural effects and the presence of defects may give rise to the required pinning. Obviously, more experimental information is needed to sort this out.

Finally, we consider self-trapping of vortices. By this we mean that the vortex is trapped in a potential it generates. This is a temporal effect in the sense that the arrangement in the core of a vortex changes with time so as to minimize the difference in free energies between the superconducting and normal states. Let us consider a specific model to illustrate the mechanism. Suppose that the concentration of ions determines the order parameter. If an ion can migrate rapidly enough at low temperatures, either due to thermal activation or by a tunneling process, we might expect the ion concentration to change in the core of a vortex. If the kinetics permit this ionic motion, the energy of the vortex core is decreased and it is locally trapped. Although activated ionic motion is strongly suppressed at low temperatures, the high-T_c compounds belong to the general class of oxides in which ion mobility is high. In fact, they were initially investigated by Raveau and his colleagues for their potential as ion transport media rather than as superconductors. There are two factors that favor the possibility of self-trapping of vortices in the high-temperature superconductors. First, local charge perturbations are not screened out as rapidly in the high-T_c superconductors as they would be in conventional metallic superconductors, because of the much lower carrier concentrations in the oxides. Second, the short coherence length enables local fluctuations to be more effective. We do not have a quantitative estimate of the effectiveness of the self-trapping mechanism, which is why we have rated it "not known" in Table 1.

2. Critical Currents Across Grain Boundaries

We have measured the voltage, temperature, magnetic field, and orientation dependence of critical currents in bicrystal epitaxial films. All of these measurements have been summarized in recent publications and we shall not discuss them here [9, 10]. In this note we shall explore possible mechanisms for the orientation dependence of the critical current density across a grain boundary. For convenience we reproduce the data presented by *Dimos* et al. [10] in Fig. 1. This figure shows the ratio of the grain boundary critical current density to the average critical current density of the two adjacent grains as a function of the total misorientation angle. The critical current density of the grain boundaries varies between 10^4 and 4×10^6 A/cm^2 at 4.2 K, the normal state resistivity lies in the range of $10^{-8} \Omega$ cm^2 [10]. We note that the grain boundary critical current for all three sets of rotations follows the same general trend. The three cases correspond to rotations about the c-axis (tilt boundary), and about the other two axes, one lying in the plane of the boundary (tilt), and the other perpendicular to it (twist). In Table 2 we summarize the role of some of the possible explanations of the grain-boundary-limited critical current. Although we can eliminate a number of possibilities, there are still a few left to be sorted out experimentally.

Table 2. Possible explanations of grain-boundary-limited critical current density

Pair symmetry: s, d, p wave pairing	Not controlling
Anisotropic order parameter	Not controlling
Second phase	Not observed
Antiferromagnetic order	Not known
Grain boundary scattering	Not known
Composition	Not known

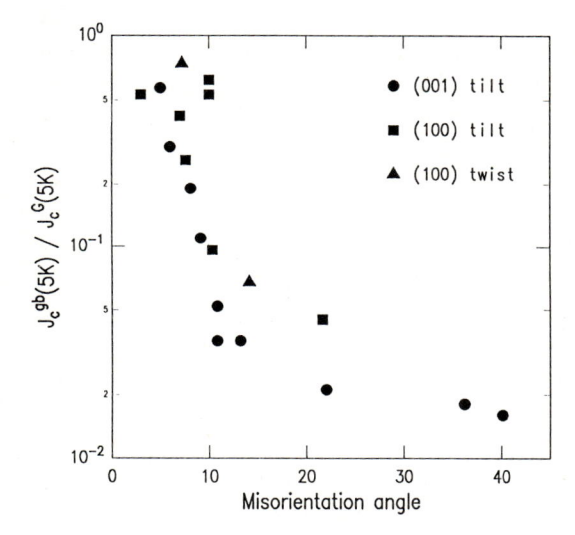

Fig. 1. Ratio of the grain boundary critical current density to the average value in the adjacent grains at 4.2 K as a function of misorientation angle. The different symbols distinguish the primary component of the misorientation with respect to the three sample geometries. (From [10])

The symmetry of the order parameter is not known unequivocally in the high temperature superconductors. Tunneling data as well as the recent experiments described by *Gough* et al. [11] suggest that the order parameter has s-wave symmetry. If the order parameter had non-s-wave symmetry, coupling across the grain boundary would almost certainly have an orientation dependence. However, if we examine in detail the shape of the orientation dependence predicted from non-s-wave pairing for the three orientations and compare them with the experimental data, we can conclude that this is not the dominant mechanism [10]. The anisotropic nature of the order parameter can also give rise to currents at interfaces [12, 13]. In particular, *Kogan* [13] has proposed a possible scenario to explain the experimental data we obtained. In this mode, currents flowing across an interface generate currents parallel to the interface due to anisotropy in the effective mass of the carriers. These currents generate a magnetic field, which breaks up into flux quanta; these quanta are poorly pinned and move in response to the current across the interface. Although interface currents of the type proposed by *Kogan* et al. must exist, the predictions of this model are not in agreement with the results for all of the boundary orientations and we conclude that the model does not describe the controlling process.

It has frequently been shown that grain boundaries in these materials can contain both discrete and continuous second phases. However, since we have not seen any evidence of a second phase in the boundaries that have been examined by electron microscopy, we conclude that grain boundary phases are not responsible for the properties of the boundaries.

The next three mechanisms are related to the structure of a grain boundary, which we will discuss before describing the specific models. Electron microscope examination of bicrystal films by us and of ceramic samples by a large number of groups show that these boundaries can be described by standard dislocation theory. As the misorientation angle increases, the dislocation density increases. Qualitatively, the variation of critical current density with misorientation angle, θ, is similar to the angular dependence of the dislocation spacing, d, which is given by $d = b/\theta$, where b is the Burgers vector. Attractive as this model seems, it is not sufficient since it does not explain the magnitude of the critical current depression. This difficulty can be appreciated readily by comparing the measured critical current density with the value expected from a simple grain boundary. Taking the width of a boundary to be a couple of unit cell dimensions and using the bulk resistivity, we can estimate the critical current across the boundary using the Ambegaokar-Baratoff equation. This estimate yields a value that is two to three orders of magnitude larger than the measured one. Alternatively, we can get the resistance of the grain boundary directly from I-V measurements. If this value is substituted into the Ambegaokar-Baratoff equation, the predicted critical current density is close to but slightly larger than the experimental value. However, this difference can be explained if the gap is depressed at the boundary.

The quest to understand the low grain boundary critical current density can then be rephrased into understanding the origin of the high grain boundary resistance. A possible explanation is related to the theoretical models which use antiferromagnetic ordering as the starting point for explaining high-temperature superconductivity. *Nabarro* [14] has suggested that this may have important implications in critical current transport across a grain boundary. We shall use a particular model to develop this theme further. It can readily be shown that a dislocation in an antiferromagnetic solid can result in local ferromagnetic order. In the case of the Cu-O planes, neighboring Cu atoms are antiferromagnetically coupled. Since a dislocation has a Burgers vector which goes from one Cu atom to a neighboring Cu atom, its presence results in a plane across which neighboring Cu atoms are ferromagnetically ordered. This implies that, apart from strain fields, dislocations can have a strong influence on the local pairing. Long-range antiferromagnetism is gradually replaced by short-range order as the oxygen content, and thus the carrier concentration, is increased. However, the topological constraint on magnetic order around the core of a dislocation would still hold. In addition, a charge can migrate and be localized at the dislocation core in order to minimize the change of energy associated with antiferromagnetic-to-ferromagnetic ordering. The charge will considerably enhance scattering of carriers, thereby increasing the local resistivity. Enhanced scattering associated with charge imbalance can also be associated with ionic bonds in these materials. Unlike insulating oxides, the charge would be screened over short distances, but would still give rise to enhanced resistivity.

The last possibility we wish to consider is connected with local deviations in stoichiometry at the grain boundary. We shall consider the role of oxygen in order to be specific, even though it is clear that all of the constituents can play a role. What if the oxygen concentration were lower at the boundary? This would have the effect of lowering the carrier concentration and increasing scattering. Clearly more experiments are needed to understand which of these mechanisms is most important.

In this note, we have summarized our current understanding of the possible mechanisms responsible for pinning of vortices in single-crystal films and for the weak-link behavior associated with grain boundaries. Although we can rule out many possibilities in both cases, a few alternative mechanisms remain. A complete understanding of these phenomena will require additional experimental and theoretical work.

References

1 T.R. Dinger, T.K. Worthington, W.J. Gallagher, R.L. Sandstrom: Phys. Rev. Lett. **58**, 2687 (1987)
2 G. Koren, A. Gupta, T.R. McGuire: Appl. Phys. Lett. **54**, 1054 (1989)
3 G. Deutscher, K.A. Müller: Phys. Rev. Lett. **59**, 1745 (1987)
4 G.J. Dolan, G.V. Chandrasekhar, T.R. Dinger, C. Feild, F. Holtzberg: Phys. Rev. Lett. **62**, 827 (1989)
5 M. Tinkham: *Introduction to Superconductivity* (McGraw-Hill, New York 1975)
6 P. Chaudhari: Jpn. J. Appl. Phys. **26**, 2023 (1987)
7 K.F. Gray, R.T. Kampwirth, J.M. Murduck, D.W. Capone II: Physica C **152**, 445 (1988)
8 M. Tachiki, S. Takahashi: Preprint, Tohoku University
9 P. Chaudhari, D. Dimos, J. Mannhart: IBM J. Res. Dev. **33**, 299 (1989)
10 D. Dimos, P. Chaudhari, J. Mannhart: Submitted for publication
11 M.N. Keene, T.J. Jackson, C.E. Gough: Nature **340**, 210 (1989)
12 A. Baratoff, F. Guinea: Preprint
13 V. Kogan: Phys. Rev. Lett. **62**, 3001 (1989)
14 F.R.N. Nabarro: Preprint

What Limits the Critical Current Density in High-T_c Superconductors?

J. Mannhart

Physikalisches Institut, Lehrstuhl für Experimentalphysik II,
Universität Tübingen, Auf der Morgenstelle 14,
D-7400 Tübingen, Fed. Rep. of Germany

Present address: IBM Research Division, Zurich Research Laboratory, Säumerstr. 4, CH-8803 Rüschlikon, Switzerland

Abstract. Whereas single crystals and epitaxial films of high-T_c superconductors support supercurrent densities well above 10^6 A/cm^2 at 77 K, the critical current densities of the technically far more interesting polycrystalline high-T_c superconductors are lower by three orders of magnitude. In this article, after a discussion of the general mechanisms which limit supercurrents, experiments will be described which allow one to study directly supercurrent limiting phenomena in high-T_c superconductors.

1. Introduction

Presently, the main challenge on the way to exciting applications of high-T_c superconductivity is the enhancement of the critical current density of the high-T_c copper oxides. While large-scale applications of high-T_c superconductors, such as resistanceless power transmission lines or computer interconnects, require critical current densities of 10^5–10^6 A/cm^2 at 77 K [1], the critical current density J_c of the best polycrystalline high-T_c superconductors lies only in the range 10^3–10^4 A/cm^2 at liquid-nitrogen temperature. In contrast, epitaxially grown films of YBa$_2$Cu$_3$O$_7$ show a considerably higher J_c: here values well above 10^6 A/cm^2 at 77 K have been reported. However, the small size of the epitaxial films hinders their use on a large scale. Consequently, much effort has been devoted to understanding the mechanisms which limit critical current densities in polycrystalline high-T_c superconductors and to finding ways of improving them [2]. In an attempt to give an introduction to this field, this paper is divided into three parts. First, the fundamental mechanisms which control J_c in superconductors will be discussed, then experiments will be presented which allow the identification of the mechanisms that control J_c in polycrystalline YBa$_2$Cu$_3$O$_7$. Finally, various approaches to improving the critical current density of high-T_c superconductors will be addressed.

2. Mechanisms that Limit Supercurrents

The critical current density of a superconductor is defined as the highest current density the superconductor can support without resistance; in other words, above the critical current density the current generates a voltage. Generally speaking, a voltage can be generated in a superconductor by two mechanisms. First, the order parameter of the superconductor can be depressed by an electric current, so that parts of the superconductor become resistive because the Cooper-pair density vanishes. Second, according to Faraday's law, a voltage is induced in a superconductor by moving magnetic flux structures. While in type I superconductors in the intermediate state [3] the voltage can be induced by moving flux tubes, in type II superconductors the voltage is generated by moving flux lines. In inhomogeneous superconductors, for example in tunnel junctions, the voltage may be generated by Josephson vortices, which start to move as soon as the critical current density is exceeded.

2.1 Depairing. If the velocity v of the Cooper pairs exceeds a critical value [4]

$$v_{cr} = \frac{\Delta(0)}{\hbar k_F} \quad (T = 0\,\mathrm{K}), \tag{1}$$

which corresponds to a critical current density

$$\tilde{J}_c^{dp} = n^*(v) e^* v_{cr}, \tag{2}$$

the gap goes to zero for some quasiparticle states. In (1), Δ is the gap energy and $\hbar k_F$ is the momentum of the quasiparticles at the Fermi level. In (2), $n^*(v)$ and e^* are the density and the charge of the Cooper pairs, respectively. Therefore, above v_{cr}, Cooper pairs quickly break up and the maximum supercurrent density J_c^{dp} is only slightly higher than $\tilde{J}_c^{dp}$. The depairing critical current density J_c^{dp} can be expressed as a simple function of the coherence length ξ and the effective quasiparticle mass m:

$$J_c^{dp} = \frac{n^* e \hbar}{\pi m \xi}. \tag{3}$$

A similar expression can be inferred from the London theory by calculating the current density at which the kinetic energy of the charge carriers equals the condensation energy, yielding

$$J_c^{dp} = \frac{B_c}{\mu_0 \lambda}, \tag{4}$$

where B_c and λ are the critical field and the London penetration depth, respectively.

Critical current densities limited by depairing have been found, for example, in narrow bridges patterned into films of tin and lead [5]. The critical current densities of such samples lie in the range of 10^7 A/cm^2 at 2 K and agree well with critical current densities calculated on the basis of (4).

For high-T_c superconductors, the depairing critical current density is estimated to be high [6] because the small coherence length ξ of the CuO-based superconductors compensates for their small carrier concentration n^* [see (3)]. Equation (4) yields for YBa$_2$Cu$_3$O$_7$ at 4.2 K a depairing critical current density of 10^9 A/cm^2 with $B_c \simeq 1$ T and $\lambda \simeq 1400$ Å [6].

2.2 Depinning. Magnetic fields with a flux density above H_{c1} enter a type-II superconductor in the form of flux lines which carry single flux quanta ϕ_0. The fields may be supplied by external sources or by an electric current which flows through the superconductor (Fig. 1). A current with a density J exerts a Lorentz force of approximately

$$F_L = J\phi_0 d \tag{5}$$

on the flux lines, where d is the thickness of the sample. If the flux lines move in response to this force with a velocity v, energy is dissipated in the superconductor because of relaxation effects and the acceleration of quasiparticles. According to Faraday's law, moving flux lines induce an electric field $\boldsymbol{E} = \boldsymbol{B} \times \boldsymbol{v}$ in the sample. This field is oriented parallel to J, so that the superconductor becomes resistive. Therefore a type-II superconductor can only support a substantial supercurrent if the flux lines are held in place by pinning forces F_p that prevent flux line movement. In this case, the critical current density is given as the current density at which the Lorentz force equals the pinning force (critical state).

Flux line pinning is a complex subject [3]. First, flux lines may be pinned by direct interactions with the superconducting phase. Such elementary pinning forces may arise, for example, if the core of a flux line is located at a defect

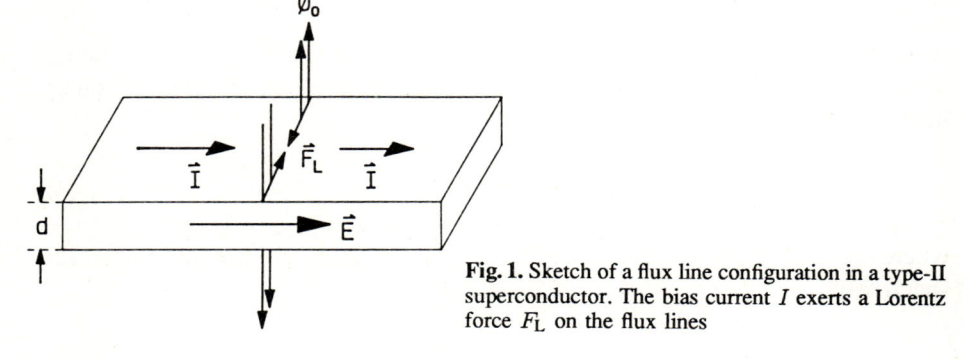

Fig. 1. Sketch of a flux line configuration in a type-II superconductor. The bias current I exerts a Lorentz force F_L on the flux lines

which has a reduced condensation energy density. This is an energetically favorable configuration, because the flux line core has to provide only for the reduced condensation energy. Because this interaction is most effective if the pinning site is of a size comparable to the coherence length ξ,

$$F_p \simeq \frac{U_p}{\xi} \simeq \frac{\xi^2 B_c^2}{2\mu_0} \tag{6}$$

is a crude measure of the pinning strength. Here, U_p is the pinning energy. Second, flux lines are held in place by interactions with the rest of the flux line lattice [3, 7]. In this case, the shear properties of the flux line lattice and the force with which the lattice is pinned determine J_c.

Besides the Lorentz force, the thermal energy kT may help to overcome the pinning energy U_p [8]. The resulting hopping process, where flux lines jump from pinning site to pinning site, is known as flux creep. A flux line which in the presence of a current density J is pinned by an effective pinning force

$$U_p^{\text{eff}}(T) = U_p(T)\left(1 - \frac{J}{J_{c0}(T)}\right) \tag{7}$$

will escape from the pinning center with a hopping rate

$$R_{\text{hop}}(T) = f \exp\left(-\frac{U_p^{\text{eff}}(T)}{kT}\right). \tag{8}$$

In (7) J_{c0} is the critical current density of the sample without thermally activated depinning, and in (8) f is the frequency with which the flux lines attempt to leave the pinning site. Following the idea that the flux lines are fixed in space whereas the atomic lattice vibrates [9], f is usually approximated by phonon frequencies of the superconductor. According to (8), for $T > 0$ a fraction of the flux quanta is not pinned, therefore, strictly speaking, there is no supercurrent in thermally activated superconductors. (However, the resistance of the sample becomes exceedingly small at low T.) In this case, J_c has to be defined by an electric field criterion. Defining J_c as the current density that produces an electric field E_c, (7) and (8) yield for high J_c

$$J_c(T) = J_{c0}(T)\left(1 - \frac{kT}{U_p^{\text{eff}}(T)} \ln \frac{Baf}{E_c}\right), \tag{9}$$

where a is the average hopping distance of the flux lines. Equation (9) should be used with care, because it is based on a number of simplifications. For example, in a more complete treatment of flux creep, interactions between flux lines, which may lead to a correlated movement of flux line bundles, and a spread of pinning energies should be considered.

Equation (9) predicts a high rate of flux creep in high-T_c superconductors, because, due to the small coherence length of these materials, the flux lines are expected to be pinned with moderate pinning energies at small defects. In combination with the high temperature T at which these superconductors can be operated, the moderate pinning leads to a high rate of thermally activated hopping (8) [10]. Indeed, flux creep has been observed at a rate unknown from low-T_c superconductors [11] in single crystals of YBaCuO and BiSrCaCuO [12].

2.3 Decoupling. The critical current density of a superconductor can be reduced drastically by thin layers of insulating material, through which the supercurrent has to tunnel. As shown by *Josephson* [13], if two superconductors are coupled by a thin insulator, the energy of the system is reduced by the so-called coupling energy. The density of the coupling energy is given by

$$E = \frac{hJ_c}{2e} \cos\gamma , \tag{10}$$

where γ is the difference of the phases of the two superconductors. Furthermore, a supercurrent tunnels through the barrier that has a density

$$J = J_c \sin\gamma . \tag{11}$$

Ambegaokar and *Baratoff* [14] have calculated $J_c(T)$ to be

$$J_c(T) = \frac{\pi \Delta_i(T)}{2eR_N} \tanh\left(\frac{\Delta_i(T)}{2kT}\right) , \tag{12}$$

where R_N is the resistance of the tunnel barrier for barrier voltages which are large compared to $2\Delta_i$, and Δ_i is the gap energy at the interface between superconductor and insulator (Fig. 2). The Josephson relation (11) implies that on increasing the phase difference γ from 0 to $\pi/2$ the supercurrent density increases to J_c. At J_c the coupling energy vanishes. For even higher current densities, the phases of the superconductors decouple and a voltage develops across the barrier.

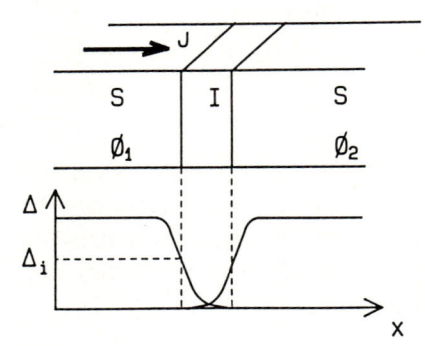

Fig. 2. Sketch of an SIS-type weak link. *Bottom*: Illustration of the depression of the order parameter Δ at the weak links

Because the value Δ_i of the gap at the interference between superconductor and insulator enters the Ambegaokar-Baratoff equation (12), J_c is indirectly a function of the coherence length ξ. While in superconductors with large ξ the gap Δ_i corresponds to the value of the gap deep inside the superconductor, in superconductors where the coherence length ξ is comparable to the size of a unit cell, Δ_i can be depressed by defects and chemical variations of the superconductor at the interface. Furthermore, as shown by *Deutscher* and *Müller* [15], even for perfect superconductors Δ_i is lowered if ξ is small, because of restricting boundary conditions on $\Delta(r)$ at the interface. Hence, for superconductors with a small coherence length, J_c will be lowered further. This implies that in high-T_c superconductors weak links effectively reduce J_c.

Critical current densities which are controlled by the Josephson effect are highly sensitive to magnetic fields, in contrast to critical current densities which are limited by depinning or by depairing. Magnetic fields of a few gauss are generally strong enough to insert a flux quantum into the tunnel barrier and to shift the phase difference along the tunnel barrier, so that J_c is reduced markedly [16].

Weak links, in which the insulating layer is replaced by a normal metal or by a weak superconductor (SNS- or SS'S-weak links), may also have severe effects on J_c. In these cases, the normal state resistance of the links is small, so that the coupling energy can be large. (This is the case if $\xi_n \gtrsim d_n$, where ξ_n and d_n are the coherence length and the thickness of the normal layer.) Hence, critical current densities of SNS- and SS'S-type weak links may be limited by depinning at the barrier layer or by depairing in the weak link region. For a more detailed discussion of these types of weak links, the reader is referred to [17].

3. Experimental Investigation of Critical Current Limiting Mechanisms in YBa$_2$Cu$_3$O$_7$

A variety of effects may be responsible for the low critical current density of polycrystalline high-T_c superconductors compared to typical values of single-crystalline materials. Certainly, in polycrystals J_c is lowered by voids, cracks and second phases which are formed during the fabrication process. Furthermore, in polycrystalline high-T_c superconductors, the grains are usually randomly oriented, so that the current has to flow in some grains along the c-direction, which has a relatively low J_c. Finally, grain boundaries have been suspected to be responsible for the low J_c, although in some low-T_c superconductors grain boundaries enhance J_c, because they act as pinning centers [18]. In the following, experiments will be presented which have been designed to identify the mechanisms that limit J_c in YBa$_2$Cu$_3$O$_7$ and to clarify which role the grain boundaries play in this context.

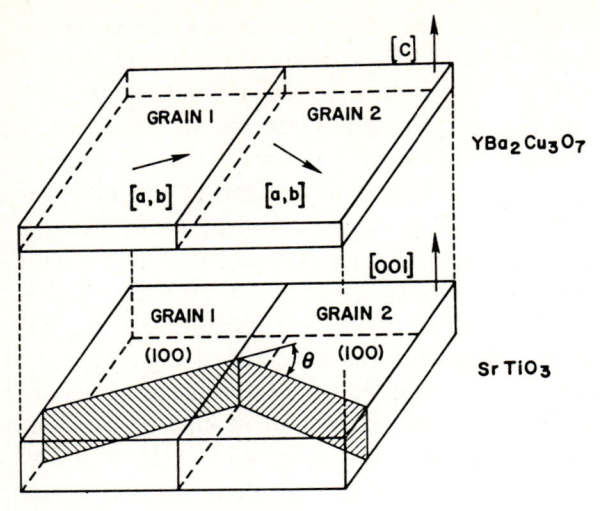

Fig. 3. Schematic diagram of the bicrystalline samples. The c-axis oriented bicrystalline YBaCuO-films were grown epitaxially on SrTiO$_3$ bicrystals. (After [5])

YBa$_2$Cu$_3$O$_7$ has been chosen for these studies, because it is well characterized in general and high quality films of this material can be prepared. As will be discussed later, the behavior of the BiSrCaCuO and TlBaCaCuO systems resembles that of YBa$_2$Cu$_3$O$_7$.

To measure directly the transport properties of individual grain boundaries and of their adjacent grains, c-axis oriented bicrystalline films of YBa$_2$Cu$_3$O$_7$ were grown epitaxially on SrTiO$_3$ bicrystals, as shown in Fig. 3 [19]. This technique allowed the fabrication of millimeter-sized films with single, well-defined grain boundaries. With an excimer-laser ablation technique [20], three superconducting bridges were patterned into each film, as shown in Fig. 4. Two of the bridges were located within the grains and one bridge straddled

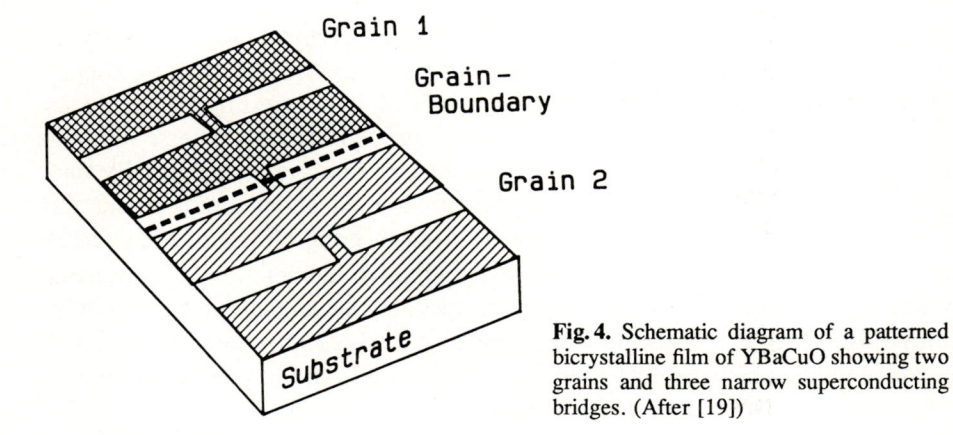

Fig. 4. Schematic diagram of a patterned bicrystalline film of YBaCuO showing two grains and three narrow superconducting bridges. (After [19])

the grain boundary. Typically, the bridges were 10–20 μm wide and 40 μm long. This configuration allowed direct comparison of the transport properties of the grain boundary with the behavior of the adjoining grains.

In the samples investigated, the critical current density of the bridges which straddled the grain boundary was significantly lower than the critical current density of the lines located within the grains [19, 21]. Furthermore, as discussed in the article by P. Chaudhari in these proceedings, the critical current density of the line across the grain boundary has been found to be closely coupled to the structure of the grain boundary [19]. Therefore, these measurements clearly demonstrate that grain boundaries limit the critical current density in polycrystalline $YBa_2Cu_3O_7$ samples.

To gain further insight into supercurrent transport in high-T_c materials, the transport properties of the bicrystalline films were investigated with high spatial resolution using Low Temperature Scanning Electron Microscopy (LT-SEM) [23]. This technique allows one to study transport phenomena in superconductors, such as the spatial distribution of the critical current density, with a spatial resolution of about 1 μm. The principle of LTSEM is quite simple [23]. The specimen film is mounted into the sample chamber of a scanning electron microscope and is cooled to the temperature range where it becomes superconducting. Simultaneously, an electric bias current is applied to the film and the electron beam induced change $\delta V(x, y)$ of the voltage V along the sample is recorded as a function of the coordinate point (x, y) of the electron beam focus on the film surface (Fig. 5). The electron beam (energy and current

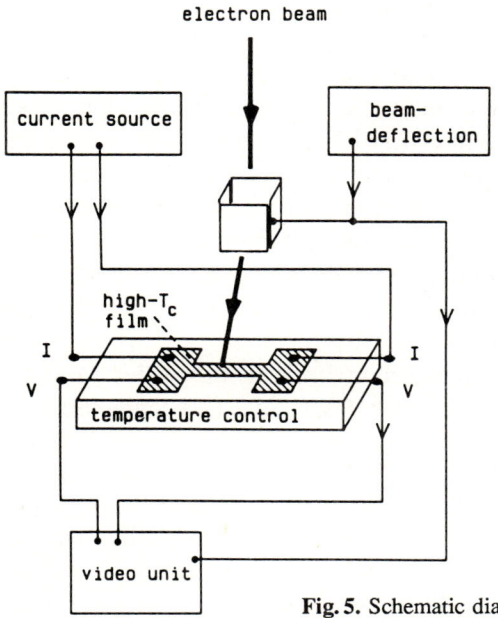

Fig. 5. Schematic diagram of the sample arrangement in LTSEM

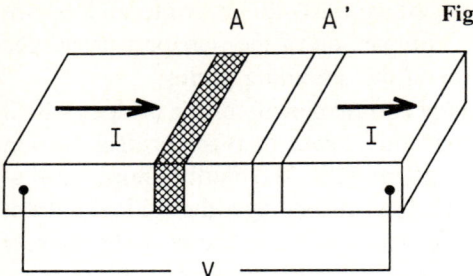

Fig. 6. Illustration of signal generation in LTSEM

typically 10–30 keV and 10-100 pA, respectively) heats the film locally. For a beam power of 10^{-7} W, a temperature rise of a fraction of a kelvin is observed in an area of about 1 μm^2 around the beam focus. To employ this local heating for imaging the supercurrent transport, the sample is biased with a current that is higher than its critical current, so that at least one cross section A of the film is in the resistive state and a voltage V is obtained along the sample (Fig. 6). If the electron beam is focused on a point in a cross section A' different from A, it heats the superconducting film. However, as long as the critical current at A' is higher than the bias current, no change δV of V is measured. On the other hand, if the electron beam is focused on a point located on A, it increases the resistance of A. Because the sample is current biased, this increase of resistance leads to an increase δV of V. Therefore, by mapping δV as a function of the position of the electron beam focus, the resistive areas of a superconductor can be imaged.

This technique has been used to study the distribution of critical current densities in single crystalline, bicrystalline and polycrystalline YBa$_2$Cu$_3$O$_7$ films [23–25]. As shown in Fig. 7 in the case of a bicrystalline film, it has been found that, within the spatial resolution of LTSEM, the critical current density is indeed limited by the grain boundary. Presumably, part of the structure of δV shown in Fig. 7 directly reflects inhomogeneities of the grain boundary, so that either the current density or the normal state resistance varies along the boundary plane. Some of the inhomogeneities which can be seen in the voltage response signal δV in Fig. 7 can be traced back to precipitates on the surface of the specimen.

Further information about the supercurrent limiting properties of grain boundaries can be obtained by measuring their $I(V)$ and $I_c(B)$ characteristics, where I_c is the critical current of the grain boundary. The $I(V)$ characteristic (Fig. 8a) is typical for strongly coupled weak links (e.g. SNS-type weak links), and the $I_c(B)$ characteristic (Fig. 9) corresponds to the behavior of inhomogeneous Josephson junctions. Backed by the observation of microwave-induced steps in the $I(V)$ characteristics and by studies of $I_c(T)$, we can conclude [26] that the grain boundaries act like SNS-type weak links. However, it re-

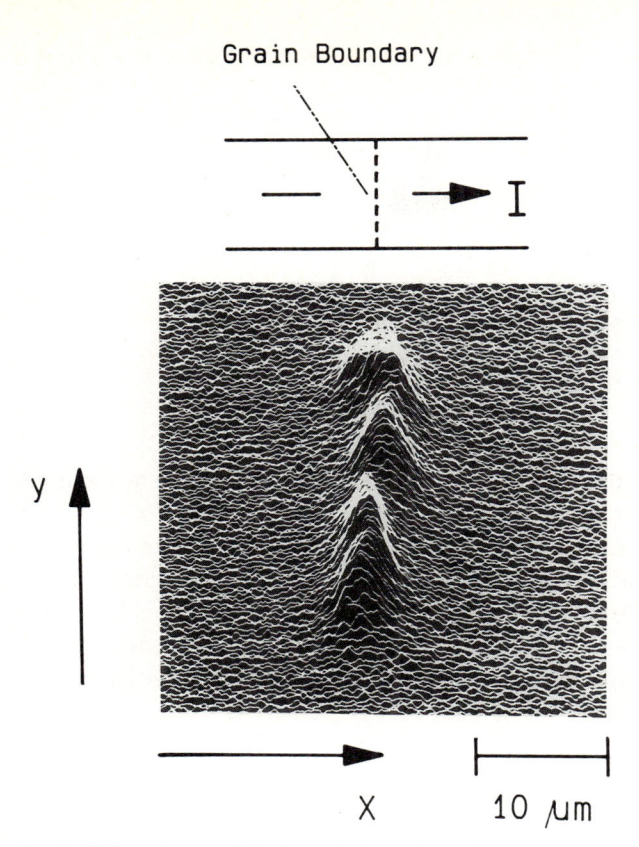

Fig. 7. Voltage image δV of a grain boundary taken with a bias current of 39 mA at 14 K. (After [25])

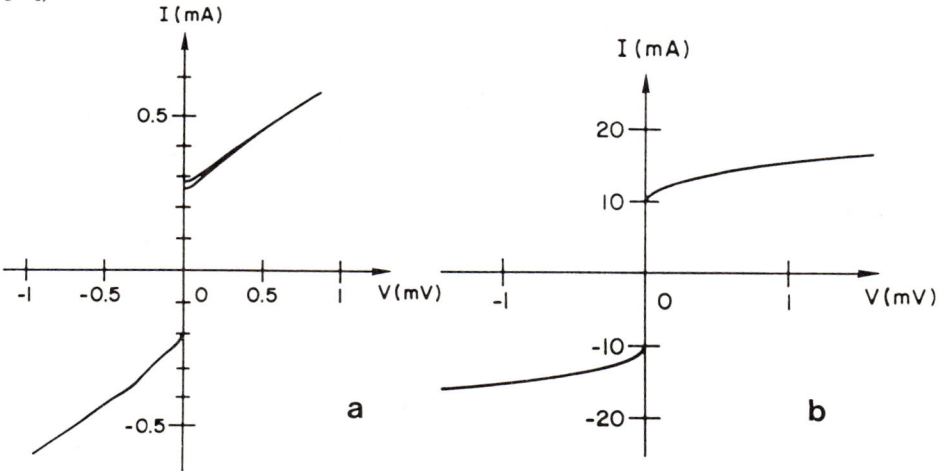

Fig. 8. (a) Current-voltage characteristic of a 4.5 μm wide line straddling a grain boundary (after [21]) at 5 K [34]. (b) Current-voltage characteristic of a 8 μm wide line at 5 K inside one of the grains that form the grain boundary above (after [21]) [34]

217

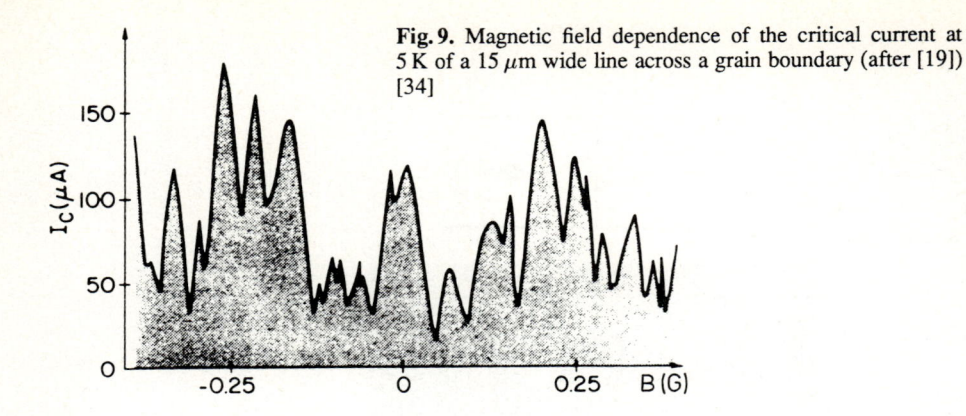

Fig. 9. Magnetic field dependence of the critical current at 5 K of a 15 μm wide line across a grain boundary (after [19]) [34]

mains to be clarified whether the critical current is limited at the weak links be depinning, depairing or decoupling.

In comparison to the critical current density across grain boundaries, the critical current density of the grains depends much less on magnetic fields (Fig. 10) and the $I(V)$ characteristic is more rounded (Fig. 8b). This behavior, combined with an analysis of $J_c(T)$, allows the conclusion that the critical current density of the grains is controlled by flux creep [26]. As shown by LTSEM, also inside the grains, the critical current density is usually lowest in

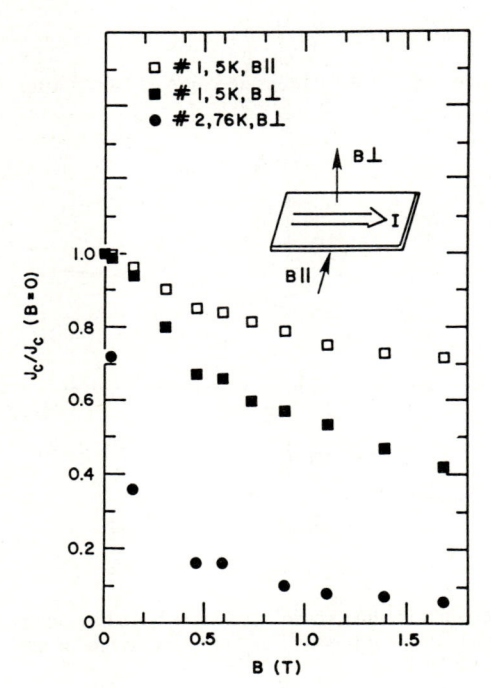

Fig. 10. Normalized magnetic field dependence of the critical current of two grains at 5 K and 76 K. (After [26])

well-defined local areas [24]. For higher currents, these areas grow and thereby increase the sample resistance. Because these spatial effects conceal the basic flux creep properties, the $I(V)$ characteristic and the $I_c(T)$ characteristic near T_c have not been quantitatively analyzed in terms of thermally activated flux motion.

4. Concluding Remarks

Although critical current densities of single grain boundaries in BiSrCaCuO and TlBaCaCuO superconductors have not yet been measured, a number of reports indicate that in both systems, similar to $YBa_2Cu_3O_7$, the intragrain critical current density is considerably higher than the intergrain J_c [27]. This suggests that also in these superconductors the critical current density is limited by grain boundaries. This appears plausible, in view of the small coherence lengths of these systems.

The fact that grain boundaries limit J_c of polycrystalline high-T_c superconductors does not imply, however, that the critical current density of the polycrystals is identical with the critical current density of an average grain boundary [28]. First, as shown by the experiments with bicrystals described above, the critical current density of the grain boundaries depends nonlinearly on the misorientation of the grains [19]. It has been demonstrated that grain boundaries between well-aligned grains have critical current densities that are comparable to those of the grains themselves [19]. In addition, the aspect ratio of the grains will affect J_c. A polycrystal which consists of long grains oriented parallel to the direction of the total current has a large grain boundary area (Fig. 11), and may therefore support a large J_c. Relatively high critical current densities have in fact been achieved in $YBa_2Cu_3O_7$ fibers that consisted of aligned, needle-shaped grains [29]. Second, structural defects like cracks and voids will affect J_c and most likely lower it [30]. Furthermore, supercurrent flow in polycrystalline high-T_c superconductors is a percolative process. Therefore the grain arrangement influences J_c. Finally, the magnetic field of the sample current may reduce the critical current density of the grain boundaries, so that a specimen is only able to carry a supercurrent that does not create a critical magnetic field on the surface [31].

Aside from the approach to enhancing J_c of high-T_c polycrystals by optimizing the grain configuration, the critical current density may be increased

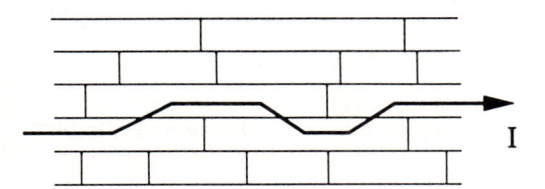

Fig. 11. Sketch of a polycrystal with a large effective grain boundary area

by improving the grain boundaries, for example by doping or by thermal processing. In this context, it should be mentioned that for Ag-coated BiSrCaCuO wires a critical current density of 1.5×10^4 A/cm^2 at 4.2 K and 26 T has already been reported [32]. This current density lies well above the corresponding critical current density of Nb$_3$Sn. Finally, it may be possible to increase the intragrain critical current density by creating artificial defects to strengthen flux pinning [33].

Acknowledgements. This work was performed in collaboration with P. Chaudhari, D. Dimos, and C.C. Tsuei of the IBM Research Division at the T.J. Watson Research Center and R. Gross, K. Hipler and R.P. Huebener of the University of Tübingen. In addition to these colleagues, the author is grateful to J. Hagerhorst, M.M. Oprysko, and M.L. Fenske for their assistance as well as to Vakuumschmelze GmbH for their support. Part of this work was supported by the Bundesministerium für Forschung und Technologie (Projekt Nr. 13N5482).

References

1 F.J. Edeskuty et al.: "High Temperature Superconductors for Power Transmission Applications", in Proc. Conf. on Electrical Applications of Superconductivity, Orlando, FL, 1988;
A.P. Malozemoff: Physica C **153–155**, 1049 (1988);
T.D. Schlaback: "High-T_c Superconductor Wire-Fabrication and Prospects for Use in Large Scale Applications" in *Superconductivity Applications and Development*, MD-Vol. II (ASME, New York) p. 29
2 See for example Proc. Int. Conf. on Critical Currents in High-Temperature Superconductors, Snowmass, CO, in Cryogenics **29** (March Supplement), (1989)
3 R.P. Huebener: *Magnetic Flux Structures in Superconductors*, Springer Ser. Solid-State Sci., Vol. 6 (Springer, Berlin, Heidelberg 1979)
4 M. Tinkham: *Introduction to Superconductivity* (McGraw-Hill, New York 1975)
5 T.K. Hunt: Phys. Rev. Lett. **151**, 325 (1966)
6 C.C. Tsuei, J. Mannhart, D. Dimos: "Limitation on Critical Currents in High Temperature Superconductors", in Proc. Topical Conf. on High T_c Superconducting Thin Films, Devices and Applications, AIP Conf. Proc. No. 182, ed. by G. Margariton, R. Joynt, M. Onellion (1989) pp. 194–297
7 A.I. Larkin, Yu. N. Ovchinnikov: J. Low Temp. Phys. **34**, 409 (1979)
8 P.W. Anderson: Phys. Rev. Lett. **9**, 309 (1962)
P.W. Anderson, Y.B. Kim: Rev. Mod. Phys. **36**, 39 (1964)
9 P. Kes: Private communication
10 D. Dew-Hughes: Cryogenics **28**, 674 (1988);
M. Tinkham: Helv. Phys. Acta **61**, 443 (1988)
11 M.R. Beasley, R. Labusch, W.W. Webb: Phys. Rev. **181**, 682 (1969)
12 Y. Yeshurun et al.: Cryogenics **29**, 259 (1989)
13 B.D. Josephson: Phys. Lett. **1**, 251 (1962)
14 V. Ambegaokar, A. Baratoff: Phys. Rev. Lett. **10** 486 (1963) and (errata) Phys. Rev. Lett. **11**, 104 (1963)
15 G. Deutscher, K.A. Müller: Phys. Rev. Lett. **59**, 1745 (1987);
G. Deutscher: IBM J. Res. Dev. **33**, 293 (1989)
16 R.L. Peterson, J.W. Ekin: Phys. Rev. B **37**, 9848 (1988)
17 J. Clem, B. Bumble, S.I. Raider, W.J. Gallagher, Y.C. Shik: Phys. Rev. B **35**, 6637 (1987);
K.K. Likharev: Rev. Mod. Phys. **51**, 101 (1979)

18 A. DasGupta et al.: Philos. Mag. B **38**, 367 (1978)
19 D. Dimos, P. Chaudhari, J. Mannhart, F.K. LeGoues: Phys. Rev. Lett. **61**, 219 (1988)
20 J. Mannhart, M. Scheuermann, C.C. Tsuei, M.M. Oprysko, C.C. Chi, C.P. Umbach, R.H. Koch, C. Miller: Appl. Phys. Lett. **52**, 1271 (1988)
21 P. Chaudhari, J. Mannhart, D. Dimos, C.C. Tsuei, C.C. Chi, M.M. Oprysko, M. Scheuermann: Phys. Rev. Lett. **60**, 1653 (1988)
22 R.P. Huebener: Adv. Electron. Electron Phys. **70**, 1 (1988)
23 R. Gross, J. Bosch, R.P. Huebener, J. Mannhart, C.C. Tsuei, M. Scheuermann, M.M. Oprysko, C.C. Chi: Nature **332**, 818 (1988)
24 R. Gross, K. Hipler, J. Mannhart, R.P. Huebener, P. Chaudhari, D. Dimos, C.C. Tsuei, J. Schubert, U. Poppe: Appl. Phys. Lett. **55**, 2132 (1989)
25 J. Mannhart, R. Gross, K. Hipler, R.P. Huebener, C.C. Tsuei, D. Dimos, P. Chaudhari: Science **245**, 839 (1989)
26 J. Mannhart, P. Chaudhari, D. Dimos, C.C. Tsuei, T.R. McGuire: Phys. Rev. Lett **61**, 2476 (1988)
27 See for example J.R. Thompson, J. Brynestad, D.M. Kroeger, Y.C. Kim, S.T. Sekula, D.K. Christen, E.D. Specht: Phys. Rev. B **39**, 6652 (1989);
J.W.C. de Vries, G.M. Stollman, M.A.M. Gijs: "Analysis of the Critical Current Density in High-T_c Superconducting Films", to be published in Physica C; and [2]
28 J. Mannhart, C.C. Tsuei: Z. Phys. B **77**, 53 (1989)
29 S. Jin, T.H. Tiefel, R.C. Sherwood, M.E. Davis, R.B. van Dover, G.W. Kammlot, R.A. Fastnacht, H.D. Keith: Appl. Phys. Lett. **52**, 2074 (1988)
30 The possibility of pinning at cracks has been studied by T. Matsushita: Jpn. J. Appl. Phys. **27**, L1712 (1988)
31 H. Dersch, G. Blatter: Phys. Rev. B **38**, 391 (1988)
32 K. Heine, J. Tenbrink, M. Thöner: Appl. Phys. Lett. **55**, 2441 (1989)
33 B. Roas, B. Hensel, G. Saemann-Ischenko, L. Schultz: Appl. Phys. Lett. **54**, 1051 (1989)
34 Here, the YBaCuO film was grown epitaxially on a $SrTiO_3$ polycrystal, as described in [21]

Muon Spin Rotation Experiments in High-T$_c$ Superconductors

H. Keller

Physik-Institut der Universität Zürich, Schönberggasse 9,
CH-8001 Zürich, Switzerland

Abstract. The muon spin rotation (μSR) method is a unique tool for investigating the local magnetic field distribution at the muon site, $p(B_\mu)$, in a superconductor, yielding important information on the *microscopic* magnetic flux line structure in the bulk of the sample. In uniaxial superconductors, such as the high-T$_c$ oxides, the shape of $p(B_\mu)$ strongly depends on the direction of the external field with respect to the symmetry axis, reflecting the anisotropic screening in these layered superconductors. A precise knowledge of the second moment of $p(B_\mu)$ allows a determination of the London penetration depths λ_{ab} and λ_c. As an example, recent μSR experiments on Y-Ba-Cu-O are presented. The possibilities and limitations of the μSR technique for investigating $p(B_\mu)$ in high-T$_c$ superconductors are discussed.

1. Introduction

The determination of the pairing mechanism in the superconducting copper oxides has stimulated an enormous number of experimental and theoretical investigations of these oxides and closely related antiferromagnetic materials. A knowledge of the magnetic properties of these novel compounds is of special importance. In contrast to *macroscopic* magnetization measurements, the muon spin rotation (μSR) technique provides a sensitive *microscopic* probe of the local magnetic field distribution (at the muon). The μSR technique is a unique method for investigating on a microscopic scale various magnetic properties (Meissner-Ochsenfeld effect, diamagnetic shielding, perfect/imperfect vortex structure, magnetic penetration, superconducting glass state) of the superconducting oxides. In the related antiferromagnetic materials, on the other hand, μSR is a sensitive method for detecting small frozen local magnetic moments. These experiments are of special interest, since in some theories it is assumed that magnetism plays a relevant role in the pairing mechanism of these CuO$_2$-based materials [1,2].

So far, µSR experiments have made many relevant contributions which are not easily obtained with other experimental techniques to the microscopic understanding of the superconducting copper oxides, as well as of the related magnetically ordered copper oxide systems. As a result, a large number of papers has appeared in the literature [1,2]. In most experiments the London penetration depth, one of the fundamental parameters of a superconductor, has been determined from the temperature dependence of the µSR depolarization rate in the superconducting oxides. In addition, µSR experiments on sintered Y-Ba-Cu-O have been performed by the University of Zurich group [3,4] and show characteristic features (irreversible and hysteresis effects, irreversibility phase line) that are consistent with the existence of a superconducting glass state [5-7]. Several investigations were made in order to search for a possible interplay between magnetism and superconductivity in the CuO_2-based systems by studying the magnetic phase diagram of these materials with µSR [1,2].

Since the field is growing so fast, it would be beyond the scope of this paper to give a *complete* review of all the µSR work on high-T_c superconductors. The purpose of this paper is to demonstrate the potential applications of the µSR technique to the microscopic magnetic properties of high-T_c superconductors. As an example, recent experiments on Y-Ba-Cu-O performed by the University of Zurich group [8,9] are discussed in some detail and compared with other experiments. In particular, it will be shown how the London penetration depths in uniaxial type-II superconductors, such as the superconducting oxides, can be extracted from the local magnetic field distribution measured by µSR.

2. The Muon Spin Rotation (µSR) Technique

The positive muon (μ^+) is a lepton with a mass of $m_\mu \simeq 207 m_e$ (electron mass m_e) and spin $S = 1/2$. In a magnetically ordered solid or superconductor the positive muon serves as a *microscopic* magnetic probe of the local magnetic field B_μ at the muon site. The basic principle of a standard µSR experiment is illustrated schematically in Fig. 1. Spin-polarized positive muons are implanted into a sample, which is located in an external magnetic field B_{ext} perpendicular to the initial muon-spin polarization. During a short thermalization process ($\sim 10^{-13}$ s) the muons retain their polarization and subsequently precess in the local magnetic field B_μ (at the muon site) with a Larmor frequency $\omega_\mu = \gamma_\mu B_\mu$, where $\gamma_\mu = 2\pi \times 135.5 \, \text{MHz/T}$ is the gyromagnetic ratio of the muon. After a mean lifetime of $\tau_\mu = 2.2 \mu s$, the muon decays into a positron (e^+) and two neutrinos ($\nu_e, \bar{\nu}_\mu$) as shown in Fig. 2. Because of parity violation the

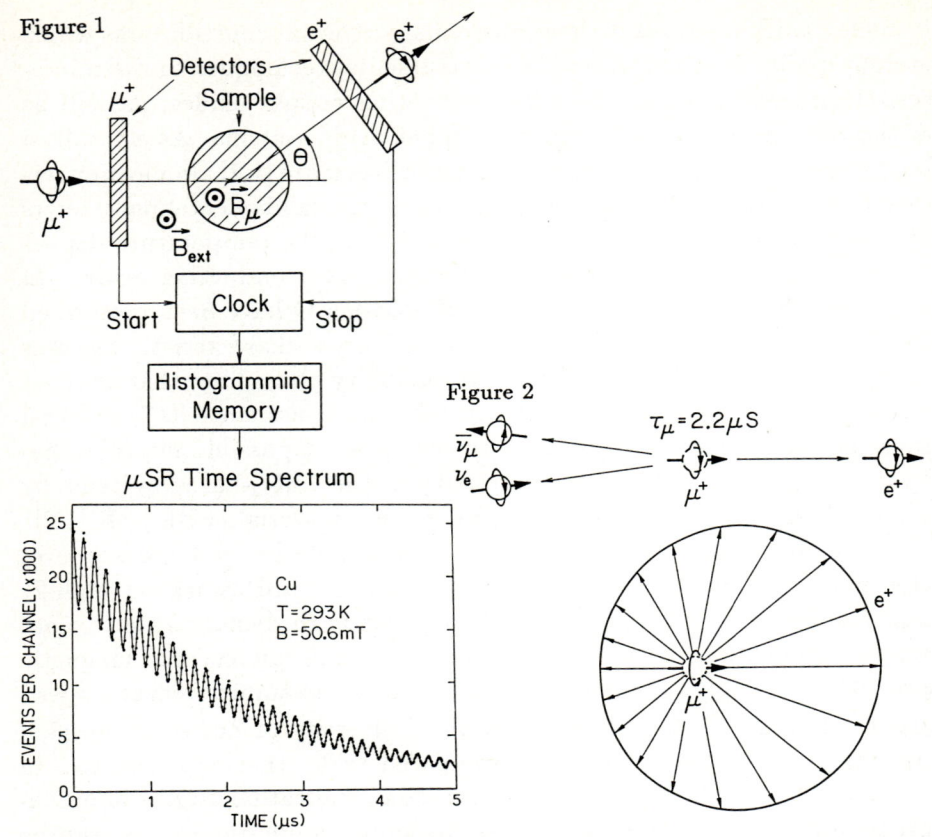

Figure 1: Schematic diagram of a time-differential, transverse-field μSR apparatus. The μ^+ detector registers the incoming muon and the e^+ detector the corresponding decay positron. The lower part of the figure shows a typical μSR time spectrum of copper. The oscillations superimposed on the exponential muon-decay are due to muon-spin precession.

Figure 2: Muon-decay scheme and angular distribution pattern of the decay positrons with respect to the muon-spin direction.

decay positrons are emitted preferentially along the muon-spin direction. The muon-spin precession is observed by detecting the number of decay positrons as a function of time after a muon has stopped in the sample (see Fig. 2). The basic ingredients of a standard μSR experiment are as follows (see Fig. 1): When a muon passes through the μ^+ detector a clock is started. The muon stops in the sample and then precesses in the local field B_μ at the muon site. After the muon decays the corresponding decay positron is registered in the e^+ detector and stops the clock. The resulting time-histogram is stored in memory. As an example, a typical μSR time spectrum taken on a copper sample in a transverse field of 50.6 mT is shown at the bottom of Fig. 1. One can clearly observe oscillations due to muon-spin precession, superimposed on the exponential muon decay. A detailed description of the μSR technique is given in Ref. [10].

The observed μSR time-histogram may be written in the form:

$$N(t) = N_o \exp(-t/\tau_\mu)[1 + A\, P(t)] + b, \qquad (1)$$

$$P(t) = R(t)\cos(2\pi\nu_\mu t + \theta), \qquad (2)$$

where N_0 is a normalization constant, τ_μ is the muon lifetime (2.2 μs), A is the maximum experimental decay asymmetry (precession amplitude), $P(t)$ is the time evolution of the muon-spin polarization, $\nu_\mu = \omega_\mu/(2\pi)$ is the muon-spin precession frequency, θ is the initial phase, and b is a time-independent background. The relaxation function $R(t)$ describes the damping of the precession signal. In the case of a Gaussian distribution of static internal fields $p(B_\mu)$, the relaxation function has the form [10]:

$$R(t) = \exp(-\sigma^2 t^2), \qquad (3)$$

where σ is the muon-spin depolarization rate. The precession amplitude A is a measure of the fraction of muons precessing in the field, and the depolarization rate σ yields information on the width of $p(B_\mu)$. In general, the data are analyzed by performing least-squares fits to the time-histogram $N(t)$ defined in Eqs. (1-3).

In the case of a type-II superconductor in the vortex phase, $p(B_\mu)$ is in general *not* Gaussian [11], and the time evolution of the muon-spin polarization $P(t)$ [Eq. (2)] may be written in a more general form:

$$P(t) = \int dB_\mu\, p(B_\mu)\cos(\gamma_\mu B_\mu t + \theta). \qquad (4)$$

Here it is assumed that the implanted muons are immobile (no muon diffusion). The field distribution $p(B_\mu)$ is then obtained from the real part of the (complex) Fourier transform of $P(t)$ or by numerically solving the integral equation defined by Eq. (4) [9]. The average local magnetic field $<B_\mu>$ [first moment of $p(B_\mu)$] is given by

$$<B_\mu> = \int dB_\mu\, p(B_\mu) B_\mu = B_{ext} + \mu_0(1-N)M, \qquad (5)$$

where B_{ext} is the external field, N is the demagnetization factor, and M is the magnetization of the sample. Eq. (5) indicates that the magnetization M of a superconductor can be determined from the first moment of $p(B_\mu)$ measured by μSR [3,4]. The second moment of $p(B_\mu)$,

$$<\Delta B_\mu^2> = \int dB_\mu\, p(B_\mu)(<B_\mu> - B_\mu)^2, \qquad (6)$$

is a measure of the width of the local magnetic field distribution $p(B_\mu)$.

For a Gaussian $p(B_\mu)$ the second moment is given by [10]

$$< \Delta B_\mu^2 >= 2\sigma^2/\gamma_\mu^2, \qquad (7)$$

where σ is the muon-spin depolarization rate. A critical discussion concerning the evaluation of $p(B_\mu)$ in high-temperature superconductors from μSR time-spectra is given in a article by Brandt [11].

3. Magnetic Flux Distribution and Penetration Depth

There are two fundamental microscopic lengths that characterize the magnetic properties of a superconductor: the coherence length ξ and the London penetration depth λ. In the framework of the London and Ginzburg-Landau theories the magnetic penetration depth for an isotropic superconductor is given by [12]

$$\lambda(0) = \sqrt{m^*/\mu_0 e^2 n_s} \propto \sqrt{m^*/n_s}, \qquad (8)$$

where m^* is the effective mass of the superconducting carriers and n_s is the superconducting carrier density which, for ordinary superconductors at $T = 0$ is identical to the normal-state carrier density n. A knowledge of $\lambda(0)$ gives the ratio m^*/n_s, and a precise measurement of the temperature dependence of λ yields crucial information on the type of pairing mechanism (s-wave or p-wave pairing, weak- or strong-coupling). The temperature dependence of λ for various s-wave BCS superconductors (weak- or strong-coupling, clean or dirty limit) is discussed in the literature [12-14]. For a conventional s-wave superconductor the temperature dependence of λ is generally described by the *empirical* formula (two-fluid model):

$$\lambda(T) = \lambda(0)[1 - (T/T_c)^4]^{-1/2}. \qquad (9)$$

Eq. (9) usually holds for s-wave BCS-like superconductors in the extreme anomalous ($\xi \gg \lambda$) and weak-coupling limit [12]. This condition is certainly not fulfilled for the short coherence length superconducting oxides ($\xi \ll \lambda$). However, it has been shown by Rammer [14] that Eq. (9) also represents an appropriate approximation of $\lambda(T)$ for a local ($\xi \ll \lambda$) s-wave superconductor with strong-coupling.

The μSR technique provides an ideal tool for determining λ in an elegant way. In the Shubnikov phase of a type-II superconductor, the external magnetic field ($B_{ext} > B_{c_1}$) penetrates the sample in the form of a regular vortex lattice, each vortex having an elementary flux quantum $\Phi_0 = h/2e$. This results in a broad distribution $p(B)$ of local fields (at the muon site), and the width of $p(B)$ is determined by λ. In an isotropic extreme type-II superconductor ($\lambda/\xi \gg 1$) the second moment of the magnetic flux distribution $p(B)$ for a perfect triangular vortex lattice may be calculated using simple London theory [11]. For $2B_{c_1} < B_{ext} < B_{c_2}/4$ the second moment $<\Delta B^2>$ of $p(B)$ is given by [11]:

$$<\Delta B^2> = 0.00371\Phi_0^2\lambda^{-4}. \quad (10)$$

Note that in this case the field variation $<\Delta B^2>^{1/2}$ in a perfect flux lattice is field-independent and much smaller than the average field $<B>$ and $<B> \approx B_{ext}$. For the superconducting oxides this has been confirmed by μSR experiments [1,2]. For a very homogeneous sample, where flux pinning effects are small, it is found that $<\Delta B^2>$ is indeed constant for $B_{ext} > 0.1\text{T} \approx 2B_{c_1}$ (see Fig. 3), and Eq. (10) can in principle be used to determine λ by means of a μSR experiment.

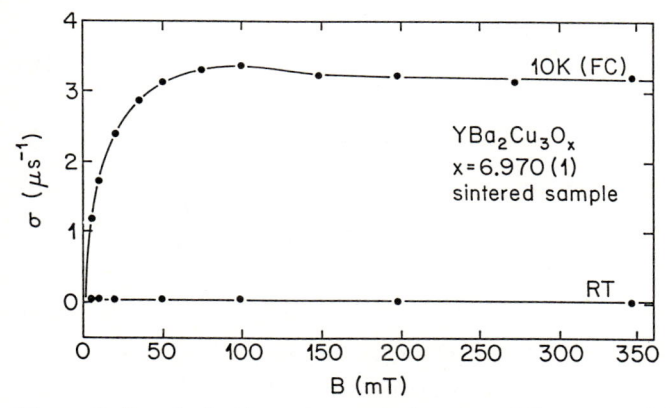

Figure 3: Depolarization rate σ as a function of the external field B_{ext} (FC) for a high quality sintered $YBa_2Cu_3O_x$ sample [$x = 6.970(1)$] at 10 K and at room temperature (RT), respectively. The lines are guides to the eye.

So far, we have neglected the extreme electronic anisotropy of the superconducting oxides. In these layered superconductors the supercurrents flow predominantly in the CuO_2-planes (ab-planes) giving rise to an *anisotropic* penetration depth. In a uniaxial superconductor the magnetic screening of an external magnetic field depends on whether the supercurrents flow in the ab-plane (λ_{ab}) or along the c-axis (λ_c), and the London formula in Eq. (8) has the more general form [15,16]:

$$\lambda_{ab,c}(0) = \sqrt{m^*_{ab,c}/\mu_0 e^2 n_s} \propto \sqrt{m^*_{ab,c}/n_s}, \quad (11)$$

where m^*_{ab} (m^*_c) is the effective mass of the superconducting carriers associated with λ_{ab} (λ_c). For the superconducting oxides the anisotropy is

227

large and $\lambda_c/\lambda_{ab} \approx 3-10$. Schneider and Frick [16] developed a tight-binding BCS-type model (singlet pairing) of layered high-temperature superconductors, consistent with experimental constraints. As a result, they obtained values for $\lambda_{ab}(0)$ and $\lambda_c(0)$ that are in the range of current experimental values (see Sec. 4).

The μSR lineshape of an anisotropic type-II superconductor in the vortex phase was calculated by Celio et al. [17] using a simple model assuming $\lambda_c = \infty$. Schopohl and Baratoff [18] used a more sophisticated approach to calculate the local magnetic field distribution $p(B)$ in an anisotropic extreme type-II superconductor. Barford and Gunn [15] derived expressions for the second moment of $p(B)$ in a uniaxial type-II superconductor, allowing a determination of the principal penetration depths λ_{ab} and λ_c. For a single crystal with the external field B_{ext} parallel to the c-axis ($B_{ext} \parallel c$), the second moment of $p(B)$ is given by [15]

$$<\Delta B^2>_\parallel = 0.00371\Phi_0^2 \lambda_{ab}^{-4}, \tag{12}$$

whereas for $B_{ext} \perp c$

$$<\Delta B^2>_\perp = 0.00371\Phi_0^2 (\sqrt{\lambda_{ab}\lambda_c})^{-4}. \tag{13}$$

Eq. (13) indicates that for $B_{ext} \perp c$ the effective penetration depth is the geometric mean of λ_{ab} and λ_c. An interesting quantity from an experimental point of view is the ratio

$$<\Delta B^2>_\parallel / <\Delta B^2>_\perp = (\lambda_c/\lambda_{ab})^2. \tag{14}$$

Using single crystals λ_{ab} and λ_c can thus be determined independently from a μSR experiment with the external field parallel and then perpendicular to the c-axis. However, large single crystals of high quality are required and are not readily available. Therefore, most μSR experiments are performed on powder samples which are available in large sizes of high purity. In a polycrystalline sample one measures an effective penetration depth λ_{eff} (average over all orientations), and λ_{ab} and λ_c cannot be determined independently unless the anisotropy ratio λ_c/λ_{ab} is known [15]. It was shown by Barford and Gunn [15] that for a large anisotropy ratio ($\lambda_c/\lambda_{ab} \geq 5$), λ_{eff} is independent of the anisotropy ratio and is solely determined by λ_{ab}:

$$\lambda_{eff} \simeq 1.23\lambda_{ab}. \tag{15}$$

This is a very useful result since it allows a direct determination of λ_{ab} in a uniaxial large anisotropy superconductor from a μSR spectrum taken on a polycrystalline sample.

So far, we have assumed that the spatial variation of the local magnetic field is that of the perfect flux lattice. However, the superconducting oxides have extremely *short coherence lengths*, giving rise to flux pinning at twin boundaries and defects. Moreover, it is not clear whether the regular Abrikosov flux lattice observed in Y-Ba-Cu-O single crystals at low temperatures and low fields [19] is also stable at higher temperatures and fields. Gammel et al. [20] found experimental evidence that the flux lattice in these short coherence length superconductors may melt above a temperature $T_m(B)$. They believe that at $T_m(B)$ a transition to a vortex-liquid state occurs, similar to that observed in two-dimensional superconducting films. Shortly after the discovery of the superconducting oxides, it was suggested by Müller et al. [5] that these materials may have glasslike superconducting properties. In the superconducting glass state, flux lines at low temperatures might be randomly frozen at fixed positions and on short time scales [6,7]. Several experiments, including μSR experiments [3,4], were reported and show several characteristic magnetic features of a superconducting glass [5-7]: (a) irreversible and hysteresis effects, (b) the existence of a phase line separating reversible from irreversible behavior, and (c) the unconventional time dependence of the magnetization. Direct experimental evidence of the existence of a glassy superconducting state was reported by Rossel et al. [21]. In the time dependence of the magnetization of a Y-Ba-Cu-O single crystal they observed the so-called 'memory effect', a phenomenon that was first seen in spin glasses. Schneider et al. [22] considered a model of weakly-coupled superconducting grains where the magnetic screening effects are introduced self-consistently within the framework of mean-field theory. They found that this inhomogeneous system exhibits 'glassy' behavior and that under certain conditions an Abrikosov lattice is formed at low temperatures. Here the question arises: What does the magnetic phase diagram (B-T-diagram) of the short coherence length copper oxide superconductors look like? In order to answer this question in detail more experimental and theoretical work has to be done.

In conclusion, any microscopic field inhomogeneity not caused by the regular vortex lattice will tend to increase the overall μSR line width $<\Delta B^2>$, and since $<\Delta B^2> \propto \lambda^{-4}$ [Eq. (10)] values for the penetration depth determined by μSR should only be considered as lower estimates.

4. Experiments and Results

The experiments described here were performed at the Paul Scherrer Institute (PSI), Switzerland, with a standard time-differential, transverse-field μSR apparatus [23] using low-momentum muons (29 MeV/c)(see Fig. 1). Special attention was paid to assure that *all* the muons stopped in the sample, avoiding misleading background signals from the sample holder and cryostat windows. For this reason the samples were mounted on an Fe_2O_3 sample holder, and very thin windows in the beam line and the cryostat were used. Experiments on a high quality sintered $YBa_2Cu_3O_x$ sample [24] were performed in order to deduce the local magnetic field distribution in the vortex phase, allowing an accurate measurement of the temperature dependence of the effective penetration depth λ_{eff}. In addition, complementary measurements on a c-axis-oriented $YBa_2Cu_3O_x$ polycrystal [25] were used to obtain information on the anisotropic magnetic properties of this layered superconductor. The samples were:

(1) a high quality sintered $YBa_2Cu_3O_x$ sample [$x = 6.970(1)$] which was prepared at the ETH Zurich [24]. The method for the accurate determination of the oxygen concentration has been described elsewhere [26]. Magnetization measurements exhibit a very sharp transition at $T_c = 89.5(5)$ K with a width of 8 K (10-90%) and a Meissner fraction of approximately 60% in a field of 1.42 mT. The disk-shaped sample has a diameter of 14 mm and a thickness of 3 mm.

(2) a c-axis-oriented $YBa_2Cu_3O_x$ polycrystal [$x \approx 6.9$] which was prepared at the University of Frankfurt [25]. The sample ($7.5 \times 4 \times 3.5$ mm^3) with $T_c = 89(1)$ K showed a Meissner fraction of 23% in a field of 20 mT.

In order to attain a 'regular' vortex lattice in the sample, the measurements must be done on slowly field-cooled (FC) samples in high external magnetic fields ($B_{ext} > 2B_{c_1} \approx 100$ mT) [27]. In this case the magnetic flux density is almost constant above and below T_c, and flux pinning effects in these inhomogeneous materials are small. Flux pinning effects are in general temperature-dependent and give rise to inhomogeneous line broadening. In order to check the homogeneity of the high quality sintered Y-Ba-Cu-O sample, a field scan between 5 and 350 mT was performed at 10 K. In this experiment each data point was obtained after field-cooling the sample slowly to 10 K from a temperature well above T_c. The μSR time-spectra were analyzed assuming a Gaussian distribution of local magnetic fields $p(B_\mu)$ [Eqs. (1-3)], yielding the depolarization rate σ, which according to Eq. (7), is a measure of the second moment of $p(B_\mu)$. Figure 3 shows the corresponding field dependence of σ at 10 K. Note that at 10 K the depolarization rate σ increases with increasing field, has a slight maximum at around 100 mT and remains constant above 150 mT ($B_{ext} > 2B_{c_1} \approx 100$ mT). This is exactly the behavior expected for a regular vortex lattice [see Eq. (10)], except for the small 'bump' observed around 100 mT, which most probably is due to weak flux pinning [27]. This finding indicates that the sample is of high quality and that the applied fields are large enough to make use of Eq. (10) for a determination of the London penetration depth. The constant behavior of σ above 150 mT does not establish, however, that a regular vortex lattice is formed. A randomly distorted flux lattice exhibits the same behavior, although with a larger second moment (depolarization rate) due to additional field inhomogeneities [11]. For comparison, the corresponding field dependence of σ measured at room temperature (RT) is also displayed in Fig. 3. As expected, only a small and almost temperature-independent $\sigma_0 \approx 0.1 \mu s^{-1}$ is observed at RT, arising most likely from copper nuclear moments.

In the next step, a FC temperature scan in an external field of 350 mT was performed. In such a high field 'glassy' effects, which arise only at low fields, may be ruled out. It is not at all clear at present what the vortex structure in this field at low temperature really looks like. Nonetheless, under the assumptions that a 'regular triangular vortex lattice' is formed in the sample at 350 mT and that the local magnetic field distribution is Gaussian-like, Eqs. (7) and (10) can be used to deduce the effective penetration depth $\lambda_{eff}(T)$ from the measured depolarization rate $\sigma(T)$. Since $\sigma(T) \propto \lambda_{eff}(T)^{-2}$, the temperature dependence of σ for the two-fluid model [Eq. (9)] is given by the *empirical* expression:

$$\sigma(T) = \sigma(0)[1 - (T/T_c)^4]. \qquad (16)$$

Figure 4 shows the temperature dependence of σ measured in a field of 350 mT (FC), after subtraction of the small temperature-independent contribution ($\sigma_0 \approx 0.1 \mu s^{-1}$) from the copper nuclear moments. The solid

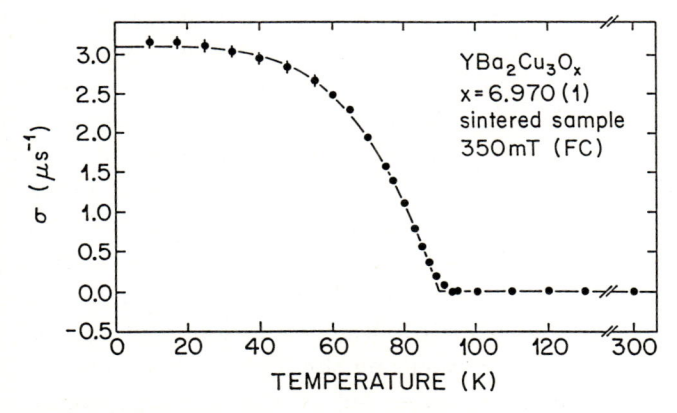

Figure 4: Depolarization rate σ as a function of temperature for a high quality sintered YBa$_2$Cu$_3$O$_x$ sample [$x = 6.970(1)$] measured in a field of 350 mT (FC). The solid curve corresponds to a fit using the empirical temperature dependence of the two-fluid model [Eq. (16)].

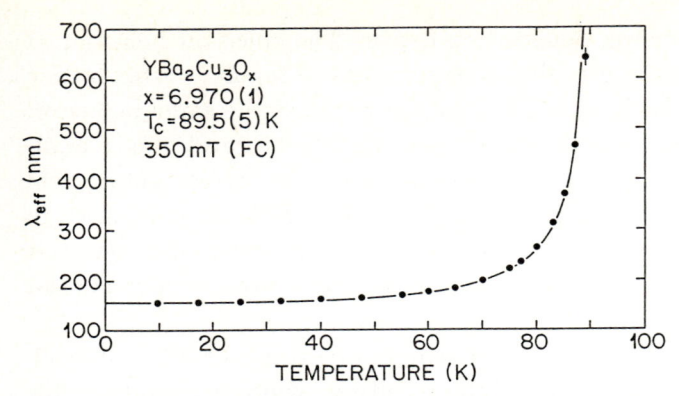

Figure 5: Effective penetration depth λ_{eff} as a function of temperature for a high quality sintered YBa$_2$Cu$_3$O$_x$ sample [$x = 6.970(1)$] measured in a field of 350 mT (FC). The solid curve is a fit using the empirical expression given in Eq. (9).

curve represents a fit to the power law defined in Eq. (16). The fit is excellent, indicating that the observed temperature dependence of σ at 350 mT is consistent with conventional s-wave pairing and suggesting evidence for strong-coupling [14], in agreement with other μSR results [28-30]. The temperature dependence of the effective penetration depth λ_{eff}, deduced from Eqs. (7) and (10), is shown in Fig. 5. It is obvious that $\lambda_{eff}(T)$ is well described by the empirical temperature dependence of the two-fluid model [Eq. (9)] as indicated by the solid curve in Fig. 5.

The data were also analyzed by calculating the *real part* of the complex Fourier transform [11] of the μSR time-spectra [Eq. (4)]. This procedure allows a direct and model-independent determination of the second moment $< \Delta B_\mu^2 >$ of the local magnetic field distribution $p(B_\mu)$. Both analysis methods were used to determine an accurate value for $\lambda_{eff}(0)$ by extrapolating the measured $< \Delta B_\mu(T)^2 >^{1/2}$ to $T = 0$ K using the empirical expression given in Eq. (16). As a result, $\lambda_{eff}(0) = 155(10)$ nm was obtained. The error, which is mainly from systematic effects [11], was estimated from two completely different analysis methods. The present value of $\lambda_{eff}(0)$ is consistent with those obtained by other μSR groups [27-31].

Although λ_{eff} in a sintered sample is an average over all orientations according to Eq. (15), λ_{ab} can be extracted directly from λ_{eff} if the anisotropy ratio $\lambda_c/\lambda_{ab} \geq 5$. In order to obtain an estimate of λ_c/λ_{ab} in Y-Ba-Cu-O, experiments similar to those on the high quality sintered sample were performed on the c-axis-oriented polycrystal. Figure 6 shows the temperature dependence of the depolarization rate σ taken in an external magnetic field $B_{ext} = 350$ mT (FC) for $B_{ext} \parallel c$ and $B_{ext} \perp c$, respectively. Note that the temperature dependence of σ for both field

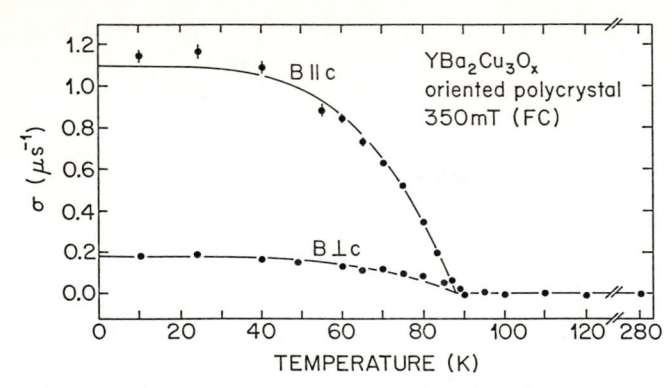

Figure 6: Depolarization rate σ as a function of temperature for a c-axis-oriented polycrystalline YBa$_2$Cu$_3$O$_x$ sample measured in 350 mT (FC) for $B_{ext} \parallel c$ and $B_{ext} \perp c$, respectively. The solid lines represent fits to the data using the empirical expression defined in Eq. (16).

directions is well described by the two-fluid model [Eq. (16)], although the fits are not as good as for the high quality sintered sample (see Fig. 4). The different values of σ obtained for $B_{ext} \parallel c$ and $B_{ext} \perp c$ at low temperatures clearly reflect the anisotropic magnetic screening in this layered superconductor. At low temperatures (10 K), the depolarization rate $\sigma_\parallel$ for $B_{ext} \parallel c$ is 4 to 6 times larger than $\sigma_\perp$ for $B_{ext} \perp c$, i.e. $\sigma_\parallel/\sigma_\perp \simeq 5(1)$. Combining Eqs. (14) and (7) one gets for the anisotropy ratio $\lambda_c/\lambda_{ab} = \sigma_\parallel/\sigma_\perp \simeq 5(1)$, which is in fair agreement with results obtained by other groups (see Table 1). According to Eq. (11), this value of the anisotropy ratio of the penetration depths implies an anisotropy ratio of the effective masses $m_c^*/m_{ab}^* \approx 25(10)$. Note that the value of $\sigma_\parallel(0)$ for the oriented polycrystal (see Fig. 6) is smaller than $\sigma(0)$ for the sintered sample (see Fig. 4). This is because the large oriented polycrystal is not very homogeneous, and no conclusive values for the penetration depths λ_{ab} and λ_c could be determined for this sample. More reliable values of λ_{ab} and λ_c can be deduced from the effective penetration depth $\lambda_{eff}(0) = 155(10)$ nm obtained for the high quality sintered sample and the known anisotropy ratio $\lambda_c/\lambda_{ab} \simeq 5(1)$. According to Eq. (15), the following values for the penetration depths at $T = 0$ were calculated: $\lambda_{ab}(0) = 130(10)$ nm and $\lambda_c(0) \simeq 500 - 800$ nm. As pointed out earlier in this paper, these values should be considered as lower limits of the 'true' intrinsic values. For comparison, a selected list of present values of the penetration depths and the anisotropy ratio of Y-Ba-Cu-O measured with various methods is given in Table 1. We note that our results are consistent with those obtained with other experimental techniques

Table 1: Values of the penetration depths λ_{ab} and λ_c and the anisotropy ratio λ_c/λ_{ab} for Y-Ba-Cu-O measured by various experimental techniques. Samples: sintered sample (S), c-axis-oriented polycrystalline sample (OP), single crystal (SC), c-axis-oriented film (OF).

Method (Sample)	λ_{ab} (nm)	λ_c (nm)	λ_c/λ_{ab}	Reference
µSR (S/OP)	130(10)	500-800	5(1)	this work [8,9]
µSR (SC)	141.5(3.0)	> 700	> 5	Harshman et al. [30]
Magnetic Torque (SC)	68(20)	500(70)	≈ 7	Fruchter et al. [32]
Kinetic Inductance (OF)	150	-	-	Fiory et al. [33]
Magnetization (SC)	140(50)	-	-	Krusin-Elbaum et al. [34]
Magnetization (SC)	130	450	3	Krusin-Elbaum et al. [35]
Magnetization (OP)	140(20)	610(100)	≈ 4.5	Scheidt et al. [36]
Microwave Absorption (SC)	-	420	-	Blazey [37]

and for different types of samples (sintered samples, oriented polycrystalline samples, single crystals and oriented thin films). Moreover, the present experimental values for the penetration depths λ_{ab} and λ_c are in fair agreement with those calculated by Schneider and Frick [16] using a tight-binding BCS-type model for a layered superconductor. According to Eq. (11), the large penetration depth and the anisotropy can be understood in terms of the low carrier density n_s and the large effective mass m^* [16].

Some typical Fourier spectra (local field distributions) for a sintered sample and a c-axis-oriented polycrystalline sample taken in an external field of 350 mT (FC) at 10 K are displayed in Fig. 7. The sharp peak observed in the Fourier spectrum in Fig. 7a is an artifact, arising from a small fraction (≈ 5%) of muons stopping in the cryostat windows and precessing in the external magnetic field. Celio [17,31] calculated the local magnetic field distribution for an anisotropic type-II superconductor in the mixed phase and obtained field distributions that are very similar to the experimental Fourier spectra shown in Fig. 7. Note that the powder-averaged field distribution (Fig. 7a) for a uniaxial superconductor is considerably more symmetric than for the external magnetic field parallel (Fig. 7b) or even perpendicular (Fig. 7c) to the symmetry axis (c-axis). This confirms our previous assumption that the field distribution in the high quality sintered sample is well approximated by a Gaussian. Having a closer look at Fig. 7a, one observes a slight asymmetry in the Fourier spectrum, which is also seen in the calculated spectra [17,31], although much more pronounced. This is because any field inhomogeneity present in the sample 'smears out' the details of the magnetic flux distribution expected for a perfect vortex lattice. On the other hand,

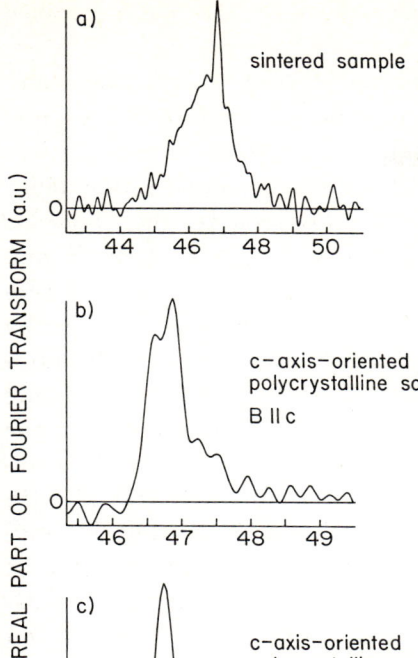

Figure 7: Fourier spectra (real part of complex Fourier transform) of μSR time-histograms taken on two Y-Ba-Cu-O samples in an external field of 350 mT (FC) at 10 K: (a) high quality sintered sample (average over all orientations), (b) c-axis-oriented polycrystalline sample ($B_{ext} \parallel c$), (c) c-axis-oriented polycrystalline sample ($B_{ext} \perp c$).

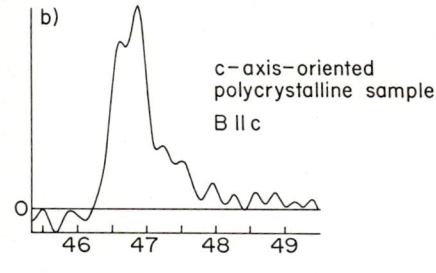

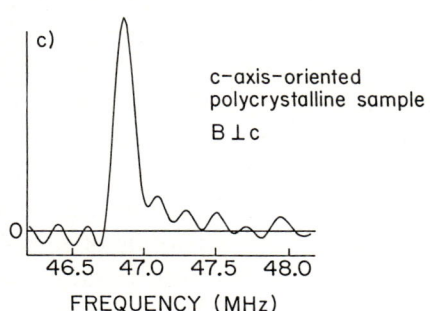

the long tail on the high-field side seen in the Fourier spectrum of the c-axis-oriented polycrystalline sample (Fig. 7b) is a characteristic feature of the magnetic flux distribution of an Abrikosov lattice [11,31].

5. Conclusions

In conclusion, we have demonstrated that the μSR technique is in many respects a unique tool for investigating the local magnetic field distribution (at the muon site) $p(B_\mu)$ in high-T_c superconductors, allowing a determination of the principal-axis penetration depths λ_{ab} and λ_c and their temperature dependence from the second moment of $p(B_\mu)$ in the vortex phase. A precise knowledge of $\lambda_{ab}(T)$ and $\lambda_c(T)$ is of special importance, since it yields crucial information on the type of the pairing mechanism (*s*- or *p*-wave, weak- or strong-coupling) [12-14,16] involved in these uniaxial superconductors.

As an example, recent µSR measurements [8,9] on a high quality sintered sample [24] and on a *c*-axis-oriented polycrystalline Y-Ba-Cu-O sample [25] were discussed in detail. For the sintered sample, it was found that the temperature dependence of the effective penetration depth (average over all orientations) λ_{eff} is well described by the *empirical* expression [Eq. (9)] of the two-fluid model with $\lambda_{eff} = 155(10)$ nm. This observation is consistent with conventional *s-wave* pairing and suggests evidence for strong-coupling [14]. Similar results were reported by other µSR groups [27-31]. On the other hand, Fiory et al. [33] determined $\lambda_{ab}(T)$ in *c*-axis-oriented epitaxial Y-Ba-Cu-O films from kinetic inductance measurements. They found that $\lambda_{ab}(T)$ follows closely the temperature dependence of an *s*-wave BCS superconductor with weak-coupling [$\Delta(0) = 1.76 k_B T_c$] [13]. This finding is in contradiction with the µSR results, which are also consistent with *s*-wave pairing but suggest strong-coupling [14]. Therefore, whether the coupling is weak or strong remains an open question. From measurements on the *c*-axis-oriented polycrystalline sample, the anisotropy ratio of the penetration depth along the principal axes was found to be $\lambda_c/\lambda_{ab} \simeq 5(1)$. From this value and $\lambda_{eff}(0) = 155(10)$ nm obtained for the high quality sintered sample, $\lambda_{ab}(0) = 130(10)$ nm and $\lambda_c(0) \simeq 500 - 800$ nm were estimated. These values are in good agreement with those listed in Table 1, measured with various experimental techniques and on different kinds of samples. In addition, the observed local magnetic field distributions (Fourier spectra) for the two samples (see Fig. 7) exhibit several features that are characteristic of the magnetic flux distribution in the vortex phase of a uniaxial type-II superconductor [11,15,17,31]. It has not been confirmed for the superconducting oxides that a regular Abrikosov flux lattice is indeed formed in high magnetic fields and at moderate temperatures. More experimental work, especially on high quality single crystals, is required to get a better understanding of the microscopic magnetic flux distribution in these novel superconducting materials.

Acknowledgments

I would like to thank K.A. Müller, J.G. Bednorz and G. Benedek for organizing this Course on Earlier and Recent Aspects of Superconductivity, and for the opportunity to contribute. The kind collaboration with W. Assmus, J.G. Bednorz, E. Kaldis, J. Kowalewski, W. Kündig, Y. Maeno, I. Morgenstern, K.A. Müller, W. Odermatt, B. Pümpin, C. Rossel, S.

Rusiecki, I.M. Savić, J.W. Schneider, T. Schneider, H. Simmler and P. Zimmermann is gratefully acknowledged. We wish to thank J.P. Blaser and the staff of the Paul Scherrer Institute (PSI) for technical assistance. This work was partly supported by the Swiss National Science Foundation.

References

[1] See, e.g., *Proceedings of the International Conference on High-Temperature Superconductors and Materials and Mechanism of Superconductivity*, Interlaken, Switzerland, Physica C **153-155** (1988).

[2] See, e.g., H. Keller, IBM J. Res. Develop. **33**, 314 (1989), and references quoted therein.

[3] H. Keller, B. Pümpin, W. Kündig, W. Odermatt, B.D. Patterson, J.W. Schneider, H. Simmler, S. Connell, K.A. Müller, J.G. Bednorz, K.W. Blazey, I. Morgenstern, C. Rossel, and I.M. Savić, Physica C **153-155**, 71 (1988).

[4] B. Pümpin, H. Keller, W. Kündig, W. Odermatt, B.D. Patterson, J.W. Schneider, H. Simmler, S. Connell, K.A. Müller, J.G. Bednorz, K.W. Blazey, I. Morgenstern, C. Rossel, and I.M. Savić, Z. Phys. B - Condensed Matter **72**, 175 (1988).

[5] K.A. Müller, M. Takashige, and J.G. Bednorz, Phys. Rev. Lett. **58**, 1143 (1987).

[6] I. Morgenstern. K.A. Müller, and J.G. Bednorz, Z. Phys. B - Condensed Matter **69**, 33 (1987).

[7] I. Morgenstern, IBM J. Res. Develop. **33**, 307 (1989), and this course.

[8] B. Pümpin, H. Keller, W. Kündig, W. Odermatt, I.M. Savić, J.W. Schneider, H. Simmler, P. Zimmermann, J.G. Bednorz, Y. Maeno, K.A. Müller, C. Rossel, E. Kaldis, S. Rusiecki, W. Assmus, and J. Kowalewski, *Proceedings of the International Conference on Superconductivity*, Stanford, 1989, to be published in Physica C.

[9] B. Pümpin, H. Keller, W. Kündig, W. Odermatt, I.M. Savić, J.W. Schneider, H. Simmler, P. Zimmermann, J.G. Bednorz, Y. Maeno, K.A. Müller, C. Rossel, E. Kaldis, and S. Rusiecki, to be published.

[10] See, e.g., A. Schenck, *Muon Spin Rotation Spectroscopy: Principles and Applications in Solid State Physics* (Adam Hilger, Bristol 1985).

[11] E.H. Brandt, Phys. Rev. B **37**, 2349 (1988).

[12] M. Tinkham, *Introduction to Superconductivity* (McGraw-Hill, New York 1975).

[13] B. Mühlschlegel, Z. Phys. **155**, 313 (1959).

[14] J. Rammer, Europhys. Lett. **5**, 77 (1988).

[15] W. Barford and J.M.F. Gunn, Physica C **156**, 515 (1988).

[16] T. Schneider and M. Frick, *Proceedings of the IBM Japan International Symposium on Strong Correlations and Superconductivity*, Mt Fuji, Japan, 1989, to be published by Springer-Verlag, and this course.

[17] M. Celio, T.M. Riseman, R.F. Kiefl, J.H. Brewer, and W.J. Kossler, Physica C **153-155**, 753 (1988).

[18] N. Schopohl and A. Baratoff, Physica C **153-155**, 689 (1988).

[19] P.L. Gammel, D.J. Bishop, G.J. Dolan, J.R. Kwo, C.A. Murray, L.F. Schneemeyer, and J.V. Waszczak, Phys. Rev. Lett. **59**, 2592 (1987).

[20] P.L. Gammel, L.F. Schneemeyer, J.V. Waszczak, and D.J. Bishop, Phys. Rev. Lett. **61**, 1666 (1988).

[21] C. Rossel, Y. Maeno, and I. Morgenstern, Phys. Rev. Lett. **62**, 681 (1989).

[22] T. Schneider, D. Würtz, and R. Hetzel, Z. Phys. B - Condensed Matter **72**, 1 (1988).

[23] J.W. Schneider, Hp. Baumeler, H. Keller, W. Odermatt, B.D. Patterson, K.A. Müller, J.G. Bednorz, K.W. Blazey, I. Morgenstern, and I.M. Savić, Phys. Lett. A **124**, 107 (1987).

[24] E. Kaldis and S. Rusiecki, to be published.

[25] J. Kowalewski, D. Nikl, and W. Assmus, Physica C **153-155**, 429 (1988).

[26] K. Conder, S. Rusiecki, and E. Kaldis, Materials Research Bulletin, in press.

[27] R.F. Kiefl, T.M. Riseman, G. Aeppli, E.J. Ansaldo, J.F. Carolan, R.J. Cava, W.N. Hardy, D.R. Harshman, N. Kaplan, J.R. Kempton, S.R. Kreitzman, G.M. Luke, B.X. Yang, and D.Ll. Williams, Physica C **153-155**, 757 (1988).

[28] D.R. Harshman, G. Aeppli, E.J. Ansaldo, B. Batlogg, J.H. Brewer, J.F. Carolan, R.J. Cava, and M. Celio, Phys. Rev. B **36**, 2386 (1987).

[29] Y.J. Uemura, V.J. Emery, A.R. Moodenbaugh, M. Suenaga, D.C. Johnston, A.J. Jacobson, J.T. Lewandowski, J.H. Brewer, R.F. Kiefl, S.R. Kreitzman, G.M. Luke, T. Riseman, C.E. Stronach, W.J. Kossler, J.R. Kempton, X.H. Yu, D. Opie, and H.E. Schone, Phys. Rev. B **38**, 909 (1988).

[30] D.R. Harshman, L.F. Schneemeyer, J.V. Waszczak, G. Aeppli, R.J. Cava, B. Batlogg, L.W. Rupp, E.J. Ansaldo, and D.Ll. Williams, Phys. Rev. B **39**, 851 (1989).

[31] T.M. Riseman, R.F Kiefl, Y.J. Uemura, E.J. Ansaldo, J.H. Brewer, J.F. Carolan, M. Celio, W.N. Hardy, H. Hart, R. Kadono, N. Kaplan, W.J. Kossler, J.R. Kempton, S.R. Kreitzman, G.M. Luke, B. Statt, B. Sternlieb, C.E. Stronach, D.Ll. Williams, and B.X. Yang, to be published.

[32] L. Fruchter, C. Giovannella, G. Collin, and I.A. Campbell, Physica C **156**, 69 (1988).

[33] A.T. Fiory, A.F. Hebard, P.M. Mankiewich, and R.E. Howard, Phys. Rev. Lett. **61**, 1419 (1988).

[34] L. Krusin-Elbaum, R.L. Greene, F. Holtzberg, A.P. Malozemoff, and Y. Yeshurun, Phys. Rev. Lett. **62**, 217 (1989).

[35] L. Krusin-Elbaum, A.P. Malozemoff, Y. Yeshurun, D.C. Cronemeyer, and F. Holtzberg, Phys. Rev. B **39**, 2936 (1989).

[36] E.-W. Scheidt, C. Hucho, K. Lüders, and V. Müller, Solid State Commun., in press.

[37] K.W. Blazey, Physica Scripta, in press.

(Spin) Glass Behavior in High-T_c Superconductors

I. Morgenstern

IBM Research Division, Zurich Research Laboratory,
CH-8803 Rüschlikon, Switzerland, and
Physik-Institut der Universität Zürich, Schönberggasse 9,
CH-8001 Zürich, Switzerland

Abstract. This paper considers application of the (spin) glass theory to high-T_c superconductivity on two levels. The first is concerned with the macroscopic glass picture as a generalization of the traditional flux creep picture. In this case, a hierarchy of energy barriers dominates the physical behavior. A spin-glass-like memory effect supports this picture for high-T_c superconductors. An important technical aspect is the influence of the macroscopic glassiness on critical current densities. In contrast, the second level deals with a microscopic spin glass in the insulating phase for lower hole concentrations. Numerical simulations predict the occurrence of a spin-glass antiphase. A gap in the spin excitation could allow for a nonmagnetic superconducting mechanism in high-T_c superconductors.

1. Introduction

In this contribution I would like to consider glassy effects in high-T_c superconductors on two different levels. I distinguish between macroscopic glass behavior and the microscopic (spin) glass in the insulating phase of the new materials. To clarify the situation let us turn to Figure 1. On level 1 we have a disordered array of grains — a ceramic. Inside the grain we see the macroscopic glass consisting of weakly coupled two-dimensional (2-D) planes. Each plane (level 3) consists of an array of domains, probably caused by the heavy twinning of the system, but other defects may also cause these domains (e.g. oxygen deficiencies). The resulting weak links are essential for the glass behavior. Level 4 inside the domain is related to the microscopic mechanism. Here we find in the insulating phase the high-T_c spin glass for lower hole concentrations. The macroscopic glass behavior is clearly not the origin of high-T_c superconductivity and probably only indirectly related to the microscopic mechanism. The microscopic spin glass, in contrast, may even coexist with the superconducting phase and may therefore be related to the mechanism.

(1) 3-D ARRAY OF GRAINS

(2) 2-D PLANES: INSIDE GRAINS

(3) PLANE: 2-D ARRAY OF DOMAINS (TWINS)

(4) INSIDE DOMAIN: MICROSCOPIC MECHANISM

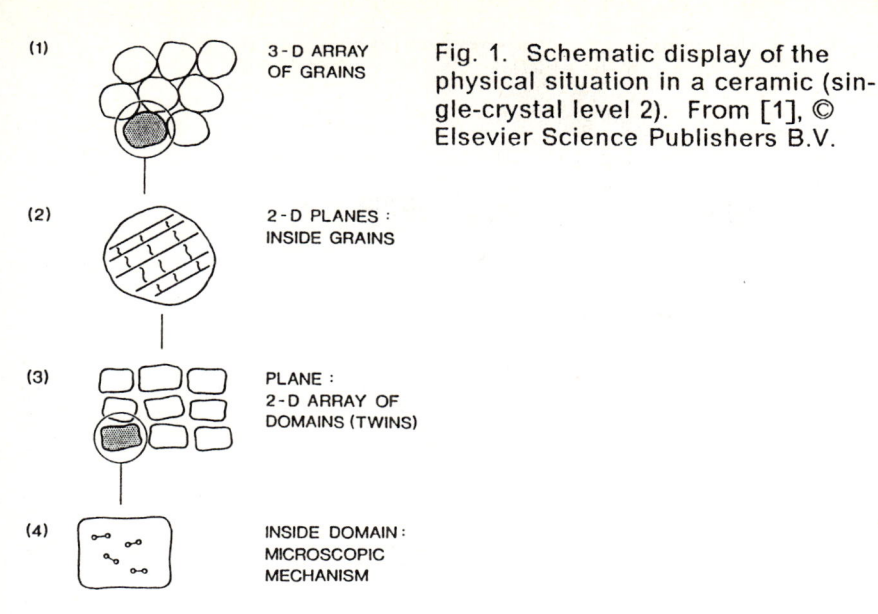

Fig. 1. Schematic display of the physical situation in a ceramic (single-crystal level 2). From [1], © Elsevier Science Publishers B.V.

2. Macroscopic Glass

The first experimental findings of glassy behavior in high-T_c materials were made by Müller et al. [2]. In the spirit of spin-glass experiments [3,4], they considered zero-field cooling (ZFC) versus field-cooling behavior (FC) for a ceramic La-Ba-Cu-O system. Measuring the susceptibility (Figure 2), they found reversible behavior after field cooling. In sharp contrast, metastability appeared in the irreversible behavior after ZFC. The two curves met at a temperature later denoted as $T_c^*(H)$, H = magnetic field, above which only reversible behavior was detected. Thus, $T_c^*(H)$ is the temperature below which metastable behavior occurs just as in the corresponding spin-glass experiment.

Taking measurements in different magnetic fields clarified the analogy to spin glasses even further. A "quasi" de Almeida-Thouless line [5] was found, i.e.

$$H^{2/3} \sim T_c^*(0) - T_c^*(H) \tag{1}$$

just as in spin glasses (Figure 3). The prefix "quasi" had to be added, because further theoretical work [6] showed that the underlying mechanism for the $H^{2/3}$ behavior is different from that in spin glasses. Furthermore, experiments [2] confirmed nonexponential decay of magnetization or susceptibility (insert of Figure 2), which is also reminis-

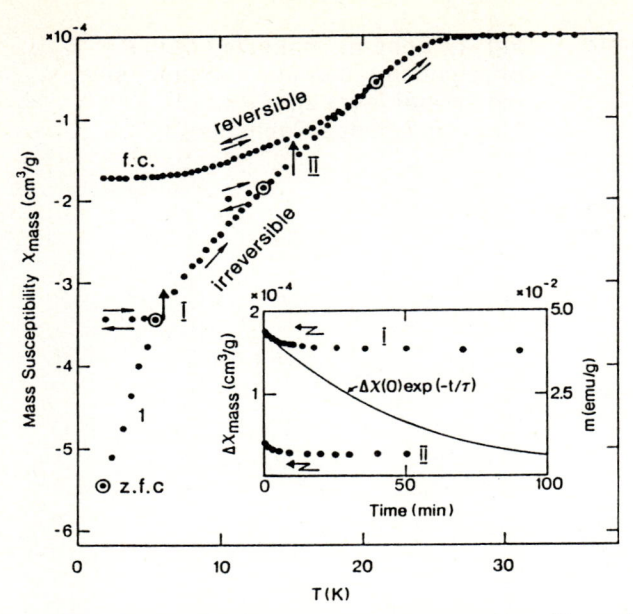

Fig. 2. Susceptibility versus temperature. Experimental situation for La-Ba-Cu-O. Field-cooling and zero-field cooling measurements. Insert: nonexponential time decay. From [2], © 1987 The American Physical Society

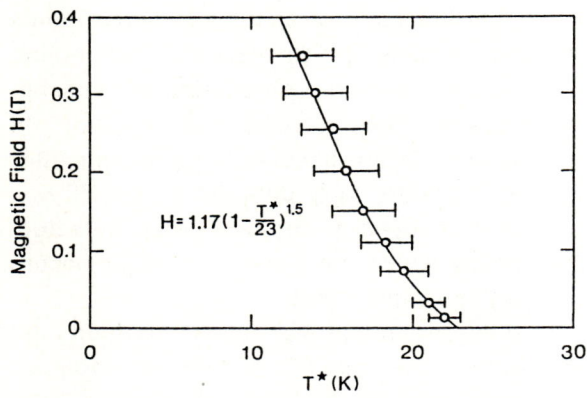

Fig. 3. "Quasi" de Almeida-Thouless line experiment. From [2], © 1987 The American Physical Society

cent of spin glasses. Therefore, a natural theoretical approach to the problem had to consider various (spin) glass models.

Carrying out numerical simulations for an XY-spin-glass model based on earlier work of Ebner and Stroud [7], we were able to repeat the experimental results. Figure 4 shows the FC and ZFC susceptibilities for

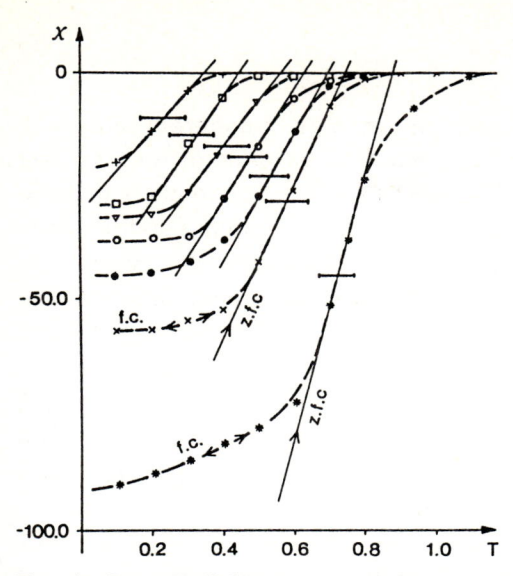

Fig. 4. Susceptibility versus temperature; numerical simulation; various magnetic fields from $H = 0.05$ (lowest curve) to $H = 0.40$ in steps of $\Delta H = 0.05$ in units of $2\pi/\phi_0$ (except $H = 0.35$). ϕ_0 is the elementary flux quantum. From [6], © Springer-Verlag 1987

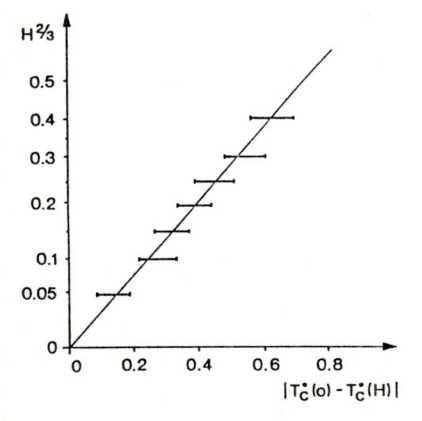

Fig. 5. "Quasi" de Almeida-Thouless line numerical simulation. From [6], © Springer-Verlag 1987

various magnetic fields H. The resulting error bars denote $T_c^*(H)$. Figure 5 shows nothing unexpected: plotting $H^{2/3}$ as a function of $T_c^*(0) - T_c^*(H)$ results in a straight line. Furthermore, the model was surprisingly successful in describing the features of high-T_c glass experiments [8]. We refer the reader to [6] for details.

The theoretical model of the XY-spin glass of Ebner and Stroud [7] can be understood as a disordered array of Josephson junctions. The following Hamiltonian describes the system:

$$-\beta \mathcal{H} = J \sum_{<ij>} \cos(\phi_i - \phi_j - A_{ij}) \ . \qquad (2)$$

ϕ_i (the XY-spins) describes the regions or domains of coherent phases in the system, originally the (physical) grains of the ceramic or the granular superconductor, as considered in [7]. But it soon became clear that the La-Ba-Cu-O ceramic behaved differently — the domains of the glass model had to be located *inside* the grains. Deutscher and Müller [9] solved the resulting puzzle. The extremely short coherence length in the system and the existence of twin boundaries lead to Josephson junctions or, more generally, to weak links inside the grains. Therefore we predicted glassy behavior also in single crystals, especially the occurrence of a $H^{2/3}$ line which was later found by Malozemoff's group [10]. But they gave a different explanation based on their flux creep picture. I will comment on this later on.

Returning to the XY-spin-glass model we may consider it an approximation to the Landau-Ginzburg theory omitting higher-order terms but including disorder as considered in [7]. It should be noted that the model is only relevant at higher temperatures close to T_c, since it does not include supercurrents. A careful study in T. Schneider's group later on showed no substantial differences to the results presented especially concerning the AT-line [11]. The resulting disorder is described by the phase factor A_{ij} in the model which is given by the line integral

$$A_{ij} \sim \int_i^j \vec{A} \, d\ell \qquad (3)$$

from site i to site j over the vector potential $\vec{A}$. Sites i and j reflect the positional disorder in the system.

The main feature of Hamiltonian (2) is frustration. Figure 6 shows the effect well known to spin-glass physicists [3]. Spin 4 simply does not know whether to follow spin 1 in accordance with phase factor A_{41} or spin 3 due to A_{34}. It is frustrated. The origin of frustration is disorder leading to "randomized" A_{ij}. The interaction in the system via the weak links leads to cooperative effects — the glass behavior. Glass behavior is best described by considering the energy landscape of the system (Figure 7). Again, it is well known that we have a hierarchy of barriers, as shown in the 2-D section through the multidimensional phase space (denoted by coordinate P). Most of the behavior results from the fact that the system has to hop over all these little but also those larger barriers, i.e. the hierarchy of barriers in the system.

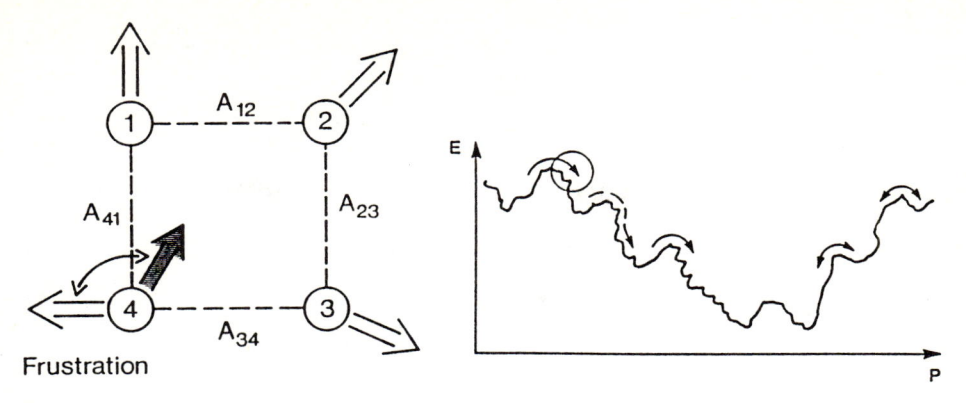

Fig. 6. Frustration in theoretical model (see text). From [6], © Springer-Verlag 1987

Fig. 7. Energy landscape of glassy system. Hierarchy of barriers. From [12]

Mota et al. [13] realized that the decay of the magnetization with time showed $M(t) \sim \ln t$. They compared their results with the traditional flux creep picture [14], which indeed predicted this type of decay. Malozemoff et al. [10] even went a step further: they argued that the traditional flux creep picture leads to a sufficient description of their experimental findings, considering just a single or only a few barriers originating from the pinning forces acting on the flux lines. It is important to note that these experiments were carried out at low temperatures and that only "relatively" short measurements were taken. In Mota et al.'s case [13], the measurements were taken for up to 24 hours, but at "relatively" low temperatures (about helium temperature) no deviation from the $\ln t$ behavior was found. Here it has to be noted that the emphasis is on "relatively", as we shall see in the next paragraph.

Back to Figure 7, our hierarchy of barriers. By simply considering the slope of a hill, we know from spin-glass research that this slope consists of a "rough surface" [15], shown in Figure 8. Then it immediately becomes clear that the flux creep theory, based on a similar phenomenological physical picture, only describes the decay of the system down one single slope. Considering the average size of a barrier U, Malozemoff et al.'s experiments [10] show that it is (a) relatively too low for the traditional flux creep picture, but (b) also relatively too high for the glass picture, again in the traditional sense. Malozemoff's reasoning leads to the introduction of the giant flux creep picture, which merely allows more flux to creep over the relatively small barriers [10]. But what happens at higher temperatures when the hopping probabilities are

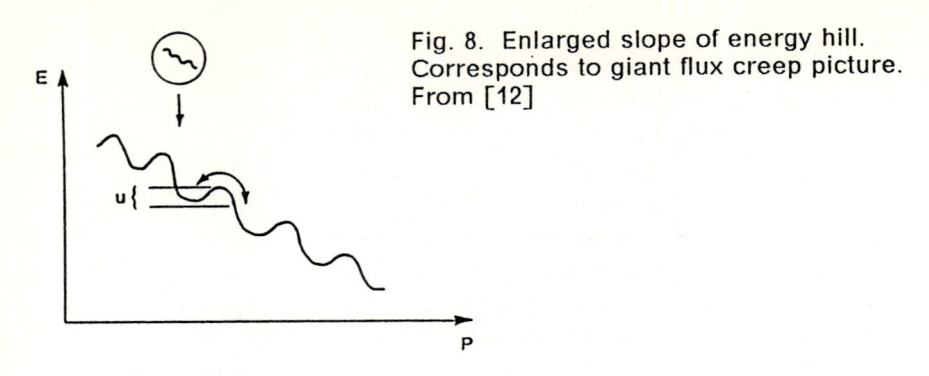

Fig. 8. Enlarged slope of energy hill. Corresponds to giant flux creep picture. From [12]

higher? Is the system then able to experience our nice hierarchy of barriers? The answer is certainly yes! Giant flux creep or the influence of only a few barriers is seen at temperatures below about 70% T_c and relatively short times which can be up to several days or weeks at helium temperatures. Here the phenomenological approach to the problem, the flux creep picture, is certainly valid; yet, we always have to keep in mind that we are dealing with a glass, but with such long relaxation times that they do not show up in ordinary low-temperature experiments. Above about 70% T_c, however, glass behavior can be detected on relatively reasonable time scales (Figure 9). The appropriate magnetic fields are roughly in the range from 300 G to 2000 G. Too close an approach to T_c destroys the glassy behavior again because we experience critical effects. Magnetic fields outside the above range take the system out of the glass regime for the temperatures mentioned [6].

The fundamental message of this paper is the following: in the regime of about 70%-95% T_c, we will find glass behavior and rich new physics in the field of superconductivity. In the following I would like to comment on some typical glass features in the new superconductors.

GIANT FLUX CREEP

$M(t) \sim \ell n\, t$

ANDERSON - KIM

GLASS BEHAVIOR

$M(t) \sim \dfrac{e^{-(t/\tau)^\beta}}{t^x}$

KOHLRAUSCH

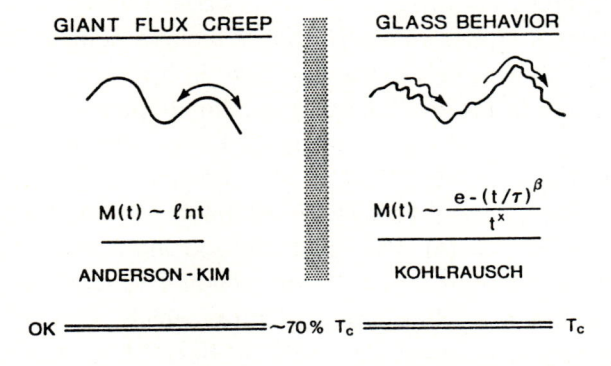

OK ══════════ ~70% T_c ══════════ T_c

Fig. 9. Schematic view of experimental situation. From [12]

2.1. Glassy Dynamics

Glassy systems are usually considered to exhibit Kohlrausch (type) relaxation, for example the magnetization $M(t)$ should decay as follows

$$M(t) \sim \frac{e^{-(t/\tau)^\beta}}{t^x}, \qquad (4)$$

with $0 < \beta < 1$ and $0 < x < 1$; τ is the relaxation time. Note that the short-term behavior of $M(t)$ can be very well fitted to $\ln t$. Therefore, only very time-consuming experiments can solve (4), because knowledge of the entire decay is necessary to unambiguously show Kohlrausch behavior. Nevertheless, Kohlrausch decay does not prove glass behavior, as it also exists in an Ising-ferromagnet below T_c. It could, however, disprove it. A particular complication also has to be taken into account: the large anisotropy in the superconductors which allowed the implementation of 2-D simulations. Numerical simulations of a sufficient quality are currently not available but recent experiments on 2-D spin glasses by Dekker et al. [16] show activated dynamic behavior:

$$M(t) \sim e^{-(\ln(t/\tau))^\delta} \qquad (5)$$

with $\delta > 1$. Activated dynamics was first introduced by Fischer and Huse [17] for 2-D spin glasses. For higher temperatures the form expressed in (5) could also be expected in high-T_c superconductors.

2.2. Memory Effect

In the search for a critical experiment which favors the glass picture because it cannot be explained in terms of the giant flux creep [10], we tried to repeat a memory effect known hitherto only in spin glasses, where it was first discovered by Lundgren et al. [18]. The experiment was carried out in collaboration with C. Rossel and Y. Maeno. After field-cooling the crystal in $H_0 = 500$ G and aging t_w up to 24 hours, the decay of the magnetization measured after switching the field by $\Delta H = 1000$ G showed a characteristic shift of shape immediately following the same time t_w. Such an effect can only be described by a hierarchy of barriers or the resulting broad distribution of relaxation times [19].

2.3. "Quasi" de Almeida-Thouless line

As this line is found only in the "glass regime", its origin is clearly related to the hierarchy of barriers. Other approaches which take only the flux creep single barrier picture into account are bound to fail. At this point, it becomes clear why the certainly "oversimplified" XY-spin-glass model was so successful. It contains the essential physics — the hierarchy of barriers — and therefore reproduces the (universal) $H^{2/3}$ behavior. Both the flux creep approach and the XY-model describe different sides of the same coin. The XY-model can be considered a rather basic approach to the Landau-Ginzburg theory, because it is limited to the physics essential for a more qualitative description of the glass phase. Further intensive work will certainly be necessary to obtain a quantitative theory for the glass behavior.

Summarizing at this point, I wish to emphasize that a whole new world of glassy effects is waiting to be discovered in high-T_c experiments. In view of applications, it should be noted that, when we deal with nitrogen temperature, glass behavior becomes dominant and has to be understood. Especially in view of reaching higher critical currents in the new superconductors it obviously plays an important role.

From the theoretical point of view, our current situation is the following. We do not find a real contradiction between the (giant) flux creep and the glass picture. Flux creep is merely a phenomenological macroscopic approach to the glass behavior for low temperatures and relatively short times. As already pointed out, the XY-spin-glass model has to be considered a basic approach to the Landau-Ginzburg theory including disorder. The XY-spins are related to the different phases in the theory, and not to any physical spins. The term "spin glass" should thus be understood with the emphasis on glass. In contrast, we will discuss a real spin glass in the next chapter, the microscopic spin glass with spins on the copper sites.

In general I wish to emphasize that for the temperatures of technical interest around 77 K, critical currents are influenced by cooperative (glassy) pinning effects, in contrast to the traditional picture. For technical applications when higher critical currents are required, the cooperative (glassy) behavior is of great importance.

3. Microscopic Spin Glass

In this chapter I consider a model recently introduced by Aharony et al. [20] for the insulating phase of high-T_c superconductors. The model is

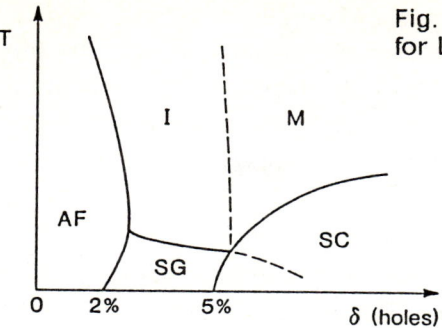

Fig. 10. Experimental phase diagram for La-Sr-Cu-O

referred to as the Aharony model. Figure 10 schematically shows the experimental situation for La-Sr-Cu-O [21,22]. With increasing hole concentration δ, we observe a breakdown of the antiferromagnetic phase (AF). Then a spin-glass (SG) or spin-glass-like phase occurs before superconductivity (SC) sets in for higher doping. It is unclear at the moment whether a spin-glass phase coexists at lower temperatures with the superconducting phase. The observed effects could be due to mixed phases in the samples. The Aharony model [20] tries to describe the physical situation with emphasis on (a) the insulating phase (antiferromagnet, spin glass) and (b) the superconducting mechanism. They considered the following Hamiltonian:

$$\mathcal{H} = -\sum_{<ij>} J_{ij} S_i S_j \tag{6}$$

describing a 2-D spin-1/2 Heisenberg model with random interactions J_{ij} which depend on the hole concentration. Figure 11 describes the phys-

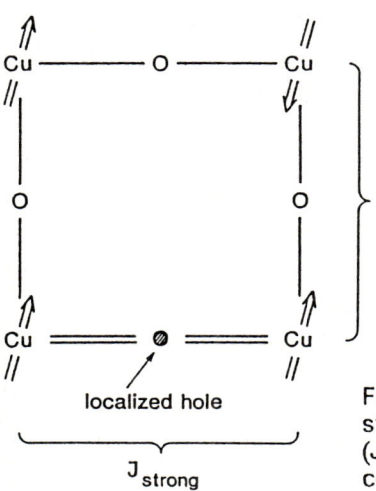

Fig. 11. Localized hole causes strong ferromagnetic coupling (J_{strong}) between adjacent copper spins

ical situation considered by Aharony and coworkers. We have Heisenberg spins on the copper sites antiferromagnetically coupled via the oxygen sites. Holes are now assumed to be localized in the π orbital on the oxygen sites, leading to strong ferromagnetic couplings between the adjacent copper spins. Therefore we have

$$J_{ij} = \begin{cases} -J_{strong} & \text{if hole} \\ J & \text{otherwise.} \end{cases} \quad (7)$$

The localization of the holes could be caused by variable range hopping and the disorder in the superconductor. The Aharony Model is currently under attack by new experimental results. I refer the reader to the article of Mehring et al. in these proceedings, who found the holes in the σ rather than in the π-orbital as originally proposed by Aharony et al. The σ-orbital has been considered not to lead to frustrated bonds. But recent diagonalization results of Cu_2O_7 clusters precisely lead to frustrating ferromagnetic bonds as in the Aharony Model [23]. I do not wish to go more deeply into detail and refer the reader to the paper of Aharony et al. [20]. Here I intend to concentrate on numerical simulations of Hamiltonian (6) with different degrees of hole concentration (i.e. ferromagnetic bonds).

3.1. Quantum Monte Carlo Simulations

For the quantum Monte Carlo simulations, the Trotter-Suzuki transformation [6] was applied. The transformation maps a d-dimensional quantum onto a $d+1$-dimensional classical system. The resulting classical simulation is usually more complex than standard Monte Carlo work. The main reason are the forbidden configurations in the $d+1$-dimensional system. A way around this problem is the "world-line" formalism introduced by Hirsch et al. [24]. This algorithm usually works well for a "bosonic" system, including the Heisenberg antiferromagnet [25]. But for fermionic systems an additional extremely severe complication occurs, the minus-sign problem [25]. For frustration in a Heisenberg model (induced by the ferromagnetic bonds), we also have negative contributions to the partition function and therefore also a minus sign problem. In the case of model (6), the problem could be circumvented by the "barrier trick" first published in [26] and recently worked out in more detail in [27]. It is important to note that the choice of

$$J_{strong} = -5 \cdot J \quad (8)$$

and the relatively small amount of randomness allow efficient simulations for the model.

3.2. Numerical Results

3.2.1. Spin-glass antiphase. The simulations were carried out mainly for 16×16 and 32×32 lattices. The temperature was $T = 0.05 \times J$ with 400 Trotter slices. Figure 12 shows the staggered magnetization M_{st} versus the hole concentration δ for 32×32 lattices. First of all I wish to emphasize that the $\delta = 0$ case, i.e. the Heisenberg spin-1/2 antiferromagnet, does not show a breakdown of the staggered magnetization with increasing system size, at least not up to 32×32 as anticipated by part of the scientific community. It should be noted that this does not exclude the possibility that for much larger system sizes a breakdown could occur caused by a "turnover" of clusters (i.e. topological excitations) much larger then the present system size. But a value of $M_{st} = 0.30 \pm 0.01$ can be reported for 32×32 lattices. Increasing δ leads to a decrease of M_{st} until it vanishes at about $\delta_c = 2.5\%$. Then the staggered magnetization is almost zero. Due to finite size effects and the dynamical symmetry breaking in the z-direction, it is not exactly zero in the simulation. It should be noted that δ_c is a relatively rough estimate based on only two 32×32 lattices. Averaging over ten 16×16 lattices yields a similar value.

In this paper, however, I wish to concentrate on a qualitative description of the physical situation which allows experimental tests. A value of δ_c depends on too many experimentally unknown parameters,

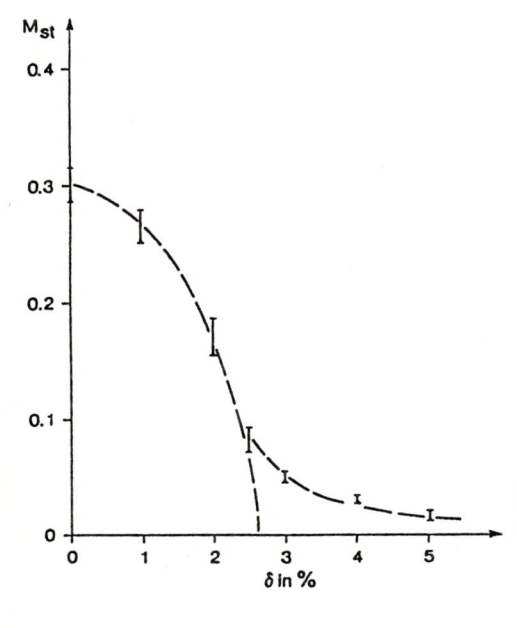

Fig. 12. Staggered magnetization M_{st} versus hole concentration δ

for instance J_{strong}. But the qualitative picture is clear: increasing the number of "localized" holes leads to a breakdown of the staggered magnetization for an experimentally tolerable hole concentration. Especially if one considers the suggestion of Aharony et al. [20] that the holes are not precisely localized on the oxygen sites but are still somewhat mobile, it is possible to decrease δ_c considerably leading to a more reasonable value compared to experiment. It should be noted that mobile holes are much more difficult to treat in a numerical simulation due to the more severe minus-sign problem. Such simulations have not been carried out by the author.

Holes are always localized in the present simulation. In Fig. 13 we switch to the measurement of the Edwards-Anderson (EA) order parameter for spin glasses. It is defined as

$$q_{EA} = \frac{1}{N} \sum_i <S_i>^2.$$

Figure 13 shows the square root of q_{EA}, because this figure is meant to be compared to μSR experiments [13]. In the antiferromagnetic phase, q_{EA} is obviously also different from zero. Increasing the hole concentration, however, does not produce a breakdown at $\delta_c \simeq 2.5\%$ but rather an almost constant value of q_{EA} for increasing δ up to $\delta \simeq 6\%$, where it drops to zero. Thus we deal with the fact that $M_{st} = 0$ for $\delta > 2.5\%$ while $q_{EA} > 0$ in that regime up to $\delta = 6\%$. To explain the origin of this

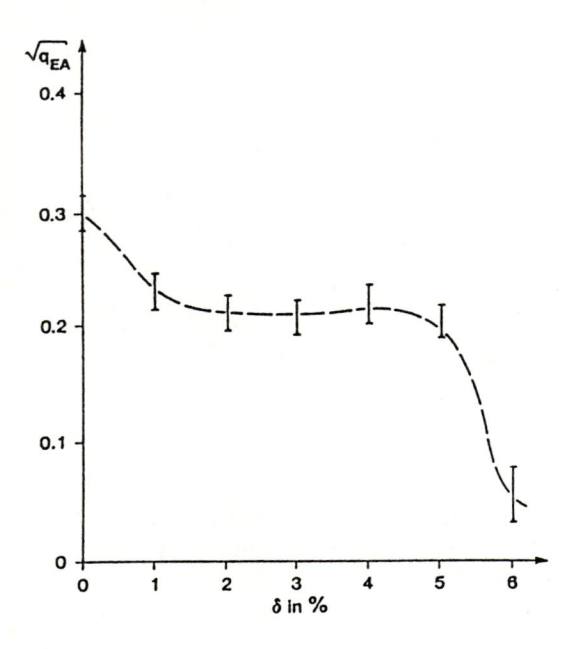

Fig. 13. Edwards-Anderson order parameter versus hole concentration δ

behavior, I show in the following the local magnetizations in z-direction (S_i^z) for various δ for the corresponding 32×32 lattices. In Fig. 14 for $\delta = 1\%$ the numbers n_i denote

$$n_i = \text{integer}\{10 \cdot (-1)^i \cdot \langle S_i^z \rangle\}, \tag{9}$$

i.e. the local magnetizations are projected onto the Néel state to make deviations visible by a minus sign. The number 3 means a local magnetization $S_i^z = \pm 0.3$, with the sign corresponding to its lattice position. In Fig. 14 (for $\delta = 1\%$), we notice numerous 3's, which is reminiscent of the $\delta = 0$ case. But due to the randomness, weaker regimes are also found in the larger cluster of 1's near the center. A complete turnover is even seen in the left corner, denoted by the minus sign. This means that the Néel order is shifted by one lattice spacing. The position of the strong bonds leads to pairs of the type $-5,5$ or $-4,4$. The origin of the weaker clusters or the turned-over clusters is therefore connected to the occurrence of the (strong) ferromagnetic bonds or, more precisely, to the occurrence of local frustration. The behavior shown Fig. 14 is reminiscent of a Swiss cheese. I therefore call the situation for $\delta < \delta_c \simeq 2.5\%$ Swiss-cheese behavior.

Figure 15 shows $\delta = 4\%$, where we have $M_{st} = 0$ but $\sqrt{q_{EA}} \simeq 0.20$. In this case we find large clusters of either positive or negative n_i. That means that the antiferromagnetic (AF) correlation length ξ_{AF} is finite but large. In contrast, the spin glass (SG) ξ_{SG} diverges. A situation like this is well known for 2-D spin glasses. It was first reported by Maynard et

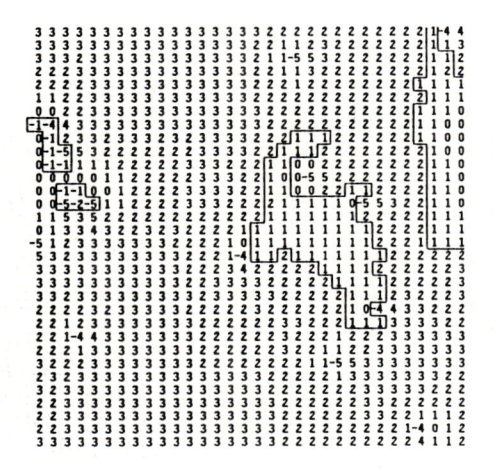

Fig. 14. Swiss-cheese configuration, $\delta = 1\%$

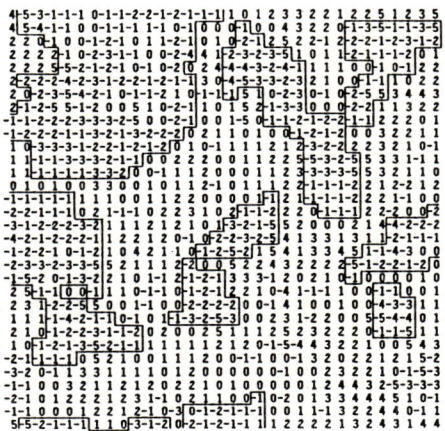

Fig. 15. Spin-glass antiphase, $\delta = 4\%$

al. [28] and was later the subject of a controversy also involving myself [29]. In spin glasses, the classical $\pm J$ model for Ising spins was considered. Increasing the concentration of negative ($-J$) bonds leads to a breakdown of the ferromagnetic order according to a Swiss-cheese effect as described here in contrast to Fig. 14. In the spin-glass antiphase, we have large clusters of up and down spins rigidly coupled to each other, leading to a divergence of the spin-glass correlation length ξ_{SG}. The existence of an antiphase in the 2-D $\pm J$ model has recently been supported by the numerical findings of Ozeki and Nishimori [30].

One of the major results discussed in the present paper is the existence of a spin-glass antiphase between the antiferromagnetic and superconducting phase in the high-T_c lamellar compounds. Increasing δ to 6% in Fig. 16, we see a breakdown of the up or down clusters (always projected on the Néel state). The mechanism again seems to be similar to the $\pm J$ spin-glass model. As described in an overview by the author [17], the nonexistence of a phase transition in 2-D spin glasses is due to the existence of finite clusters of relatively rigidly coupled spins, which can be turned against each other with no cost of energy. These clusters were called zero-energy loops. Thermal fluctuations now destroy any long-range order for finite temperatures, while we still have a $T=0$ phase transition. A similar picture evolves here. We also have the type of finite zero-energy clusters, which turn against each other during the simulation. But here it is clearly seen that the quantum fluctuations also destroy long-range order at zero temperature. Note that the simulations had to be carried out for $T = 0.05 \cdot J$, but for the finite system size (correlation length/system size) we can suppose ground state properties.

Figure 16 shows numerous zeros because of the movement of the zero-energy clusters against each other. The remaining clusters of larger numbers are short-term or freezing effects. The dynamical averaging of the clusters to zero takes a substantial amount of additional simulation time. In equilibrium all numbers have to be zero. This is almost the case for $\delta = 8\%$ (Fig. 17), where only a small cluster on the left-hand side survived so far. Thus I conclude that at about $\delta \simeq 6\%$ the antiphase breaks down and even at $T=0$ we do not have a spin-glass phase. The mechanism is comparable to the classical case. Finite clusters with zero-energy surfaces turn against each other (due to quantum fluctuations) and destroy the long-range order. Summarizing the situation at this point, I wish to emphasize that the current simulations are quite similar to a classical Heisenberg spin-glass simulation supporting Aharony et al.'s [20] point of view, which considers only the classical

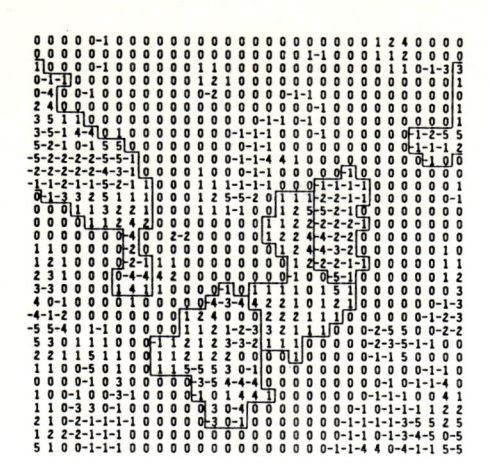

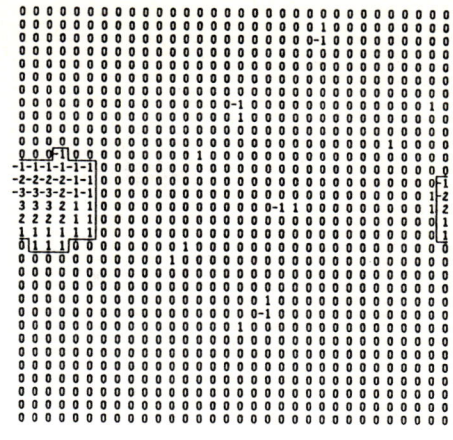

Fig. 16. Spin-glass phase, $\delta = 6\%$ Fig. 17. Spin-glass phase, $\delta = 8\%$

case. An experimental prediction is that the current situation for $T > 0$ can be understood in terms of a 2-D classical Heisenberg spin glass. This includes the occurrence of typical spin-glass effects in experiments such as cusp in the susceptibility, specific heat, divergence of nonlinear susceptibility, field-cooled versus zero-field-cooled measurements etc. In particular, typical freezing effects should be observed in future experiments.

3.2.2. Microscopic pairing. The Aharony model does not only consider the antiferromagnetic and the spin-glass phases in the high-T_c superconductors, but more importantly also considers a microscopic mechanism for superconductivity which is embedded in the framework of the phases mentioned earlier. Due to the frustration effects caused by the localized holes, an attractive interaction between two holes should be established according to Aharony. I performed simulations for two holes in an antiferromagnetic background (no further ferromagnetic bonds) for increasing distances r.

Figure 18 shows the difference in energy ΔE versus $1/r^2$, not confirming the prediction of Aharony et al. [20] of a dipolar interaction between the two holes. Figure 18 shows relatively large error bars for the 24×24 runs. Due to these error bars, it is impossible to state whether two holes attract each other even for very short distances. Therefore it is truly impossible to state whether the Aharony mechanism works at all. These findings are supported by recent exact diagonalizations of small lattices by von der Linden and coworkers [31].

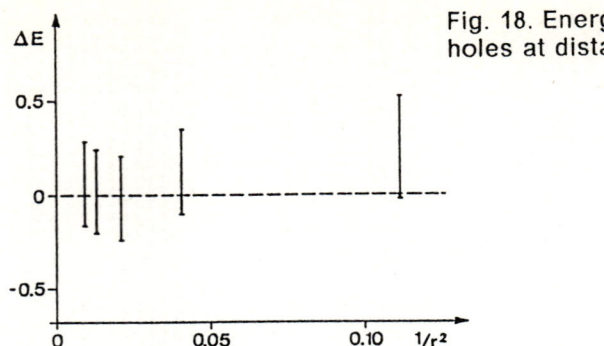

Fig. 18. Energy difference ΔE for two holes at distance r, $J_{strong} = -5 \cdot J$

For numerical simulations of the present type it is almost impossible to rule out a very small dipolar attraction because of the strong fluctuations in the energy caused by the minus-sign problem. But to allow for a reasonable attraction for high-T_c superconductivity, the remaining possible energy differences are too low in any case. Furthermore the calculations for Fig. 18 were carried out for a complete antiferromagnetic background. As we have seen before this is certainly not true for the hole concentration where the real system becomes superconducting. In that case we already have the (pure) spin-glass phase, where antiferromagnetic clusters have already disappeared completely. Just when the hole concentration is sufficient for superconductivity in a real sample, the essential antiferromagnetic background is gone and therefore also the attraction between holes mediated by this background.

Hence it is difficult to see how an overall picture for the insulating and superconducting phases should hold. The only solution seems to be a scenario where increased doping first localizes all holes and then causes more and more holes to stay mobile. This means that we do not have a breakdown of the spin-glass antiphase, which contains relatively large antiferromagnetic clusters. Nonetheless the doubts as to an attractive interaction in the quantum case remain. Thus the Aharony model — while nicely describing antiferromagnetic and spin-glass phases — has considerable difficulties to survive as a model for the microscopic mechanism. But all other mechanisms based on an antiferromagnetic background such as Schrieffer's spin bag theory are still compatible with the present scenario.

At this point we can formulate an important experimental question. Are the holes at low doping localized and are they becoming more and more mobile with increased doping? A fraction of localized holes may lead to a spin-glass antiphase, which could provide the necessary background for some kind of microscopic mechanism. Considering the Aharony model, the present simulations provide evidence for its validity for low doping. But exactly this nice agreement for low doping almost rules out its validity for the microscopic mechanism.

In the following, I will discuss the possibility that magnetic correlations do not play an essential role in high-T_c superconductivity. Considering again the above spin-glass-antiphase concept, I first want to consider the temperature dependence of the spin-glass correlations. We have relatively rigid antiferromagnetic clusters. They show comparably strong dynamic correlations at relatively high temperatures. This is due to the movement of the clusters as a whole. As the clusters are rather large, we obtain a commensurate signal in neutron scattering measurements. The spin-glass antiphase should not open a gap at lower temperatures. The dynamic correlations should already exist at a temperature not significantly lower than the Néel temperature. The temperature dependence should be very weak. On increasing doping, the antiferromagnetic clusters should become smaller and smaller, therefore an incommensurate signal has to occur. A gap in the spin excitations could occur for the pure quantum spin-glass state. The occurrence of the pure spin-glass state is at only $\delta \simeq 6\%$, where the system becomes superconducting. This gap could open at relatively high temperatures well above the critical temperature of the superconductor. Superconductivity could evolve just inside this gap. This means that the frustration in the model leads to a spin-glass state which just opens a gap to allow some kind of nonmagnetic mechanism without being influenced by strong magnetic correlations. This scenario may be an explanation of the recently published neutron scattering experiments by Shirane et al. [32]. Following the above reasoning, it is possible that all the nice antiferromagnetic and spin-glass phases have no direct connection with the superconducting mechanism. Moreover it is possible that the spin-glass state merely has the property of opening a sufficiently large gap in the spin excitations to stay out of the way of superconductivity.

Considering only the microscopic mechanism, I want to add a further figure for the 2-D single-band Hubbard model, Fig. 19. It is the preliminary result of a collaboration with W. von der Linden and H. de Raedt, using a modified Sorella algorithm for the quantum Monte Carlo simulations. For different repulsive interactions U, it shows the resulting energies versus the number of holes for 4×4 lattices. For small U, we notice almost straight lines indicating the absence of binding of holes. For large U, we even see repulsive interaction between holes as a result of

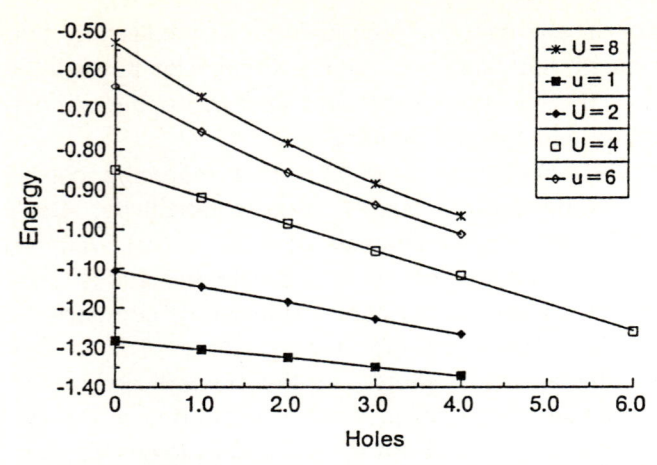

Fig. 19. Ground state energy of 2-D single-band Hubbard model versus number of holes for different repulsive interactions U (4×4 system)

the finite lattice size. Although we only have 4×4 lattices so far, we interpret this preliminary result as evidence against the pairing of holes in the 2-D single-band Hubbard model. Simulation for larger systems are in progress.

3.3. Summarizing Remarks

In conclusion I would like to recap the scenario leading to the various phases when doping is increased in the Aharony model. In Fig. 20 the three phases are schematically shown. For $0 < \delta \lesssim 2.5\%$, we have the Swiss-cheese phase with holes of weaker clusters on an antiferromagnetic background. Above 2.5%, the spin-glass antiphase exists up to about 6%. It is characterized by large but finite antiferromagnetic clusters leading to a commensurate signal in neutron scattering experiments. Above 6% we have a pure spin-glass phase with very short antiferromagnetic correlations and also finite spin-glass correlations. This state opens a gap in spin excitations already at relatively high temperatures caused by the zero-energy clusters in the system. The clusters lead to an incommensurate signal. Further doping should lead to mobile holes. This point could not be clarified by the present simulations. Considering the pairing mechanism in the Aharony model, it is relatively unlikely according to the measurements of the hole-hole attraction and the vanishing of an antiferromagnetic background for increased doping. Therefore, the Aharony model provides a consistent description of the insulating phase for low doping in terms of the Swiss-

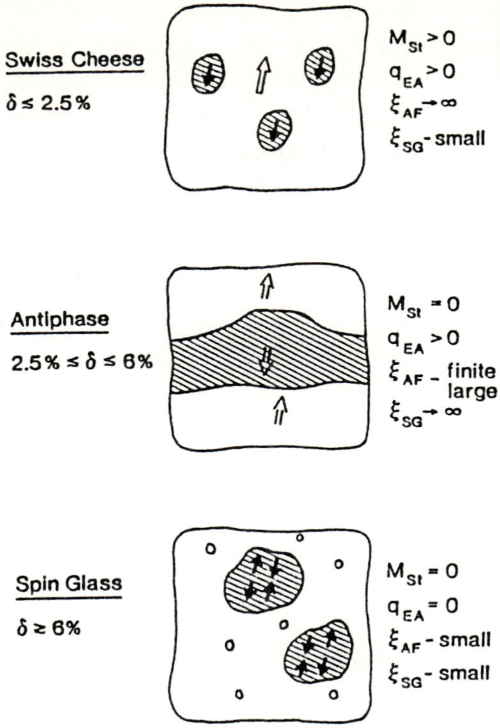

Swiss Cheese
$\delta \leq 2.5\%$

$M_{St} > 0$
$q_{EA} > 0$
$\xi_{AF} \to \infty$
ξ_{SG} - small

Fig. 20. Schematic scenario for increased doping in the Aharony model. See text

Antiphase
$2.5\% \leq \delta \leq 6\%$

$M_{St} = 0$
$q_{EA} > 0$
ξ_{AF} - finite large
$\xi_{SG} \to \infty$

Spin Glass
$\delta \gtrsim 6\%$

$M_{St} = 0$
$q_{EA} = 0$
ξ_{AF} - small
ξ_{SG} - small

cheese and the spin-glass antiphases. But depending on the degree of localization for further doping it allows (a) other magnetic mechanisms which depend on an antiferromagnetic background and even (b) non-magnetic mechanisms since the spin-glass state opens a gap in the spin excitation at high enough temperatures to avoid any disturbance of the superconducting state. The predictions of the Aharony model have to be tested in careful experiments.

4. Conclusion

In conclusion of these two chapters, taking macroscopic and microscopic glass together, we clearly recognize the usefulness of (spin) glass concepts within the framework of high-T_c superconductivity. While in the first case the analogy leads to experimental findings such as the famous (quasi) de Almeida-Thouless line, in the second case we have a "real" spin glass with "real" spins on the copper sites. At first glance, there is — apart from the overall concept — no real connection between the two levels, but a somewhat speculative conjecture may be presented at the end. For macroscopic glass, an essential ingredient are the domains or clusters of level 3 in Fig. 1. These clusters may simply be related to the clusters in the spin-glass phase or the antiphase. Clusters in spin glasses are surrounded by a cloud of "fast flipping" spins which could provide the weak links so essential to the macroscopic glass. Considering the possibility that we have a nonmagnetic mechanism and a gap opening in the spin excitations to allow for just this mechanism, this gap could be suppressed in the "fast-flipping" cloud to produce the weak links. But at this point, I will end this speculation which on the other hand could become the subject of extremely complex experimental ventures.

As a more realistic conclusion, I wish to stress once again the usefulness of the glass picture on the macroscopic level in the direction of technical applications concerning critical current densities and in the microscopic case considering more fundamental questions.

References

1. I. Morgenstern, K. A. Müller and J. G. Bednorz, *Physica C* **153-155**, 59 (1988).
2. K. A. Müller, M. Takashige and J. G. Bednorz, *Phys. Rev. Lett.* **58**, 1143 (1987).
3. For a recent review, see: *Lecture Notes in Physics*, **275**, Heidelberg Colloquium on Glassy Dynamics, J. L. van Hemmen and I. Morgenstern, eds. Springer-Verlag, Berlin Heidelberg, 1987.
4. For a recent review on spin glasses in particular, see: K. Binder and A. P. Young, *Rev. Mod. Phys.* **58**, 801 (1986).
5. J. R. L. de Almeida and D. J. Thouless, *J. Phys. A* **11**, 983 (1978).
6. I. Morgenstern, K. A. Müller, and J. G. Bednorz. *Z. Phys. B* **69**, 33 (1987).
7. C. Ebner and D. Stroud, *Phys. Rev. B* **31**, 165 (1985).
8. H. Keller, B. Pümpin, W. Kündig, W. Odermatt, B. D. Patterson, J. W. Schneider, H. Simmler, S. Connell, K. A. Müller, J. G. Bednorz, K. W. Blazey, I. Morgenstern, C. Rossel, and I. M. Savic, *Physica C* **153**, 71 (1988).
9. G. Deutscher and K. A. Müller, *Phys. Rev. Lett.* **59**, 1745 (1987).
10. L. Krusin-Elbaum, A. P. Malozemoff, Y. Yeshurun, D. C. Cronemeyer, and F. Holtzberg, *Physica C* **153**, 1469 (1988) and references therein.
11. R. Hetzel, M. Vanhimbeeck, and T. Schneider, *Z. Phys. B* **76**, 259 (1989).
12. I. Morgenstern, K. A. Müller and J. G. Bednorz, in: Proceedings of the 2nd Yukawa International Seminar (YKIS'88), Springer-Verlag, in press.
13. A. C. Mota, A. Pollini, P. Visani, K. A. Müller, and J. G. Bednorz, *Physica C* **153**, 67 (1988).

14. See for example: M. Tinkham, *Introduction to Superconductivity*, McGraw-Hill Inc., New York, 1975.
15. I. Morgenstern, in: *Lecture Notes in Physics*, **192**, Heidelberg Colloquium on Spin Glasses, J. L. van Hemmen and I. Morgenstern, eds., Springer-Verlag, Berlin Heidelberg, 1983.
16. C. Dekker, A. F. Arts, H. W. de Wijn, A. J. van Duyneveldt, and J. A. Mydosh, preprint, submitted to Phys. Rev. B
17. D. S. Fischer and D. A. Huse, *Phys. Rev. Lett.* **56**, 1601 (1986).
18. L. Lundgren, P. Svedlindh, P. Nordblad, and O. Beckman, *Phys. Rev. Lett.* **51**, 911 (1983).
19. C. Rossel, Y. Maeno, I. Morgenstern, *Phys. Rev. Lett.* **62**, 681 (1989).
20. A. Aharony, R. J. Birgenau, A. Coniglio, M. A. Kastner and H. E. Stanley, *Phys. Rev. Lett.* **60**, 1330 (1988).
21. R. J. Birgenau et al., *Phys. Rev. Lett.* **59**, 1329 (1987).
22. G. Shirane et al., *Phys. Rev. Lett.* **59**, 1613 (1987).
23. D. Baeriswyl, private communication.
24. J. E. Hirsch, D. I. Scalopino, R. L. Sugar and R. Blancenbecler, *Phys. Rev. Lett.* **47**, 1628 (1981).
25. For a recent review see, *Quantum Monte Carlo Methods*, M. Suzuki, ed., Springer-Verlag, Berlin Heidelberg (1986).
26. I. Morgenstern, *Z. Phys. B* **70**, 291 (1988).
27. I. Morgenstern, preprint, submitted to *Z. Phys. B*.
28. R. Maynard and R. Rammal, *J. Physique (France) Lett.* **43**, L347 (1982).
29. I. Morgenstern, *Phys. Rev. B* **25**, 6071 (1982).
30. Y. Ozeki and H. Nishimori, *J. Phys. Soc. Jpn.* **56**, 3265 (1987).
31. W. von der Linden and P. Horsch, preprint.
32. G. Shirane, R. J. Birgenau, Y. Endoh, P. Gehring, M. A. Kastner, K. Kitazawa, H. Kojima, I. Tanaka, T. R. Thurston and K. Yamada, preprint, submitted to *Phys. Rev. Lett.*.

Microwave Absorption in Granular Superconductors

K.W. Blazey

IBM Research Division, Zurich Research Laboratory,
CH-8803 Rüschlikon, Switzerland

Abstract. The results of modulated microwave absorption experiments are described in terms of two kinds of loss processes. Viscous fluxon motion in the microwave field is the most widely observed source of dissipation. Another source of losses is junction breakdown which in certain cases can induce regular series of narrow absorption lines in a quantum interferometer structure such as in single crystal $YBa_2Cu_3O_{7-\delta}$.

1. Introduction

On cooling a homogeneous material through its metal to superconductor transition a rapid drop of the surface impedance occurs as the resistance due to the normal electrons freezes out when they couple to form Cooper pairs. Microwave absorption was proposed as a technique for observing this [1] and studied extensively in conventional superconductors during the 1960's [2]. However this effect does not cause the surface resistance to go to zero at very low temperatures and there is a residual resistance which is a sensitive indicator of the surface quality. This is particularly noticeable in the ceramic samples of the new high T_c cuprate superconductors [3] due to their granularity but is also apparent in thin films and single crystals of the same materials. In fact most superconductors will show microwave losses at low temperatures and magnetic fields.

The microwave absorption of superconductors is usually studied by measuring the reflectivity or the quality factor of a resonant cavity containing a sample of the material under study [2,4]. Recently, there have been many such studies on the high T_c cuprate superconductors [5-28] some of which used a conventional electron spin resonance (ESR) spectrometer to measure the modulated microwave absorption. An ESR spectrometer measures the microwave absorption of the sample in a resonant cavity, where its temperature can be varied, with an external magnetic field that can be swept. A modulation field is applied parallel to the external field and the microwave signal detected synchronously with it which normally gives the differential of the absorption. This is not the case with a superconducting sample. The field dependent microwave absorption of a superconductor depends on its magnetisation which is not always reversible due to flux trapping even at low fields. Another complication is that the signals are related to the superconducting properties of the surface which are not always the same as those of the bulk. Oxidation of the surface creates Josephson junctions at grain boundaries and other defects or even forms a surface phase.

2. Conventional Superconductors Pb and Nb

The modulated microwave absorption of Pb and Nb are shown in Figure 1 as a function of the external field. Both samples have oxidised surfaces, the Pb sample through exposure to the air and the Nb has been oxidised electrolytically. The spectra are typical of all superconductors in that the principal feature is an open hysteresis loop where the signal changes sign upon field sweep reversal at any field within the cycle. The spectrum for Pb is similar to that reported by Rubins et al. [29] who also noted that sample heating by large modulation fields can distort the spectrum. Similar distortions were obtained here also with increasing microwave power. The essential information of the Pb spectrum is that despite it being a type I superconductor there is a lot of hysteretic microwave absorption at all fields not only up to the critical field, $H_c \sim 520$ G at 4.3 K but also beyond even the critical field for surface superconductivity, $H_{c3} \sim 880$ G [30]. The Nb spectrum also shows hysteretic absorption over a 10 kG field range determined principally by a surface superconducting layer in such an oxidised sample [31]. The maximum in the increasing field and the minimum in the decreasing field sweep both occur near 3.5 kG and correspond to the H_{c2} of the surface layer [31]. In decreasing field there is very little absorption until near 6 kG which corresponds to the critical field for nucleation of surface superconductivity, $H_{c3} = 1.7\ H_{c2}$ [30]. With either large modulation or microwave fields the value of H_{c3} was depressed to lower fields probably due to sample heating [29]. The small minimum near 3.3 kG is due to a defect electron spin resonance in the oxidised surface with $g = 2.3$ and is not related to the superconductivity or the conduction electron spin resonance of Nb which has $g = 1.84$ [32].

Viscous fluxon motion is the usual source of dissipation in bulk superconductors even at microwave frequencies [33]. The oxidation of the surface of conventional superconductors creates a layer with a very low

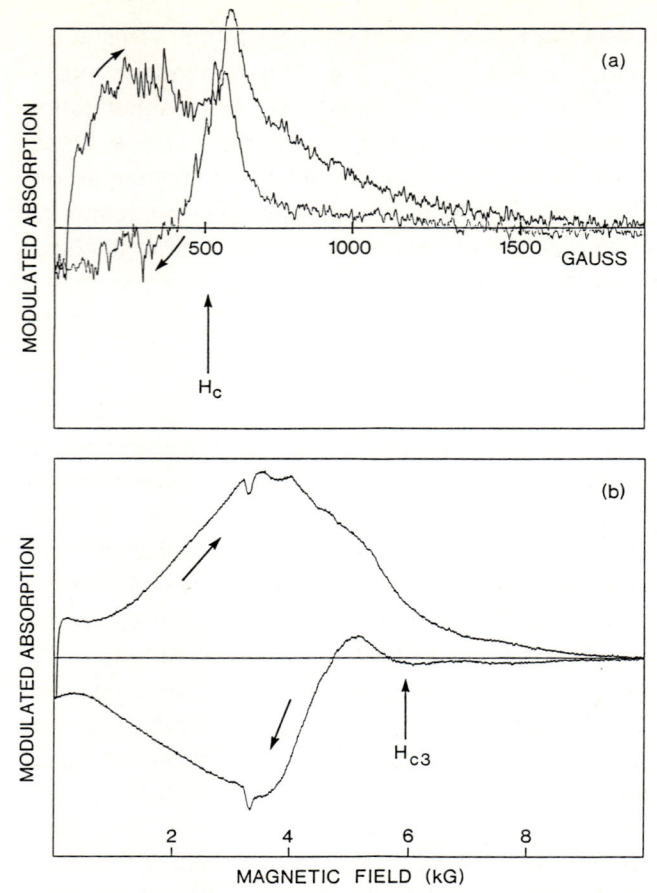

Fig. 1. Field dependence of the modulated microwave absorption of (a) a small lead plate at 4.3 K (microwave power 0.1 mW, frequency 9.45 GHz, field modulation 0.2 Gauss) and (b) a small anodically oxidised niobium plate at 4.4 K (microwave power 0.5 mW, frequency 9.44 GHz, field modulation 1 Gauss).

effective H_{c1} easing fluxon entry already at low fields and causes the microwave absorption [2]. This surface layer is not composed of a continuous superconductor but is granular in that it probably consists of an array of islands connected by weak links where magnetic flux may penetrate. In this respect the surfaces of these bulk superconductors resemble thin Al films which also show microwave loss due to junctions [34]. Granularity has been shown to give rise to disorder in the system [35] and may cause it to show spin glass behavior [36].

3. High T_c Cuprate Ceramics

The high T_c cuprate superconductors also have a large microwave absorption at very low magnetic fields [5-28]. For example, the modulated spectrum over 100 G is shown in Figure 2 for $YBa_2Cu_3O_{7-\delta}$. Most authors now agree that it is related to the many Josephson junctions connecting the grains of the ceramic allowing flux to enter at a very low effective H_{c1}. This strong absorption is easily seen without using modulation techniques [37], which has the advantage of observing the change of slope in the absorption which occurs when fluxons start entering the bulk where they contribute less. This is because losses due to damped fluxon motion are inversely proportional to the fluxon viscosity [33] and this quantity is much smaller for fluxons in Josephson junctions than in the bulk by several orders of magnitude [38], causing a knee in the absorption spectrum on passing through H_{c1} of the bulk.

The modulated absorption signal is rich in fluctuations, some of which may be removed by screening, showing they are due to environmental magnetic noise [27]. However, the principal feature of the spectrum of a ceramic sample cooled in zero magnetic field where the signal is zero is

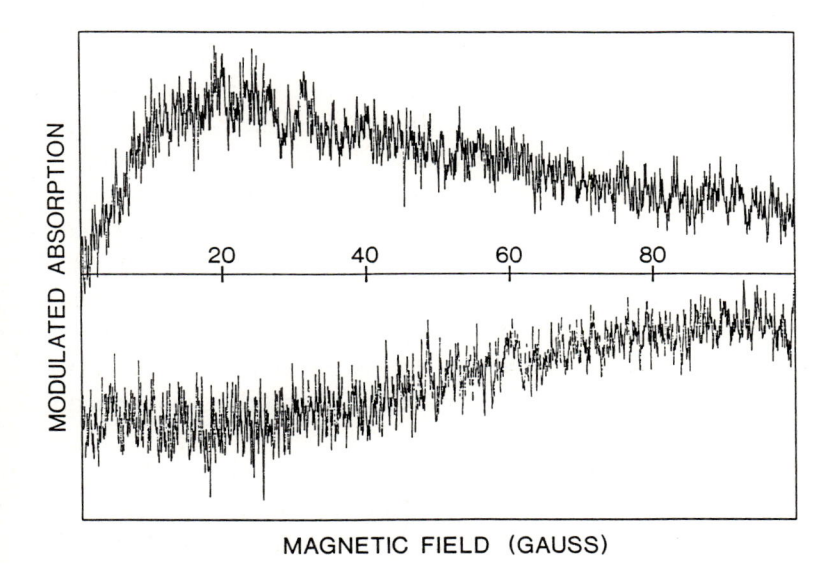

Fig. 2. Field dependence of the modulated microwave absorption of a $YBa_2Cu_3O_{7-\delta}$ ceramic at 4.4 K. The microwave power was 20 μW and the frequency was 9.45 GHz. The modulation field was 0.2 Gauss.

a maximum at very low fields, typically 0.1-10 G for $YBa_2Cu_3O_{7-\delta}$ but it can be as high as 100 G, followed by a long tail. With small modulation fields the signal changes sign whenever there is a change in field sweep direction as in Figure 2. The magnitude of the signal on returning to zero field shows a memory effect in that it depends on the range of the field swept [7,10,12].

The change of sign of the modulated signal on changing the field sweep direction is related to the resultant change of the critical state at the surface [39,40]. A magnetic field applied to a granular superconductor (or a type II superconductor above H_{c1}) induces a critical state at the surface where the critical current is flowing already at very low fields. The critical current is the maximum current sustainable by the superconductor and it is just sufficient to depin the fluxons which enter with increasing external field. In this state the depinned fluxons [39] and the critical current contribute to the microwave absorption [40]. Upon reversing the field sweep the critical current is reversed at the surface only after a certain field interval $2H_{c1}^* = 4\pi/c\lambda'J_c$, where λ' is a flux relaxation distance and J_c the critical current. Within this interval, the current flowing at the surface is less than the critical current, causing a reduced microwave absorption which recovers to its original value again with the reversed critical current which flows after the field has been reduced by $2H_{c1}^*$. Exactly this same change of the critical current occurs over a modulation field cycle. Thus, if one looks at the microwave absorption on a scope synchronously with a modulation field $\sim 2H_{c1}^*$ then a double V shaped signal is observed [41,42]. The minimum in each V corresponds to zero critical current at the surface. Since the slopes at each side of the V have opposite signs, the signal has different sign in increasing and decreasing external field for small modulation fields less than H_{c1}^*. For modulation fields in excess of $2H_{c1}^*$, additional fluxons are swept in and out of the sample. These fluxons are depinned by the critical current and their concentration is linearly proportional to the modulation field causing a linear increase of absorption, whose sign does not change with field sweep direction.

This is illustrated in Figure 3 where the modulated absorption signal is shown as a function of the modulation field amplitude. For temperatures well below T_c, the modulated absorption shows two nearly linear regions [43]. These are the critical current signal at small amplitudes where $H_m < H_{c1}^*$ and the changing fluxon density signal for $H_m < 2H_{c1}^*$ at large amplitudes. Between these two linear regions is a minimum where $H_m \sim 2H_{c1}^*$. The results in Figure 4 show that $2H_{c1}^*$ decreases almost linearly with temperature and vanishes near T_c.

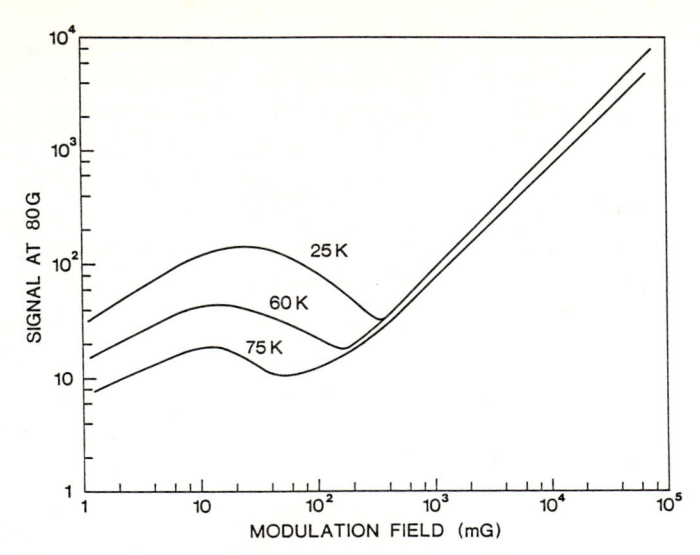

Fig. 3. Variation of the modulated microwave absorption signal with modulation field for a YBa$_2$Cu$_3$O$_{7-\delta}$ ceramic at 80 Gauss.

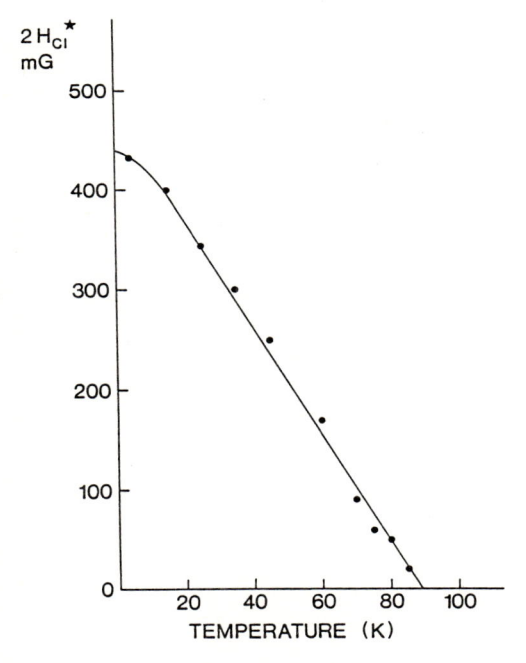

Fig. 4. Temperature variation of $2H_{c1}$ for a YBa$_2$Cu$_3$O$_{7-\delta}$ taking the modulation field where the modulated microwave absorption goes through a minimum in Fig. 3.

Similar hysteretic absorption due to granularity at the surface has been seen in conventional superconductors Nb [39] and PbMo$_6$S$_8$ [44]. In the former case the oxidised surface layer is not homogeneous and behaves as a granular superconductor and the sulphide is made as a hot sintered compressed powder which gives it similar properties to the oxides just discussed. Since these modulated absorption signals are determined by the junctions and the critical current, their field dependence does not show significant structure at the bulk critical field H_{c1}.

4. Single Crystal YBa$_2$Cu$_3$O$_{7-\delta}$

The principal difference between the absorption spectrum of ceramics and single crystals is the appearance of line spectra in the latter like that shown in Figure 5. Such lines can be seen superimposed on the signals obtained from ceramics, but due to the many grains and incorporated junctions they are not always clearly resolved [13,18,21]. Only with the preparation of single crystals could these spectra be studied in more detail [45-51]. Moreover, they are not a peculiarity of the high T_c superconductors but have also been found in the conventional superconductors Nb [50], Figure 6, and PbMo$_6$S$_8$ [45]. Their origin is thought to be microwave current induced nucleation and annihilation of fluxons within

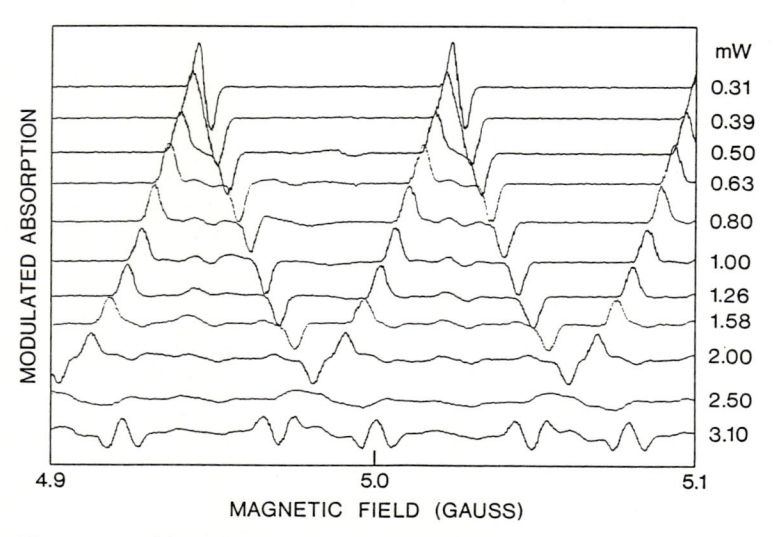

Fig. 5. Modulated microwave absorption of a YBa$_2$Cu$_3$O$_{7-\delta}$ single crystal at 27 K. The microwave power in mW is marked on each field sweep. The modulation field was 5 mGauss.

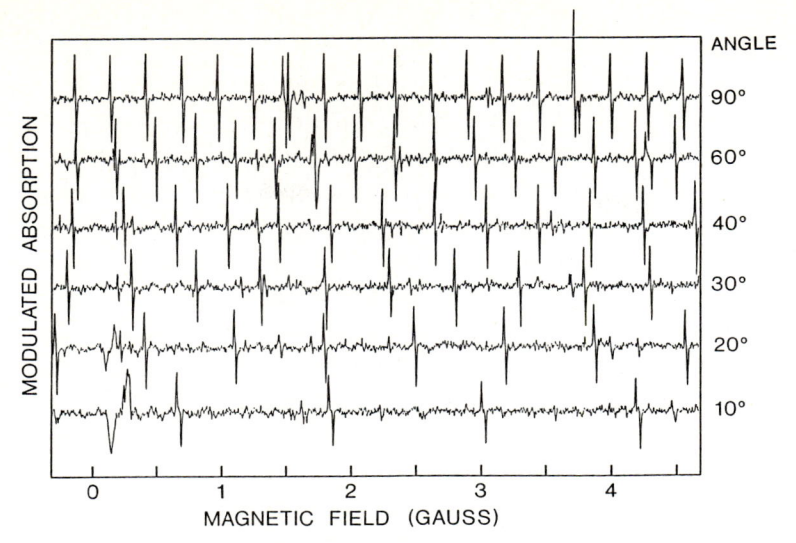

Fig. 6. Field dependence of the modulated microwave absorption of a single irregular niobium particle at 4.4 K for various field orientations. The circuit of the junction and its surrounding superconducting particle is nearly perpendicular to the field at 90° and moves towards being parallel with the field with decreasing angle.

sections of a Josephson junction in a superconducting loop [52] intersecting an applied magnetic field.

When a crystal with the c axis perpendicular to the microwave field is measured, numerous superimposed spectra are seen due to a multitude of different junctions which apparently exist in a "single crystal." This makes it difficult to study the anisotropy of such spectra. This is less of a problem when the crystal is mounted with the microwave field parallel to the c axis although overlapping spectra are still seen at moderate microwave power levels ~ 0.1 mW. On rotating the external field H_0 in the a,b-plane the line spacing is found to pass through a minimum when H_0 is parallel to a [110] direction and goes to infinity when H_0 is rotated 90° to the orthogonal [1$\bar{1}$0] direction. The line spectrum is determined by the external field according to

$$H_0 \cos \phi = \pm \left(p + \frac{1}{2} \right) \Delta H ,$$

where p is an integer, ϕ the angle between H_0 and the [110] direction, and ΔH the minimum line separation according to

$$\Delta H = \frac{\phi_0}{S},$$

where ϕ_0 is the flux quantum and S the area of cross section intersecting the magnetic field. This purely regular behavior indicates that flux is nucleated over a limited distance of a Josephson junction at field values corresponding to the regularly spaced level crossings of fluxon states in a quantum interferometer [50]. Since the responsible junctions are parallel to a [110], twin boundaries are probably involved and fluxons are created at the top and bottom ends in the c direction. Striations associated with these twin boundaries are visible through the microscope with 1 μm separation. Looking along the axis of one of these striations the field H_0 sees an area S equal to 1 μm times the crystal thickness. Originally the crystal was 100 μm thick which gives a calculated $\Delta H = 200$ mG in reasonable agreement with the observed line separation. Later the crystal cleaved in halves and the period ΔH doubled. Not every set of twin planes gives a line sequence or the spectrum would be more confused. Apparently only a few junctions with appreciable microwave absorption exist in the crystal shortly after growth and even these change with time, possibly due to surface decomposition which occurs on exposure to the air at extended planar defect [53].

Near zero external field the line absorptions can be extinguished simply by reducing the microwave power showing that there exists a certain threshold power for their generation. Above this threshold power all lines broaden as the excess microwave field in the cavity over the threshold value according to

$$\delta H = H_1 - H_{c1J},$$

where H_1 is the microwave field, H_{c1J} the junction critical field and δH the measured linewidth as illustrated in Figure 5. Thus, absorption begins when the microwave field is large enough to induce the critical current across the junction when fluxon creation and annihilation occurs. As the critical current of a Josephson junction decreases with increasing temperature, the threshold for microwave absorption also begins at lower powers. Actually, upon warming different line series with different spacings, showing they are due to different junctions, are observed to replace each other. This is particularly evident in crystals shortly after growth. Only after about six months exposure to the air did the crystal change in such a way that a single junction dominated over the entire temperature range up to T_c.

Also evident in Figure 5 is additional structure within the absorption bands in the higher power absorption spectra. This secondary fluxon excitation bandwidth extrapolates back to the same threshold as the primary absorption process [51].

When the series of lines extends over a wide field range, e.g. 100 G, the microwave threshold power for their generation also varies with the applied field. This is another facet of the quantum interferometer behavior [50,51] and in the YBaCuO crystal indicates that fluxon generation occurs over an active length which is only about one thousandth of the total length of the junction [51].

As already mentioned the absorption line interval, ΔH, is related to the cross section, S, of one of the [110] oriented twin boundary related striations. This area is the product of the crystal thickness, t, and the junction width, w, including the penetration depth λ_L into the bulk on either side of the junction

$$S = t(w + 2\lambda_L) = \frac{\phi_0}{\Delta H}.$$

The observation of different spectra with different field spacings ΔH shows there is some variation in the paths connecting active regions of the junctions at the surfaces. The spectra observed over the entire temperature range up to T_c has a period of $\Delta H = 82$ mG at low temperatures indicating a larger effective cross section, S, than that originally observed to correspond to the striations seen on the crystal surface. As the crystal remained 50 μm thick this is attributed to another junction configuration.

Any temperature variation of ΔH may be reasonably attributed to the temperature dependence of the penetration depth. The temperature variation of the area S thus derived is shown in Figure 7, where the upward curvature on approaching T_c shows the increasing λ_L. Substituting the BCS temperature dependence of λ_L near T_c in the above equation gives

$$S = tw + \frac{\sqrt{2}\, t\, \lambda(0)}{\left(1 - \frac{T}{T_c}\right)^{\frac{1}{2}}}.$$

The plot of S against $(1 - T/T_c)^{1/2}$ in Figure 7 shows this is obeyed over the entire temperature range and not just near T_c which is 92 K for this crystal.

From the gradient and intercept of the BCS line in Figure 7 one obtains $\lambda_L(0) = 0.42$ μm for screening currents perpendicular to the a,b-plane and $w = 4.5$ μm which is exceptionally wide for a Josephson

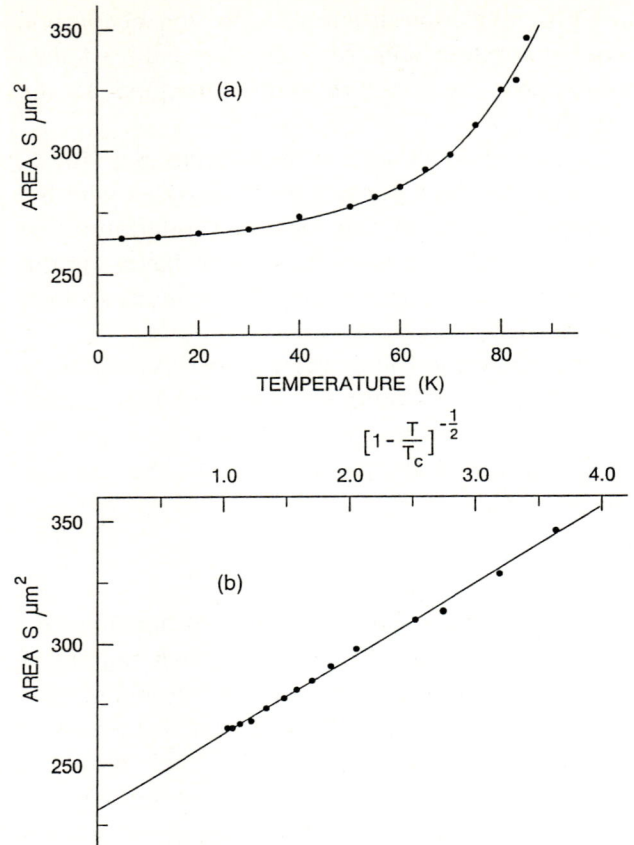

Fig. 7. (a) Temperature dependence of the effective area, S, given by the field spacing of the absorption lines of a $YBa_2Cu_3O_{7-\delta}$ single crystal. (b) Comparison of the same data with the BCS formula for the temperature dependence of the penetration depth near T_c.

junction, suggesting that this structure may be due to microcracking [54] or an accumulation of twin boundaries. Due to the uncertainty in the active thickness of the crystal, the value of $\lambda_L(0)$ should be considered a lower limit. Nevertheless its value is in reasonable agreement with that estimated from other single crystal determinations. Analysis of low field magnetization [55] and μSR [56] measurements on single crystal $YBa_2Cu_3O_{7-\delta}$ find $\lambda_L(0) = 0.14$ μm for screening currents in the a,b-plane, i.e. one third the same quantity in the orthogonal direction. These latter experiments do not yield directly $\lambda_L(0)$ for screening currents perpendicular to the a,b-plane. But low field magnetisation [57] does yield the two corresponding values of the lower critical field, H_{c1}, as 180 G for $H \perp c$ and 530 G for $H \parallel c$, which should scale as the inverse of the penetration depth. The ratio of these two values is 2.9 which is in good agreement with the ratio of the two penetration depths. The same anisotropy factor of 3 has been found for H_{c2} as well [58]. The remaining curvature of $\lambda(T)$ down to 4 K observed in Figure 7a is not seen in the temperature dependence of the orthogonal penetration depth which is very flat at low temperatures [55,56]. This may be a consequence of an anisotropic energy gap which is smaller along the c-axis [59].

Each absorption line in Figures 5 and 6 represents microwave absorption as the induced microwave current equals or exceeds the junction critical current. This happens at the regular level crossings of successive fluxon states as a function of the external field H_0. Measuring the linewidth δH as a function of the microwave field H_1 allows the determination of the junction threshold microwave field H_{c1J} by extrapolation of δH to zero as a function of the square root of the microwave power. H_{c1J} varies with the external applied field just as the critical current minima of a quantum interferometer [60]. For a given value of the applied field this threshold also depends upon temperature through the penetration depth and the junction critical current [50]

$$J_c(T) \propto H_{c1J}^2(T) \lambda_L(T) .$$

Since the penetration depth has already been shown to follow the BCS equation over a wide temperature range the temperature dependence of the junction critical current may be obtained by plotting the log of the square root of the threshold microwave power against $\log(1 - T/T_c)$ as shown in Figure 8. From about 40 K up to T_c this is

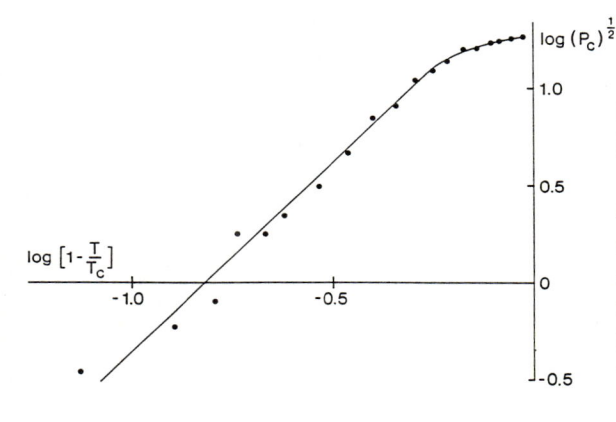

Fig. 8.
Plot of $\log(P_c)^{1/2}$ against $\log[1 - T/T_c]$ used to obtain the temperature exponent of the critical current of the junction causing the line spectrum in single crystal $YBa_2Cu_3O_{7-\delta}$.

linear and the gradient shows $J_c \propto (1 - T/T_c)^{3.4\pm0.6}$ after subtracting 0.5 from the exponent for the temperature variation of λ_L. This exponent of between 3 and 4 for the temperature dependence of a junction critical current is much larger than 1 for a SIS and 2 for a SNS usually found, again emphasizing the uniqueness of the junctions causing the line structure in $YBa_2Cu_3O_{7-\delta}$ single crystals. Particularly as the same experiment on small irregular Nb particles gives a temperature dependence consistent with a normal SIS junction which presumably forms by oxidation of surface cracks [50].

5. Summary

The microwave loss experiments described here illustrate the effects of including Josephson junctions at the surface by oxidation of the surface layer, formation of a ceramic, or crystal growth defects. These loss processes occur in addition to those predicted by the two fluid models for surface impedance of bulk superconductors and cause new features in the field dependent modulated microwave absorption spectra.

Acknowledgments

Stimulating discussions with A. M. Portis and A. Baratoff are gratefully acknowledged.

References

1. H. London, Nature 133 (1934) 497.
2. M. Cardona, J. Gittleman and B. Rosenblum, Phys. Lett. 17 (1965) 92; B. Rosenblum, M. Cardona and G. Fischer, RCA Rev. 25 (1964) 491.
3. J.G. Bednorz and K.A. Müller, Z. Phys. B 64 (1986) 189.
4. H. Piel, M. Hein, N. Klein, U. Klein, A. Michalke, G. Müller and L. Ponto, Physica C 153-155 (1988) 1604.
5. S.V. Bhat, P. Ganguly and C.N.R. Rao, Pramana-J. Phys. 28 (1987) L425.
6. S.V. Bhat, P. Ganguly, T.V. Ramakrishnan and C.N.R. Rao, J. Phys. C 20 (1987) L559.
7. K.W. Blazey, K.A. Müller, J.G. Bednorz, W. Berlinger, G. Amoretti, E. Buluggiu, A. Vera and F.C. Matacotta, Phys. Rev. B 36 (1987) 7241.

8. M. Peric, B. Rakvin, M. Prester, N. Brnicevic and A. Dulcic, Phys. Rev. B 37 (1988) 522.
9. A. Dulcic, B. Leontic, M. Peric and B. Rakvin, Europhys. Lett. 4 (1987) 1493.
10. K. Khachaturyan, E.R. Weber, P. Tejedor, A.M. Stacy and A.M. Portis, Phys. Rev. B 36 (1987) 8309.
11. S.H. Glarum, J.H. Marshall and L.F. Schneemeyer, Phys. Rev. B 37 (1988) 7491.
12. R.S. Rubins, J.E. Drumheller, S.L. Hutton, G.V. Rubenacker, D.Y. Jeong and T.D. Black, J. Appl. Phys. 64 (1988) 1312.
13. J. Stankowski, P.K. Kahol, N.S. Dalal and J.S. Moodera, Phys. Rev. B 36 (1987) 7126.
14. M.D. Sastry, R.M. Kadam, Y. Babu, A.G.I. Dalvi, I.K. Gopalakrishnan, P.V.P.S.S. Sastry, G.M. Phatak and R.M. Iyer, J. Phys. C — Solid State Phys. 21 (1988) L607.
15. M.D. Sastry, Y. Babu, R.M. Kadam, A.G.I. Dalvi, I.K. Gopalakrishnan, J.V. Yakhim and R.M. Iyer, Solid State Commun. 66 (1988) 1219.
16. W.J. Tomasch, H.A. Blackstead, S.T. Ruggiero, P.J. McGinn, J.R. Clem, K. Shen, J.W. Weber and D. Boyne, Phys. Rev. B 37 (1988) 9864.
17. R. Durny, J. Hautala, S. Ducharme, B. Lee, O.G. Symko, P.C. Taylor, D.J. Zhang and J.A. Xu, Phys. Rev. B 36 (1987) 2361.
18. Somdev Tyagi, M. Barsoum and K.V. Rao, J. Phys. C — Solid State Phys. 21 (1988) L827.
19. R. Sobolewski, A. Konopka and J. Konopka, Physica C 153-155 (1988) 1431.
20. C. Rettori, D. Davidov, I. Belaish and I. Felner, Phys. Rev. B 36 (1987) 4028.
21. V.F. Masterov, A.I. Egorov, N.P. Gerasimov, S.V. Kozyrev, I.L. Likholit, I.G. Savel'ev, A.V. Fedorov and K.F. Shtel'makh, J.E.T.P. Lett. 46 (1987) 365.
22. A.R. Harutyunyan, L.S. Grigoryan, M. Baran and S. Peichota, Phys. Lett. A 133 (1988) 339.
23. R. Karim, S.A. Oliver, C. Vittoria, A. Widom, G. Balestrino, S. Barbanera and P. Paroli, Phys. Rev. B 39 (1989) 797.
24. S. Tyagi, M. Barsoum, K.V. Rao, V. Skumryev, Z. Yu and J.L. Costa, Physica C 156 (1988) 73.
25. A.I. Tsapin, S.V. Stepanov and L.A. Blumenfeld, Phys. Lett. A 132 (1988) 373.
26. R. Marcon, R. Fastampa, M. Giura and C. Matacotta, Phys. Rev. B 39 (1989) 2796.

27. S. Tyagi and M. Barsoum, Supercond. Sci. Technol. 1 (1988) 20.
28. Z. Yu, N. Shvachko, D.Z. Khusainov, A.A. Romanyukha and V.V. Ustinov, Solid State Commun. 69 (1989) 611.
29. R.S. Rubins, S.L. Hutton and J.E. Drumheller, Phys. Rev. B 39 (1989) 4666.
30. D. Saint-James and P.G. De Gennes, Phys. Lett. 7 (1963) 306.
31. J. Halbritter, Appl. Phys. A 43 (1987) 1.
32. Y. Yafet, D.C. Vier and S. Schultz, J. Appl. Phys. 55 (1984) 2022.
33. A.R. Strnad, C.F. Hempstead and Y.B. Kim, Phys. Rev. Lett. 13 (1964) 794.
34. K. A. Muller, M. Pomerantz, G. M. Knoedler and D. Abraham, Phys. Rev. Lett. 45 (1980) 832.
35. A. Raboutou, J. Rosenblatt and P. Peyral, Phys. Rev. Lett. 45 (1980) 1035.
36. S. John and T.C. Lubensky, Phys. Rev. B 34 (1986) 4815; C. Ebner and D. Stroud, Phys. Rev. B 31 (1985) 165.
37. E.J. Pakulis and T. Osada, Phys, Rev. B 37 (1988) 5940; Y. Maniwa, A. Grupp, F. Hentsch and M. Mehring, Physica C 156 (1988) 755.
38. A.M. Portis and K.W. Blazey, Solid State Commun. 68 (1988) 1097; O.G. Symko, D.J. Zheng, R. Durny, S. Ducharme and P.C. Taylor, Phys. Lett. 134 (1988) 72.
39. K.W. Blazey. A.M. Portis and J.G. Bednorz, Solid State Commun. 65 (1988) 1153.
40. A. Dulcic, B. Rakvin and M. Pozek, preprint.
41. B.L. Walton, B. Rosenblum and F. Bridges, Phys. Rev. Lett. 32 (1974) 1047.
42. M. Stalder, G. Stefanicki, M. Warden, A.M. Portis and F. Waldner, Physica C 153-155 (1988) 659.
43. K.W. Blazey, Proc. of the 9th General Conf. of the Condensed Matter Division of the EPS, Nice, France, March 6-9, 1989, Physica Scripta, in press.
44. A.M. Portis, K.W. Blazey, C. Rossel and M. Decroux, Physica C 153-155 (1988) 633.
45. K.W. Blazey, A.M. Portis, K.A. Müller and F.H. Holtzberg, Europhys. Lett. 6 (1988) 457.
46. V. Foukis, O. Dobbert, K.P. Dinse, M. Lehnig, T. Wolf and W. Goldaker, Physica C 156 (1988) 467.
47. D. Hoffmeister, O. Dobbert, K.P. Dinse, W. Gold-acker and T. Wolf, Europhys. Lett. 8 (1989) 369.

48. A.A. Bugai, A.A. Bush, I.M. Zaritskii, A.A. Konchits, N.I. Dashirina and S.P. Kolesnik, Pis'ma. Zh. Eksp. Teor. Fiz. 48 (1988) 209; English translation JETP Lett. 48 (1988) 228.
49. A. Dulcic, R.H. Crepeau and J.H. Freed, Phys. Rev. B 39 (1989) 4249.
50. K.W. Blazey, A.M. Portis and F.H. Holtzberg, Physica C 157 (1989) 16.
51. K.W. Blazey and F.H. Holtzberg, IBM J. Res. Develop. 33 (1989) 324.
52. H. Vichery, F. Beuneu and P. Lejay, Physica C, in press.
53. H.W. Zandbergen, R. Gronsky and G. Thomas, Phys. Stat. Sol. (a) 105 (1988) 207.
54. K.L. Keester, R.M. Housley and D.B. Marshall, J. Crystal Growth 91 (1988) 295.
55. L. Krusin-Elbaum, R.L. Greene, F. Holtzberg, A.P. Malozemoff and Y. Yeshurun, Phys. Rev. Lett. 62 (1989) 217.
56. D.R. Harshman, L.F. Schneemeyer, J.V. Waszczak, G. Aeppli, R.J. Cava, B. Batlogg, L.W. Rupp, E.J. Ansaldo and D. Ll. Williams, Phys. Rev. B 39 (1989) 851.
57. L. Krusin-Elbaum, A.P. Malozemoff, Y. Yeshurun, D.C. Cronemeyer and F. Holtzberg, Phys. Rev. B 39 (1989) 2936.
58. K. Nakao, N. Miura, K. Tatsuhara, H. Takeya and H. Takei, Phys. Rev. Lett. 63 (1989) 97.
59. G. Deutscher, private communication.
60. A. Barone and G. Paterno, Physics and Applications of the Josephson Effect, John Wiley Sons Inc., New York, 1982.

Microwaves and Superconductivity: Processes in the Intergranular Medium

A.M. Portis

Department of Physics, University of California at Berkeley,
Berkeley, CA 94720, USA

Abstract. Models are developed for intergranular microwave absorption in superconductors. Loss mechanisms in bulk superconductors are first considered, followed by related processes in the intergranular medium. Sources of microwave loss may be identified from their frequency dependence. Loss from bulk flux flow is independent of frequency, normal-phase transport loss increases as the square-root of the frequency, hysteretic loss is linear in frequency and two-fluid transport loss, within and between grains, is quadratic in frequency.

1. Introduction

1.1. Background

The electrical and magnetic properties of the new superconductors are largely dominated by processes in the intergranular medium--the region between grains. This is certainly true of the ceramics but also of nominal single crystals and epitaxial films.

The resistivity of the ceramics is percolative as a result of the partial isolation of grains. The rf and microwave surface impedance of type II superconductors, the main subject of this paper, is largely determined by intergranular coupling with the surface resistance at these frequencies increased by the penetration of rf and microwave magnetic fields into the intergranular medium. Additionally, static and low frequency magnetic flux readily penetrates the intergranular medium at fields well below H_{c1} [1.1.1-1.1.3].

The history of the improvement of the microwave properties of high-field superconductors has been one of the control and elimination of grain boundaries. This has been just as true of the new superconductors with the potential usefulness of these promising materials depending on the extent to which intergranular processes can be understood and controlled.

1.2. Microwave Penetration and Absorption

The surface reactance measures the penetration of microwave fields into the superconducting medium while the surface resistance measures dissipation processes. Early rf and microwave studies of type I superconductors close to their transition temperature T_c provided a useful measure of transport phenomena [1.2.1-1.2.7] and of excitation across the superconducting energy gap [1.2.8-1.2.10]. Although such studies are possible in type II and granular superconductors, rf and microwave losses in these materials appear to be dominated by processes in the intergranular medium rather than by bulk transport.

2. Electrodynamics

2.1. Type I Superconductors

Normal metals and classical surface impedance. From the Maxwell equations (in Gaussian units)

$$\nabla \times \mathbf{E} = -(1/c)\partial \mathbf{B}/\partial t, \tag{2.1.1}$$

$$\nabla \times \mathbf{H} = (4\pi/c)\mathbf{J} + (\varepsilon/c)\partial \mathbf{E}/\partial t, \tag{2.1.2}$$

with $\nabla \cdot \mathbf{E} = 4\pi\rho = 0$ the wave equation is written as

$$\nabla^2 \mathbf{E} - (\varepsilon/c^2)\partial^2 \mathbf{E}/\partial t^2 = (4\pi/c^2)\partial \mathbf{J}/\partial t. \tag{2.1.3}$$

So long as the electric field changes slowly over a carrier mean free path l, the current may be described by the local relation $\mathbf{J} = \sigma \mathbf{E}$. Assuming

$$\mathbf{E}(x, t) = \mathbf{E}\, e^{i(kx - \omega t)}, \tag{2.1.4}$$

the dispersion relation, which connects the wavevector k with the frequency ω, is

$$k^2 = \varepsilon\omega^2/c^2 + 4\pi i\omega\sigma/c^2. \tag{2.1.5}$$

The ac surface impedance is defined as

$$Z_s = R_s + iX_s = E_s/\int J_{tot}\, dx = (4\pi/c)\, E_s/H_s = (4\pi/c^2)\, \omega/k, \tag{2.1.6}$$

where

$$J_{tot} = J - (i\omega\varepsilon/4\pi)E \tag{2.1.7}$$

is the sum of the conduction and displacement currents. From (2.1.5) the surface impedance is

$$Z_s = (4\pi/c)/(\varepsilon + 4\pi i\sigma/\omega)^{1/2}. \tag{2.1.8}$$

For $4\pi\sigma \ll \omega\varepsilon$ the surface impedance is largely resistive as the result of wave propagation into the medium with

$$Z_s \approx (4\pi/\varepsilon^{1/2}c)(1 - 2\pi i\sigma/\omega\varepsilon). \tag{2.1.9}$$

For $4\pi\sigma \gg \omega\varepsilon$, the surface impedance is complex with real and imaginary parts equal

$$Z_s \approx (1-i)/\sigma\delta \tag{2.1.10}$$

with the classical skin depth $\delta = (c^2/2\pi\omega\varepsilon)^{1/2}$.

Note that for σ independent of ω, Z_s increases as $\omega^{1/2}$ as a result of the reduction in δ.

The surface resistance R_s is related to the absorption density P_s by

$$P_s = 1/2 |\int J_{tot}\, dx|^2 R_s = 1/2 |(c/4\pi)H_s|^2 R_s. \tag{2.1.11}$$

Alternatively, the Poynting vector gives

$$P_s = (c/8\pi)\, \mathrm{Re}\, (E_s H_s^*). \tag{2.1.12}$$

With the definition of the surface impedance from (2.1.6) the absorption

$$P_s = 1/2 |(c/4\pi)H_s|^2 \mathrm{Re}\, (Z_s) \tag{2.1.13}$$

is equivalent to (2.1.11).

The Londons and flux exclusion. Following the discovery of the exclusion of magnetic flux by a superconductor, the Meissner effect [2.1.1], F. and H. London [2.1.2] developed the appropriate electrodynamics from the expression for the rate of change of a lossless current

$$(d/dt)\, \mathbf{J} = (\varepsilon/4\pi)\omega_p^2\, \mathbf{E} \tag{2.1.14}$$

with the plasma frequency $\omega_p^2 = 4\pi n e^2/\varepsilon m$. With the definition of the vector potential

$$\mathbf{B} = \nabla \times \mathbf{A} \tag{2.1.15}$$

(2.1.1) leads to

$$E = -\nabla \phi - (1/c)\partial A/\partial t. \qquad (2.1.16)$$

So long as there are no electrostatic fields, time-integration leads to the London equation

$$(4\pi/c)\, \mathbf{J} = -(1/\lambda_L^2)\, \mathbf{A} \qquad (2.1.17)$$

with $\lambda_L = c/\varepsilon^{1/2}\omega_p = (mc^2/4\pi n e^2)^{1/2}$. Differential equations equivalent to (2.1.17) are

$$\nabla^2 \mathbf{A} = -(1/\lambda_L^2)\, \mathbf{A} \qquad (2.1.18)$$

and similar equations with **A** replaced by **B** or **J** so long as these vectors are solenoidal.

At a plane surface the field decays into the superconductor as

$$B(x) = H_s \exp(-x/\lambda_L) \qquad (2.1.19)$$

and the current similarly decays into the medium as

$$J(x) = -(c/4\pi\lambda_L)B(x). \qquad (2.1.20)$$

Gorter, Casimir and the two-fluid model. Gorter and Casimir [2.1.3] originally developed a model of normal and superconducting electrons to account for the thermal properties of superconductors. Taking for the carrier concentration $n = n_n + n_s$ with a concentration of normal carriers $n_n = xn$ and a concentration of superconducting carriers $n_s = (1-x)n$, they were able to fit the thermal properties with a Helmholz free energy

$$F = x^{1/2} F_n(T) + (1-x)F_s(T). \qquad (2.1.21)$$

Note for a mixed phase to be stable that (2.1.21) must be nonlinear in x. The thermodynamics of a free electron Fermi gas leads to an energy $U_n = 1/2\, \gamma T^2$, a heat capacity $C_n = \gamma T$, an entropy $S_n = \int C\, dT/T = \gamma T$ and a consequent free energy $F_n = U_n - TS_n = -1/2\, \gamma T^2$. Assuming that C_s is appreciable over only a limited range of temperature leads to the simplification $F_s \approx -\beta$. Minimizing (2.1.21) with respect to x for this choice of free energies yields for the fraction of normal carriers

$$x = (F_n/2F_s)^2 = t^4 \qquad (2.1.22)$$

with $t = T/T_c$ and $T_c = (\beta/\alpha)^{1/2}$. The two-fluid model suggests for the electrodynamics $\mathbf{J} = \mathbf{J}_n + \mathbf{J}_s$ with $\mathbf{J}_n = \sigma\mathbf{E}$ and $\mathbf{J}_s = -(c/4\pi\lambda_L^2)\mathbf{A}$ for $\sigma = x(ne^2/m)\tau$ and $1/\lambda_L^2 = 4\pi(1-x)ne^2/mc^2$. Substituting into (2.1.3) gives for the dispersion relation

$$k^2 - \varepsilon\omega^2/c^2 = 4\pi i\sigma/c^2 - 1/\lambda_L^2 \qquad (2.1.23)$$

or $k = (i/\lambda_L)(1 - \omega^2/\omega_L^2 - 4\pi i\omega\sigma/c^2)^{1/2}$ with the London plasma frequency $\omega_L^2 = c^2/\varepsilon\lambda_L^2$. The surface impedance of the two-fluid model is $Z_s = R_s + iX_s = (4\pi/c^2)\omega/k$ with surface resistance and reactance, respectively

$$R_s = (16\pi^2/c^4)\omega^2\sigma\lambda_L, \qquad (2.1.24)$$

$$X_s = -(4\pi/c^2)\omega\lambda_L(1 + \omega^2/\omega_L^2). \qquad (2.1.25)$$

The measurements of H. London [1.2.1] showed early the presence of substantial loss just below T_c as expected from (2.1.24).

Pippard and nonlocal conductivity. The classical model of the rf surface impedance assumes a local relation between current and field of the form $\mathbf{J}(\mathbf{r}) = \sigma(\omega)\mathbf{E}(\mathbf{r})$. From a series of experiments on pure metals at low temperature, Pippard [1.2.2] recognized that the use of a local conductivity $\sigma(\omega)$ gives incorrect results and must be replaced by a nonlocal relation of which the simplest, neglecting retardation, is the Chambers formula [2.1.4, 2.1.5]

$$\mathbf{J}(\mathbf{r}) = \sigma(\omega)\int(3/4\pi\rho^4 l)\boldsymbol{\rho}\,[\boldsymbol{\rho}\cdot\mathbf{E}(\mathbf{r}+\boldsymbol{\rho})]\,e^{-\rho/l}\,dV. \qquad (2.1.26)$$

For a superconductor [1.2.6, 1.2.7], the carrier mean-free-path l is replaced in the exponent by the coherence length ξ

$$(4\pi/c)\mathbf{J}(\mathbf{r}) = -(1/\lambda_L^2)\int(3/4\pi\rho^4\xi_0)\boldsymbol{\rho}\,[\boldsymbol{\rho}\cdot\mathbf{E}(\mathbf{r}+\boldsymbol{\rho})]\,e^{-\rho/\xi}\,dV \qquad (2.1.27)$$

where ξ is given by $1/\xi = 1/\xi_0 + 1/l$ with ξ_0 the intrinsic coherence length. *Note* for $l < \xi_0$ that $\mathbf{J}(\mathbf{r})$ is depressed by a factor ξ/ξ_0. Ginzburg [2.1.6] has discussed the superconducting surface impedance when the current is anomalous and concludes that it is always possible to represent the surface impedance by an effective dielectric function $\varepsilon(\omega, T)$ as

$$Z_s(\omega, T) = (4\pi/c)[\varepsilon(\omega, T)]^{-1/2}. \qquad (2.1.28)$$

Bardeen, Cooper, Schrieffer and the microscopic theory.
The theory of Bardeen Cooper and Schrieffer (BCS) [2.1.7] has provided a model of the superconducting state in which paired carriers interact over substantial distances to produce a coherent state. The removal of a pair from the BCS state requires an energy 2Δ, which in weak-coupling theory is related to the transition temperature T_c by

$$\Delta(0) = 1.764\, k_B T_c \ . \tag{2.1.29}$$

Close to T_c the half-gap varies as

$$\Delta(T) \approx 1.74\, \Delta(0)(1 - t)^{1/2} \ . \tag{2.1.30}$$

BCS have shown that the Meissner effect [2.1.1] is a consequence of the gap in the energy spectrum. They obtain an intrinsic coherence length

$$\xi_0 \approx h/2\pi\Delta p = h/(2\pi\Delta/v_F) \tag{2.1.31}$$

with just the electrodynamics anticipated by Pippard [1.2.6, 1.2.7]. The new superconductors, with Δ large and v_F small, have unusually short coherence lengths, of the order of atomic distances. BCS obtained from their theory a temperature dependence of the penetration depth $\lambda(T)$ that closely follows the two-fluid model in the limit $\xi_0 \ll \lambda_L$ where the electrodynamics are local. In the opposite limit $\xi_0 \gg \lambda_L$, the Pippard ineffectiveness concept with $1/\lambda^2 \approx (\lambda/\xi_0)(1/\lambda_L^2)$ yields $\lambda \approx (\lambda_L^2 \xi_0)^{1/3}$.

Mattis, Bardeen and anomalous rf conductivity. Mattis and Bardeen [2.1.8] extended the BCS derivation of the Meissner effect to $\omega > 0$ through the development of a quantum mechanical form of the Chambers integral [2.1.4, 2.1.5]. They obtain the surface impedance in the extreme anomalous limit $\xi_0 \gg \lambda_L$ in terms of which a complex conductivity may be defined

$$Z_s(\omega, T) = \{\sigma_n/[\sigma_1(\omega, T) + i\, \sigma_2(\omega, T)]\}^{1/2} Z_n \tag{2.1.32}$$

with Z_n the surface impedance in the normal state and $\sigma_n = ne^2\tau/m$ the normal-state conductivity. At $T = 0$, Z_s may be evaluated in terms of complete elliptic integrals. Above $T = 0$, numerical integration is required.

2.2. Superconducting Junctions

Josephson current-phase relation. In a remarkable theoretical contribution that was stimulated by P. W. Anderson, Josephson [2.2.1-

2.2.4] developed a theory of lossless tunneling between superconductors with the pair-tunneling current

$$J = J_0 \sin \delta \qquad (2.2.1)$$

where

$$\delta = -(2\pi e/hc) \int dr \cdot A(r) \qquad (2.2.2)$$

is the difference in phase between superconductors. For tunneling at T=0, Anderson had calculated for the maximum Josephson current

$$I_0(0) = \pi \Delta(0)/2eR_n \qquad (2.2.3)$$

where R_n is the normal-state tunneling resistance between conductors. Ambegaokar and Baratoff [2.2.5] have obtained at elevated temperatures

$$I_0(T) = [\pi \Delta(T)/2eR_n] \tanh \Delta(T)/2k_B T. \qquad (2.2.4)$$

Expansion of (2.2.4) near T_c gives

$$I_0(T) \approx (\pi/4ek_B T_c R_n) \Delta^2(T) \approx 4.19 [\Delta(0)/eR_n] (1 - t) \qquad (2.2.5)$$

where (2.1.30) has been used.

The presence of normal-phase material within a junction suppresses the gap parameter $\Delta(T)$, leading near T_c to a reduced current density $J_0(T) \propto (1 - t)^2$. This reduction is accentuated by the short coherence length of the new superconductors [2.2.6].

The ac Josephson effect. Josephson also found that a potential between superconductors leads to a rate of change of phase. From the expression for the electric field (2.1.6) and (2.2.2) the rate of change of phase is

$$\partial \delta/\partial t = (2\pi e/h) \int dr \cdot E(r) = 2\pi eV/h \qquad (2.2.6)$$

When integrated around the junction (2.1.2) gives

$$(4\pi/c) J d + (\varepsilon/c) \partial V/\partial t = 0 \qquad (2.2.7)$$

where $d = w + 2\lambda_L$ is the magnetic thickness of the junction with w the physical separation between conductors. Combining (2.2.7) with (2.2.1) and (2.2.6) gives the differential equation

$$\partial^2 \delta/\partial t^2 = -\omega_J^2 \sin \delta \qquad (2.2.8)$$

with $\omega_J^2 = 8\pi^2 edJ_0/\varepsilon h$ the Josephson plasma frequency of the junction.
Note that the dielectric function $\varepsilon(\omega)$ is a complex function of frequency that is characteristic of the inter-granular medium. If the medium is an electrical conductor, for example, we can expect to write

$$\varepsilon(\omega) = 1 + 4\pi i\sigma/\omega \qquad (2.2.9)$$

and $\delta(t)$ will be damped exponentially in time.

3. Flux Quantization and Type II Superconductivity

3.1. Quantization in Bulk

Ginzburg, Landau and the macroscopic theory. Ginzburg and Landau [3.1.1, 3.1.2] applied to superconductivity the Landau theory of phase transitions [3.1.3]. Where for ferroelectrics, the polarization density P is the order parameter and for ferromagnets it is the magnetization density M, for superconductors it is the complex gap function Δ [3.1.4], which is represented by a pseudo-wavefunction ψ in the Ginzburg-Landau theory.

In the Landau theory, the free energy in a magnetic field H takes the form

$$F_{sH} = F_{s0} + (1/8\pi)H^2 + (1/2m)|(-ih/2\pi)\nabla\psi - (q/c)A\psi|^2 \qquad (3.1.1)$$

with the charge q a phenomonological quantity equal to 2e in BCS theory. The free energy in zero field is

$$F_{s0} = F_{n0} + \alpha|\psi|^2 + 1/2\,\beta|\psi|^4 \,. \qquad (3.1.2)$$

The Ginzburg-Landau theory thus contains three free parameters, α, β and q. Minimizing the free energy in a field F_{sH} with respect to the conjugate pseudo-wavefunction ψ^* and with respect to the vector potential **A** leads to the two Ginzburg-Landau differential equations

$$(1/2m)|(-ih/2\pi)\nabla - (q/c)\mathbf{A}|^2\psi + \alpha\psi + \beta|\psi|^2\psi = 0 \qquad (3.1.3)$$

$$\nabla^2\mathbf{A} = -(4\pi/c)\,\mathbf{J}_s = (-ihq/mc)(\psi^*\nabla\psi - \psi\nabla\psi^*) + (4\pi q^2/mc^2)|\psi|^2\mathbf{A}$$
$$= (-4\pi q/mc)|\psi|^2\,[(h/2\pi)\nabla\phi - (q/c)\,\mathbf{A}] \qquad (3.1.4)$$

The angle ϕ is the phase of the wavefunction ψ. The equation

$$(h/2\pi)\nabla \phi = m \, v_s + (q/c) \, A = (q/c)[(4\pi/c)\lambda_L^2 \, J + A] \qquad (3.1.5)$$

which for ϕ constant, as expected in weak fields, is the London equation (2.1.17). Ginzburg-Landau theory leads to a breakdown of the London equation in strong fields as a result of the spatial dependence of ψ.

In place of α, β and q, the parameters of the theory may be taken as H_c, λ_L and $\kappa = (1/\sqrt{2})(4\pi q/hc)H_c\lambda_L^2$. Agreement with the observed properties of metals like Pb and Sn is obtained for $\kappa \approx 0.1$. Ginzburg and Landau noticed early however [3.1.1] that for $\kappa > 1/\sqrt{2}$ their differential equations lead to anomalous behavior, which was further investigated by Abrikosov.

Abrikosov and flux quanta. Abrikosov [3.1.5] studied the Ginzburg-Landau theory for variable κ and concluded that there are two distinct classes of superconductors, type I with $\kappa < 1/\sqrt{2}$ and type II with $\kappa > 1/\sqrt{2}$. Ginzburg and Landau had found that the surface energy of a type II superconductor is negative, allowing for the stabilization of superconductivity up to an upper critical field $H_{c2} = \sqrt{2} \, \kappa \, H_c$. Abrikosov investigated the properties of type II superconductors at superconductivity up to an upper critical field $H_{c2} = \sqrt{2} \, \kappa \, H_c$. Abrikosov investigated the properties of type II superconductors at fields close to H_{c2} and found a regular lattice of flux lines to be stable with $|\psi|^2 = 0$ at their cores. At low fields the flux lines are distinct and satisfy the equation

$$\nabla^2 A - (1/\lambda_L)^2 A = (\phi_0/2\pi\lambda_L^2) \, \nabla \phi \qquad (3.1.6)$$

with $\phi_0 = hc/2e$ the quantum of flux and $1/\lambda_L^2 = (4\pi e^2/mc^2)|\psi|^2$. Well outside the core of the flux line, λ_L is constant as assumed in the London theory. The phase, however, increases by 2π around the core with $\nabla \phi = (1/r) \, \theta$. Taking the curl of (3.1.6) gives

$$\nabla^2 B - (1/\lambda_L)^2 B = -(\phi_0/\lambda_L^2) \, z \, \delta(r) . \qquad (3.1.7)$$

The solution of (3.1.7) is

$$B(r) = (\phi_0/2\pi\lambda_L^2) \, K_0(r/\lambda_L) \, z \qquad (3.1.8)$$

where $K_0(r/\lambda_L)$ is the modified Bessel function. For $r \gg \lambda_L$ the function falls off exponentially as

$$K_0(r/\lambda_L) \approx (\pi \lambda_L/2r)^{1/2} \exp(-r/\lambda_L) . \tag{3.1.9}$$

For $r \ll \lambda_L$ the function has a logarithmic singularity, which may be taken down to a core radius λ_L/κ. The flux enclosed within a cylinder of radius r is

$$\int d\mathbf{S}\cdot\mathbf{B} = \phi_0 - (4\pi\lambda_L^2/c) \int d\mathbf{l} \cdot \mathbf{J}, \tag{3.1.10}$$

which approaches the quantum of flux ϕ_0 for $r \to \infty$. From an examination of (3.1.6) and the ideas of Pippard, $\nabla^2\mathbf{A}$, which is the curvature of $\mathbf{A}$, is limited for anomalous conduction to $(1/\xi^2)$ $\mathbf{A}$ which gives $\kappa = \lambda_L/\xi$ with, as assumed by Pippard, $1/\xi = 1/\xi_0 + 1/l$.

The Ginzburg-Landau theory assumes near T_c that α and $|\psi|^2$ go to zero as $(1 - t)$ with β assumed constant. With coefficients obtained from the BCS theory, the Ginzburg-Landau macroscopic theory leads near T_c to the lengths [1.1.3]

$$\xi(T) = 0.74\, \xi_0/(1 - t)^{1/2} \tag{3.1.11}$$

$$\lambda_L(t) = 0.71\, \lambda_L(0)/(1 - t)^{1/2} \tag{3.1.12}$$

and to the fields [1.1.3]

$$H_{c1}(T) \approx H_c^2(T)/H_{c2}(T) = 1.64\, (2\pi\xi_0^2/\phi_0)\, H_c^2(0) \tag{3.1.13}$$

$$H_{c2}(T) = [\phi_0/2\pi\xi^2(T)] = 1.83\, (\phi_0/2\pi\xi_0^2)(1 - t) \tag{3.1.14}$$

The theory also leads to a thermodynamic critical current [1.1.3]

$$J_c = (1/3\pi\sqrt{6})\, H_c(T)/\lambda_L(T) \propto (1 - t)^{3/2} \tag{3.1.15}$$

In Sections 3.2 and 4.2 are discussed the related critical currents in Josephson junctions. Studies of the temperature dependence of microwave absorption near T_c may be useful in identifying loss mechanisms since these mechanisms depend variously on characteristic lengths, fields and critical currents.

Current-driven flux flow. The equation of motion of a damped flux line or fluxon driven by a force per unit length $F(t)$ is

$$m\, d^2x/dt^2 + \eta\, dx/dt + kx = F(t) \tag{3.1.16}$$

where m is the mass per unit length [3.1.6], η is the viscosity [3.1.7-3.1.9], and k is the force constant associated with pinning. If the force is produced by a static current with $F = (1/c) J \phi_0$, fluxons are displaced by $x = (\phi_0/c) J/k$. As J is increased there is a field-dependent critical current J_c at which fluxons break away from their pinning centers and flow at a velocity

$$v \approx (\phi_0/\eta c) (J - J_c) \tag{3.1.17}$$

inducing an electric field

$$E = -n (v/c) \phi_0 = -(v/c) B \tag{3.1.18}$$

with n the fluxon concentration. The medium thus develops a resistivity

$$\rho = - E/J = (\phi_0 B/\eta c^2)(1 - J/J_c). \tag{3.1.19}$$

Microwave surface impedance. Taking the force in (3.1.16) to arise from a periodic transverse current density

$$F(t) = (\phi_0/c) J_1 e^{-i\omega t} \tag{3.1.20}$$

the fluxon displacement is $x(t) = x_1 e^{-i\omega t}$ with

$$x_1 = (1/c) J_1 \phi_0/(-\omega^2 m - i\omega\eta + k) . \tag{3.1.21}$$

At microwave frequencies [3.1.10, 3.1.11] with $k/\eta < \omega < \eta/m$, viscous damping dominates with a resultant fluxon velocity

$$v_1 = -i\omega x_1 = \phi_0 J_1/\eta c \tag{3.1.22}$$

and a reaction electric field [3.1.12]

$$E_r = -(n\phi_0/c)v_1 = -B\phi_0 J_1/\eta c^2 \tag{3.1.23}$$

Writing for the superconducting carriers

$$dJ/dt = (\varepsilon\omega_L^2/4\pi)(E + E_r) \tag{3.1.24}$$

with $\omega_L^2 = 4\pi n e^2/m\varepsilon$ leads to

$$dJ/dt + (1/\tau) J = (\varepsilon\omega_p^2/4\pi) E \tag{3.1.25}$$

with $\tau = \eta c^2/\phi_0 B$. The conductivity σ is

$$\sigma = J/E = (1/4\pi)\,\varepsilon\,\omega_L^2 \tau/(1 - i\omega\tau). \qquad (3.1.26)$$

As seen from (2.1.8) the surface impedance is

$$Z_s = (4\pi/c)\,(\varepsilon + 4\pi i\sigma/\omega)^{-1/2}. \qquad (3.1.27)$$

Taking (3.1.26) for the conductivity in the presence of flux flow yields

$$Z_s = (4\pi/\varepsilon^{1/2}c)[1 + i\omega_L^2 \tau/\omega(1 - i\omega\tau)]^{-1/2} \qquad (3.1.28)$$

Although this equation may be solved for R_s and X_s for arbitrary ω, ω_L, and τ, it is simpler to neglect the displacement current, obtaining

$$R_s = [4\pi/(2\varepsilon)^{1/2}c]\,(\omega/\omega_L)[-1 + (1 + 1/\omega^2\tau^2)^{1/2}]^{1/2} \qquad (3.1.29)$$

$$X_s = -[4\pi/(2\varepsilon)^{1/2}c]\,(\omega/\omega_L)[1 + (1 + 1/\omega^2\tau^2)^{1/2}]^{1/2}. \qquad (3.1.30)$$

Expanding (3.1.29) and (3.1.30) for $\omega\tau > 1$ with $\tau = \eta c^2/\phi_0 B$ gives

$$R_s \approx (4\pi/\varepsilon^{1/2}c)(\phi_0 B/\omega_L \eta c^2)(1 - \phi_0^2 B^2/8\omega_L^2 \eta^2 c^4) \qquad (3.1.31)$$

$$X_s \approx -(4\pi/\varepsilon^{1/2}c)(\omega/\omega_L)(1 + \phi_0^2 B^2/8\omega_L^2 \eta^2 c^4). \qquad (3.1.32)$$

Note that to this order, the surface resistance R_s is independent of ω.

3.2. Magnetic Flux in Junctions

Classical transmission lines. Before discussing the propagation of micro-wave radiation within a Josephson junction, it is useful to review the theory of a classical distributed-element transmission line. In series with the line are an inductance L and parallel resistance R per unit length. Across the line are shunt capacitance C, conductance G and reciprocal inductance S per unit length. The equation for the voltage across the line is

$$dV/dx = -L\,dI_L/dt = -R\,I_R \qquad (3.2.1)$$

where I_L is the current flowing through the inductance and I_R the current flowing through the resistance. The equation for the current along the line is

$$(d/dx)\, dI/dt = -C\, d^2V/dt^2 - G\, dV/dt - S\, V \, . \tag{3.2.2}$$

Combining (3.2.1) and (3.2.2) gives

$$[1 + (L/R)d/dt]\, d^2V/dx^2 - LC\, d^2V/dt^2 - L\, G\, dV/dt = LS\, V \, . \tag{3.2.3}$$

For a time-dependence $e^{-i\omega t}$ it is useful to write (3.2.3) as

$$(1/\mathbf{L})\, d^2V/dx^2 - \mathbf{C}\, d^2V/dt^2 = S\, V \tag{3.2.4}$$

with $\mathbf{L} = L/(1 - i\omega L/R)$ and $\mathbf{C} = C + iG/\omega$ where $\mathbf{L}$ is a lossy inductance and $\mathbf{C}$ a lossy capacitance. *Note* that the term on the right of (3.2.4) is a shunt susceptance. The characteristic impedance of the line

$$Z = R + iX = V/I = [\mathbf{L}/(\mathbf{C} - S/\omega^2)]^{1/2} \tag{3.2.5}$$

is reduced by large susceptance. At low frequency the capacitance C, which represents displacement current, may be neglected and for $S \gg \omega G$ the impedance becomes

$$Z \approx \tfrac{1}{2}\, \omega^2(L/S)^{1/2}(L/R + G/S) - i\omega(L/S)^{1/2} \, . \tag{3.2.6}$$

The first term on the right is resistive and increases with ω^2 as in bulk transport (2.1.24). The losses arise from conductive current in shunt with L and with S. The reactance $X = -\omega(L/S)^{1/2}$ is inductive. In the opposite limit $S \ll \omega G$, the impedance is $(1-i)(\omega L/2G)^{1/2}$.

The Josephson transmission line. As discussed in Section 2.2, the current across a junction is given by the Josephson relation $J = J_0 \times \sin \delta$ where J_0 is the Josephson critical current and δ is the phase across the junction, taken to be in the xy-plane. Differentiating (2.2.2) with respect to x gives

$$\partial \delta/\partial x = -(2\pi/\phi_0) \int dy\, B_z = (2\pi d/\phi_0)\, B_z \, . \tag{3.2.7}$$

Differentiating (3.2.7) with respect to x gives

$$\partial^2 \delta/\partial x^2 = -(2\pi d/\phi_0)[(4\pi/c)\, J_y + (\varepsilon/cd)\, \partial V/\partial t] \, . \tag{3.2.8}$$

Using (2.2.6) finally gives the sine-Gordon equation for wave propagation in the junction

$$\partial^2 \delta/\partial x^2 - (\varepsilon/c^2)\, \partial^2 \delta/\partial t^2 = (1/\lambda_J^2) \sin \delta \tag{3.2.9}$$

with $1/\lambda_J^2 = 8\pi^2 J_0 d/\phi_0 c$. *Note* that taking the junction to be short leads to (2.2.8) with $\omega_J^2 = c^2/\epsilon\lambda_J^2$. For a sufficiently weak oscillating field $H(t) = H_1 e^{-i\omega t}$, δ will be small and (3.2.9) becomes

$$\partial^2\delta/\partial x^2 - (\epsilon/c^2)\partial^2\delta/\partial t^2 = (1/\lambda_J^2)\,\delta \ . \tag{3.2.10}$$

Assuming a wave of the form $\delta(x, t) = \delta e^{i(kx - \omega t)}$ leads to the dispersion relation

$$k^2 = -1/\lambda_J^2 + \epsilon\omega^2/c^2 = (\epsilon/c^2)(\omega^2 - \omega_J^2) \ . \tag{3.2.11}$$

For ω less than the junction plasma frequency ω_J, the wave decays exponentially into the junction. The weak-field condition using (3.2.7) is $\delta = (2\pi d/\phi_0) H_1 \ll 1$ or at frequencies $\omega < \omega_J$

$$H_1 \ll H_{c1J} = \phi_0/2\pi\lambda_J\,d \tag{3.2.12}$$

where H_{c1J} is the lower critical field within the junction. This definition of H_{c1J} differs slightly from that of Clem [1.1.1] and Tinkham and Lobb [1.1.2], which follow from Ginzburg-Landau theory. The surface impedance within the junction is

$$Z_s = R_s + iX_s = V/I = (4\pi/c)\,E_s/H_s = -(4\pi/c^2)(\partial\delta/\partial t)/(\partial\delta/\partial x) \tag{3.2.13}$$

where (2.2.6) and (3.2.7) have been used. Using (3.2.11) gives

$$Z_s = 4\pi\omega/c^2 k = -(4\pi/c^2)(i\omega\lambda_J)(1 - \omega^2/\omega_J^2)^{-1/2}. \tag{3.2.14}$$

Expanding (3.2.14) for $\omega^2 \ll \omega_J^2$ yields

$$Z_s \approx -(4\pi i\omega\lambda_J/c^2)(1 + \omega^2/2\omega_J^2). \tag{3.2.15}$$

Using (2.2.9) for the junction medium leads to

$$R_s = (8\pi^2/c)(\omega^2\sigma/\omega_J^3) \tag{3.2.16}$$

$$X_s = -(4\pi/c)(\omega/\omega_J)(1 + \omega^2/2\omega_J^2) \ . \tag{3.2.17}$$

Microwave absorption and granular decoupling. We now consider the simultaneous application of a weak microwave field satisfying (3.2.12) and a static field H_0 such that the boundary condition on δ at the surface is

$$\partial\delta/\partial x = (2\pi d/\phi_0)(H_0 + H_1 e^{-i\omega t}) \ . \tag{3.2.18}$$

The junction is assumed to be sufficiently weak and sufficiently regular that the Josephson penetration depth λ_J is much longer than the scale of structural irregularities. Substituting the integral of (3.2.18)

$$\delta = (2\pi d/\phi_0)\{H_0 x + [1/ik(0)]H_1 \, e^{i(\int k(x)dx - \omega t)}\} \tag{3.2.19}$$

into the right hand side of (3.2.9) gives

$$\partial^2\delta/\partial x^2 - (\varepsilon/c^2) \partial^2\delta/\partial t^2 = $$
$$(1/\lambda_J^2) \sin (2\pi d/\phi_0)\{H_0 x + [1/ik(0)]H_1 \, e^{i(\int k(x)dx - \omega t)}\} \ . \tag{3.2.20}$$

Expanding (3.2.20) for H_1 small leads to

$$\partial^2\delta/\partial x^2 - (\varepsilon/c^2) \partial^2\delta/\partial t^2 = $$
$$(1/\lambda_J^2) \sin K_0 x + (1/\lambda_J^2)[K_1/ik(0)] \, e^{i(\int k(x)dx - \omega t)} \cos K_0 x \tag{3.2.21}$$

with $K_0 = 2\pi d H_0/\phi_0$ and $K_1 = 2\pi d H_1/\phi_0$. *Note* for $K_0 = 0$ that (3.2.9) is obtained as expected with δ given by (3.2.19). To obtain the surface impedance Z_S from (3.2.13) it is sufficient to expand (3.2.21) for small x and retain the time-dependent part of the right side of (3.2.21)

$$[K_1/ik(0)\lambda_J^2](1 - \tfrac{1}{2} K_0^2 x^2 + \cdots) \, e^{i(\int k(x)dx - \omega t)} \ . \tag{3.2.22}$$

A solution for k of the form

$$k(x) = k_0 + k_1 x + \tfrac{1}{2} k_2 x^2 + \tfrac{1}{6} k_3 x^3 + \cdots \tag{3.2.23}$$

has for the leading term

$$k_0^2 = (1 - \omega^2/\omega_J^2)(-1/\lambda_J^2 + K_0^2/4) \ . \tag{3.2.24}$$

The components of the surface impedance are

$$R_S \approx (8\pi^2/c)(\omega^2\sigma/\omega_J^3)[1 + (\pi^2/2) H_0^2/H_{c1J}^2], \tag{3.2.25}$$

$$X_S \approx -(4\pi/c)(\omega/\omega_J)[1 + (\pi^2/2) H_0^2/H_{c1J}^2]. \tag{3.2.26}$$

The above two equations can be modeled approximately as an inductor and resistor in parallel with admittance

$$Y_S = 1/Z_S = 1/R + i/\omega L \ . \tag{3.2.27}$$

For R >> ωL the surface impedance is approximately

$$Z_s \approx \omega^2 L^2/R - i\omega L \ . \tag{3.2.28}$$

Comparing this equation with (3.2.25) and (3.2.26) leads to

$$L \approx (4\pi/\omega_J c)[1 + (\pi^2/2) H_0^2/H_{c1J}^2] \tag{3.2.29}$$

$$R \approx (2/\sigma\lambda_J)[1 + (\pi^2/2) H_0^2/H_{c1J}^2] \tag{3.2.30}$$

suggesting that the additional loss is associated with an increased junction inductance that results from the decoupling of grains and not from a reduced resistance. *Note* that for $H_0 = 0$ these expressions become (3.2.16) and (3.2.17) to lowest order in ω.

The ratio of the field-dependent parts of the surface impedance is

$$\Delta X_s/\Delta R_s = (1/2\pi)(\omega_J^2/\omega\sigma) \ . \tag{3.2.31}$$

Taking [3.2.1] $H_{c1J} \approx 0.1$ Oe for $La_{1-x}Sr_xCuO_4$ ($x \approx 0.15$) and $d = 2\lambda_L \approx 0.4$ μm from muon depolarization [3.2.2, 3.2.3] gives $\omega_J = (cd/\phi_0)H_{c1J} \approx 6 \times 10^{11}$ Hz. The observed value [3.2.1] of $\Delta X_s/\Delta R_s \approx 0.5$ indicates a junction conductivity $\sigma_J \approx 1.0$ $(\Omega\cdot cm)^{-1}$, which is to be compared with the measured [3.2.1] bulk resistivity $\rho = 25$ mΩ·cm, indicating an effective grain size $a_0 = d/\sigma\rho \approx 15$ μm separating the weakest junctions.

At low applied fields $H_0 < H_{c1J}$ microwave flux is expected to penetrate the intergranular medium to a distance λ_J. As the magnetic field H_0 is increased above H_{c1J}, the Josephson currents become periodic and the junction critical current is reduced. With the weakening of intergranular coupling, microwave absorption increases to the normal value at a frequency-dependent upper junction critical field. Clem [1.1.1] and Tinkham and Lobb [1.1.2] designate a closely related decoupling field by H_{c2J}.

As the magnetic field is increased, the Josephson currents become less and less effective in limiting the penetration of microwave fields. At high fields using (3.2.29) we estimate for the inductance [3.2.4]

$$L \approx (4\pi/c^2)\lambda_J (H/H_{c1J}). \tag{3.2.32}$$

We obtain H_{c2J} from the applied field H_0 at which the inductive reactance of surface Josephson currents is of the order of the surface resistance of a transmission line without distributed Josephson currents

$$\omega L = (4\pi\omega/c^2)\lambda_J (H_{c2J}/H_{c1J}) \approx 1/\sigma\delta_J . \qquad (3.2.33)$$

With the junction skin-depth $\delta_J = (c^2/2\pi\omega\sigma)^{1/2}$ we obtain

$$H_{c2J} \approx (\delta_J/2\lambda_J) H_{c1J} = (4\pi^2/c)J_0 \delta_J . \qquad (3.2.34)$$

For a range of junction skin-depths 10-100 µm and Josephson current densities 10^4-10^6 A/cm^2, we estimate a range for $H_{c2J} \approx 0.3$ to 300 kOe. The lower limit is appropriate to weak intergranular junctions and the upper limit to strong intragranular junctions. At fields above H_{c2J}, the surface resistance of the junction is the same as that in the normal state and increases as $\sqrt{\omega}$.

Studies of microwave absorption as a function of magnetic field support the view that mobile flux makes the major contribution to the field-dependent microwave loss at high fields [3.2.1]. As the magnetic field is increased from H = 0, the absorption at first increases quadratically with field and then more slowly. If the field is reversed at a maximum below the bulk H_{c1}, quantized flux is not expected to penetrate the grains. For the same value of field, the absorption is larger with decreasing fields as a result of trapped intergranular flux.

If the field is reversed at a maximum above the bulk H_{c1}, the absorption with decreasing fields is observed to lie below the absorption at the same increasing field [3.2.5, 3.2.6]. This is presumably because quantized flux has been trapped within grains, reducing the concentration of mobile flux. The absorption with decreasing field is observed to go through a minimum and then increase with further reduction of the field, crossing the absorption measured with increasing field. The observed minimum is at around the bulk H_{c1} and suggests that the concentration of mobile fluxons crosses zero around this field. Similar crossings have been observed in microwave absorption studies of type II metallic alloys [3.2.7] and support the general view that absorption in these materials is similarly associated with the fraction of mobile fluxons.

Flux flow in junctions. Multifluxon and single fluxon profiles have been observed in Josephson transmission lines and agree well with soliton theory [3.2.8-3.2.10]. Fluxon propagation is well described by a modification of (3.2.9) with dissipative terms included explicitly in the sine-Gordon equation

$$(1 + \beta\partial/\partial t) \partial^2\delta/\partial x^2 - (\varepsilon/c^2) \partial^2\delta/\partial t^2 - \alpha \partial\delta/\partial t = (1/\lambda_J^2) \sin\delta . \qquad (3.2.35)$$

By comparison with (3.2.3) we see that the term in α represents shunt losses and the term in β represents series losses. The dominant loss is evidently associated with the surface of the Josephson transmission line and not with quasiparticle excitation [3.2.11, 3.2.12]. In the description of small-amplitude periodic waves, ε in (3.2.9) may be replaced with the complex dielectric function

$$\varepsilon(\omega) = \varepsilon + (ic^2/\omega)\alpha \qquad (3.2.36)$$

to represent the parallel loss. We may introduce in addition a complex magnetic permeability

$$1/\mu(\omega) = 1 - i\omega\beta \qquad (3.2.37)$$

to represent the series loss.

4. Critical State Phenomena

4.1. Bulk Critical Phenomena in Type II Superconductors

Bean and the critical state model. The central idea of the bulk critical state [4.1.1-4.1.4] is that magnetic flux is transported against a macroscopic force

$$\mathbf{F} = -(1/c) \langle \mathbf{B} \times \mathbf{J}_c \rangle \qquad (4.1.1)$$

where $\mathbf{J}_c$ is a phenomenological and not a true current. This critical current, which inhibits fluxon transport, is associated with pinning forces in bulk type II superconductors. As discussed in Section 4.2, it is associated in weak links with the Josephson currents that connect grains. These forces are opposed by the gradient of a magnetic pressure that gives

$$\nabla P = (1/4\pi) \langle \mathbf{B} \times (\nabla \times \mathbf{B}) \rangle \qquad (4.1.2)$$

and a true current

$$\langle \nabla \times \mathbf{B} \rangle = -(4\pi/c) \mathbf{J}_c . \qquad (4.1.3)$$

For an applied field H_0 parallel to the surface, the flux density is

$$B(x) = H_0 \pm (4\pi/c) J_c x \qquad (4.1.4)$$

where x is directed into the superconductor. The negative sign in the above equation is for fluxons forced into the superconductor by an increasing external field. The positive sign is for fluxons forced out of the superconductor by a decreasing external field.

The critical current J_c normally represents flux pinning. A position-dependent critical current density $J_c(x)$ may represent a surface barrier [4.1.5] and a critical current $J_c(x, H)$ that depends on both position and magnetic field may represent the interface between the intergranular medium and the interior of grains.

From studies of axial flux penetration through thin cylindrical tubes, Kim, Hempstead and Strnad [4.1.6] have found that the critical current varies with magnetic field as

$$J_c(H) = J_c/(1 + H/H^*) \tag{4.1.5}$$

where H^* is of the order of the thermodynamic critical field.

Hysteretic ac loss. As Bean has shown [4.1.3], (4.1.3) leads to hysteretic loss over the period of an alternating current-produced (ac) field. The work performed per cycle and per unit area of surface is

$$W_s = (1/4\pi) \int H_s \, d\phi \tag{4.1.6}$$

where $d\phi = \int dx \, dB$ is the integrated flux. Using (4.1.3) gives per cycle

$$W_s = (c/24\pi^2) H_s^3/J_c \tag{4.1.7}$$

with power dissipation per unit area

$$P_s = \nu W_s = (\nu c/24\pi^2) H_s^3/J_c \, . \tag{4.1.8}$$

Taking $P_s = \frac{1}{2}(c/4\pi)^2 H_s^2 R_s$ for the dissipation (2.1.11) gives a surface resistance

$$R_s = \frac{4}{3} (\nu/c)(H_s/J_c) \tag{4.1.9}$$

linear in frequency ν and peak magnetic field H_s applied to the surface.

We show that (4.1.9) is exact for the Kim form of the critical current. From (4.1.5) the field gradient into the medium is

$$dH/dx = -(4\pi/c) J_c/(1 + H/H^*) \tag{4.1.10}$$

and the integrated flux is

$$\phi = \int H \, dx = (c/4\pi J_c) \int (1 + H/H^*) \, dH = $$
$$(c/4\pi J_c)(\frac{1}{2}H^2 + \frac{1}{3}H^3/H^*) \, . \tag{4.1.11}$$

The integral used to compute the work done around a cycle is

$$\int \phi \, dH = (c/24\pi J_c)(H^3 + {}^1\!/_2 H^4/H^*) \, . \tag{4.1.12}$$

The total work is the area included within the hysteresis curve that is enclosed by a rectangle of width $2H_s$ and height $(c/4\pi J_c)(H_s^2 + {}^2\!/_3 H_s^4/H^*)$ from (4.1.11) for a total area

$$S = (c/8\pi^2 J_c)(H_s^3 + {}^2\!/_3 H_s^4/H^*). \tag{4.1.13}$$

It is easiest to find the area within the hysteresis loop by subtracting the corner areas, each of which is (4.1.12) with both H_s and J_c doubled

$$\Delta S = (1/16\pi^2)(2c/3J_c)(H_s^3 + H_s^4/H^*) \, . \tag{4.1.14}$$

Finally, the work per cycle is

$$W = (1/16\pi^2)(S - 2\Delta S) = (c/24\pi^2 J_c) \, H_s^3 \tag{4.1.15}$$

which is identical with (4.1.7) and independent of H^*.

Anderson, Kim and flux creep. The Bean critical state is metastable and over long times shows logarithmic relaxation associated with thermally activated flux creep. Anderson and Kim [4.1.7, 4.1.8] have constructed a theory in which the relaxation slows as relaxation proceeds because the activation energy for further creep increases. The theory may be written in terms of relaxation of the local current density, which is proportional to the field-gradient

$$dJ/dt = -(1/\tau)J \tag{4.1.16}$$

with a thermally activated relaxation rate

$$1/\tau = \omega \exp(J/\alpha k_B T) \tag{4.1.17}$$

where as the current relaxes, the activation energy for further relaxation increases and the relaxation rate becomes slower. Combining the last two equations gives

$$-\exp(-J/\alpha k_B T] \, dJ = \omega J \, dt \, . \tag{4.1.18}$$

For $J > \alpha k_B T$ the left-hand-side of (4.1.18) changes rapidly with J while the right-hand-side is relatively fixed. Integrating with the approximation on the right-hand-side $J(t) \approx J(0)$ gives

$$J(t) \approx -\alpha k_B T \ln\{\omega J(0)t/\alpha k_B T + \exp[-J(0)/\alpha k_B T]\} \, . \tag{4.1.19}$$

For times sufficiently long that the first term in the logarithm dominates, the above equation becomes, in agreement with experiment [4.1.8, 4.1.9]

$$J(t) \approx -\alpha k_B T \ln t + const. \qquad (4.1.20)$$

4.2. Critical Phenomena in Junctions

Microwave absorption and granular decoupling. A gradient of magnetic field in a junction is equivalent to a junction bias-current and leads to absorption that increases when the applied magnetic field is increasing or decreasing [4.2.1]. This problem is treated most simply for fields $H > H_{c1J}$ by absorbing the spatial variation of δ into a field-dependent Josephson current as proposed by Dulcic *et al.* [4.2.2, 4.2.3]

$$(\varepsilon J_0/\omega_J^2)\partial^2\delta/\partial t^2 + J_0(H) \sin \delta = J_c + J_s e^{-i\omega t} \qquad (4.2.1)$$

where $J_0(H)$ is the Josephson current reduced by flux in the junction. Expanding $\sin \delta = \sin[(\delta - \delta_0) + \delta_0] \approx (\delta - \delta_0) \cos \delta_0 + \sin \delta_0$

$$(\varepsilon J_0/\omega_J^2)\partial^2\delta/\partial t^2 + J_0(H) (\delta - \delta_0) \cos \delta_0 = J_s e^{-i\omega t} \qquad (4.2.2)$$

with $J_0(H) \sin \delta_0 = J_c$. For $\delta = \delta_0 + \delta_1 e^{-i\omega t}$, the time-dependent phase difference across the junction is

$$\delta_1 = [J_s/J_0(H)]/(\cos \delta_0 - \varepsilon\omega^2/\omega_J^2). \qquad (4.2.3)$$

The power dissipation is

$$P_s = \tfrac{1}{2} \operatorname{Re} (J_s^* V) = (h/4\pi q) \operatorname{Re} [J_s^* (\partial\delta/\partial t)]. \qquad (4.2.4)$$

Substituting from (4.2.2) the power is

$$P_s = (h\omega/4\pi q) [J_s^2/J_0(H)] \operatorname{Re} [i (\cos \delta_0 - \varepsilon\omega^2/\omega_J^2)]^{-1}. \qquad (4.2.5)$$

Writing $\varepsilon = 1 + 4\pi i\sigma/\omega$ gives for the surface resistance at frequencies ω well below $\omega_J = c/\lambda_J$

$$R_s \approx 2(h/q)(\omega^2\sigma/\omega_J^2)/[J_0(H) \cos^2\delta_0] \ . \qquad (4.2.6)$$

The reduction in $J_0(H)$ and the presence of bias-current both increase R_s, leading to additional microwave absorption. The striking observation of the effect of critical-state current is that when the magnetic field is reversed, the resistance initially decreases and then recovers [3.2.10, 3.2.11, 4.2.4-4.2.6].

Early studies identified this process variously with a modification in fluxon density [3.2.11] or with a modification of fluxon response [4.2.4]. The analysis of Dulcic *et al.* [4.2.2, 4.2.3] offers a convincing argument that the origin of the additional loss is the junction decoupling that results from critical-state current in the surface.

The microwave critical state. As shown in Section 4.1, ac loss leads to a surface resistance

$$R_S = (16\pi/3)\,(v/c)\,(H_S/J_c)\;. \tag{4.2.7}$$

Surface resistance linear in H_S has been observed in ceramics [4.2.7, 4.2.8], suggesting that ac losses persist to microwave frequencies and indicating that flux penetrates the intergranular medium in nanoseconds. Particularly at low frequencies and high microwave fields, this process may be an important source of microwave loss and should be investigated.

Acknowledgements

I am grateful to J. G. Bednorz and K. A. Müller for inviting me to participate in the *16th Course of the International School of Materials Science and Technology: Earlier and recent aspects of superconductivity* and for their hospitality at the Zürich Research Laboratory of the IBM Research Division, where this paper was written. I owe much to my colleague K. W. Blazey with whom I have had the good fortune to collaborate on microwave studies of granular superconductors.

References

1.1.1. J. R. Clem, *Physica C* **153-155** (1988) 50.

1.1.2. M. Tinkham and C. J. Lobb, "Solid State Physics, Volume 42," Academic Press, Inc., New York, 1989. Pages 91-134. This review is keyed to the following extremely useful reference.

1.1.3. Michael Tinkham, *Introduction to Superconductivity,* McGraw-Hill, New York, 1975. Reprinted 1980 by Robert E. Krieger Publishing Company, Inc., Krieger Drive, Malabar, FL 32950, USA.

1.2.1. H. London, *Proc. Roy. Soc. A* **176** (1940) 522.

1.2.2. A. B. Pippard, *Proc. Roy. Soc. A* **191** (1947) 370, 385, 399.

1.2.3. W. M. Fairbank, *Phys. Rev.* **76** (1949) 1106.

1.2.4. E. Maxwell, P. M. Marcus and J. C. Slater, *Phys. Rev.* **76** (1949) 1332.

1.2.5. I. Simon, *Phys. Rev.* **77** (1950) 384.

1.2.6. A. B. Pippard, *Proc. Roy. Soc. A* **203** (1950) 98, 195.

1.2.7. A. B. Pippard, *Proc. Roy. Soc. A* **216** (1953) 547.

1.2.8. R. E. Glover and M. Tinkham, *Phys. Rev.* **104** (1956) 844.

1.2.9. R. E. Glover and M. Tinkham, *Phys. Rev.* **108** (1957) 243.

1.2.10. A. T. Forrester, *Phys. Rev.* **110** (1958) 769; M. Tinkham and R. E. Glover, *Phys. Rev.* **110** (1958) 771.

2.1.1. W. Meissner and R. Ochsenfeld, *Naturwiss.* **21** (1933) 787.

2.1.2. F. and H. London, *Proc. Roy. Soc. A* **149** (1935) 71.

2.1.3. C. J. Gorter and H. B. G. Casimir, *Phys. Z.* **35** (1934) 963.

2.1.4. See A. B. Pippard in "Advances in electronics," Academic Press, New York, 1954. Edited by L. Marton, Volume 6, page 1.

2.1.5. See J. M. Ziman, "Principles of the theory of solids, Second edition" Cambridge Press, 1972. Pages 283, 402.

2.1.6. V. L. Ginzburg, *Nuovo Cimento* **2** (1955) 1234.

2.1.7. J. Bardeen, L. N. Cooper and J. R. Schrieffer, *Phys. Rev.* **108** (1957) 1175.

2.1.8. D. C. Mattis and J. Bardeen, *Phys. Rev.* **111** (1958) 412.

2.2.1. B. D. Josephson, *Phys. Lett.* **1** (1962) 251.

2.2.2. B. D. Josephson, *Rev. Mod. Phys.* **34** (1964) 216.

2.2.3. B. D. Josephson, *Advan. Phys.* **14** (1965) 419.

2.2.4. B. D. Josephson, *Rev. Mod. Phys.* **46** (1974) 251.

2.2.5. V. Ambegaokar and A. Baratoff, *Phys. Rev. Lett.* **10** (1963) 486.

2.2.6. G. Deutscher and K. A. Müller, *Phys. Rev. Lett.* **59** (1987) 1745.

3.1.1. V. L. Ginzburg and L. D. Landau, *J. Exptl. Theor. Phys., USSR* **20** (1950) 1064, *Soviet Phys.JETP* **5** (1957) 1442.

3.1.2. V. L. Ginzburg, *Soviet Phys. JETP* **3** (1956) 621, **7** (1958) 78.

3.1.3. L. D. Landau and I. M. Lifshitz, "Course of Theoretical Physics, Volume 5: Statistical Physics," Pergamon Press, London, 1958, Chapter IV.

3.1.4. L. P. Gor'kov, *Soviet Phys. JETP* **9** (1959) 1364.

3.1.5. A. A. Abrikosov, *Dokl. Akad. Nauk., SSSR* **86** (1952) 489, *Soviet Physics JETP* **5** (1957) 1174.

3.1.6. H. Suhl, *Phys. Rev. Lett.* **14** (1965) 226.

3.1.7. C. Caroli, P. G. de Gennes and J. Matricon, *Phys. Lett.* **9** (1964) 307.

3.1.8. M. Tinkham, *Phys. Rev. Lett.* **13** (1964) 804.

3.1.9. M. J. Stephen and J. Bardeen, *Phys. Rev. Lett.* **14** (1965) 112; J. Bardeen and M. J. Stephen, *Phys. Rev.* **140** (1965) A1197.

3.1.10. J. Gittleman and B. Rosenblum, *Phys. Lett.* **20** (1966) 453.

3.1.11. J. le G. Gilchrist, *Proc. Roy. Soc. A* **295** (1966) 399. J. le G. Gilchrist and P. Monceau, *Phil. Mag.* **18** (1968) 237; *J. Phys. C* **3** (1970) 1399.

3.1.12. J. Baixeras and G. Fournet, *Phys. Lett.* **20** (1966) 226.

3.2.1. A. M. Portis, K. W. Blazey, K. A. Müller and J. G. Bednorz, *Europhys. Lett.* **5** (1988) 467.

3.2.2. G. Aeppli, R. J. Cava, E. J. Ansaldo, J. H. Brewer. S. R. Kreitzman, G. M. Luke, D. R. Noakes and R. F. Kiefl, *Phys. Rev. B* **35** (1987) 7129.

3.2.3. W. J. Kossler, J. R. Kempton, X. H. Yu, H. E. Schone, Y. J. Uemura, A. R. Moodenbaugh, M. Suenaga and C. E. Stronach, *Phys. Rev. B* **35** (1987) 7133.

3.2.4. R. A. Ferrell and R. E. Prange, *Phys. Rev. Lett.* **10** (1963) 479.

3.2.5. E. J. Pakulis and T. Osada, *Phys. Rev. B* **37** (1988) 5940.

3.2.6. Y. Maniwa, A. Grupp, F. Hentsch and M. Mehring, *Physica C* **156** (1988) 755.

3.2.7. B. L. Walton, B. Rosenblum and F. Bridges, *Phys. Rev. Lett.* **32** (1974) 1047.

3.2.8. A. Barone and G. Paterno,"Physics and Applications of the Josephson Effect," Wiley-Interscience, New York, 1982.

3.2.9. N. F. Pedersen and D. Welner, *Phys. Rev. B* **29** (1984) 2551.

3.2.10. S. Sakai and N. F. Pedersen. *Phys. Rev. B* **34** (1986) 3506.

3.2.11. A. Matsuda and T. Kawakami, *Phys. Rev. Lett.* **51** (1983) 694.

3.2.12. W. R. McGrath, H. K. Olsson, T. Claeson, S. Erikkson and L. -G. Johansson, *Europhys. Lett.* **4** (1987) 357.

4.1.1. C. P. Bean, *Phys. Rev. Lett.* **8** (1962) 250.

4.1.2. H. London, *Phys. Lett.* **6** (1963) 162.

4.1.3. C. P. Bean, *Rev. Mod. Phys.* **36** (1964) 31.

4.1.4. C. P. Bean, *J. Appl. Phys.* **41** (1970) 2482.

4.1.5. C. P. Bean and J. D. Livingston, *Phys. Rev. Lett.* **12** (1964) 14.

4.1.6. Y. B. Kim, C. F. Hempstead and A. R. Strnad, *Phys. Rev. Lett.* **9** (1962) 306; *Phys. Rev.* **129** (1963) 528; *Rev. Mod. Phys.* **36** (1964) 43;*Phys. Rev.* **139** (1965) A1163.

4.1.7. P. W. Anderson, *Phys. Rev. Lett.* **9** (1962) 309.

4.1.8. P. W. Anderson and Y. B. Kim, *Rev. Mod. Phys.* **36** (1964) 39.

4.1.9. M. R. Beasley, R. Labusch and W. W. Webb, *Phys. Rev.* **181** (1969) 682.

4.2.1. Y. W. Kim, A. M. de Graaf, J. T. Chen, E. J. Friedman and S. H. Kim, *Phys. Rev. B* **6** (1972) 887.

4.2.2. M. Pozek, A. Dulcic and B. Ravkin, *Solid State Commun.* **70** (1989) 889.

4.2.3. A. Dulcic, B. Rakvin and M. Rozek, to be published.

4.2.4. K. W. Blazey, A. M. Portis and J. G. Bednorz, *Solid State Commun.* **65** (1988) 1153.

4.2.5. A. M. Portis, K. W. Blazey and F. Waldner, *Physica C* **153-155** (1988) 659.

4.2.6. A. M. Portis, M. Stalder. G. Stefanicki, F. Waldner and M. Warden, *J. de Physique* **12** Colloque C8 (1988) 2231.

4.2.7. H. Piel, M. Hein, N. Klein, U. Klein, A. Michalke, G. Müller, and L. Ponto, *Physica C* **153-155** (1988) 1604.

4.2.8. S. M. Rezende and F. M. de Aguiar, *Phys. Rev. B* **39** (1988) 1604.

Part III

**Electronic and
Magnetic Properties**

Superconductivity of Strongly Correlated Electrons: Heavy-Fermion Systems

F. Steglich

Institut für Festkörperphysik, Technische Hochschule Darmstadt,
D-6100 Darmstadt, Fed. Rep. of Germany

Abstract. The Cooper pairs in the superconductors $CeCu_2Si_2$, UBe_{13}, UPt_3, and URu_2Si_2 are formed by quasiparticles with effective mass m* of order 100 m_{el}. m* originates in very strong correlations between f-electrons which become weakly delocalized well below a characteristic temperature $T^* \sim 5\text{-}50$ K via a Kondo-type mechanism. These "heavy fermions" dominate the low-temperature properties of a number of lanthanide- and actinide-based compounds, spanning the full range between coherent Fermi liquid, band magnetism and superconductivity. The pairing interaction in the "heavy-fermion superconductors" is likely to differ from the BCS-type phonon-mediated coupling. In addition, their order parameters appear to be highly anisotropic. Most of the experimental evidence points to even-parity ("singlet") rather than odd-parity ("triplet") pairing.

1. Introduction

Traditionally, superconductivity and magnetism are considered antagonistic phenomena: Superconducting ground states are frequently observed in metals containing delocalized (s-, p-, 5d- and 4d-) conduction electrons. According to Bardeen, Cooper and Schrieffer (BCS) [1], a net attraction between such electrons via the exchange of virtual phonons leads to the formation of the Cooper pairs that carry the superconducting state. Magnetic order preferentially occurs in metals which contain more localized (3d-, 5f- and 4f-) electrons, whose motion is dominated by the intra-atomic correlation energies, notably their on-site Coulomb repulsion. The magnetic moments formed in the free atoms according to Hund's rules can often be retained in a metallic environment where either direct wave-function overlap or hybridization with ligand orbits tends to delocalize these "magnetic electrons". Among elemental metals, magnetic order is found in the Fe-group metals, in the actinides and lanthanides. The unique case of coexistence between superconducting and antiferromagnetically ordered states was established for some of the Chevrel phases and related ternary compounds, in which the interaction between superconducting and magnetic electrons is sufficiently weak [2]. Superconductivity and magnetic order are also close and can be transformed into each other via changes of physical (chemical) parameters in both the organic [3] and the cuprate high-T_c superconductors [4]. Similar phase diagrams, containing adjacent antiferromagnetically ordered and superconducting states, sometimes even coexisting, are reported for the heavy-fermion compounds, i.e. certain Ce- and U-based intermetallics. Unlike the Chevrel-phase superconductors, however, "heavy-fermion superconductors" (HFS) owe their magnetic as well as superconducting effects to the <u>same</u> (4f- or 5f-) electrons and, thus, exhibit "superconductivity in a system of strongly correlated electrons" [5].

The relationship between superconductivity and electron correlations is illustrated in figure 1. For the "classical" superconductors the critical

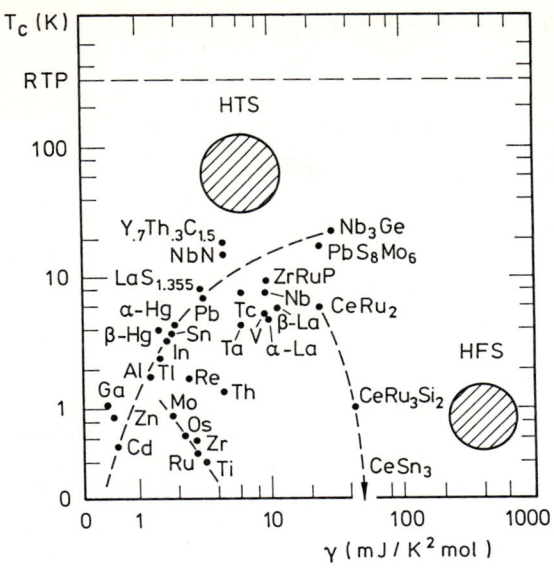

Fig. 1. Superconducting transition temperature T_c vs γ, the coefficient of the electronic specific heat in the normal state, γT, for several superconductors. Hatched areas mark approximate locations of high-T_c superconductors (HTS) and heavy-fermion superconductors (HFS).

temperature, T_c, continuously increases upon increasing γ, the Sommerfeld coefficient of the electronic specific heat which is proportional to the normal-(n-)state density of states (DOS) at the Fermi energy, E_F, or the effective carrier mass, m^*. For these simple metals, γ is typically of order mJ/K²mole. The observed $T_c(\gamma)$ dependence reflects the gain in thermodynamic stability (condensation energy) of the superconducting state, when the density of Cooper-pair states becomes increased. However, the tendency of $T_c(\gamma)$ to saturate for the A15 superconductors between 20 and 30 K indicates the action of a competing effect. The latter apparently dominates within the series of the transition metals Mo, Os, Zr, Ru and Ti. Likewise, Ce-based intermediate-valence compounds display a disastrous T_c-depression upon increasing γ [6]. For example, the canonical intermediate-valence system $CeSn_3$ is not a superconductor, although its La homolog superconducts below $T_c \approx 6K$. We attribute this precipitous drop of T_c to an increasing importance of the correlation energies relative to the kinetic energy of the carriers, measured by the band width or the inverse DOS. This is accompanied by a strong reduction of the typical frequency, ν_{sf}, of spin fluctuations. When long lived ($\nu_{sf} \to \nu_c = h^{-1}k_BT_c$), spin fluctuations act as efficient pair breakers [7,8].

In $CeCu_2Si_2$ and its U-based counterparts, all showing T_c's of order 1K and being found well beyond the stability range of the intermediate-valence superconductors, the correlations seem to help constitute rather than to break the Cooper pairs. For $CeCu_2Si_2$, an estimate of the effective mass from the measured γ, based upon a reasonable valence-electron concentration and a parabolic band, yields $m^* \approx 300\ m_{el}$, where m_{el} is the free-electron mass. The huge m^* has the meaning of a large mass of inertia, which originates in the strong correlations between the 4f electrons and can likewise be expressed by a very small Fermi velocity, $v_F^* \approx 5 \cdot 10^3$ m/s.

Since cuprate high-T_c superconductors show comparatively small γ's, they are also well separated, in figure 1, from the classical superconductors. Despite their low γ-values, however, and owing to their reduced carrier concentrations, a substantial mass enhancement, $m^*/m_{el} \sim 5$, is inferred [9]. It is similar to the one found for strongly intermediate-valent compounds like $CeRu_2$ and is assumed to result mainly from the correlations on the Cu-3d shells [10]. Therefore, pronounced correlations appear to be a common property of heavy-fermion and high-T_c superconductors.

We begin these notes with some historical remarks concerning HFS, which are followed by a brief discussion of the characteristic properties of heavy-fermion materials, before a few selected topics of the current research on HFS will be addressed. The interested reader is referred to existing review articles [11-17].

2. Historical remarks

An investigation of the low-T state of thoroughly prepared, nearly single-phase polycrystalline $CeCu_2Si_2$ material revealed a number of anomalies near 0.6 K, indicating the transition into a bulk-superconducting phase (cf. figure 2) [5]. These included a drop in the electrical resistivity, a large diamagnetic signal in the ac-susceptibility, a substantial volume fraction from which the magnetic flux became expelled upon field cooling in a dc-magnetization measurement ("Meissner-Ochsenfeld effect"), as well as a pronounced discontinuity in the specific heat (figure 2a).

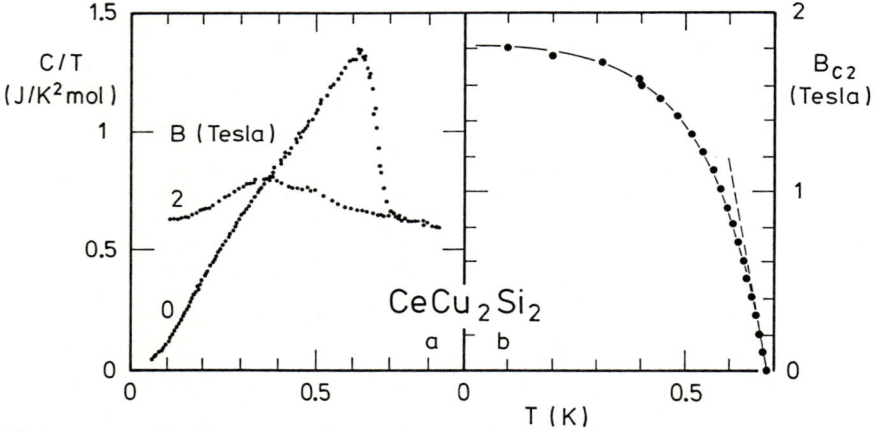

Fig. 2. Specific heat as C/T vs T (a) and upper critical field as B_{c2} vs T (b), for $CeCu_2Si_2$ [5,27]. Dashed line represents $B_{c2}(T)$ slope at T_c (-13 T/K). Solid line is guide to the eye.

Before we focus on the implications of the superconducting phase transition, let us discuss the n-state properties of $CeCu_2Si_2$. Figure 3 shows the body-centered tetragonal unit cell of the $ThCr_2Si_2$ structure. The Ce ions reside in the basal and center planes which are separated by a pair of Cu and Si planes. While the lattice parameter c ($\simeq 9.94$ Å) appears to be determined by a dominantly covalent Si-Si bond, the lattice parameter a equals the Ce-Ce

CeCu$_2$Si$_2$

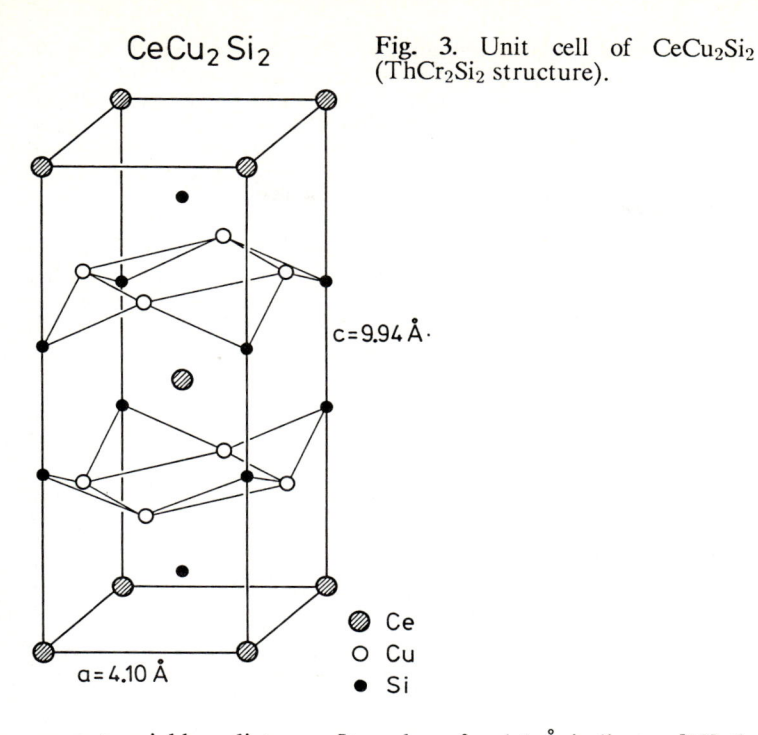

$a = 4.10$ Å

$c = 9.94$ Å

⊘ Ce
○ Cu
● Si

Fig. 3. Unit cell of CeCu$_2$Si$_2$ (ThCr$_2$Si$_2$ structure).

nearest-neighbor distance. Its value of $\simeq 4.1$ Å indicates [18] that the valence of Ce is close to 3, according to a (Xe)4f^1 configuration with a J = 5/2 Hund's rule ground state. Magnetic susceptibility, $\chi(T)$, measurements [19] reveal a Curie-Weiss law for T > 150 K involving an effective moment of 2.68 μ_B which exceeds the free-ion value by approximately 5%. Paramagnetic, 4f-derived moments can also be identified through the magnetic relaxation rate as determined by the half-width of a quasi-elastic line, $\Gamma/2$, in the magnetic neutron scattering cross section [20]. $\Gamma(T)/2$ approaches a constant value $\Gamma(0)/2 \simeq k_B \cdot 10$ K as T → 0, while $\chi(T)$ turns from the Curie-Weiss to a Pauli-like behavior and becomes essentially temperature independent below T $\simeq$ 6 K [21,19]. These observations demonstrate the lack of a magnetically ordered ground state, which might have been anticipated in view of the dense lattice of Ce^{3+} ions. The low-temperature (as T → 0) normal state of CeCu$_2$Si$_2$ is characterized by (1) a linear specific-heat term γT, with $\gamma(0)$ being of order 1000 mJ/K^{2}mole (figure 2a), and (2) a Pauli susceptibility $\chi(0)$ of approximately $8 \cdot 10^{-8}$ m^3/mole [19].

Both numbers exceed the corresponding values for simple metals by 2-3 orders of magnitude and point to the existence of a "heavy-Fermi-liquid" phase. In this respect the system appears to be closely related to CeAl$_3$ [22] and a few other Ce intermetallics like CeCu$_6$ [23] and CeRu$_2$Si$_2$ [24] which, however, are not superconductors. Inherent to all these compounds is the existence of a characteristic scaling temperature T*, which ranges from 4K (CeAl$_3$) to 25 K (CeRu$_2$Si$_2$) as determined, e.g. by measurements of the entropy [25,12] and by $\Gamma(0)/2$ [26]. Well above T*, the material contains a collection of 4f-derived magnetic moments, weakly coupled to ordinary conduction electrons (with effective masses m* $\lesssim$ 10 m$_{el}$), while well below T*, a non-magnetic state with strongly renormalized masses develops. The low-lying excitations (quasiparticles) of this non-magnetic metallic ground state have been labeled "heavy fermions" or, alternatively, "heavy electrons".

The specific-heat anomaly at the superconducting transition, when replaced by an idealized one, is of the same gigantic size as the n-state specific heat at T_c, $\gamma(T_c) \cdot T_c$. This, in fact, proves that the term γT must be of electronic origin. Moreover, it indicates that the Cooper pairs in $CeCu_2Si_2$ are formed by those heavy fermions concluded from the n-state thermodynamic properties. Consequently B'_{c2}, the initial slope of the upper critical magnetic field curve, $B_{c2}(T)$, is of a gigantic size too (see figure 2b). Since $B_{c2}(T)$ at T_c results from the "diamagnetic pair breaking" by the external magnetic field acting on the orbital state of the Cooper pair (via the Lorentz force), the large B'_{c2} clearly reflects the very small renormalized Fermi velocity v_F^* of the quasiparticles [27]. Heavy fermions constitute a necessary, though not sufficient ($CeAl_3$!), condition for this kind of superconductivity: The homologs $LaCu_2Si_2$ (with no 4f electron and no heavy fermions) as well as any other Rare-EarthCu_2Si_2 (with 4f electrons, but with no heavy fermions) are not superconductors [5].

Superconductivity of $CeCu_2Si_2$ was initially considered with substantial scepticism, in particular by experimentalists [28,29]. One argument was the apparent dependence of superconducting properties on the "quality", namely - as was learnt from subsequent investigations [30-32,12] - the stoichiometry of the samples. In particular, $CeCu_2Si_2$ single crystals do not superconduct at ambient pressure [33-35], but do so under a surprisingly low external pressure, $p \gtrsim 1$ kbar [33]. Also, a few at% excess of Cu in the "starting" melt was found to lead to bulk superconductivity in single crystals already at ambient pressure [31,35]. Furthermore, in view of the disastrous pair-breaking effect of Ce^{3+} ions when diluted in a BCS superconductor, it seemed most puzzling that an intermetallic containing 100 at% Ce^{3+} ions should be a superconductor. For example, about 0.6 at% Ce^{3+} replacing La^{3+} in $LaAl_2$ ($T_c \simeq 3.3$ K) suffices to suppress superconductivity [36]. The origin of this paramagnetic pair breaking is an exchange scattering of the carriers from the local Ce spin which is associated by spin-flip processes [8] and, thus, destroys the spin pairing of the Cooper pair. This effect is particularly pronounced since the Ce^{3+} dopants in $LaAl_2$ give rise to a Kondo effect [36].

Many theorists, however, liked heavy-fermion superconductivity in $CeCu_2Si_2$ as a new kind of "high-T_c superconductivity": the ratio of T_c/T_F^*, where T_F^* is the low degeneracy temperature of the heavy fermions, exceeded the ratio T_c/T_F for all hitherto known superconductors. In particular, the renormalized Fermi velocity is of the same order of magnitude as the velocity of sound, $v_s \simeq 2 \cdot 10^3$ m/s [37], in contrast to BCS superconductors, exhibiting $v_F/v_s \simeq 1000$. This gave rise to speculations concerning "non-conventional" superconductivity, in that a non-phononic mechanism might be needed to provide the net attraction between heavy fermions [5]. In fact, $CeCu_2Si_2$ reminds one of the case of liquid ^{3}He, being the prototype of a heavy Fermi-liquid which, in the mK range, becomes unstable against formation of a superfluid state [38]. Here a non-phononic mechanism, i.e. the exchange of virtual (ferromagnetic) spin fluctuations, is commonly considered to greatly contribute to the Cooper-pair formation [38]. Since spin-triplet (S = 1) pairing had been established for superfluid ^{3}He [38], $CeCu_2Si_2$ was regarded a promising candidate for triplet superconductivity. Subsequent investigations, namely measurements of $B_{c2}(T)$ [39] and of the dc-Josephson effect [40], were devoted to this issue and will be discussed briefly below.

The $B_{c2}(T)$ curve in figure 2b shows a very pronounced flattening below $T \simeq 0.5$ K. This proves the presence of an additional pair-breaking mechanism caused by the magnetic field. In fact, regardless of the actual Cooper-pair state, diamagnetic pair breaking alone should result in a low-T value, $B_{c2}(T \rightarrow 0)$, exceeding the experimental one by a factor 4-6 [39]. A meaningful fit of the

$B_{c2}(T)$ data is possible, on the other hand, if a strong "Pauli-limiting" is included. This paramagnetic pair breaking effect by the B-field, acting on the spin state of the Cooper pair, shows almost no anisotropy, i.e. is almost independent of the orientation of the field relative to the crystal axes [35]. From these experimental results it was concluded [39] that $CeCu_2Si_2$ cannot be in an "equal-spin pairing" state, and that any other triplet state is at least unlikely.

Figure 4 shows I-V characteristics of a weak link between $CeCu_2Si_2$ and Al at different temperatures [40]. A critical current I_c is detected upon cooling to below the T_c of $CeCu_2Si_2$. $I_c(T)$ increases with decreasing temperature but becomes suppressed if, at 0.14 K, a small magnetic field of 25 mT is applied which is overcritical for Al, but only of the order of the lower critical field, $B_{c1}(0)$, of $CeCu_2Si_2$. By these observations it is assured that both metals are involved in forming the weak link. The main result of the experiments is a critical current $I_c(T\rightarrow 0)$ as large as for usual Josephson contacts between two BCS superconductors. In fact, $I_c(T\rightarrow 0)$ was found to assume 80% of I_{cmax}, the upper theoretical limit of such a dc-Josephson current [41]. This has to be compared with the prediction that in leading order no Josephson current should occur for a triplet/singlet junction [42]. Therefore the experimentally observed I_c values, being much too large for a higher-order effect, support the conclusion from the $B_{c2}(T)$ measurements that $CeCu_2Si_2$ is a spin-singlet (even-parity) superconductor. This does not necessarily invoke s-wave ($\ell = 0$) pairing. By contrast, the Cooper pair-wave function in $CeCu_2Si_2$ as well as in the other HFS UBe_{13} [43], UPt_3 [44] and URu_2Si_2 [45-47] appears to be highly anisotropic, see Sec. 4.

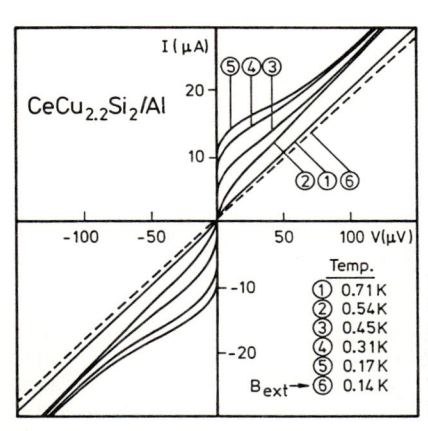

Fig. 4. Current-voltage characteristics of a $CeCu_{2.2}Si_2/Al$ weak link at different temperatures [40]. Solid lines are for B = 0T, dashed line holds for B = 25 mT, a field which is overcritical for Al and comparable to $B_{c1}(T\rightarrow 0)$ of $CeCu_2Si_2$.

3. Characteristic properties of heavy-fermion materials

In figure 5, the 4f-derived specific heat of $CeCu_2Si_2$ is shown as a function of temperature between 0.04 K and 400 K [25]. The plot $C_{4f}(T)/T$ displays the T-dependence of the Sommerfeld coefficient, $\gamma(T)$. On cooling the system to below the characteristic temperature $T^* \simeq 15K$, at which the majority of the Ce's exist in their crystal-field (Kramers-doublet) ground-state, $\gamma(T)$ is found to increase by almost one order of magnitude within about one decade in temperature. This demonstrates that the formation of the large effective masses is a low-temperature phenomenon which cannot be explained in the frame of

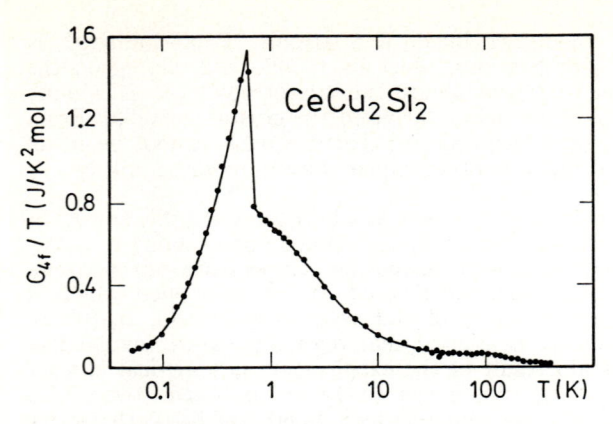

Fig. 5. 4f-derived specific heat as C_{4f}/T vs T (on a logarithmic scale) for CeCu$_2$Si$_2$ [25]. Solid line is a guide to the eye.

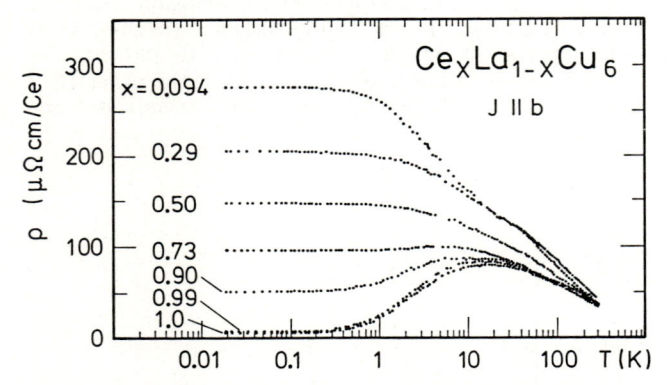

Fig. 6. Ce increment of the electrical resistivity as a function of temperature (on a logarithmic scale) for several Ce$_x$La$_{1-x}$Cu$_6$ systems [51].

one-electron theory. Similar specific-heat behavior was already discovered in 1971 [48] for canonical dilute Kondo alloys, i.e. CuFe and CuCr [49]. For example, the Fe-increment to the specific heat, when plotted as C_{Fe}/T vs T, increases monotonically below the "Kondo temperature" $T_K \simeq 27$ K and saturates for $T \lesssim 1$K. If the actual Fe-concentration (83 ppm) is scaled to 100 at%, the low-T value $\gamma(0) = C_{Fe}(0)/T$ is found to be of the same gigantic size as for the heavy-fermion compounds ($\simeq 0.9$ J/K^{2}mole-Fe). Further phenomenological similarities between the two classes of materials concern, e.g. the temperature dependences of the electrical resistivity sufficiently far above the respective scaling temperatures T_K and T^* (cf. figure 6) and of the magnetic susceptibility.

The Kondo-impurity effect is by now well understood. It is a true many-body phenomenon which originates in both the dynamic nature of the conduction electron-local moment scattering and the sharpness of the Fermi-Dirac distribution function for the conduction electrons at low temperatures. Exact theoretical results exist, through Bethe-Ansatz calculations, for the thermodynamic properties of dilute Kondo alloys [50]. The well-known

Kondo anomaly in the electrical resistivity is displayed in figure 6 for a $La_{1-x}Ce_xCu_6$ alloy containing a Ce concentration of 9.4 at% [51]. It is explained by a resonance scattering of the conduction electrons from the partially filled 4f shell of Ce, i.e. by a phase shift for the $\ell = 3$ component being close to $\pi/2$. As a result of the resonance scattering, the 4f electrons become weakly delocalized and form a "local band of heavy fermions". In addition, their spins screen the 4f spin and, thus, give rise to the non-magnetic ground state, typical of Kondo alloys for $T \ll T_K$ [49].

For most of the Ce^{3+}-based intermetallic compounds, however, the single-ion Kondo interaction is predominated at low temperatures by the interionic magnetic correlations mediated by the Ruderman, Kittel, Kasuya, Yosida (RKKY) polarization of the conduction electrons. Therefore, these systems adopt a (usually complex antiferro-) magnetically ordered ground state. More interesting are those compounds like $CeCu_6$ which are lacking magnetic ordering between local moments, but approach a Fermi-liquid state instead. As figure 6 indicates, increasing importance of inter-site couplings (upon increasing Ce concentration) results in a depression of the low-T resistivity. In the "Kondo-lattice" system $CeCu_6$, $\rho(T)$ becomes maximum at a finite temperature T_g (being of the same order as T^*), rather than at $T = 0K$. This essential difference between the dilute alloy and the intermetallic compound derives from the fact (known as "Bloch's theorem") that (as $T \rightarrow 0$) the eigenstates of an electron system contained in a perfectly periodic lattice of ions are Bloch states, which do not cause resistivity. The decrease of $\rho(T)$ upon cooling indicates decreasing incoherent scattering [53], in that the phases of the conduction electrons scattered off the periodic Ce^{3+} sites become coupled. The low-T resistivity behavior is given by $\rho = \rho_0 + AT^2$, where an "electron-electron scattering" derived term, AT^2, adds to the ordinary residual resistivity, ρ_0. Since the coefficient A is proportional to $(m^*)^2$, it exceeds the one in simple metals by several orders of magnitude and, thus, supports the existence of extended (Bloch) states with very large effective mass.

The validity range of the $\rho - \rho_0 = AT^2$ law (below $T_{coh} \lesssim 0.1\, T^*$) is often called the "coherence regime", since incoherent scattering is replaced by (inelastic) scattering processes between coherent quasiparticles. The formation of the heavy Fermi liquid, for $T < T_{coh}$, can be inferred not only from other transport coefficients, e.g. Hall number [54] magneto-resistivity [55] and Seebeck coefficient [56], but also from the specific heat. This was first recognized for $CeAl_3$ [57], for which the Sommerfeld coefficient $\gamma(T)$ was found to assume a broad maximum near $T \simeq T_{coh}$ - in striking contrast to the experimental [48,49] and theoretical [50] results for dilute Kondo impurities which reveal a continuous increase of $\gamma(T)$ upon lowering T. Very similar $\gamma(T)$ peaks have been observed, e.g. for n-state $CeCu_2Si_2$ [30] (see figure 2a), $CeRu_2Si_2$ [40] and n-state URu_2Si_2 [45]. For $CeCu_2Si_2$ it was shown that this peak can be removed upon doping with ~ 10 at% of La or Y, which proves that this distinct feature must derive from the periodicity of the Ce sublattice [58]. It was proposed [58] to reflect interference of the narrow band of coherent 4f states with the ordinary conduction bands of the metal, which may lead to the opening of a hybridization-derived pseudo gap at E on top of the resonant 4f-DOS. Such structure in the DOS has also been obtained theoretically [59-61]. An alternative explanation for the observed maximum in $\gamma(T)$ invokes the vicinity of a spin-density-wave (SDW) transition [62].

The most direct proof for a band structure constituted by strongly renormalized states near E_F is provided by the de-Haas-van-Alphen (dHvA) oscillations observed, e.g. for $CeCu_6$ [63] and UPt_3 [64]. For heavy-fermion compounds, dHvA measurements require very low temperatures, in addition to high magnetic fields and high-quality samples: Firstly, the large effective mass drastically reduces the dHvA amplitude, and secondly, the strong inelastic electron-electron scattering process has to be frozen out in order to achieve sufficiently large mean free paths (> 1000 Å). Effective masses up to $40m_{el}$ and $90m_{el}$ have been identified in the dHvA results for $CeCu_6$ and UPt_3, respectively. For the latter compound, these data can be compared with results of band-structure calculations performed within the local-density approximation [65]. While there is a surprisingly good agreement as far as the topology of the Fermi surface is concerned, one-electron theory fails to predict the observed large, correlation-derived electron masses by a factor of approximately 20 [66].

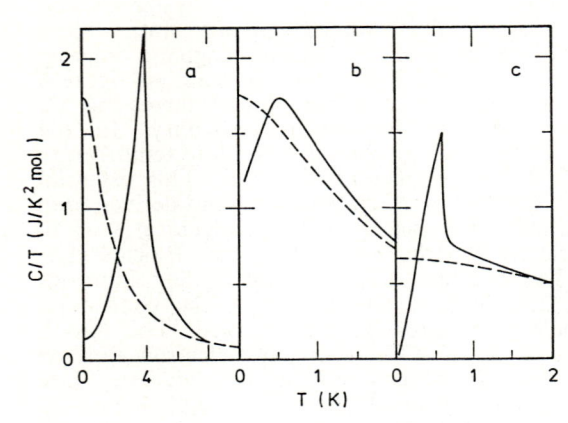

Fig. 7. Specific-heat behavior for exemplary heavy-fermion magnet $CeAl_2$ [68], heavy-Fermi-liquid system $CeAl_3$ [58] and heavy-fermion superconductor $CeCu_2Si_2$ [5]. Dashed curves indicate Bethe-Ansatz results for $S = 1/2$ Kondo impurity with $T_K = T^*$ [50].

We conclude this section by comparing, in figure 7, the three prototypical heavy-fermion systems $CeAl_2$ [67,68], $CeAl_3$ [22,58] and $CeCu_2Si_2$ [5,30]. $CeAl_2$ is representative for the majority of heavy-fermion compounds, i.e. those which exhibit long-range magnetic order between local moments, mediated by the RKKY interaction. We mention a ("Kondo"-)reduced ordered f-derived moment and an enhanced Sommerfeld coefficient γ, typically larger than 100 mJ/K^2mole-(f-ion), as hallmarks of these heavy-fermion local-moment magnets like CeB_6 [69,70], $CePb_3$ [71], $CeCu_2$ [72], $CeAg_2Si_2$ [73], $CeCu_2Ge_2$ [74], U_2Zn_{17} [75], UCd_{11} [76], $NpSn_3$ [77] and $NpBe_{13}$ [78]. $CeAl_3$ is the archetype of a heavy Fermi-liquid system. Apparent deviations from the Kondo-impurity behavior are due to the <u>periodicity</u> (rather than just the high concentration) of the Ce's. Therefore, $CeAl_3$ and its counterparts $CeCu_6$ [23], $CeRu_2Si_2$ [24,79], $CeCu_4Al_8$ [80], $CePtSi$ [81] and $YbCuAl$ [82] may be labeled "Kondo-lattices" in a strict sense. This term is adequate also for the two superconductors $CeCu_2Si_2$ [5] and UBe_{13} [43], although an alternative mechanism, i.e. a "quadrupolar Kondo effect", has been proposed to be responsible for the heavy fermions in the latter [83]. The superconductors UPt_3 [64,65] and URu_2Si_2 [47], on the other hand, are often discussed in terms of models starting from the outset (i.e. at high temperatures) with itinerant 5f electrons. Two issues of major timely interest in heavy-fermion superconductivity, concerning (1) the nature of the pairing interaction and (2) the type of the Cooper-pair state, will be addressed in the following section.

4. Current problems in heavy-fermion superconductivity

4.1 Pairing mechanism

Most physicists in this field believe in a Cooper pairing of heavy fermions via the exchange of some intermediate bosons, although a Bose-Einstein condensation by "heavy bosons" existing already at high temperatures has also been proposed [84]. Since the discovery of HFS, magnetic couplings, namely the exchange of virtual (antiferromagnetic) spin fluctuations, have been considered essential for the Cooper-pair formation. Additional arguments for a magnetic mechanism may be derived from the recent observation, through sensitive microscopic probes like muon-spin rotation (μSR), nuclear magnetic resonance (NMR) and neutron diffraction, that magnetic correlations dominate the low-T "Fermi-liquid" properties of compounds like $CeCu_6$ and $CeRu_2Si_2$ [85], and even lead to a cooperative antiferromagnetic state in other heavy-fermion systems. This new kind of "heavy-fermion band magnetism" [86] seems to be characterized by extremely small magnetic moments of order $10^{-2}\mu_B$/f-ion.

All of the superconducting heavy-fermion systems show magnetic phenomena at sufficiently low temperatures: A SDW transition at T_N = 17 K was found for URu_2Si_2 to be followed by a superconducting one at $T_c \simeq$ 1.5K [45-47]. It is assumed in this case that the SDW is gapping the Fermi surface only partly so that superconductivity can form on the remaining portion. Neutron-diffraction studies [87] suggest a similar scenario for UPt_3, where long-range antiferromagnetism seems to develop below $T_N \simeq$ 5K. Upon cooling the system to below $T_c \simeq$ 0.5 K, the magnetic order parameter levels off [87]. For UBe_{13}, an anomaly in the specific heat detected for external fields $B \gtrsim$ 2 T near T = 0.15K has been attributed to an antiferromagnetic phase transition as well [88]. Even in the case of $CeCu_2Si_2$, μSR data at B = 0T appear to be consistent with the onset of long-range antiferromagnetism at $T_N \simeq$ 0.8 K [89], whereas NMR [90] and magneto-resistivity [91] anomalies have led to the suggestion [92] of a low-T phase diagram with superconductivity below $B_{c2}(0) \simeq$ 2T, antiferromagnetism up to B $\simeq$ 7T and probable coexistence of the two between 1 T and 2 T. Neutron-diffraction experiments are in preparation to check this phase diagram.

Alternatively, an electron-phonon coupling unique for Kondo-lattices had been proposed to explain heavy-fermion superconductivity with $T_c \lesssim$ 1 K in $CeCu_2Si_2$ [93,94]. It derives from the pronounced volume dependence of the characteristic energy of the heavy fermions, $k_B T^*$, as expressed by a corresponding "electronic" Grüneisen parameter, $\Gamma = - d\ln(k_B T^*)/d\ln V \simeq$ 100 [95,96]. Thus, the "breathing modes" associated with substantial changes of the atomic volume will strongly couple to the heavy-fermion degrees of freedom. Experimental evidence for this "Grüneisen-parameter coupling" is found, among others, in striking anomalies of the elastic constants [97], the longitudinal ultrasound attenuation both in the normal [98] and superconducting [99] state and the thermal expansion [100]. Further work has to show how such a phonon-mediated pairing mechanism can encounter the small ratio $v_F^*/v_s \simeq$ 2 [37]. In this connection, it should be remarked that preliminary results obtained on $CeCu_2Si_2$ samples with differing values of v_F^* (assumed to be proportional to T*) highlight the existence of a critical ratio, $v_F^*/v_s \simeq$ 1, for which superconductivity becomes suppressed [30,91].

In summary, we can presently not rule out that superconductivity in heavy-fermion compounds originates in some interplay between spin-fluctuation- and phonon-mediated pairing mechanisms. More insight into this problem requires detailed information about the material properties of the four unique HFS $CeCu_2Si_2$, UBe_{13}, UPt_3 and URu_2Si_2.

4.2 Cooper-pair state

The conclusion that the Cooper pairs are formed by heavy fermions, i.e. only weakly delocalized f electrons, has led to the assumption of a node of the Cooper-pair wave function at the origin of the relative coordinate : here the on-site Coulomb repulsion is assumed to play a similar role to the hard-core repulsion between the two atoms within a Cooper pair in superfluid ^{3}He. A non-s-wave pair state would be a natural consequence. It is important to note, however, that the residual Coulomb repulsion between the heavy quasiparticles is smaller compared to the bare Coulomb energy on the 4f shell by many orders of magnitude (i.e. by $k_B T^*/U_{intra} \simeq 10^{-4}$ [101]). Still, a highly anisotropic Cooper-pair wave function (order parameter) of HFS is commonly presupposed. Experimental support for this assumption derives from the observation of distinctly non-exponential temperature dependences of, e.g., the specific heat, thermal conductivity, ultrasound attenuation and NMR spin-lattice relaxation rate. For a bibliography, see recent review articles [13,17].

Figure 8 proves [30] the existence of a considerable low-T excess specific heat of CeCu$_2$Si$_2$ compared to that of a BCS superconductor (with an isotropic order parameter). Very similar results have been obtained for UBe$_{13}$ [102]. They reflect an excitation spectrum which is strongly modified compared to the BCS spectrum, in showing a finite DOS already at rather low excitation energies. The latter is caused by correspondingly small values of the energy gap, $\Delta(\underline{k})$, in certain $\underline{k}$ directions. Isotropic potential scattering is expected to "smear out" the anisotropies of the gap [103]. If $\Delta(\underline{k})$ does not contain an isotropic contribution, i.e. if it is exhibiting zeros on certain parts of the Fermi surface, potential scattering should be "pair breaking" [8]. Figure 9 displays specific-heat results for CeCu$_2$Si$_2$ samples doped with non-magnetic La or Y substituting Ce [104]. Both T_c and ΔC are found to be depressed by the doping, with Y being more efficient than La (figure 9a) and, in addition, the reduction in ΔC to be more pronounced than that in T_c (figure 9b). These results resemble earlier ones on BCS superconductors doped with magnetic impurities, which will be briefly recalled below.

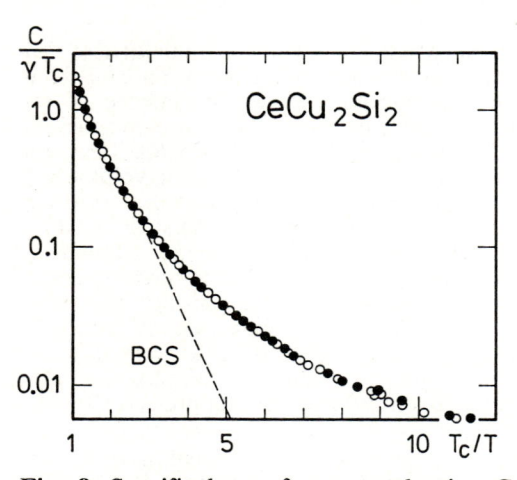

Fig. 8. Specific heat of superconducting CeCu$_2$Si$_2$ as $C/\gamma T_c$ (on a logarithmic scale) vs T_c/T [30]. Different symbols refer to different runs. Below $T \simeq 0.5\ T_c$, CeCu$_2$Si$_2$ exhibits a large excess in specific heat compared with a BCS superconductor.

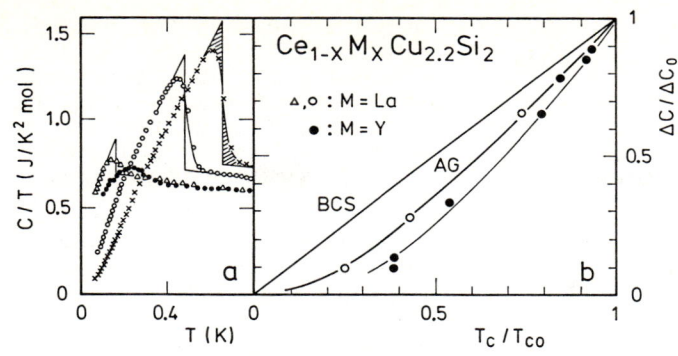

Fig. 9. a: Specific heat as $\Delta C/T$ vs T for $Ce_{1-x}M_xCu_{2.2}Si_2$ with M = 3 at% (○) and 9 at% (△)La as well as 3 at% Y (●). b: $\Delta C/\Delta C_0$ vs T_c/T_{c0} (T_{c0} and ΔC_0 refer to the undoped material) [104]. Some of the broadened transitions as measured are replaced in (a) by idealized ones. Also shown in (b) is the BCS "law of corresponding states" and the universal result of the Abrikosov-Gorkov (AG) theory [106].

The case of $\Delta C/\Delta C_0 = T_c/T_{c0}$ indicating the BCS "law of corresponding states" has been verified for certain dilute alloys like ThU [105] containing "nearly magnetic" impurities, which cause a "pair-weakening" [8]. (T_{c0} and ΔC_0 refer to the undoped system). Magnetic impurities in BCS superconductors, on the other hand, give rise to pair breaking, i.e. they break the time-reversal symmetry of the Cooper pairs. For dopants like Gd^{3+} with stable moments, $\Delta C/\Delta C_0$ vs T_c/T_{c0} follows the universal result of the Abrikosov-Gorkov (AG) theory [106], which can be ascribed to a "collision-time broadening" of the gap, i.e. to the development of excited states at energies below the gap energy. These give rise to an increased specific heat at low T and, because of simple thermodynamic arguments, to a ΔC value reduced more strongly than T_c. A finite DOS even in the entire gap regime ("gapless superconductivity") can be achieved for sufficiently high dopant concentration [106]. If the dopant dissolved in a BCS superconductor is a Kondo ion like Ce^{3+}, in addition to collision-time broadening, the formation of "localized excited states" in the vicinity of the Fermi surface causes "impurity bands" inside the gap already at minor dopant content [107]. Thus for example, "gapless superconductivity" has been inferred from specific-heat measurements on $La_{1-x}Ce_xAl_2$ for Ce concentrations as low as 0.1 at% [108]. Consequently, for this alloy the $\Delta C/\Delta C_0$ vs T_c/T_{c0} results lie well below the universal AG curve [108].

Although neither La nor Y carry a magnetic moment, their effect on the "Kondo-lattice superconductor" $CeCu_2Si_2$ is similar to that of magnetic impurities on a BCS superconductor: While La-doped $CeCu_2Si_2$ shows a $\Delta C/\Delta C_0$ vs T_c/T_{c0} curve surprisingly close to the AG result, Y doping achieves a ΔC depression comparable to the one caused by Kondo impurities in BCS superconductors. For corresponding results on doped UBe_{13}, we refer to [109]. A theoretical treatment of impurities in HFS is found in [110]. The observation of "pair breaking" by non-magnetic dopants ("Kondo holes") in $CeCu_2Si_2$ and UBe_{13} confronts one with a serious problem, hereafter called the "mean-free-path problem": Assuming an isotropic potential scattering, one estimates from the n-state residual resistivities for the two HFS elastic mean free paths, which can be considerably smaller than the Ginzburg-Landau coherence lengths derived from $B_{c2}(T)$ slopes [91]. Actual mean free paths at T_c where, in both materials, the n-state resistivity is <u>dominated</u> by inelastic scattering processes are even much shorter. Furthermore, if one estimates the (isotropic) scattering rates, τ_x^{-1}, caused by x = 1 at% dopant concentration which is quite harmless to T_c, they are found to exceed for both HFS the "gap frequency" Δ_0/h (Δ_0 = energy gap at T = 0; h = Planck's constant) by almost one order of magnitude. It is hard to understand how under these conditions an anisotropic order parameter can exist at all [103].

Based upon our present knowledge about these materials we are left with, at least, two possible solutions of the "mean-free-path problem" : If one wants to retain an anisotropic superconducting order parameter, one has to assume an anisotropic potential scattering, too, inherent to a Kondo-lattice system. Alternatively, "strong-coupling effects" may be able to explain the observations. Strong-coupling effects become important, if $k_B T_c$ is <u>not</u> small compared to the relevant n-state energy [111], no matter whether we use $k_B T^*$ or $k_B T_{coh}$ (cf. Sec. 3) for the latter. Since for both $CeCu_2Si_2$ and UBe_{13}, superconductivity sets in before the Fermi-liquid state has been fully established ($T_c/T_{coh} > 1$), a non-exponential C(T) may reflect anomalies, e.g., in the n-state DOS near E_F. In the same type of reasoning, the results of figure 9b may indicate that, in addition to the breaking of existing Cooper pairs, the <u>formation</u> of pairs is hindered by a disorder-induced destruction of coherence, i.e. a "quasi localization" of heavy fermions.

Interestingly enough, however, all UPt_3 samples found to be superconducting exhibit mean-free paths that exceed considerably the coherence length [91]. In addition, the superconducting state of UPt_3 forms out of the <u>coherent</u> Fermi-liquid phase ($T_c/T_{coh} \simeq 0.2$). Therefore this compound remains the prime candidate for an anisotropic-pairing superconductor.

No agreement can presently be stated concerning the type of anisotropy. Many researchers favor an "unconventional" order parameter, which has a symmetry lower than that of the Fermi surface [14,15,112], while others discuss arguments for a "conventional" order parameter, exhibiting the same symmetry as the Fermi surface [16]. Since the latter is highly anisotropic in the low-T heavy Fermi-liquid state, conventional order parameters in HFS should be highly anisotropic as well [16]. Among unconventional order parameters, two different general classes are distinguished according to group theoretical considerations [113] : The first class involves zeros of the energy gap along lines on the Fermi surface. A special case of such a "polar state" is the (even-parity) d-wave state. It should be manifested, well below T_c, by a T^2-dependence of the specific heat and related properties, such as the thermal conductivity and $(T_1 T)^{-1}$, where T_1^{-1} is the NMR spin-lattice relaxation rate [114]. UPt_3 is often considered a d-wave superconductor, cf. figure 10. The second class of unconventional order parameters is characterized by gap zeros at certain symmetry points on the Fermi surface. This kind of "axial state" is expected for an (odd-parity) p-wave state. Certain observations, concerning, e.g. the T-dependences of the magnetic penetration depth [115] and of the dc-Josephson current [116], were taken as evidence for an axial order parameter in UBe_{13}. These notions seem to be partly supported by simple power-law dependences of the ultrasound attenuation [99,117] in both UPt_3 and UBe_{13}. Several problems concerning so-called simple power laws in HFS should be mentioned, however : (1) Thorough analysis of the experimental data sometimes reveals no power laws at all [118,119]. (2) Whereas power laws should occur in pure HFS only way <u>below</u> T_c, they are often registered rather close to T_c (see figure 10). (3) In the presence of lattice imperfections, more complex T-dependences are predicted [120]. (4) Seemingly simple power-law dependences can sometimes be observed for classical superconductors [121].

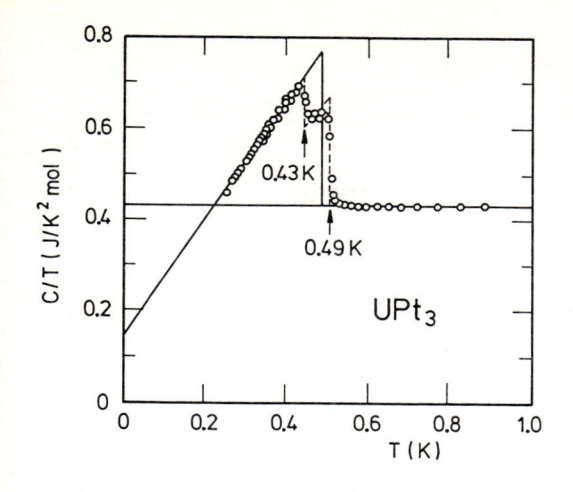

Fig. 10. Temperature dependence of the specific heat as C/T vs T for UPt$_3$ [125]. Dashed lines represent two idealized transitions at T$_a$ and T$_b$. Solid lines indicate one idealized transition, with the same total entropy at T$_c$, and extrapolations of measured n-state and superconducting data to T = 0K.

Further exciting phenomena, which may in principle be explicable in terms of both conventional and unconventional Cooper-pair states, have been discovered in HFS. We should like to mention notably the existence of more than one superconducting phase transition and the occurrence of terms linear in T in the specific heat and related properties. The first example [122] of a material exhibiting two subsequent superconducting [123] transitions was U$_{1-x}$Th$_x$Be$_{13}$ with $0.02 < x < 0.05$. It has become even more interesting by reason of the discovery of an extremely small ordered moment ($< 10^{-2}\mu_B/U$) which forms close to the lower of the two superconducting transition temperatures [124]. More recently two closely spaced T$_c$'s have been found also in the specific-heat of UPt$_3$ [125], as is shown in figure 10. The magnetic field-temperature phase diagram of this material being intensively studied by several groups apparently contains a number of different (superconducting and magnetic) phases [126].

A residual linear term, $\gamma_s T$, of considerable size in the specific heat of superconducting UPt$_3$ is inferred in figure 10 upon extrapolation of the data to T = 0. This has also been found for other (polycrystalline) samples of lower purity [127] as well as for high-purity single crystals [128], although in one case a vanishing $\gamma_s T$ term was reported [129]. For UBe$_{13}$ [13] and particularly CeCu$_2$Si$_2$ (figure 2a), γ_s is substantially smaller than for UPt$_3$. This term indicates the presence of normal carriers which may originate either in pair breaking or in a multi-band structure. Comparison with a corresponding term, $\lambda_s T$, in the thermal conductivity yields information about the effective mass of those normal carriers. They can be heavy as in UPt$_3$ or light as in CeCu$_2$Si$_2$ [130] and conceivably coexist with the Cooper pairs on different portions of the Fermi surface, i.e. in different parts of k-space. This may be compared with the behavior of cuprate high-T$_c$ superconductors, for which recent thermal conductivity results also indicate the presence of normal carriers way below T$_c$ [131]. In view of the intrinsic multi-layer structure of the cuprates it was suggested, however, that here the two kinds of carriers coexist in different parts of real, rather than k-space [131].

5. Perspective

Heavy-fermion compounds provide a rich manifold of phenomena which are unexpected within traditional concepts of metal physics. They allow one to investigate, for example, the development of band magnetism in systems that are characterized at moderate temperatures by a dense array of local moments. Heavy-fermion superconductors are a very suitable testing ground for models concerned with superconductivity in strongly correlated electron systems. It is hoped, in particular, that the role of correlations in high-T_c superconductivity may be better understood, if further progress is made in unraveling the physics of heavy fermions.

Acknowledgements

I am grateful for the continuing collaboration with the low-temperature-physics group at TH Darmstadt and for many stimulating discussions with N. Grewe and P. Fulde. This work was supported by the Deutsche Forschungsgemeinschaft, in part under the auspices of the Sonderforschungsbereich 252 Darmstadt / Frankfurt / Mainz.

References

[1] J. Bardeen, L.N. Cooper, and J.R. Schrieffer, Phys. Rev. 108, 1175 (1957).
[2] Φ. Fischer, this issue; and references cited therein.
[3] D. Jerome, this issue; and references cited therein.
[4] J.G. Bednorz, and K.A. Müller, Z. Phys. B - Condensed Matter 64, 189 (1986).
[5] F. Steglich, J. Aarts, C.D. Bredl, W. Lieke, D. Meschede, W. Franz, and H. Schäfer, Phys. Rev. Lett. 43, 1892 (1979).
[6] F. Steglich, Physica B+C 130, 145 (1985).
[7] N.Y. Rivier, and D.E. MacLaughlin, J. Phys. F : Metal Physics 1, 67 (1971).
[8] P. Fulde, this issue; and references cited therein.
[9] See, e.g.: U. Gottwick, R. Held, G. Sparn, F. Steglich, H. Rietschel, D. Ewert, B. Renker, W. Bauhofer, S. von Molnar, M. Wilhelm, and H.E. Hoenig, Europhys. Lett. 4, 1183 (1987).
[10] W. Hanke, this issue; and references cited therein.
[11] G.R. Stewart, Rev. Mod. Phys. 56, 755 (1984).
[12] F. Steglich, Springer-Series in Solid-State Sciences 62, 23 (1985).
[13] H.R. Ott, Progr. Low Temp. Phys. XI, 217 (1987).
[14] C.M. Varma, Comments on Solid State Phys. 11, 221 (1985).
[15] P.A. Lee, T.M. Rice, J.W. Serene, L.J. Sham, and J.W. Wilkins, Comments on Condensed Matter Phys. 12, 99 (1986).
[16] P. Fulde, J. Keller, and G. Zwicknagl, Solid State Phys. 41, 1 (1988).
[17] N. Grewe, and F. Steglich, in: Handbook on the Physics and Chemistry of Rare Earths, Vol. 14, K.A. Gschneidner, Jr., and L. Eyring (eds.), North Holland, Amsterdam (1990), forthcoming.
[18] For a review on intermediate-valence systems, see, e.g.: J.M. Lawrence, P.S. Riseborough, and R.D. Parks, Rep. Prog. Phys. 44, 1 (1981).
[19] W. Lieke, U. Rauchschwalbe, C.D. Bredl, F. Steglich, J. Aarts, and F.R. de Boer, J. Appl. Phys. 53, 2111 (1982).
[20] S. Horn, E. Holland-Moritz, M. Loewenhaupt, F. Steglich, H. Scheuer, A. Benoit, and J. Flouquet, Phys. Rev. B 23, 3171 (1981).
[21] B.C. Sales, and R. Viswanathan, J. Low Temp.Phys. 23, 449 (1976).
[22] K.Andres, J.E. Graebner, and H.R. Ott, Phys. Rev. Lett. 35, 1779 (1975).
[23] Y. Onuki, Y. Shimizu, and T.Komatsubara, J. Phys. Soc. Japan 53, 1210 (1984).
G.R. Stewart, Z. Fisk, and M.S. Wire, Phys. Rev. B 30, 482 (1984).

[24] L.C. Gupta, D.E. MacLaughlin, Cheng Tien, C. Godart, M.A. Edwards, and R.D. Parks, Phys. Rev. B 28, 3678 (1983).
[25] C.D. Bredl, W. Lieke, R. Schefzyk, M. Lang, U. Rauchschwalbe, F. Steglich, J. Klaasse, J. Aarts, and F.R. de Boer, J. Magn. Magn. Mat. 47 & 48, 30 (1985).
[26] S. Horn, F. Steglich, M. Loewenhaupt, and E. Holland-Moritz, Physica 107B, 381 (1981).
[27] U. Rauchschwalbe, W. Lieke, C.D. Bredl, F. Steglich, J. Aarts, K.M. Martini, and A.C. Mota, Phys. Rev. Lett. 49, 1448 (1982).
[28] B.T. Matthias, in: Proceedings of the Topical Meeting on Unusual Conditions of Superconductivity and Itinerant Magnetism in d-Materials, Bad Honnef, F.R.G. (1979), unpublished.
[29] G.W. Hull, J.H. Wernick, T.H. Geballe, J.V. Waszczak, and J.E. Bernardini, Phys. Rev. B 24, 6715 (1981).
[30] C.D. Bredl, H. Spille, U. Rauchschwalbe, W. Lieke, F. Steglich, G. Cordier, W. Assmus, M. Herrmann, and J. Aarts, J. Magn. Magn. Mat. 31-34, 373 (1983).
[31] H. Spille, U. Rauchschwalbe, and F. Steglich, Helv. Phys. Acta 56, 165 (1983).
[32] M. Ishikawa, H.F. Braun, and J.L. Jorda, Phys. Rev. B27, 3092 (1983).
[33] F.G. Aliev, N.B. Brandt, V.V. Moshchalkov, and S.M. Chudinov, Solid State Commun. 45, 215 (1983).
[34] G.R. Stewart, Z. Fisk, and J.O. Willis, Phys. Rev. B28, 172 (1983).
[35] W. Assmus, M. Herrmann, U. Rauchschwalbe, S. Riegel, W. Lieke, H. Spille, S. Horn, G. Weber, F. Steglich, and G. Cordier, Phys. Rev. Lett. 52, 469 (1984).
[36] G. Riblet, and K. Winzer, Solid State Commun. 9, 1663 (1971).
[37] F. Steglich, J. Aarts, C.D. Bredl, G. Cordier, F.R. de Boer, W. Lieke, and U. Rauchschwalbe, in: Superconductivity in d-Band and f-Band Metals, W. Buckel, and W. Weber (eds.), Kernforschungszentrum, Karlsruhe (1982), p. 145.
[38] See, e.g.: A.J. Leggett, Rev. Mod. Phys. 47, 331 (1975); and references cited therein.
[39] U. Rauchschwalbe, U. Ahlheim, F. Steglich, D. Rainer, and J.J.M. Franse, Z. Phys. B - Condensed Matter 60, 379 (1985).
[40] F. Steglich, U. Rauchschwalbe, U. Gottwick, H.M. Mayer, G. Sparn, U. Poppe, and J.J.M. Franse, J. Appl. Phys. 57, 3054 (1985). U. Poppe, J. Magn. Magn. Mat. 52, 157 (1985).
[41] V. Ambegaokar, and A. Baratoff, Phys. Rev. Lett. 10, 486 (1963); 11, 104 (1963).
[42] J.A. Pals, W. van Haeringen, and M.H. van Maaren, Phys. Rev. B15, 2592 (1977).
[43] H.R. Ott, H. Rudigier, Z. Fisk, and J.L. Smith, Phys. Rev. Lett. 50, 1595 (1983).
[44] G.R. Stewart, Z. Fisk, J.O. Willis, and J.L. Smith, Phys. Rev. Lett. 52, 679 (1984).
[45] W. Schlabitz, J. Baumann, B. Pollit, U. Rauchschwalbe, H.M. Mayer, U. Ahlheim, and C.D. Bredl, Abstracts of 4th Int. Conf. on Valence Fluctuations, Köln (1984), unpublished; Z. Phys. B - Condensed Matter 62, 171 (1986).
[46] T.T.M. Palstra, A.A. Menovsky, J. van den Berg, A.J. Dirkmaat, J.G. Niewenhuys, and J.A. Mydosh, Phys. Rev. Lett. 55, 2727 (1985).
[47] M.B. Maple, J.W. Chen, Y. Dalichaouch, Y. Kohara, C. Rossel, M.S. Torikachivili, H.W. McElfresh, and J.D. Thompson, Phys. Rev. Lett. 56, 185 (1986).
[48] B.B. Triplett, and N.E. Phillips, Phys. Rev. Lett. 27, 1001 (1971).

[49] For a review, see, e.g.: G. Grüner and A. Zawadowski, Rep. Progr. Phys. 37, 1497 (1974).
[50] For a review, see, e.g.: N. Andrei, K. Furuya, and J.H. Loewenstein, Rev. Mod. Phys. 55, 331 (1983).
[51] Y. Onuki and T. Komatsubara, J. Magn. Magn. Mat. 63 & 64, 281 (1987).
[52] M.A. Ruderman, and C. Kittel, Phys. Rev. 96, 99 (1954).
T. Kasuya, Progr. Theor. Phys. 16, 45 (1956).
K.Yosida, Phys. Rev. 106, 895 (1957).
[53] N.F. Mott, Phil. Mag. 30, 403 (1974).
[54] See, e.g.: T. Penney, F.P. Milliken, S. von Molnar, F. Holtzberg, and Z. Fisk, Phys. Rev. B 34, 5959 (1986).
[55] See, e.g.: G. Remeny, A. Briggs, J. Flouquet, O. Laborde, and F. Lapierre, J. Magn. Magn. Mat. 31-34, 407 (1983).
[56] See, e.g.: F. Steglich, C.D. Bredl, W. Lieke, U. Rauchschwalbe, and G. Sparn, Physica, 126B, 82 (1984).
[57] A. Benoit, A. Berton, J. Chaussy, J. Flouquet, J.C. Lasjaunias, J. Odin, J. Palleau, J. Peyrard, and M. Ribault, in: Valence Fluctuations in Solids, L.M. Falicov, W. Hanke, and M.B. Maple (eds.), North-Holland, Amsterdam (1981), p. 283.
[58] C.D. Bredl, S. Horn, F. Steglich, B. Lüthi, and R.M. Martin, Phys. Rev. Lett. 52, 1982 (1984).
C.D. Bredl, PhD Thesis, TH Darmstadt (1985), unpublished.
[59] R.M. Martin, Phys. Rev. Lett. 48, 362 (1982).
[60] N.Grewe, Solid State Commun. 50, 19 (1984).
[61] N. d'Ambrumenil, and P. Fulde, Springer Series in Solid Sciences 62, 195 (1985).
[62] S. Doniach, in: Theoretical and Experimental Aspects of Valence Fluctuations and Heavy Fermions, L.C.Gupta and S.K. Malik (eds.), Plenum, New York (1987), p. 179.
[63] P.H.P. Reinders, M. Springford, P.T. Coleridge, R. Boulet, and D. Ravot, Phys. Rev. Lett. 57, 1631 (1986).
[64] L. Taillefer, R. Newbury, G.G. Lonzarich, Z. Fisk, and J.L. Smith, J. Magn. Magn. Mat. 63 & 64, 372 (1987).
L. Taillefer, and G.G. Lonzarich, Phys. Rev. Lett. 60, 1570 (1988).
[65] T. Oguchi, A.J. Freeman, and G.W. Crabtree, Phys. Lett. 117A, 428 (1986).
[66] See also: G. Zwicknagl, J. Magn. Magn. Mat. 76 & 77, 16 (1988).
[67] B. Barbara, J.X. Boucherle, J.L. Buevoz, M.F. Rossignol, and J. Schweizer, Solid State Commun. 24, 481 (1977).
[68] C.D. Bredl, F. Steglich, and K.D. Schotte, Z. Phys. B - Condensed Matter 29, 327 (1978).
[69] K. Winzer and W. Felsch, J. Physique 39, C6-832 (1978).
[70] M. Kawakami, S. Kunii, T. Komatsubara, and T.Kasuya, Solid State Commun. 36, 435 (1980).
[71] C.L. Lin, J. Teter, J.E. Crow, T. Mihalisin, J. Brooks, A.I. Abou-Aly, and G.R. Stewart, Phys. Rev. Lett. 54, 2541 (1985).
[72] E. Gratz, E. Bauer, B. Barbara, S. Zemirli, F. Steglich, C.D. Bredl, and W. Lieke, J. Phys. F: Met. Phys. 15, 1975 (1985).
[73] B.H. Grier, J.M. Lawrence, V. Murgai, and R.D. Parks, Phys. Rev. B29, 2664 (1984).
[74] F.R. de Boer, J.C.P. Klaasse, P.A. Verhuizen, A. Böhm, C.D. Bredl, U. Gottwick, H.M. Mayer, L. Pawlak, U. Rauchschwalbe, H. Spille, and F. Steglich, J. Magn. Magn. Mat. 63 & 64, 91 (1987).
[75] H.R. Ott, H. Rudigier, P. Delsing, and Z. Fisk, Phys. Rev. Lett. 52, 1551 (1984).

[76] Z: Fisk, G.R. Stewart, J.O. Willis, H.R. Ott, and F. Hulliger, Phys. Rev. B30, 6360 (1984).
[77] R.J. Trainor, M.B. Brodsky, B.D. Dunlap, and G.K. Shenoy, Phys. Rev. Lett. 37, 1511 (1976).
[78] G.R. Stewart, Z. Fisk, J.L. Smith, J. O. Willis, and M.S. Wire, Phys. Rev. B30, 1249 (1984).
[79] J.-M. Mignot, J. Flouquet, P. Haen, F. Lapierre, L. Puech, and J. Voiron, J. Magn. Magn. Mat. 76 & 77, 97 (1988).
[80] U.Rauchschwalbe, U. Gottwick, U. Ahlheim, H.M. Mayer, and F. Steglich, J. Less Common Metals 111, 265 (1985).
[81] L. Rebelsky, K. Reilly, S. Horn, H. Borges, J.D. Thompson, J.O. Willis, R. Aikin, R. Caspary, and C.D. Bredl, J. Appl. Phys. 63, 3405 (1988).
[82] W.C.M. Mattens, P.F. de Ch tel, A.C. Moleman, and F.R. de Boer, Physica $69B$, 138 (1979).
[83] D.L. Cox, Phys. Rev. Lett. 59, 1240 (1987).
[84] J. Ranninger, S. Robaszkiewicz, A. Sulpice, and R. Tournier, Europhys. Lett. 3, 347 (1987).
[85] J. Rossat-Mignod, L.P. Regnault, J.L. Jacoud, C. Vettier, P. Lejay, J. Flouquet, E. Walker, D. Jaccard, and A. Amato, J. Magn. Magn. Mat. 76 & 77, 376 (1988).
[86] N.Grewe, and B. Welslau, Solid State Commun. 65, 437 (1988).
[87] G. Aeppli, E. Bucher, A.I. Goldman, G. Shirane, C. Broholm, and J.K. Kjems, J. Magn. Magn. Mat. 76 & 77, 385 (1988).
[88] J.P. Brison, A. Ravex, J. Flouquet, Z. Fisk, and J.L. Smith, J. Magn. Magn. Mat. 76 & 77, 525 (1988).
[89] Y.J. Uemura, W.J. Kossler, X.H. Yu, H.E. Schone, J.R. Kempton, C.E. Stronach, S. Barth, F.N. Gygax, B. Hitti, A. Schenck, C. Baines, W.F. Lankford, Y. Onuki, and T. Komatsubara, Phys. Rev. B39, 4726 (1989).
[90] H. Nakamura, Y. Kitaoka, H. Yamada, and K. Asayama, J. Magn. Magn. Mat. 76 & 77, 517 (1988).
[91] U. Rauchschwalbe, Physica $147B$, 1 (1987).
[92] F. Steglich, J. Phys. Chem. Solids 50, 225 (1989).
[93] H. Razafimandimby, P. Fulde, and J. Keller, Z. Phys. B - Condensed Matter 54, 111 (1984).
[94] N. Grewe, Z. Phys. B - Condensed Matter 56, 111 (1984).
[95] J. Flouquet, J.L. Lasjaunias, J. Peyrard, and M. Ribault, J. Appl. Phys. 53, 2127 (1982).
[96] R. Takke, M. Niksch, W. Assmus, B. Lüthi, R. Pott, R. Schefzyk, and D.K. Wohlleben, Z. Phys. B - Condensed Matter 44, 33 (1981).
[97] See, e.g.: B. Lüthi, and M. Yoshizawa, J. Magn. Magn. Mat. 63 & 64, 274 (1987).
[98] V. Müller, D. Maurer, K. de Groot, E. Bucher, and H.E. Bömmel, Phys. Rev. Lett. 56, 248 (1986).
[99] For UBe$_{13}$, see : B. Golding, D.J. Bishop, B. Batlogg, W.H. Haemmerle, Z. Fisk, J.L. Smith, and H.R. Ott, Phys. Rev. Lett. 55, 2479 (1985).
For UPt$_3$, see : V. Müller, D. Maurer, E.W. Scheidt, C. Roth, K. Lüders, E. Bucher, and H.E. Bömmel, Solid State Commun. 57, 319 (1986).
[100] M. Ribault, A. Benoit, J. Flouquet, and J. Palleau, J. Physique 40, L-413 (1979).
M. Lang, R. Schefzyk, F. Steglich, and N. Grewe, J. Magn. Magn. Mat. 63 & 64, 79 (1987).
[101] M. Tachiki, and S. Maekawa, Phys. Rev.B 29, 2497 (1984).
[102] H.R. Ott, H. Rudigier, T.M. Rice, K. Ueda, Z. Fisk, and J.L. Smith, Phys. Rev. Lett. 52, 1915 (1984).
[103] D. Markowitz, and L.D. Kadanoff, Phys. Rev. 131, 563 (1963).

[104]   F. Steglich, U. Ahlheim, U. Rauchschwalbe, and H. Spille, Physica 148B, 6 (1987).
          U. Ahlheim,M. Winkelmann, C. Schank, C. Geibel, F. Steglich,and A.L. Giorgi, Physica B, in press.
[105]   For a review, see, e.g.: M.B. Maple, Appl. Phys. 9, 179 (1976).
[106]   S. Skalski, O. Betbeder-Matibet, and P.R. Weiss, Phys. Rev. 136A, 1500 (1964).
          A.A. Abrikosov, and L.P. Gorkov, Sov. Phys. JETP 12, 1243 (1961).
[107]   J. Zittartz, A. Bringer, and E. Müller-Hartmann, Solid State Commun. 10, 513 (1972).
          H. Shiba, Progr. Theor. Phys. 50, 50 (1973).
[108]   F. Steglich, Z. Phys. B - Condensed Matter 23, 331 (1976).
[109]   U. Ahlheim, M. Winkelmann, P. van Aken, C.D. Bredl, F. Steglich, and G.R. Stewart, J. Magn. Magn. Mat. 76 & 77, 520 (1988).
[110]   K. Maki, and X. Huang, J. Magn. Magn. Mat. 76 & 77, 499 (1988).
[111]   D. Rainer, Physica Scripta T23, 106 (1988).
[112]   P. Wölfle, J. Magn. Magn. Mat. 76 & 77, 492 (1988).
[113]   P.W. Anderson, Phys. Rev. B30, 4000 (1984).
          G.E. Volovik, and L.P. Gorkov, Sov. Phys. JETP 61, 843 (1985).
          K. Ueda, and T.M. Rice, Phys. Rev. B31, 7114 (1985).
          E.I. Blount, Phys. Rev. B32, 2935 (1985).
[114]   D.E. MacLaughlin, C. Tien, W.G. Clark, M.D. Lan, Z. Fisk, J.L. Smith, and H.R. Ott, Phys. Rev. Lett. 53, 1833 (1984).
          K. Asayama, Y. Kitaoka, and Y. Kohori, J. Magn. Magn. Mat. 76 & 77, 449 (1988).
[115]   D. Einzel, P.J. Hirschfeld, F. Gross, B.S. Chandrasekhar, K. Andres, H.R. Ott, J. Beuers, Z. Fisk, and J.L. Smith, Phys. Rev. Lett. 56, 2513 (1986).
[116]   S. Han, K.W. Ng, E.L. Wolf, A. Millis, J.L. Smith, and Z. Fisk, Phys. Rev. Lett. 57, 238 (1986).
[117]   B.S. Shivaram, Y.H. Jeong, T.F. Rosenbaum, and D.J. Hinks, Phys. Rev. Lett. 56, 1078 (1986).
[118]   U. Rauchschwalbe, U. Ahlheim, H.M. Mayer, C.D. Bredl, and F. Steglich, J. Magn. Magn. Mat. 63 & 64, 447 (1987).
[119]   A. Ravex, J. Flouquet, J.L. Tholence, D. Jaccard, and A. Meyer, J. Magn. Magn. Mat. 63 & 64, 400 (1987).
[120]   K. Scharnberg, D. Walker, H. Monien, L. Tewordt, and R.A. Klemm, Solid State Commun. 60, 535 (1986).
[121]   B. Lüthi, M. Herrmann, W. Assmus, H. Schmidt, H. Rietschel, H. Wühl, U. Gottwick, G. Sparn, and F. Steglich, Z. Phys. B - Condensed Matter 60, 387 (1985).
[122]   H.R. Ott, H. Rudigier, Z. Fisk, and J.L. Smith, Phys. Rev. B31, 1651 (1985).
[123]   U. Rauchschwalbe, F. Steglich, G.R. Stewart, A.L. Giorgi, P. Fulde, and K. Maki, Europhys. Lett. 3, 751 (1987).
[124]   R.H. Heffner, D.W. Cooke, and D.E. MacLaughlin, in: Theoretical and Experimental Aspects of Valence Fluctuations and Heavy Fermions, L.C. Gupta, and S.K. Malik (eds.), Plenum, New York (1987), p. 319.
[125]   R.A. Fisher, S. Kim, B.F. Woodfield, N.E. Phillips, L. Taillefer, K. Hasselbach, J. Flouquet, A.L. Giorgi, and J.L. Smith, Phys. Rev. Lett. 62, 1411 (1989).
[126]   V. Müller, C. Roth, D. Maurer, E.W. Scheidt, K. Lüders, E. Bucher, and H.E. Bömmel, Phys. Rev. Lett. 58, 1224 (1987).
          A. Schenstrom, M.F. Xu, Y. Hong, D. Bein, M. Levy, B.K. Sarma, S. Adenwalla, Z. Zhao, T. Tokusayu, D.W. Hess, J.B. Ketterson, J.A.

Sauls, and D.G. Hinks, Phys. Rev. Lett. 62, 323 (1989).
K. Hasselbach, L. Taillefer, and J. Flouquet, preprint (1989).

[127] J.J.M. Franse, A. Menovsky, A. de Visser, C.D. Bredl, U. Gottwick, W. Lieke, H.M. Mayer, U. Rauchschwalbe, G. Sparn, and F. Steglich, Z. Phys. B - Condensed Matter 59, 15 (1985).
A. Sulpice, P. Gandit, J. Chaussy, J. Flouquet, D. Jaccard, P. Lejay, and J.L. Tholence, J. Low Temp. Phys. 62, 39 (1986).

[128] L. Taillefer, private communication (1989).

[129] H.R. Ott, E. Felder, A. Bernasconi, Z. Fisk, J.L. Smith, L. Taillefer, and G.G. Lonzarich, Jpn. J. Appl. Phys. 26, Suppl. 26-3, 1217 (1987).

[130] F. Steglich, U. Ahlheim, J.J.M. Franse, N. Grewe, D. Rainer, and U. Rauchschwalbe, J. Magn. Magn. Mat. 52, 54 (1985).

[131] G. Sparn, M. Baenitz, S. Horn, F. Steglich, W. Assmus, T. Wolf, A. Kapitulnik, and Z.X. Zhao, Physica C, in press.

Pair Breaking in Superconductors

P. Fulde and G. Zwicknagl

Max-Planck-Institut für Festkörperforschung,
D-7000 Stuttgart 80, Fed. Rep. of Germany

Abstract. The theory of pair breaking in superconductors is reviewed. After a discussion of the standard theory, extensions are discussed which are required when impurities in superconductors have an internal frequency structure. Special attention is paid to Zeeman splitting in superconductors and spin polarized tunneling as an application of it. In addition the case of strongly anisotropic pairing is considered and the effect of impurities on it is discussed. Whenever it is of relevance, implications for the new high-T_c superconductors are pointed out.

1 Introduction

An important concept in the theory of superconductivity is that of pair breaking. Its significance was first realized, when Reif and Woolf [1] discovered by tunneling experiments on superconductors containing paramagnetic impurities that there is a regime of impurity concentrations within which the gap vanishes although the material remains superconducting. These experiments demonstrated clearly that the existence of a gap is not a prerequisite for superconductivity. This is by now common knowledge, but was a great surprise when the tunneling experiments were done. As is sometimes the case, it was realized soon after, that there existed already a theory by Abrikosov and Gorkov [2], which predicted precisely this type of behavior for superconductors with paramagnetic scattering centers. Since at that time not many people were familiar with the mathematical tools used by those authors, this theoretical work caught the attention which it deserved only after the experimental discovery of Reif and Woolf.

Since then the theory of pair breaking in superconductors has been worked out in great detail, and many experiments were done which gave an excellent overall agreement with theory. The latter was summarized by Maki [3] in 1969 in a rather complete review. Since then, a number of interesting supplements to the theory have been made, but the essential features of it have remained unchanged.

The aim of this article is to summarize some of the most important results of pair breaking theory and to point out possible relevances for the new high-T_c superconductors. We want to provide an overview for newcomers to the field of superconductivity. Specialists might find some aspects of sections 5 and 6 of interest.

2 Origin of pair breaking

In order to construct a *BCS* type of wave function for the superconducting ground state, one must combine pairwise electronic one-particle states of the same single-particle energy. In the original *BCS* paper, pairs were formed from states $(\underline{k}, \sigma)$ and $(-\underline{k}, -\sigma)$, where $\underline{k}$ and σ are the momentum and the spin of the one-electron wave function. The well-known *BCS* ground state wave function is of the form

$$|\phi_{BCS}\rangle = \prod_{\underline{k}}(u_{\underline{k}} + v_{\underline{k}}c^+_{\underline{k}\uparrow}c^+_{-\underline{k}\downarrow})|0\rangle. \tag{1}$$

The question arises how this wave function changes when $\underline{k}$ and σ are no longer good quantum numbers, like in the presence of nonmagnetic scattering centers. In many cases the electronic system still has time reversal symmetry. In that case the time-dependent Schrödinger equation for a single electron

$$i\frac{\partial \psi}{\partial t} = H\psi \tag{2}$$

remains unchanged when $t \to -t$, provided the complex conjugate of the wave function is taken and i is replaced by $-i$. The time reversal operator T_R is therefore defined as

$$T_R = C \tag{3}$$

or $T_R = -i\sigma_y C$ when the spins are taken into account, where C changes a function into its complex conjugate. In the presence of time reversal symmetry one has for each one-particle eigenstate ψ_n another one $\psi_{\bar{n}}$ with the same energy ϵ_n, i.e.

$$\begin{aligned} H\psi_n &= \epsilon_n \psi_n \\ H\psi_{\bar{n}} &= \epsilon_n \psi_{\bar{n}} \quad ; \quad \psi_{\bar{n}} = T_R \psi_n. \end{aligned} \tag{4}$$

This enables one to generalize the BCS wave function (1) to the form

$$|\phi_{SC}\rangle = \prod_n (u_n + v_n c_n^+ c_{\bar{n}}^+)|0\rangle \tag{5}$$

(Anderson's theorem). From the above it is clear that this concept breaks down when there are interactions present which break the time-reversal symmetry of the electronic system. Examples are the interaction Hamiltonians

$$H_{int} = \frac{e}{2m}\sum_i (\underline{p}_i \underline{A} + \underline{A}\underline{p}_i) - \mu_B \sum_i \underline{\sigma}_i \underline{H} \tag{6}$$

where $\underline{p}_i$ and $\underline{\sigma}_i$ are the momenta and spins of the electrons and $\underline{A}$ is the vector potential of an external field $\underline{H}$, and

$$H_{int} = -J_{ex}\sum_i \underline{\sigma}_i \underline{S}\delta(\underline{R} - \underline{r}_i) \tag{7}$$

which describes the exchange interaction of conduction electrons with a magnetic ion at site $\underline{R}$ and spin S. In both cases, changing $\underline{p}_i \to -\underline{p}_i$ and $\underline{\sigma}_i \to -\underline{\sigma}_i$ does not leave H_{int} invariant. One important question is how T_c changes when interactions are present which break time-reversal symmetry. De Gennes [4] has given a derivation of the equation for T_c which relates it nicely to properties of the time reversal operator T_R. We present directly the result and refer for its derivation to Ref. [4]. It is found that in the weak coupling limit the following relation holds:

$$1 = N(0)V \int d\epsilon d\epsilon' \frac{1 - f(\epsilon) - f(\epsilon')}{\epsilon + \epsilon'} g(\epsilon' - \epsilon). \tag{8}$$

Here V denotes the attractive interaction of the electrons which leads to Cooper pair formation and $f(\epsilon)$ is the Fermi function. The function $g(\epsilon)$ contains a correlation function of the time reversal operator and is of the form

$$g(\epsilon) = \int \frac{dt}{2\pi} \langle T_R^+(0) T_R(t) \rangle e^{-i\epsilon t}. \tag{9}$$

The time evolution of the time reversal operator T_R is according to the equation

$$\frac{dT_R}{dt} = i[H, T_R]. \tag{10}$$

When $[H, T_R] = 0$, i.e. time reversal symmetry is preserved, one obtains $g(\epsilon-\epsilon') = \delta(\epsilon-\epsilon')$ and Eq.(8) reduces to the BCS equation

$$1 = N(0)V \int_{-\omega_D}^{+\omega_D} d\epsilon \frac{1 - 2f(\epsilon)}{2\epsilon}. \qquad (11)$$

Here a cut off at the Debye energy ω_D has been introduced.

When $[H, T_R] \neq 0$, then two distinct cases arise with respect to the long time behavior of the correlation function $< T_R^+(0)T_R(t) >$. They are

$$(a) \lim_{t \to \infty} < T_R^+(0)T_R(t) > = \eta; \quad 1 > \eta > 0 \qquad (12)$$
$$(b) \lim_{t \to \infty} < T_R^+(0)T_R(t) > = e^{-t/\tau_T}.$$

In case (a) the system is said to be nonergodic while in case (b) one is dealing with an ergodic or Markovian system.

When the system is <u>nonergodic</u> the transition temperature T_c is decreased to

$$k_B T_c = 1.14 \omega_D \, e^{-\frac{1}{\eta N(0)V}}. \qquad (13)$$

The effect is due to a decrease in the electron attraction V by a factor η, and therefore one speaks of "pair weakening" instead of pair breaking. A characteristic feature of pair weakening is that the superconductor still exhibits the universal BCS behavior which follows from the law of corresponding states (strong coupling effects are neglected here). The perturbation leads to a modified critical temperature $T_{co} \to T_c$ which is the new scaling temperature. In particular, the tunneling density of states is given by the BCS expression and there is no gapless regime as found in the experiment of Reif and Woolf.

An example for pair weakening is a thin superconducting film of thickness d and with rough boundaries but no scattering centers inside the film, when it is positioned in a parallel magnetic field [4]. In that case

$$\frac{dT_R}{dt} = i[H_{int}, T_R] \qquad (14)$$
$$= -\frac{ie}{m}(pA + Ap)T_R$$
$$= -i\frac{d\phi}{dt}T_R$$

Here ϕ is the phase of the propagating electron. When an electron moves (without scattering) from one film surface to the other, its phase does not change in the constant parallel field. Therefore the only contributions to dT_R/dt result before the particle hits the boundary for the first time and after it has hit the boundary for the last time. Therefore the correlation function $< T_R(0)T_R(t) >$ remains finite even for large t. The finite limiting value, however, decreases with increasing thickness d.

A second example is an antiferromagnetic superconductor. It also serves as an example that magnetic scattering need not necessarily result in pair breaking. In an antiferromagnet the magnetic ions form a periodic lattice. Although $[H, T_R] \neq 0$, the Hamiltonian has the property that it commutes with the time-reversal operator when it is followed by a lattice translation, i.e.

$$[H, Y]_- = 0 \; with \; Y = RT_R. \qquad (15)$$

The operator R translates the system by a vector, which connects the two sublattices. Thus, when $\psi_{k\sigma}(\underline{r})$ is an eigenfunction of the mean field Hamiltonian with eigenvalue $\epsilon_\sigma(\underline{k})$, so is $Y\psi_{k\sigma}(\underline{r}) = \psi_{-\underline{k},-\sigma}$. Therefore those states can be used to form pairs and with them a BCS like ground state. Another interesting example of pair weakening is encountered in section 4 when we reconsider magnetic impurities in a metallic host.

Next ergodic systems are considered. In that case one obtains from Eqs. (9,12 b).

$$g(z) = \frac{1}{2\pi} \frac{\tau_T}{1 + z^2 \tau_T^2/4} \quad (16)$$

and Eq. (8) becomes

$$1 = N(0)V \sum_{n \geq 0} \frac{1}{n + 1/2 + 1/(2\pi T \tau_T)} \quad (17)$$

after the integrations over $d\epsilon, d\epsilon'$ have been performed. The sum in the last equation is divergent, because the required cut off at ω_D has not yet been introduced. After this is done the sum in Eq. (17) can be expressed in terms of the digamma function $\psi(z)$, i.e.

$$\ell n \frac{T_c}{T_{c0}} + \psi(\frac{1}{2} + \frac{1}{2\pi T_c \tau_T}) - \psi(\frac{1}{2}) = 0. \quad (18)$$

Hereby T_{c0} denotes the transition temperature in the absence of those terms in the Hamiltonian H which break time reversal symmetry. One finds that T_c drops continuously with increasing τ_T^{-1} until it vanishes at a critical value of

$$(\frac{1}{\tau_T})_{crit} = \frac{\pi T_{c0}}{2\gamma}; \quad \gamma = 1.78. \quad (19)$$

Eq. (18) is one of the most important results of pair breaking theory. A plot of it is contained in Fig. 6. It demonstrates that pair breaking is always connected with a reduction of the superconducting transition temperature. However, the inverse is not true, i.e. a decrease of T_c in the presence of a perturbation does not automatically imply pair breaking. As Eq. (13) shows, a reduction of the critical temperature can also reflect a weakened mutual attraction of the quasiparticles. Examples of ergodic pair breaking processes are discussed in the following Section.

3 Standard theory of pair breaking

In the following a number of different time-reversal symmetry breaking mechanisms are discussed, which are all ergodic. Furthermore, superconductivity is described within the weak coupling limit and it is assumed that impurities are elastic scattering centers, i.e. that they do not have an internal frequency structure. This results in the standard theory of pair breaking. Extensions of this theory are discussed in the next section, where impurities are considered which do have an internal frequency structure.

The mechanisms discussed here can be divided into two groups. To group 1 belong all those systems for which one can develop a theory of the breaking of Cooper pairs for all temperatures. Group 2 comprises the cases which can be treated within (standard) pair breaking theory only in the regime where the order parameter $\Delta(\underline{r})$ is small. In that case one can start from the Ginzburg-Landau equation which is written in the form

$$[\psi(\frac{1}{2} + \rho) - \psi(\frac{1}{2}) - \ell n \frac{T_{c0}}{T}] <|\Delta(\underline{r})|^2> + \frac{1}{2(2\pi T)^2} f_1(\rho) <|\Delta(\underline{r})|^4> = 0 \quad (20)$$

with $f_1(\rho)$ to be determined. The quantity $\rho = (2\pi T \tau_T)^{-1}$ characterizes the pair breaking strength. The coefficient in front of the term $<|\Delta(\underline{r})|^2>$ is given by Eq. (18) and therefore is universal for all ergodic systems. This does not hold true as regards the function $f_1(\rho)$. The latter may differ for different systems as discussed later. At lower temperatures the spatial variation of the order parameter requires a more refined treatment. We begin by discussing mechanisms belonging to group 1.

(a) <u>Paramagnetic impurities</u>:
The interaction Hamiltonian of the randomly distributed scattering centers is of the form of Eq. (7). The scattering of conduction electrons is included to second order only (Born approximation). One finds that

$$\frac{1}{\tau_T} = 2\pi n_I N(0) S(S+1) J_{ex} \qquad (21)$$

where n_I is the magnetic impurity concentration and $N(0)$ is the conduction electron density of states (per spin). The dependence of T_c on n_I follows from Eq. (18). In Eq. (21) the spin of the impurity is treated classically. An extension which leads to the Kondo effect is discussed in section 4.

(b) <u>Thin film in a parallel magnetic field</u> (local electrodynamics):
The interaction is of the form of Eq. (6) and the effect of the field on the spins is neglected. The film thickness d is assumed to fulfil the inequality $\lambda, (\ell \xi_0)^{\frac{1}{2}} \gg d \gg \ell$, where χ and ξ_0 are the London penetration depth and the coherence length and ℓ is the mean-free path. Furthermore $(\ell/\xi_0) \ll 1$ (dirty limit). In that case

$$\frac{1}{\tau_T} = \frac{\tau_{tr}}{18}(v_F e d H)^2 \qquad (22)$$

where τ_{tr} is the transport mean-free time and v_F is the Fermi velocity.

(c) <u>Thin film in a parallel magnetic field</u> (nonlocal electrodynamics):
In distinction to case (b) d can be of order ℓ or less, but otherwise the situation is the same as before. Then

$$\frac{1}{\tau_T} = \frac{\tau_{tr}}{18}(v_F e d H)^2 \, g\!\left(\frac{\pi \ell}{d}\right) \qquad (23)$$

with $g(x) = (3/2x^3)\{(1+x^2) arc\, tan x - x\}$. This expression for the pair breaking parameter is a generalization of (22) to which it reduces in the local limit $\ell/d \ll 1$.

(d) <u>Uniform current:</u>
When a uniform current is flowing through a superconductor with mean free-path $\ell \ll \xi_0$ we have

$$\frac{1}{\tau_T} = \frac{2}{3}\tau_{tr} v_F^2 q^2, \qquad (24)$$

where $2q$ is the momentum of the Cooper pairs. In the center of mass system, time reversal symmetry is broken due to the scattering centers.

(e) <u>Pauli paramagnetism</u>
When the electron mean free path is extremely small, the effect of the magnetic field on the electron spins may dominate the one on the electron orbits. In that case $H_{int} = -\mu_B \sum_i \underline{\sigma_i} \underline{H}$. When in addition the spin-orbital mean-free path ℓ_{so} of the electrons fulfills the condition $\ell_{so} \ll \xi_0$ (dirty limit) one finds

$$\frac{1}{\tau_T} = \frac{\ell_{so}}{2v_F}(\mu_B H)^2. \qquad (25)$$

The case $\ell_{so} \gg \xi_0$ is treated separately in Section 5.

As pointed out before, for systems of the form (a-e) a theory has been worked out in closed form not only for T_c but for all temperatures. For example, it is found that also the one-particle Green's function and therefore the quasiparticle excitation energy spectrum is of a universal form, depending on τ_T only. In distinction to the *BCS* case, in which the density of states per spin is given by

$$N_S(E) = N(0) Im \frac{\omega/\Delta}{\sqrt{1-(\omega/\Delta)^2}} \qquad (26)$$

one obtains in the presence of pair breaking from the poles of the Green's function an expression of the form

$$N_S(E) = N(0) Im \frac{u}{\sqrt{1-u^2}}. \qquad (27)$$

The function $u(E)$ is the solution of the equation

$$\frac{\omega}{\Delta} = u\left(1 - \frac{1}{\tau_T \Delta} \frac{1}{\sqrt{1-u^2}}\right). \qquad (28)$$

This equation has to be solved numerically. When this is done and the solutions are set into Eq. (27) one finds that for parameter values

$$1 > (\tau_T)_{crit}/\tau_T > 0.912 \qquad (29)$$

(or $\tau_T \Delta > 1$) the system has no energy gap (gapless regime) in the excitation spectrum. This property can be again related to the behavior of the one-particle states $|m>$ under time reversal symmetry [4]. For small order parameter Δ it turns out that the excitation energy E_n can be expanded in the form

$$E_n = |\epsilon_n| + <|\Delta|^2> \sum_m P \frac{|<n|T_R|m>|^2}{\epsilon_n + \epsilon_m} \qquad (30)$$

where P implies taking the principal value. When $[H, T_R] = 0$ the energies of time-reversed states $|m>$ and $|n> = T_R|m>$ are equal, $\epsilon_n = \epsilon_m$ (Kramers degeneracy) and the expression diverges for ϵ_n near the Fermi energy, i.e. $\epsilon_n \simeq 0$. This implies that E_n cannot be expanded in terms of Δ^2, because there is an energy gap in the spectrum. However, when $[H, T_R] \neq 0$, then $<n|T_R|m> \neq 0$ in an energy range $|\epsilon_m - \epsilon_n| \simeq \tau_T^{-1}$. Then one may write in analogy to Eq. (8)

$$E_p = |\epsilon_p| + <|\Delta|^2> P \int \frac{d\epsilon'}{\epsilon_p + \epsilon'} g(\epsilon' - \epsilon_p) \qquad (31)$$
$$= |\epsilon_p| + \frac{2<|\Delta|^2>|\epsilon_p|}{(2\epsilon_p)^2 + (\tau_T)^{-2}}.$$

An expansion in terms of $<|\Delta|^2>$ is therefore possible and E_p does not show a gap. From the last equation one obtains the density of states

$$N_S(E) = N(0)\left[1 + \frac{\Delta^2}{2} \frac{E^2 - (1/\tau_T)^2}{(E^2 + (1\tau_T)^2)^2}\right] \qquad (32)$$

for small values of Δ. A plot of $N_S(E)$ as obtained from Eq. (27) is shown in Fig. 1 for various values of τ_T.

As regards the various transport coefficients, one can show [6] that they again depend only on τ_T, except when the corresponding correlation function contains an s wave spin triplet vertex with a momentum transfer $q \ll (\ell \xi_0)^{-\frac{1}{2}}$. This implies that the electrodynamics, thermal conduction, ultrasonic attenuation and nuclear spin relaxation are of the same form in all cases (a-e), but that the spin susceptibility is not universal. It differs e.g. in the case of magnetic impurities from that of a thin film in a parallel magnetic field. Furthermore, it can be shown that in the presence of several pair breaking mechanisms, the corresponding parameters τ_T^{-1} are additive.

Next, various cases of pair breaking are considered which belong to group 2, as defined above. They are:

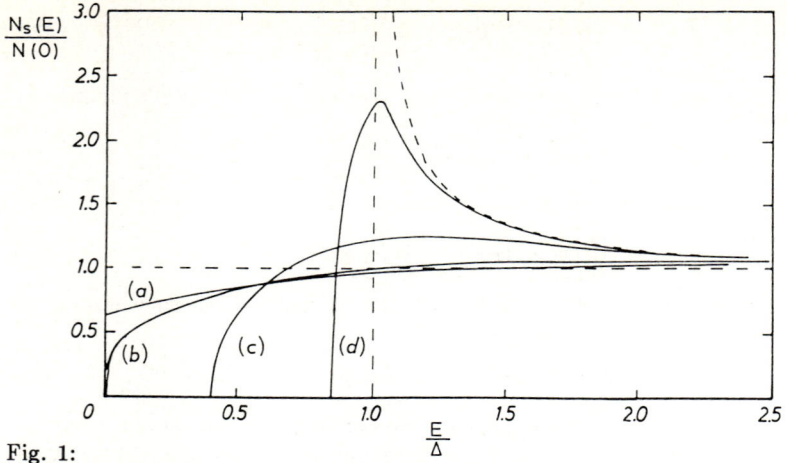

Fig. 1:
Tunneling density of states (per spin) as a function of energy E/Δ for various values of τ_T^{-1}. Curves (a) - (d) correspond to $(\tau_T\Delta)^{-1} = 1.31; 1; 0.30$ and 0.032, respectively. From Ref. [5].

(a) <u>Type II superconductors</u>:
By taking into account the effect of the magnetic field on the electron orbits (but not spins) one finds that in the dirty limit $(\ell/\xi_0 \ll 1)$

$$\frac{1}{\tau_T} = \frac{\tau_{tr} v_F^2 eH}{3} \tag{33}$$

(b) <u>Surface sheath</u>:
When a type II superconductor is in an applied magnetic field, nucleation starts at the surface. The corresponding τ_T^{-1} is given by

$$\frac{1}{\tau_T} = \frac{0.59\tau_{tr}}{3} v_F^2 eH \tag{34}$$

when the dirty limit condition is fulfilled. The pair breaking effect of the magnetic field is reduced close to the surface.

(c) <u>Contact between a superconducting and a paramagnetic film</u>:
When a superconducting film is in metallic contact with a paramagnetic one, the proximity effect requires, via a boundary condition, that the superconducting order parameter vanishes at the interface. Close to T_c the order parameter varies across the film of thickness d_s like $\Delta(x) \sim \cos\frac{\pi x}{2d_s}$, with $0 < x < d_s$. Then τ_T^{-1} is given by

$$\frac{1}{\tau_T} = \frac{\tau_{tr} v_F^2}{6}\left(\frac{\pi}{2d_s}\right)^2, \tag{35}$$

provided one is in the dirty limit, and moreover $d_s > (\ell\xi_0)^{1/2}$.

For the systems contained in group 2 the coefficient in front of the term $<|\Delta(\underline{r})|^2>$ in the Ginzburg-Landau equation (see Eq. (20)) is universal. However, the function $f_1(\rho)$ is not. Without going into details we merely mention that for systems belonging to group 1

$$f_1(\rho) = -\frac{1}{2}\psi^{(2)}(\frac{1}{2}+\rho) - \frac{1}{6}\psi^{(3)}(\frac{1}{2}+\rho) \tag{36}$$

where $\psi^{(2)}(x), \psi^{(3)}(x)$ are higher derivatives of the digamma function. As regards group 2, in cases (a) and (b) the term $\psi^{(3)}(\frac{1}{2}+\rho)$ is missing, while in case (c) it has a prefactor of $+1/18$. The function $f_1(\rho)$ enters

$$<|\Delta(\underline{r})|^2> = 2(2\pi T_c)^2 \frac{1-\rho\psi^{(1)}(\frac{1}{2}+\rho)}{\beta f_1(\rho)} \left(1-\frac{T}{T_c}\right) \tag{37}$$

as found by expansion of Eq. (20). Here $\beta = <\Delta^4>/<\Delta^2>^2$. Therefore $f_1(\rho)$ affects such quantities as the zero bias tunneling conductance near T_c, the jump of the specific heat at T_c or the slope of the magnetization near H_{c2} in a type II superconductor [3].

4 Scattering centers with internal frequencies

In this section an extension of the standard theory of pair breaking is discussed. It deals with the effect of random impurities with internal frequencies. This is required in order to interpret a number of important experimental results. It also leads to new qualitative insight. The Abrikosov-Gorkov theory [2] described in the last section relates the pair breaking by magnetic impurities to conduction electron scattering by classical spins. This simplified model qualitatively explains many observed phenomena like the depression of T_c, gap smearing etc. There are, however, fascinating phenomena which cannot be described within this picture. Two examples are considered. One deals with rare earth impurities with a pronounced crystalline electric field (CEF) splitting of the energy levels. This splitting is usually of the order of a few meV and can be measured by inelastic neutron scattering. An important new feature is that pair breaking can be appreciable even when the rare earth ion is in a singlet, i.e. nonmagnetic ground state. The second example are Kondo ions, in particular Ce ions. In that case an extended theory must explain why the pair breaking strength of a Ce ion can strongly depend on temperature and on the properties of the nonmagnetic host.

The extensions of the standard theory of pair breaking to be discussed here try to give a more realistic description of the interacting system consisting of conduction electrons and impurity ions. When an isotropic scattering center is placed into a singlet superconductor one can associate with it a one-particle scattering matrix and a scattering matrix for two particles in time reversed states. Near T_c, where the order parameter is small, this leads to the following renormalized frequencies $\tilde{\omega}_n$ and frequency-dependent order parameter $\tilde{\Delta}_n$

$$\tilde{\omega}_n = \omega_n - n_I \, Im \, T_1(\tilde{\omega}_n) \tag{38}$$

$$\tilde{\Delta}_n = \Delta + n_I \pi N(0) T_c \sum_m T_2(\tilde{\omega}_n, \tilde{\omega}_m) \frac{\tilde{\Delta}_m}{|\tilde{\omega}_m|}.$$

As before, n_I is the concentration of the random scattering centers and $T_1(\tilde{\omega}_n), T_2(\tilde{\omega}_n, \tilde{\omega}_m)$ are the one- and two-particle scattering matrices, respectively. The corresponding diagrams are shown in Fig. 2. In the dilute limit one may neglect "effective medium" effects and compute T_1 and T_2 with unrenormalized Matsubara frequencies $\omega_n = 2\pi T_c(n+\frac{1}{2})$. The self-consistency equation for T_c becomes

$$\ell n \frac{T_c}{T_{c0}} = \pi T_c \sum_{n=-\infty}^{\infty} \left\{ \frac{1}{\Delta} \frac{\tilde{\Delta}_n}{|\tilde{\omega}_n|} - \frac{1}{|\omega_n|} \right\} \tag{39}$$

(compare with Eqs. (17,18)). In the limit $n_I \to 0$ one finds from Eqs. (38, 39) for the depression of T_c the following slope

$$-\frac{dT_c}{dn_I}\bigg|_{T_{c0}} = -T_{c0}^2 \pi \sum_{n=-\infty}^{+\infty} \left\{ \frac{ImT_1(\omega_n)}{\omega_n^2} sgn\omega_n + \pi N(0) T_{c0} \sum_{m=-\infty}^{+\infty} \frac{1}{|\omega_n|} T_2(\omega_n, \omega_m) \frac{1}{|\omega_m|} \right\}. \tag{40}$$

These equations will be used in the following.

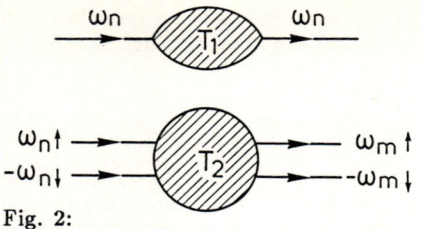

Fig. 2:
Diagrammatic representation of the one- and two-particle scattering matrices

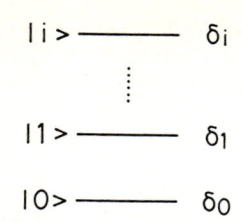

Fig. 3:
Schematic plot of the different CEF states $|i\rangle$ and energies δ_i.

The scattering matrices in the normal state, T_1 and T_2, contain specific information on the scattering mechanism. They have to be evaluated by starting from a microscopic model. The important new point of view here is that other low-energy scales in addition to the superconducting transition temperature T_{c0} come into play. It is the ratio of T_{c0} and the characteristic energy scale of the scattering mechanism which determines the pair breaking strength.

We start out by discussing the effect of CEF splitting on the pair breaking properties of rare earth (RE) ions. For that purpose we assume a given CEF splitting of the ionic states as indicated in Fig. 3. The CEF eigenstates $|i\rangle$ and energies δ_i are assumed to be known, e.g. from inelastic neutron scattering. The interaction of the states $|i\rangle$ with the conduction electrons is according to Eq. (7). We rewrite this Hamiltonian in the form

$$H_{int} = -J_{ex}(g-1) \sum_{kk'} c^+_{k'\alpha} \sigma_{\alpha\beta} c_{k\beta} \underline{J} \tag{41}$$

appropriate for RE ions. $\underline{J}$ is the total angular momentum of the $4f$ shell of a RE ion and g is its Landé factor. The scattering matrices T_1 and T_2 can be expressed in terms of a frequency dependent magnetic susceptibility

$$R(\omega_n, \omega_m) = -\sum_{ij} \frac{|\langle i|\underline{J}|j\rangle|^2 \, \delta_{ij}(n_i - n_j) sgn\,\omega_m}{(\omega_n - \omega_m)^2 + \delta_{ij}^2}. \tag{42}$$

Matsubara frequencies have been used, so that the energy transfer from the conduction electrons to the impurity is $\omega_m - \omega_n$. Furthermore, $\delta_{ij} = \delta_i - \delta_j$ and n_i denotes the thermal population of the different CEF states $|i\rangle$. In terms of $R(\omega_n, \omega_m)$ the matrices T_1 and T_2 take the form shown in Fig. 4. We find

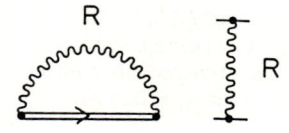

Fig. 4:
One-particle self-energy and two-particle interactions due to the interaction of the conduction electrons with the CEF split scattering centers.

$$N(0) Im T_1(\omega_n) = -\pi\gamma^2 T_c \sum_m R(\omega_n, \omega_m) \tag{43}$$

$$N(0)^2 T_2(\omega_n, \omega_m) = -\gamma^2 R(\omega_n, \omega_m) sgn\,\omega_m$$

where $\gamma = (g-1)N(0)J_{ex}$ is the dimensionless constant for spin exchange scattering. In the absence of any CEF splitting

$$R(\omega_n, \omega_m) = \frac{1}{T} J(J+1) \delta(\omega_n - \omega_m) sgn\,\omega_n \tag{44}$$

and from Eqs. (43), (38) and (39) the earlier result (18) is recovered with $N(0)/\tau_N = 2\pi n_I J(J+1)\gamma^2$. When one writes Eq. (40) in the form

$$-\frac{dT_c}{dn_I}\Big|_{T_{c0}} = s\{\delta_i/T_{c0}\} \quad (45)$$

one obtains

$$N(0)s\{\delta_i/T_{c0}\} = \frac{\pi^2}{2}J(J+1)\gamma^2 \quad (46)$$

in the limit of vanishing CEF splitting. Of particular interest are CEF split non-Kramers ions like Pr^{3+}, Tm^{3+} etc. In that case the CEF ground state can be a singlet and pair breaking is due to virtual excitation of higher CEF levels. The simplest case is that of a level scheme consisting of two singlets only, i.e. the other levels are neglected. Then $R(\omega_n, \omega_m)$ is of the form

$$R(\omega_n, \omega_m) = \frac{2\delta \cdot tanh(\delta/2T)}{(\omega_n - \omega_m)^2 + \delta^2} \quad (47)$$

where δ is the splitting energy of the two singlets and the matrix element $|<i\mid \underline{J}\mid j>|^2$ has been set equal to unity. The factor $tanh(\delta/2T)$ describes the different populations of the two levels. The form of Eq. (47) shows that the susceptibility $R(\omega_n, \omega_m)$ acts in a similar way on the electronic system as a localized phonon (Einstein oscillator) except that it is pair breaking and not pair forming. The function $s\{\delta_i/T_{c0}\}$ is written as

$$N(0)s(\delta/T_{c0}) = \frac{\pi^2}{4}y(\delta/2T_{c0})J(J+1)\gamma^2 \quad (48)$$

where $y(x)$ is plotted in Fig. 5 [7]. It determines the pair breaking strength which varies as function of the CEF splitting δ. It is seen that even for $\delta \gg T_{c0}$ the pair breaking effect of an ion can be appreciable. The solutions of Eq. (39) are plotted in Fig. 6 for various values of δ/T_{c0}, with a normalized slope at T_{c0}. One notices strong deviations from the Abrikosov-Gorkov theory ($\delta/T_{c0} = 0$). With higher concentration n_I the pair breaking effect of a single ion decreases continuously. This is so, because the ratio (δ/T_c) increases with decreasing T_c and a virtual excitation of the upper CEF singlet becomes correspondingly unlikely (compare with Fig. 5).

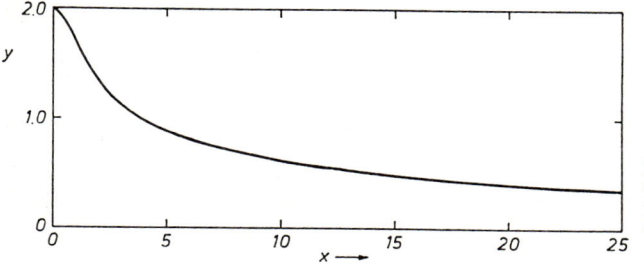

Fig. 5:
Plot of the function $y(x)$.
From Ref. [7]

In passing we note that e.g. other interactions like the aspherical Coulomb charge scattering of conduction electrons off RE ions lead to pair formation instead of pair breaking, because time reversal symmetry is not broken in that case.

The second extension of the theory to be discussed here is that to Kondo ions in general and to Ce ions in particular. Figure 7 displays the reduction of the superconducting transition temperature T_c of various $(La_{1-x}Th_x)Ce$ alloys as a function of Ce concentration and Th content. The new features are:

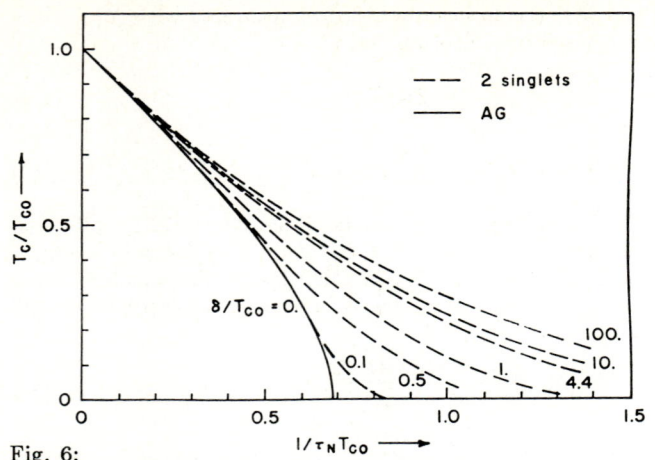

Fig. 6:
T_c vs. impurity concentration for scattering centers consisting of two singlets with energy separation δ. The quantity $(\tau_N T_{c0})^{-1}$ agrees up to a factor with $y(\delta/2T_{c0})$. The prefactor is chosen such that all curves have the same initial slope. The curve $\delta/T_{c0} = 0$ corresponds to the Abrikosov-Gorkov theory [2].

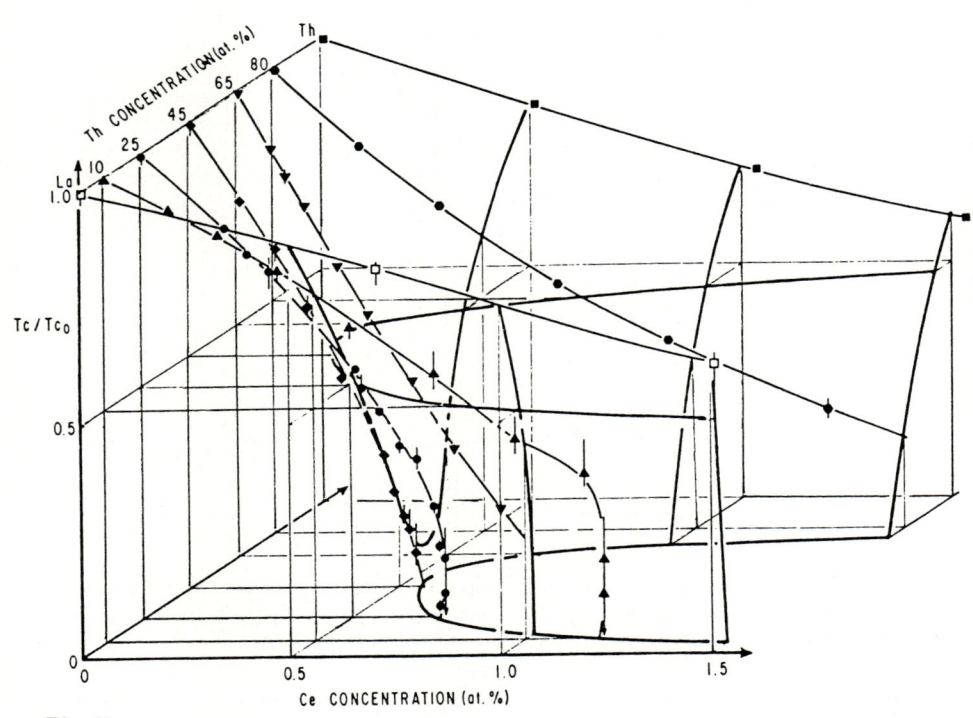

Fig. 7:
Dependence of T_c on Ce concentration for the system $(La, Th)Ce$. With increasing Th content the Kondo temperature sweeps through T_{c0}. From Ref. [9].

(a) Certain alloy systems are "re-entrant" [10]. There are ranges of impurity concentration for which the system becomes superconducting below a critical temperature T_c, but turns normal again below $T_{c2} < T_{c1}$. There are conjectures that in some systems even a third critical temperature T_{c3} was found.

(b) The reduction of the superconducting transition temperature depends on the composition of the matrix (e.g. on the Th content).

We mention briefly in this context that in the heavy fermion superconductor $CeCu_2Si_2$ superconductivity is caused by the $4f$ electrons. Therefore instead of breaking pairs the Ce ions generate them.

Let us briefly review some characteristic properties of Kondo systems which will be reflected in the depression of the superconducting transition temperature.

In the normal, i.e. nonsuperconducting, state, the interaction between the conduction electrons and the impurity ion leads to typical low-temperature anomalies in many physical properties. The most prominent example is the electrical resistivity which exhibits a minimum at some characteristic temperature T_0. This behaviour highlights very strong scattering mechanisms which develop at decreasing temperatures $T < T_0$. The characteristic temperature is (almost) independent of the impurity concentration. It sets the scale on which anomalies occur, i.e. on which the alloy changes its behavior. Of particular interest here is the magnetic susceptibility $\chi(T)$ which shows a Curie-Weiss behavior at $T \gg T_0$ as expected from the presence of magnetic moments. At low temperatures $T \ll T_0$, however, it crosses over to a Pauli paramagnetic susceptibility indicating a singlet ground state. Magnetic moments as defined by

$$\mu_{eff}^2(T) = 3T\chi(T) \tag{49}$$

are no longer present.

This behavior is explained by a microscopic theory based on the Anderson model

$$H = \sum_{\underline{k}\sigma} \epsilon(\underline{k}) c_{\underline{k}\sigma}^+ c_{\underline{k}\sigma} + \epsilon_f \sum_m n_m^f + \frac{U}{2} \sum_{m \neq m'} n_m^f n_{m'}^f + \sum_{\underline{k}m\sigma} (V_{m\sigma}(\underline{k}) f_m^+ c_{\underline{k}\sigma} + V_{m\sigma}^*(\underline{k}) c_{\underline{k}\sigma}^+ f_m). \tag{50}$$

The operators $c_{\underline{k}\sigma}^+(c_{\underline{k}\sigma})$ create (annihilate) conduction electrons with energy dispersion $\epsilon(\underline{k})$ in momentum and spin states $\underline{k}, \sigma$. All energies will be measured with respect to the Fermi level E_F. The operators for the f electrons at the impurity site are $f_m^+(f_m)$, respectively, with m=1, ...ν_f, the number operators $n_m^f = f_m^+ f_m$. The position ϵ_f of the f level is supposed to be well below the Fermi energy. The Coulomb repulsion between two f electrons is U. Since we want to exclude that there is more than one f electron at the impurity site, we shall assume later that $U \to \infty$. The last term on the right hand site describes the hybridization.

We are mainly interested in the "local moment regime" where the f-occupancy of the Ce ion is nearly one (despite broadening) and where the Ce ion possesses a magnetic moment at high temperatures. This regime is characterized by the condition

$$| \epsilon_f | \gg \nu_f \Gamma = \nu_f \pi N(0) V_{hyb}^2; \quad V_{hyb} = V(k_F) \tag{51}$$

where $N(0)$ is again the density of states per spin of the conduction electrons at the Fermi level. This model is used to calculate T_1 and T_2 in the dilute limit by means of diagrammatic techniques. For the details we refer to Ref. [11]. The dominant contribution to pair breaking comes from single-particle scattering. It can be reformulated in terms of the spectral function

$$-T_{c0}^2 \pi \sum_n \frac{1}{\omega_n^2} sgn\, \omega_n\, Im\, T_1(\omega_n, T_{co}) = T_{c0}^2 \pi \sum_n \frac{1}{|\omega_n|} \int_{-\infty}^{+\infty} d\omega \frac{1}{\omega_n^2 + \omega^2} (-\frac{1}{\pi} Im\, T_1(\omega, T_{co})). \tag{52}$$

The scattering rate for conduction electrons $-\frac{1}{\pi} Im\, T_1(\omega, T_{co})$ is proportional to the

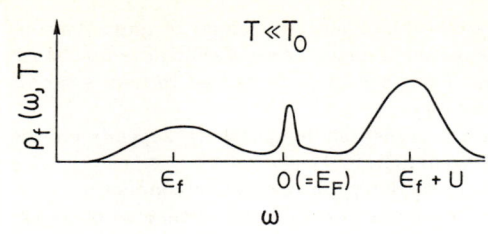

Fig. 8:
Schematic plot of the f density of states $\rho_f(\omega, T)$ at low temperatures $T \ll T_0$.

many-body spectral density for adding or removing an f electron of energy ω (see e.g. Ref.[11,12]),

$$-\frac{1}{\pi} Im T_1(\omega, T_{c0}) = \frac{\nu_f}{2} V_{hyb}^2 \rho_f(\omega, T_{c0}). \tag{53}$$

This spectrum $\rho_f(\omega, T)$ is usually referred to as the f density of states and may be obtained from photoemission and inverse photoemission. It is shown schematically in Fig. 8 for temperatures far below the characteristic Kondo temperature mentioned above. The weight at negative energies corresponds to removing an f electron while the weight at positive energies to adding one. The broad peak near the energy of the magnetic f^1 multiplet in the noninteracting system, ϵ_f, corresponds to removing an f electron from the ground state. This feature is present also at high temperatures $T \gg T_0$. It can be modelled by a Lorentzian of width $\sim \nu_f \Gamma$ and height $\frac{1}{\pi \nu_f} \frac{1}{\nu_f \Gamma}$. The corresponding contribution to the T_c-depression $N(0)(-\frac{dT_c}{dn_I})|_{T_{c0}}$ is

$$N(0)T_{c0}^2 V_{hyb}^2 \frac{\nu_f}{2} \sum_n \frac{1}{|\omega_n|} \int_{-\infty}^{+\infty} d\omega \frac{1}{\omega_n^2 + \omega^2} \frac{\Gamma}{(\epsilon_f - \omega)^2 + (\nu_f \Gamma)^2} \simeq \frac{\pi^2}{4} \nu_f \gamma^2 \tag{54}$$

where we introduced again the dimensionless coupling constant $\gamma = N(0) J_{ex}$ with $J_{ex} = -\frac{V_{hyb}^2}{|\epsilon_f|}$. This expression should be compared with the Abrikosov-Gorkov result in Eq. (46).

Note that this derivation does <u>not</u> assume that the f^1 multiplet of the noninteracting system is degenerate. The magnetism of the f-shell is not essential for this classical pair breaking. From a microscopic point of view it is the "charge fluctuations" $f^1 \to f^0$ which are modelled by the Abrikosov-Gorkov theory.

At temperatures far below the characteristic Kondo scale a narrow peak at positive energy T_0 appears, which dominates the low-temperature behavior. It is called the Abrikosov-Suhl or Kondo resonance and has no counterpart in a noninteracting system. It measures the probability of the transition $f^0 \to f^1$ when an f electron is added to the many-body ground state. In dilute magnetic alloys the ground state (as well as the low-lying excitations) is formed by coherent superposition of states containing one and no f electrons. The formation of this complicated many-body ground state which is the origin of the narrow resonance results from the magnetic degeneracy of the f^1 multiplet. It can be understood from a simple model as shown in [14]. There is no comparable feature for impurities with nondegenerate f states which can occur for ions with an even f electron number in the presence of CEF splitting. For completeness we included in Fig. 8 the broad peak at $\epsilon_f + U$ which is observed in inverse photoemission. It corresponds to the transition $f^1 \to f^2$. In Ce systems, the corresponding excitation energies are rather high. It therefore has no influence on superconducting properties. The resonant scattering of conduction electrons is reflected in the normal state transport properties. Its growth with decreasing temperature is illustrated in Fig. 9 for a typical choice of parameters. The point we want to emphasize here is the universal behavior of the dimensionless quantity $-N(0)\frac{\nu_f}{2\pi} Im T_1(\omega + i0^+)$ which depends only on the scaled frequency (ω/T_0) and the scaled temperature (T/T_0).

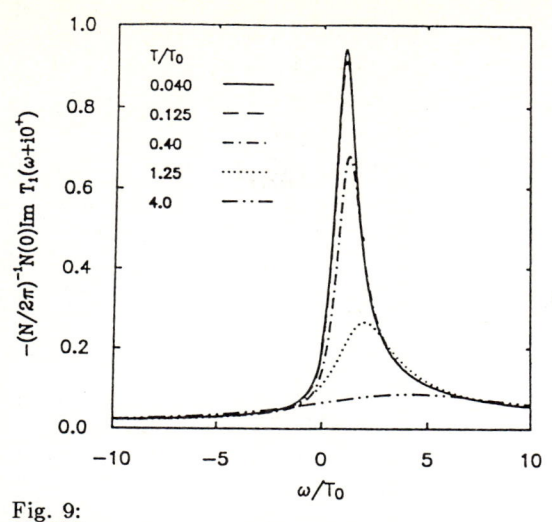

Fig. 9:
One-electron scattering amplitude as function of energy for various temperatures. At low temperatures scattering becomes very strong due to the Kondo effect. The position of the scattering resonance sets the scale for the low temperature scattering properties. From Ref. [11].

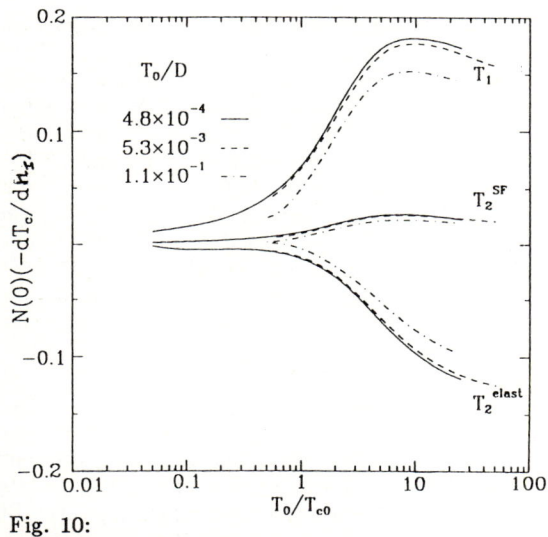

Fig. 10:
Different contributions of the one- and two-particle scattering matrices to the depression of T_c for various values of the characteristic temperature T_c measured in units of the conduction electron band width $2D$. From Ref. [11].

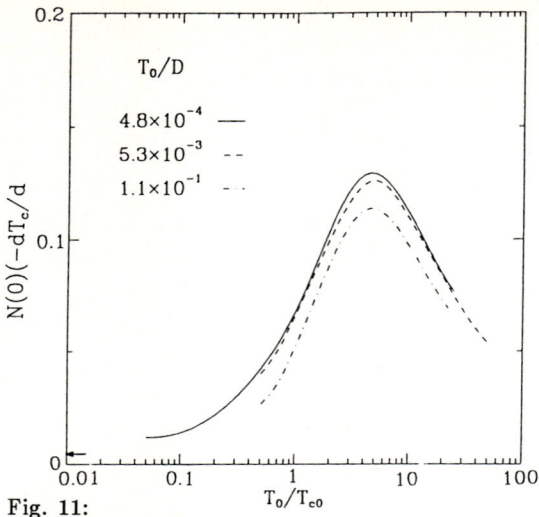

Fig. 11:
Initial depression of T_c with impurity concentration n_I as function of (T_0/T_{c0}), i.e. the ratio of the Kondo temperature and the superconducting transition temperature. The pair breaking effect of an impurity is largest for $T_0/T_{c0} \simeq 5$. This may lead to reentrant behavior as predicted by Müller-Hartmann and Zittartz [10] and also explains the findings in Fig. 7. From Ref. [11].

The effective interaction between opposite spin electrons, $T_2(\omega_n, \omega_n')$, also varies with temperature and depends strongly on energy. Microscopically it results from virtual polarization of the impurity. We shall focus here on "elastic" ($\omega_{n'} = \omega_n$) and "elastic spin flip" ($\omega_{n'} = -\omega_n$) scattering. The corresponding scattering amplitudes are denoted by T_2^{elast} and T_2^{SF}. In this case the dimensionless quantities $[N(0)]^2 Im T_2^{elast}(\omega + i0^+)$ and $[N(0)]^2 Im T_2^{SF}(\omega + i0^+)$ exhibit universal behavior as a function of (ω/T_0) and (T/T_0) in analogy with the one-particle amplitude. This fact implies that the initial depression of the superconducting transition temperature depends only on the ratio (T_0/T_{c0}) (in the limit of almost integer f valence considered here). The different contributions are shown in Fig. 10. The strong scattering of single electrons with energy of order T_0 leads to a large initial slope for the T_c depression. This effect is opposed by the elastic two-electron contribution which reflects the increasingly nonmagnetic character of impurity scattering for large (T_0/T_{c0}). The elastic spin flip amplitude is small at all temperatures and increases the pair breaking. In the "classical" limit $T_c \ll T_{c0}$ it reduces to $\frac{\pi}{8} J^2$ in analogy with the original Abrikosov-Gorkov result. The total initial slope of T_c depression displayed in Fig. 11 contains all three components. The most prominent feature is a maximum at $T_0/T_{c0} \simeq 5$ for $\nu_f = 6$. For $T_0/T_{c0} \gg 1$ the pair breaking disappears. The system goes over into a state where superconductivity results from pair condensation of quasiparticles which are formed by a coherent superposition of conduction and f electrons. The corresponding superconducting transition temperature, however, will be different from that of the pure matrix, T_{c0}.

5 Pauli paramagnetism

When the κ value of a type II superconductor is sufficiently large, or for very thin films in a parallel field, the effect of the magnetic field on the electron spins dominates that on the orbits. To lowest approximation one may therefore write for the interaction Hamiltonian

$$H_{int} = -\mu_B \sum_i \underline{\sigma}(i)\underline{H}. \tag{55}$$

The quasiparticle excitation spectrum is then simply

$$E_\sigma(p) = \sqrt{\epsilon_p^2 + \Delta^2(T)} - \mu_B \sigma_z H \tag{56}$$

and the total density of states $N_T(E)$ splits into two spin-dependent parts, one for spin up and one for spin down. In the presence of a finite spin-orbit mean free path $\ell_{s0} = v_F \tau_{s0}$ one has

$$N_T(E) = N(0)\, sgnE\, Re\{\frac{u_+}{\sqrt{u_+^2 - 1}} + \frac{u_-}{\sqrt{u_-^2 - 1}}\} \tag{57}$$

where $u_\pm(E)$ are the solutions of the following coupled equations:

$$\frac{E \mp \mu_B H}{\Delta} = u_\pm + \frac{i}{3\tau_{s0}\Delta} \frac{u_\pm - u_\mp}{\sqrt{u_\mp^2 - 1}}. \tag{58}$$

The correct branch of the square root follows from the requirement that for $E \to \pm\infty$ we have $(u_\pm^2 - 1)^{1/2} \to u_\pm sgn(Re u_\pm)$. The coupled equations (58) have to be solved numerically.

The density of states can be measured by tunneling experiments. A plot for small spin-orbit scattering is shown in Fig. 12. One notices the spin splitting by an energy $2\mu_B H$ and a small admixture of the two components, caused by the spin-orbit scattering. The spin splitting shows up in the tunneling conductance, because one can show experimentally that during the tunneling process the spin of an electron remains unchanged. The field of spin polarized tunneling was pioneered by Meservey and Tedrow [12]. By working with thin Al films in a parallel field they used spin polarized tunneling to study problems of magnetism in normal states. For example, the difference in the density of states at the Fermi level for spin ↑ and spin ↓ electrons in Ni can be measured

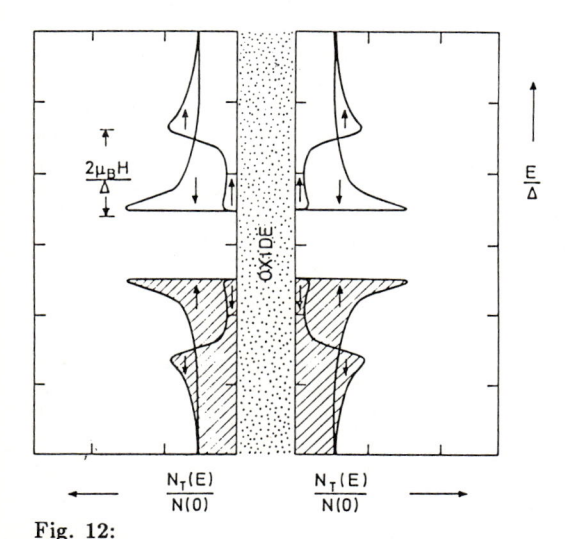

Fig. 12:
Zeeman split tunneling density of states of a $S - S$ tunneling barrier. Due to the presence of small spin-orbit scattering the two components are slightly mixed.

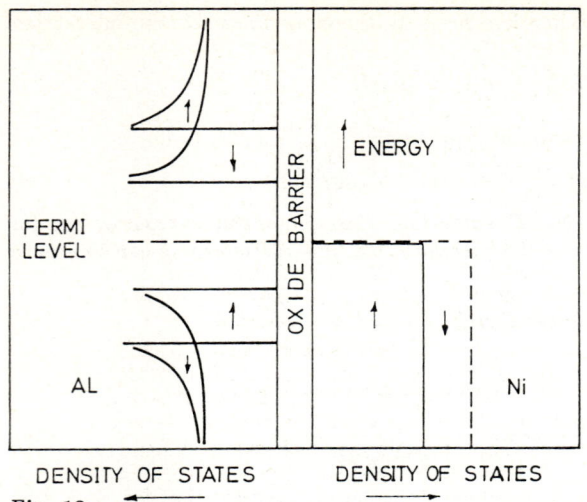

Fig. 13:
Tunneling between a superconductor with Zeeman split excitation spectrum and a ferromagnetic Ni film. The density of states for spin $\uparrow$ and $\downarrow$ electrons differ in Ni and this difference can be determined by measuring the asymmetry in the tunneling current with respect to positive and negative voltages applied to the barrier. For details see Refs. [13].

this way. This is indicated in Fig. 13. When the spin-orbit scattering rate becomes large, i.e. for $\ell_{so} \ll \xi_0$ the spin dependent density of states curves merge into one and are of the form shown in Fig. 1. One is then dealing with the situation discussed in Section 3 under case (e) of group 1.

It would be extremely interesting if spin polarization could be achieved in the high-T_c materials. For that thin films are required. Because of the small coherence length ($\xi_0^{ab} \simeq 10 - 15 \text{Å}$ in the Cu oxide planes and $\xi_0^c \simeq 2 - 5 \text{Å}$ perpendicular to them) one might be able to observe spin splitting. The constituents like Cu are rather "light" elements. We therefore do not anticipate substantial spin-orbit interaction in the ideal compounds. Due to the large values of $\Delta(T=0)$ one would have to use high fields, e.g. of the order of 20 Tesla, in order to obtain a sizeable spin splitting on the scale of Δ. Broadenings of the singularity in the BCS density of states caused by anisotropies of Δ or by the effect of the field on the orbits would not interfere with the spin splitting.

6 Strongly anisotropic pairing

During the last decade superconductivity was discovered in several new classes of materials. The most prominent examples are the new high-T_c materials and heavy fermion systems. Both classes of systems show the common feature that the strong local correlations preclude an understanding of the electronic properties in terms of a microscopic model. The superconducting properties are found to be rather unusual. There are indications that the superconducting state cannot be adequately described in terms of an isotropic order parameter. Anisotropic superconductivity, however, is rather sensitive to scattering off nonmagnetic impurities. We shall study here how the transition temperature depends on the impurity concentration.

The derivation of the T_c depression must be generalized allowing for anisotropies in such electronic properties as the effective attraction among quasiparticles and the scattering amplitudes T_1 and T_2 introduced previously. Note that in the single-electron scattering we can restrict ourselves to the forward scattering limit in the case of randomly distributed impurities since all the other contributions can be assumed to interfere destructively.

We consider here (spin-)singlet states which are characterized by a scalar order parameter. The generalization to vector order parameters encountered in triplet states is rather straightforward. Near T_c the renormalized frequencies and order parameter components $\tilde{\omega}_n$ and $\tilde{\Delta}_n$ are determined by (compare with Eq. (38))

$$\tilde{\omega}_n(\hat{k}) = \omega_n - n_I \, Im T_1(\hat{k};\omega_n) \tag{59}$$

$$\tilde{\Delta}_n(\hat{k}) = \Delta(\hat{k}) + n_I N(0) \int d\hat{k}' n(\hat{k}') T_2(\hat{k},\hat{k}';\omega_n) \frac{\tilde{\Delta}_n(\hat{k}')}{|\tilde{\omega}_n(\hat{k}')|}$$

where $\hat{k}$ is a two-dimensional variable running on the Fermi surface. The weight function $n(\hat{k})$ is normalized to unity $\int d\hat{k} n(\hat{k}) = 1$ and measures the relative contribution of the surface area $d\hat{k}$ surrounding $\hat{k}$ to the density of states at the Fermi level $N(0)$. The impurity concentration is again denoted by n_I.

For simplicity we shall adopt here the Born approximation to calculate the scattering matrices. They read

$$T_1(\hat{k};\omega_n) = -i\pi sgn\omega_n N(0) \int d\hat{k}' n(\hat{k}') \, |u(\hat{k},\hat{k}')|^2 \tag{60}$$

$$T_2(\hat{k},\hat{k}';\omega_n) = |u(\hat{k},\hat{k}')|^2$$

where $u(\hat{k},\hat{k}')$ are the matrix elements for scattering off a single impurity.

The initial slope of the variation of T_c with impurity concentration is then given by

$$\left(\frac{-dT_c}{dn_I}\right)\Big|_{T_{c0}} = \pi^2 T_{c0}^2 \sum_n \frac{N(0)}{\omega_n^2} \int d\hat{k} d\hat{k}' n(\hat{k}) n(\hat{k}') \{ |\Delta(\hat{k})|^2 |u(\hat{k},\hat{k}')|^2$$
$$- \Delta(\hat{k}) |u(\hat{k},\hat{k}')|^2 \Delta(\hat{k}')\}. \tag{61}$$

In Eq. (61) we used the normalization condition

$$\int d\hat{k} n(\hat{k}) \, |\Delta(\hat{k})|^2 = 1 \tag{62}$$

for the anisotropic gap function at T_{c0} of the undoped metal. One notices that the reduction of the superconducting transition temperature is determined by weighted Fermi surface averages of the impurity-scattering matrix elements. It is obvious that the pair breaking contribution vanishes identically for an isotropic gap, $\Delta_{iso}(\hat{k}) = const.$, as expected. As a consequence the isotropic part of the order parameter survives in the limit of short mean-free path (dirty limit). We are therefore in a pair weakening rather than pair breaking situation where the order parameter is a sum of an isotropic and an anisotropic part. For a purely anisotropic gap with

$$\int d\hat{k} n(\hat{k}) \Delta(\hat{k}) = 0 \tag{63}$$

the situation is less trivial. In that case we have to account for the anisotropies in the scattering matrix elements. Maximal pair breaking is obtained for isotropic s-wave scattering. Then the last term on the right hand side of Eq. (61) which reduces the pair-breaking effect vanishes. The pair breaking is then simply related to the measured normal state resistivity. If the scattering matrix element $|u(\hat{k},\hat{k}')|^2$ is strongly anisotropic then pair breaking is reduced. Anisotropic superconductivity can even be (almost) insensitive to sufficiently anisotropic impurity scattering. This is the case for the following example which we think is relevant for the discussion of heavy fermion superconductivity, e.g. in $U_{1-x}Th_xBe_{13}$ alloys (see [14]). Assume that superconductivity nucleates only at certain parts of the (strongly anisotropic) Fermi surface. We assume that Δ is almost constant there. If impurity scattering were to connect only those portions of the Fermi surface where Δ is finite, superconductivity would be insensitive to impurity scattering.

Among the anisotropic states we can distinguish two classes, the so-called "conventional states" when the order parameter has the symmetry of the crystal and hence the Fermi surface, and the so-called "unconventional" ones, which have a lower symmetry. Triplet states in general are unconventional. The point we want to emphasize here is that conventional superconductors seem to be less sensitive to impurity scattering than unconventional ones. We conclude this from the fact that the second term in Eq. (61) which reduces pair breaking will usually be smaller for unconventional states than it is for conventional ones.

REFERENCES

[1] F. Reif and M.A. Woolf, Phys. Rev. Letters 9, 315(1962)
[2] A. Abrikosov and L. P. Gorkov, Zh. Eksperim. i. Teor. Fiz. 39, 1781 (1960) [English transl.: Soviet Phys. - JETP 12, 1243 (1961)]
[3] K. Maki in "Superconductivity" vol. 2, edit. by R.D. Parks (M. Dekker, New York, 1969)
[4] P. de Gennes "Superconductivity of Metals and Alloys" (W.A. Benjamin, New York, 1966)
[5] S. Skalski, O. Betbeder - Matibet, and P. R. Weiss, Phys. Rev. 136 A, 1500 (1964)
[6] K. Maki and P. Fulde, Phys. Rev. 140 A, 1586 (1965)
[7] P. Fulde, L. Hirst and A. Luther, Z. Phys. 230, 155 (1970)
[8] J. Keller and P. Fulde, J. Low Temp. Phys. 4, 289 (1971)
[9] J. G. Huber, W.A. Fertig and M.B. Maple, Solid State Commun. 15, 453 (1974)
[10] E. Müller-Hartmann and J. Zittartz, Z. Phys. 234, 58 (1970); for review see also E. Müller-Hartmann in "Magnetism", (H. Suhl ed.) Vol. V, p. 353 (Academic Press, New York, 1973).
[11] N. E. Bickers and G.E. Zwicknagl, Phys. Rev. B 36, 6746 (1987)
[12] For detailed discussion see e.g. N.E. Bickers, D.L. Cox and J. W. Wilkins, Phys. Rev. B 36, 2036 (1987)
[13] P. M. Tedrow and R. Merservey, Phys. Rev. Lett. 26, 192 (1971); see also Phys. Rev. B 7, 318 (1973)
[14] P. Fulde, J. Keller and G. Zwicknagl, in "Solid State Physics", (F. Seitz, D. Turnbull, and H. Ehrenreich eds.) Vol. 41, p. 1 (Academic Press, New York, 1988)

On the Electronic Structure and Related Physical Properties of 3d Transition Metal Compounds

G.A. Sawatzky

Department of Applied and Solid State Physics, Materials Science Center,
University of Groningen, Nijenborgh 18,
NL-9747 AG Groningen, The Netherlands

1. Introduction

Compounds of the $3d$ transition metals exhibit an astonishing variety of physical properties, both electrical and magnetic [1–3]. The electric properties range from large band gap insulators (V_2O_5, Cr_2O_3, α-Fe_2O_3) and ferroelectrics ($BaTiO_3$) to semiconductors (Cu_2O, FeS_2, MnS) and via semiconductor–metal transitions (VO_2, V_2O_3, Fe_3O_4, Ti_2O_3) to metals (CrO_2, NiS, CuS) and even superconductors ($La_{2-x}Ba_xCuO_4$, $YBa_2Cu_3O_7$). In magnetism we find antiferromagnets (α-Fe_2O_3, NiO, CoO, MnO), ferrimagnets (γ-Fe_2O_3, ferrites and garnets) and ferromagnets (CrO_2). Many of these materials have been and are of great importance in various applications because their properties can be tailored and fine tuned with chemical substitutions to, for example, optimize the magnetic anisotropy or to compensate the Néel temperature. Perhaps the most astonishing example of how extremely sensitive the properties can be to the details of the chemical composition is provided by the high T_c superconductors. Here a change of x in $La_{2-x}Ba_xCuO_4$ or the stoichiometry in $YBa_2Cu_3O_{7-\delta}$ turns an antiferromagnetic insulator into a high-temperature superconductor. The reason for this wide range of properties must lie in the sensitivity of the electronic structure to the details of the crystal structure, the chemical composition and the stoichiometry.

The electronic structure of these compounds has been a controversial issue since *de Boer* and *Verwey* [4] pointed out in 1937 that compounds like NiO do not exhibit metallic behavior as predicted by the *Wilson* and *Bloch* [5] theory of solids. As a comment on the paper by *de Boer* and *Verwey*, Peierls was the first to point out that the repulsive Ni d–d Coulomb interaction could be larger than the kinetic energy gained by delocalization and therefore each Ni ion would have eight $3d$ electrons and only spin degrees of freedom would remain. These are the basic ideas behind the Mott-Hubbard theory [1, 6, 7] in which an insulating state with a correlation gap is obtained if the d–d Coulomb interaction (U) is larger than the d band dispersion width (w). Anderson, as well, in his theory of superexchange [8], recognized the importance of the d–d Coulomb interaction and used, as a starting point, the criterion $U \gg w$.

Another quantity of great importance in Anderson's theory of superexchange is the charge transfer energy (Δ), which involves the transfer of an electron from the closed shell anion to the cation. These "virtual" charge fluctuations form the intermediate states for an interaction between the cation spins. In the Mott-Hubbard picture and the Anderson theory of superexchange it was (often implicitly) assumed that $\Delta \gg U$. In this limit the anion states can be projected out, leaving us with a correlated d-band system with effective hopping integrals and renormalized Coulomb interactions.

The basic questions that one would like to answer are how large are U, Δ, and w and how large is the hybridization (T) (covalency) of the transition metal d and anion p orbitals. In this regard high-energy spectroscopies have played an important role. For example the first evidence that the above assumption ($\Delta \gg U$) was not applicable in divalent Cu compounds came from a photoelectron-Auger study of the Cu dihalides [9]. In this study it was shown that $U > \Delta$, or, in other words, the energy to remove a d electron is larger than that to remove an anion p electron. It was not until later, however, that *Fujimori* et al. [10] and *Sawatzky* and *Allen* [11] pointed out the importance of this in terms of the understanding of the electronic structure and the nature of band gaps and superexchange in transition metal compounds.

In these lectures I will review some of the more recent concepts concerning the electronic structure of transition metal compounds with special emphasis on the oxides of the late $3d$ transition metals. From this I will discuss the magnitude, character and systematics of the band gaps in insulators, the magnitude of crystal and ligand field splittings, the sign and magnitude of exchange and superexchange interactions, the influence of doping or non-stoichiometry and the character of the quasiparticle states at the Fermi level in high-T_c superconductors.

2. Electronic Structure: Basic Concepts

It is well established that the electronic structure of many $3d$ transition metal compounds (TMCs) cannot be satisfactorily described by standard band structure theory. By standard band theory I mean all the band theory approaches in which the influence of exchange and correlation are replaced by an effective one-particle potential and the many-electron wave functions are assumed to be *single* Slater determinants of one-electron Bloch states. This includes the sophisticated density functional (DF) methods [12] and local density approximations (LDAs) [13]. The problem is that the wave function used in DF calculations is a single Slater determinant of one-electron Bloch states, and therefore does not explicitly include any correlation and is not an eigenfunction of the Hamiltonian including the electron-electron interaction but is solely a mathematical tool to describe the electron density and total energy of

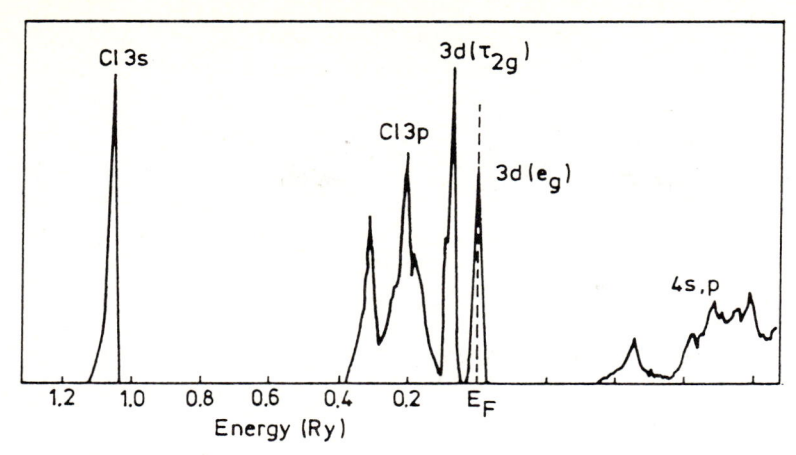

Fig. 1. The calculated density of states of NiCl$_2$ showing the full Cl$^-$ $3p$ band, the empty Ni $4s, 4p$ band and in this gap the very narrow crystal-field-split Ni $3d$ bands

the ground state. Although DF theory describes the electron density and total energy of the ground state exactly, the wave function has no physical significance formally and therefore need not give a good description of the density of states below and above the Fermi energy (E_F), which is not a ground state property.

A nice example of the breakdown of LDA theory for describing the density of states, which also serves as a basis for the discussion below, is the divalent Ni compounds NiCl$_2$. The LDA calculation of the non-magnetic phase (i.e. above the Néel temperature T_N) is shown in Fig. 1, together with the assignments of the various bands [14]. The first thing to notice is that LDA theory predicts this to be a metal, whereas it actually has a band gap of 4.7 eV [15]. The same problem is encountered in the $3d$ TM monoxides MnO, FeO, CoO, NiO [16] and, of special importance for high-T_c compounds, CuO [17] and the compounds like La$_2$CuO$_4$ [18]. This problem is a result of not including the correlation effects resulting from the strong $3d$–$3d$ Coulomb interactions *explicitly* in the many-electron wave function, as we will discuss below. This calculation however also provides us with the insight required to set up a qualitatively better approach. First we notice that the Cl $3p$ band is full, consistent with a closed shell Cl$^-$ ionic ansatz. The Ni $4s, 4p$ bands are empty and quite high in energy, consistent with an ionic Ni^{2+} ($3d^8$) ansatz. The Ni $3d$ bands are very narrow and split into t_{2g} and e_g symmetry orbitals, consistent with a ligand or crystal field approach. This also shows that the translational symmetry is not very important for the $3d$ bands so that a cluster or an impurity-type ansatz might be a good approximation. Note that the Cl $3p$ band width is quite large, so the translational symmetry is important.

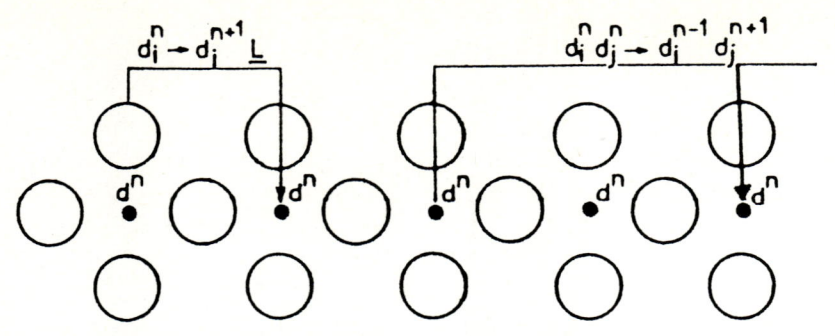

Fig. 2. Representation of an ionic lattice consisting of TM ions (d^n) closed shell anions. The most important charge fluctuation excitations are indicated

The above example gives considerable justification for starting with a localized ionic ansatz rather than a one-electron Bloch-wave-type ansatz for describing the electronic structure. Starting with such a local ionic ansatz with each TM ion having n d electrons, we are interested in the possible charge (spin is reserved for later) fluctuation excitations which will tell us about the size and nature of the band gaps. These are described in Fig. 2, and follow the procedure suggested by *Zaanen* et al. [19, 20]. In first approximation we also neglect the TM–anion hybridization (covalency) keeping in mind that this is very important, especially for the oxidic Cu compounds. We now ask for the energy for the charge fluctuations as shown and define

$$U = E_0(d^{n-1}) - E_0(d^n) + E_0(d^{n+1}) - E_0(d^n), \quad (1)$$

which is the energy required to move a d electron from one TM ion to another, and

$$\Delta = E_0(d^{n+1}\underline{L}) - E_0(d^n), \quad (2)$$

which is the energy required to move an anion p (often referred to as a ligand) electron resulting in a ligand hole ($\underline{L}$) to a TM d state. This energy Δ is usually referred to as the charge transfer energy. We now run into the problem of deciding which of the many d^n states we should take in (1) and (2). It is well known that the multiplet splitting of a d^n manifold is very large ($\approx 10\,\text{eV}$) for the 3d elements [21]. We will come back to this in a moment but for now we take in each case the lowest energy (E_0) (Hund's rule for the free ion) state as taught by *Herring* [22]. Even in the ionic picture there are two quantities, the d band (w) and anion p band (W) dispersional widths, that are important for the excited states. The states $d_i^{n-1}d_j^{n+1}$ have a dispersional width of $2w$ because both the hole (d^{n-1}) and the electron (d^{n+1}) can move freely provided they stay out of each other's way. Also the excited states $d^{n+1}\underline{L}_k$ have a dispersional width of $W + w$ where W is the ligand band width.

3. Band Gaps and Systematics

We are now in a position to draw a total energy diagram based on the ionic ansatz, as shown in Fig. 3 for $U \gg w$, $U > \Delta$, and $\Delta > W$. In Fig. 3, we can see the various types of band gaps that might occur. For $U > \Delta$, the gap is of a charge-transfer type and its magnitude is $\Delta - (W + w)/2$. So even for $U \to \infty$, we can get a metallic ground state if $\Delta < (W + w)/2$. Because generally $w \ll W$, these materials are p-type metals as, for example CuS [23]. For $\Delta > (W + w)/2$, the gap scales as the anion electronegativity for a given cation and crystal structure. This is the case for the series $NiCl_2$, $NiBr_2$, and NiI_2 with gaps of 4.7, 3.5 and 1.7 eV respectively [15] and a gap of zero for NiS.

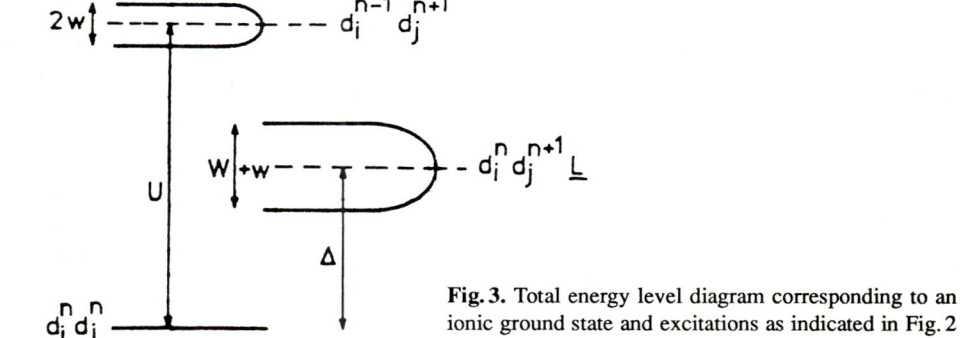

Fig. 3. Total energy level diagram corresponding to an ionic ground state and excitations as indicated in Fig. 2

For $U < \Delta$, we are in the Mott-Hubbard regime with a d–d gap for $U > w$, and a d-band metal for $U < w$. It is generally accepted that the early 3d transition-metal oxides belong to this regime.

We can put all of this information into a simple phase diagram, as shown in Fig. 4, which is a simplified version of the diagram including hybridization we have recently presented [19, 20].

As just discussed we expect the band gap to close with decreasing anion electronegativity provided we are in the charge transfer ($U > \Delta$) regime. We can get a rough idea of the systematics by looking at the trends in the ionization potentials and electron affinities of the ions involved. The simplest approach to obtaining U and Δ is to use the free ion ionization potentials (E_I), electron affinities (E_A) and Madelung potentials (V_M). Using as an example CuO, we can write in the ionic limit

$$\Delta = E_I(O^{2-}) - E_A(Cu^{2+}) + |V_M(Cu)| + |V_M(O)| + E_{corr}(\Delta) , \qquad (3)$$

$$U = E_I(Cu^{2+}) + E_A(Cu^{2+}) + E_{corr}(U) . \qquad (4)$$

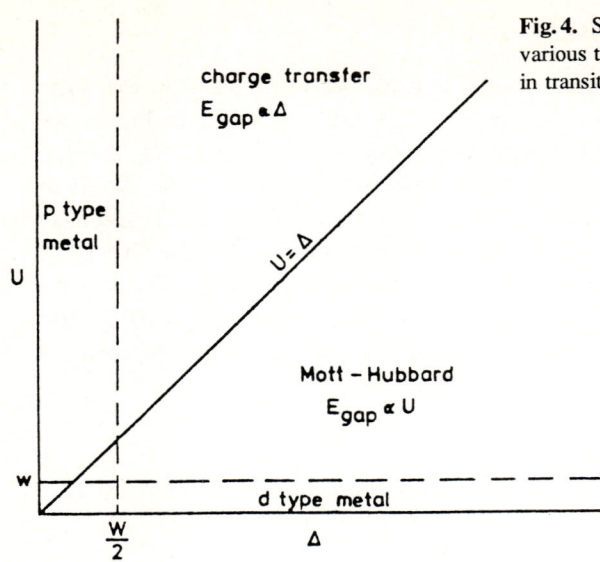

Fig. 4. Simple phase diagram showing the various types of insulating and metallic states in transition metal compounds

The problem is the large correction terms E_{corr}, which are a result of screening, polarization and covalency present in the solid. The situation is worst for Δ since the Madelung potential terms are huge ($\approx 20\,\text{eV}$) and therefore Δ is strongly dependent on the ionicity as well as the details of the crystal structure, and the ionization potentials of the anions (especially O^{2-}, S^{2-}, and Te^{2-}) are not accurately known because of the instability of these as free ions.

One of the most important corrections for both Δ and U for the ions in the solid is the result of screening. Even in an ionic insulator like NiO or CuO the effects of screening are large because of the large polarizabilities (α), especially of the anions (for example $\approx 2-3\,\text{Å}^{-3}$, for O^{2-}). The effect of screening can be understood as follows. The energy required to place a point charge in a polarizable medium is reduced by the polarization energy $E_p = (1/2)\sum_i \alpha_i F_i^2$ where α_i is the polarizability and F_i is the electric field at the position of the ith atom due to the point charge. For the $3d$ transition metal compounds $E_p \simeq 3\,\text{eV}$ [24]. Since E_p depends on F^2 it is independent of the sign of the charge therefore E_I is reduced and E_A is increased by E_p so both Δ and U are decreased by $2E_p \approx 6\,\text{eV}$ in the solid. To understand how a purely on site interaction like U could be effectively screened we simply have to realize that the polarization energy involved for two point charges well separated in space is $2E_p$ whilst that for two point charges at the same position is $4E_p$ because each point charge interacts with the polarization cloud induced by the other charge in addition to that induced by itself. This is the reason for the F^2 behavior. In fact the screening of U in the oxides is not much less than that in the metal. For example U in Cu metal for the lowest

Table 1. Theoretical estimates of the d–d Coulomb interactions U and the charge transfer energy (Δ) for (fictitious) monoxides in the rock salt structure with a NiO lattice parameter. Also given are the atomic values including the B and C Racah parameters. The atomic values are taken from Moore's tables [26]

	U_{at}	B_{at}	C_{at}	$E_A(X^{2+})$	U	Δ
Cu	16.3	0.15	0.58	20.3	5.1	4.0
Ni	18.0	0.13	0.60	18.2	7.3	6.0
Co	16.2	0.14	0.54	17.1	4.9	5.4
Fe	14.7	0.13	0.48	15.9	3.5	6.1
Mn	20.2	0.12	0.41	13.8	7.8	8.9
Cr	14.4	0.10	0.42	16.5	3.3	6.3
V	15.5	0.09	0.35	14.2	4.8	9.9
Ti	14.6	0.09	0.33	13.5	2.9	8.3

energy d^8 state is 4.7 eV whereas for the lowest energy d^8 state in CuO it is 5.3 eV, screened down from the atomic value of 16.3 eV.

In spite of these large corrections we can get at least some idea of systematics provided we keep the crystal structure and lattice parameters constant. We can then get E_{corr} from a set of experiments done on one of the compounds. This has been done in detail for NiO, for which the appropriate values for Δ and U are 6.0 eV and 7.3 eV, taking into account crystal field effects [25]. Using then the free ion values [26] for E_I and E_A as listed in Table 1 we arrive at Δ and U values for the rock salt structure TM monoxides with a NiO lattice parameter as also given in Table 1. Also given in Table 1 are the B and C Racah parameters which determine the multiplet splitting and which are not at all or hardly screened in the solid [27]. From this we see that the late 3d transition metal monoxides are expected to have charge transfer gaps $U > \Delta$ and the early ones are expected to be Mott-Hubbard systems ($U < \Delta$). These values, however, are quite strongly dependent on the crystal structure and point group symmetry. For example a detailed study of CuO has led to $U \simeq 8.8$ eV and $\Delta \simeq 2.75$–3.5 eV [17, 28]. The U value here is considerably larger because the appropriate d^8 state in this case is not the Hund's rule ground state but the low spin singlet state. It is rather interesting to note that since Δ for Cu compounds is considerably smaller than for Ni compounds we would expect CuI_2 to be a p-type metal just as CuS is. However, apparently CuI_2 would prefer to decompose to CuI+I rather than having holes in the iodine 5p band. Iodine substituted $CuBr_2$ or something similar might be rather interesting.

4. Multiplets and Ligand Fields

One of the most important aspects of TM ions in a discussion of the near ground state properties is related to multiplet structure. As is well known from atomic physics [29] a particular electronic configuration like d^n splits up into a multitude of energy levels corresponding to various possible occupations of the orbital and spin quantum number (m_l and m_s). The multiplet splitting is determined by the Slater, Coulomb and exchange integrals F_0, F_2, F_4, which are monopole, dipole and quadrupole contributions respectively, or in terms of the Racah parameters [21] $A = F_0 - 49F_4$, $B = F_2 - 5F_4$, $C = 35F_4$. The range of this multiplet splitting is as large as 10 eV and therefore of great importance. The lowest energy state of an ion in a d^n configuration is given by Hund's rule, which states that we must first maximize the spin then the orbital angular momentum, and subsequently couple this to the lowest (highest) possible total angular momentum for a less (more) than half-filled shell. In a crystal the splitting of the d levels due to the crystal fields, which we discuss below, usually quenches the orbital angular momentum since the crystal field splittings are large compared to the spin-orbit coupling, which is of the order

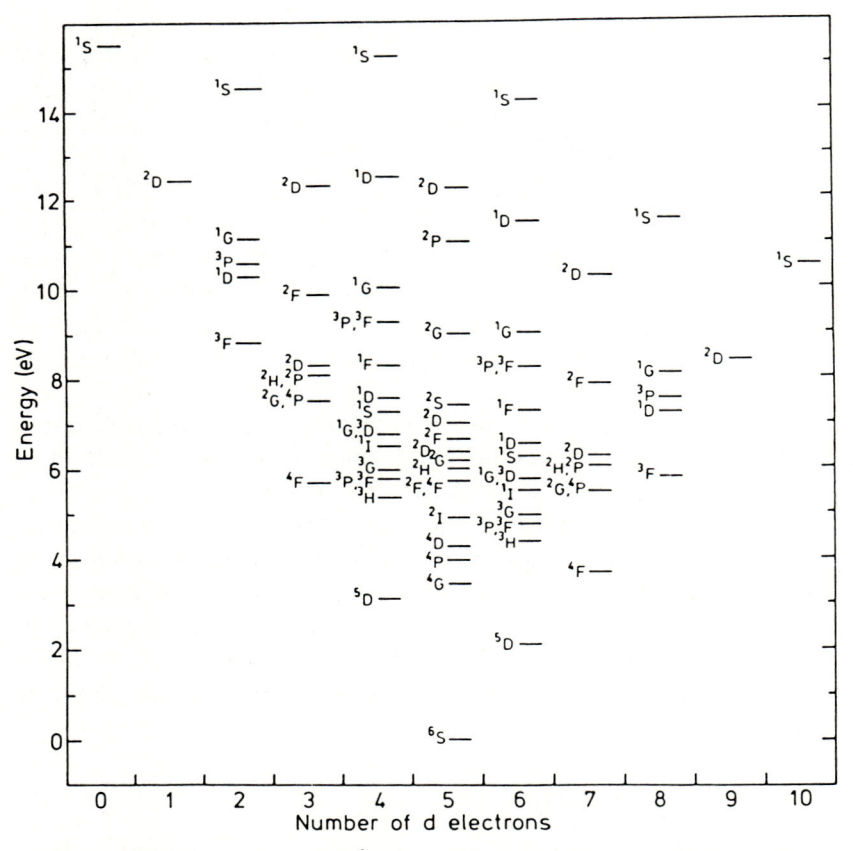

Fig. 5. Quasiatomic d states of a Mn atom in a Ag environment, calculated from Slater's atomic tables and using $F_0 = 1.04\,\text{eV}$, $F_2 = 0.18\,\text{eV}$, and $F_4 = 0.013\,\text{eV}$

of 50 meV. The effect of spin-orbit coupling then reduces to a modification of the gyromagnetic ratio (the g factor) from its spin-only value of 2 and to introducing a magnetic anisotropy. Although these effects are important for understanding the magnetic properties and the fine and hyperfine splittings in electron spin and nuclear magnetic resonance, we will here neglect the spin-orbit coupling.

To get a rough idea of the magnitude of the multiplet splittings we show in Fig. 5 the splitting for d^n taking $Mn(d^5)$ as the lowest energy state and taking the F_0, F_2 and F_4 integrals as obtained for Mn impurities in Cu in Ag hosts [27]. From the diagram we see that the ^3F state for a d^8 configuration (as in Cu^{3+}) is about 2 eV lower in energy than the first singlet. This will turn out to be rather important for the high T_c compounds. From Fig. 5 we see how important it is to specify which state one means when one specifies a U value according to (1). In general U can be written as

$$U = A + bB + cC \,, \tag{5}$$

where b and c are constants [27] depending on which multiplets are referred to in the states d^n, d^{n-1} and d^{n+1}. For the high T_c compounds we only have to worry about the multiplet of the d^8 configuration, but even here U depends strongly on the choice of multiplet. Since various types of experiments reach, because of transition matrix elements, different multiplet states it is important to do a complete analysis of the A, B and C Racah parameters and specify these rather than U.

5. Crystal (Ligand) Field Splittings

When present in a solid the d states of a TM ion are further split by crystal (ligand) field effects. In cubic symmetry the d states split into two-fold e_g $(d_{x^2-y^2}, d_{3z^2-r^2})$ and three-fold t_{2g} (d_{xy}, d_{xz}, d_{yz}) orbitals. These levels are split by the crystal field ($10\,Dq$) which is usually about 1–2 eV in the oxides. In octahedral symmetry (O_h point group) the e_g orbitals with lobes pointing towards the oxygen nearest neighbor ions are highest in energy. These crystal or ligand fields can have important consequences for the ground state properties.

In view of the high T_c materials it is interesting to look at what can in principle happen to a d^8 (Cu^{3+} or Ni^{2+}) ion in square planar symmetry (D_{4h} point group). In Fig. 6 is shown the resulting d orbital splitting as we lower the symmetry from O_h to D_{4h}. The diagram also demonstrates the exchange interaction (Hund's rule) stabilization of the state with maximum spin by including an energy difference (J) between majority and minority spin electrons. A more exact analysis will be presented below.

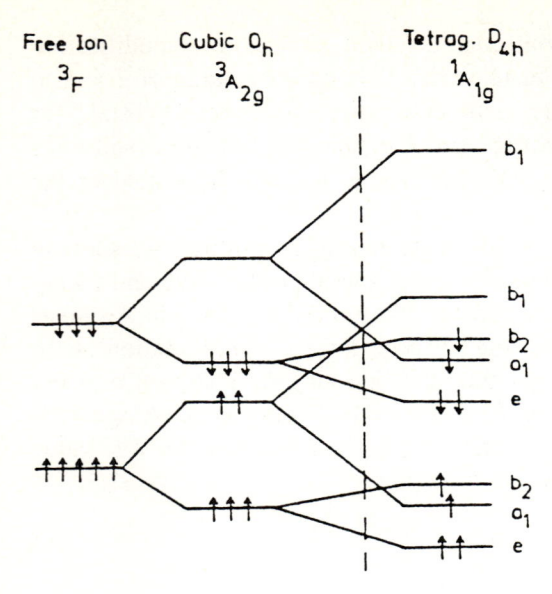

Fig. 6. The energy level diagram for a d^8 ion in spherical, cubic and tetragonal symmetry showing the change from high spin to low spin (dashed vertical line)

In D_{4h} symmetry the e_g ($d_{x^2-y^2}, d_{3z^2-r^2}$) and t_{2g} (d_{xy}, d_{yz}, d_{xz}) orbitals split up into b_1 ($d_{x^2-y^2}$), a_2 ($d_{3z^2-r^2}$) and b_2 (d_{xy}), e (d_{xz}, d_{yz}) respectively, taking the z-axis perpendicular to the plane and x- and y-axes going through the anions at the corners of a square with the cation at the center. From this diagram we see that the ground state changes from high spin ($S = 1$) to low spin ($S = 0$) when the b_1 majority spin level crosses the a_1 minority spin level, changing the electron configuration from

$$(e_g^2 b_{2g}^2 a_{1g}^1 b_{1g}^1)(^3B_{1g}) \quad \text{to} \quad (e_g^2 b_{2g}^2 a_{1g}^2)(^1A_{1g}).$$

In the limit of a strong crystal field J is approximately 2.5 eV (the energy difference between $^3B_{1g}$ and the lowest $^1A_1$ states), which shows that the non-cubic part of the crystal field must be larger than 2.5 eV for the low spin state to be stabilized.

The competition between crystal fields and Coulomb and exchange interactions is described by the *Tanabe-Sugano* diagrams [30], which have been calculated for lower symmetry cases by *König* and *Kremer* [31]. For d^8 in square planar symmetry the possible states can be calculated using the Coulomb matrix elements of Table 2 [24] together with the crystal field energies.

From the above discussion and from the experimental findings that the crystal field splittings in oxides are usually less than 2 eV, we would conclude that if Cu^{3+} were present in its usual form it would be high spin, which would present a problem for many of the high-T_c theories. Rest assured we will find a solution to this problem.

Table 2. Electrostatic matrices of d^8 in D_{4h}

$^1A_2$	b_1b_2
b_1b_2	$A+4B+2C$

$^3B_1$	a_1b_1
a_1b_1	$A-8B$

$^3B_2$	a_1b_2
a_1b_2	$A-8B$

$^3A_2$	b_1b_2	e^2
b_1b_2	$A+4B$	$6B$
e^2	$6B$	$A-5B$

$^1B_1$	a_1b_1	e^2
a_1b_1	$A+2C$	$-2B\sqrt{3}$
e^2	$-2B\sqrt{3}$	$A+B+2C$

$^1B_2$	a_1b_2	e^2
a_1b_2	$A+2C$	$-2B\sqrt{3}$
e^2	$-2B\sqrt{3}$	$A+B+2C$

^3E	eb_1	ea_1	eb_2
eb_1	$A-5B$	$-3B\sqrt{3}$	$3B$
ea_1	$-3B\sqrt{3}$	$A+B$	$-3B\sqrt{3}$
eb_2	$3B$	$-3B\sqrt{3}$	$A-5B$

^1E	eb_1	ea_1	eb_2
eb_1	$A+B+2C$	$-B\sqrt{3}$	$-3B$
ea_1	$-B\sqrt{3}$	$A+3B+2C$	$-B\sqrt{3}$
eb_2	$-3B$	$-B\sqrt{3}$	$A+B+2C$

$^1A_1$	a_1^2	b_1^2	b_2^2	e^2
a_1^2	$A+4B+3C$	$4B+C$	$4B+C$	$(B+C)\sqrt{2}$
b_1^2	$4B+C$	$A+4B+3C$	C	$(3B+C)\sqrt{2}$
b_2^2	$4B+C$	C	$A+4B+3C$	$(3B+C)\sqrt{2}$
e^2	$(B+C)\sqrt{2}$	$(3B+C)\sqrt{2}$	$(3B+C)\sqrt{2}$	$A+7B+4C$

6. Hybridization/Covalency

Up to here we have left out the hybridization between the TM d orbitals and the valence anion orbitals. This hybridization is extremely important since it determines transferred and supertransferred hyperfine interactions, superexchange, reduction of the Racah parameters and the ligand field contribution to the crystal field splittings. It is difficult, if not impossible, to take these into account even in the simple picture of Fig. 2. However, since the d band dispersional width is usually very small ($< 1\,\text{eV}$) we might in the first approximation neglect the translational symmetry of the TM but retain that of the anion. We are then left with the equivalent problem of a correlated impurity hybridizing with continuum band states, which can be treated with methods described by *Gunnarsson* and *Schönhammer* [32] but now for a semiconducting instead of metallic host [33] or with cluster methods as described by *Fujimori* [10].

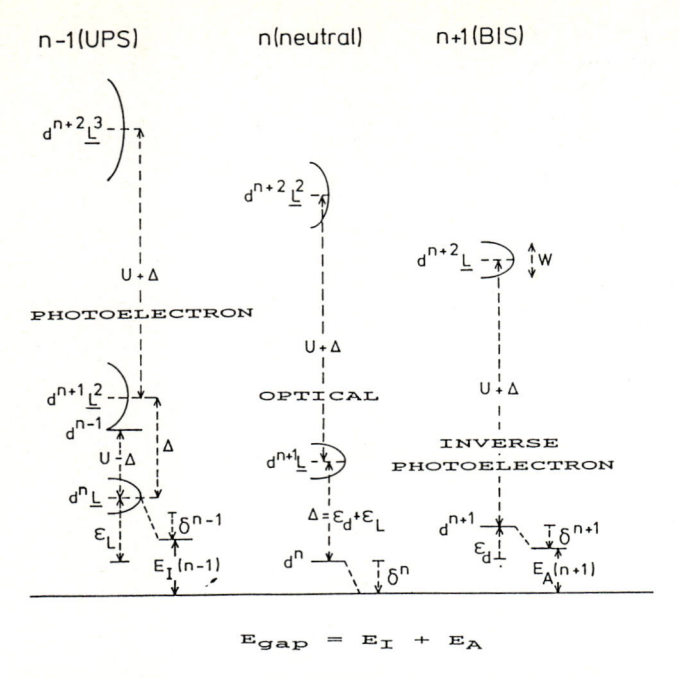

Fig. 7. Total energy diagram indicating the basis states and continua entering the model calculation as described in the text. Hybridization shifts are also indicated

Within the impurity limit we can describe the various forms of valence band spectroscopies such as optical, photoemission and inverse photoemission using a basis set of configurations as described in Fig. 7. The central diagram gives the states for the neutral system, which gives us the ground state and the local excitations as for example determined by the optical spectrum. The eigenstates of the system with one electron removed are obtained from a calculation using the configurations on the left. This, when combined with transition matrix elements, yields the photoelectron spectrum. From the right figure we obtain the eigenstates of the system with one electron added, which, together with matrix element effects, yields the inverse photoelectron or Bremsstrahlung isochromat spectrum (BIS). The band gap is determined from the minimum energy in each case as shown.

Shown in Fig. 7 are only the states corresponding to one of the many multiplets in each of the three cases. To get the total spectral distribution the calculation will have to be done for each of the multiplets or for each of the irreducible representations (IRs) of the appropriate point group. For example, to determine the ground state and optical spectrum we should calculate the eigenstates of the N-electron system for each of the IRs as done in [34] for NiO. Here I describe how one would do this for Cu^{2+} in square planar

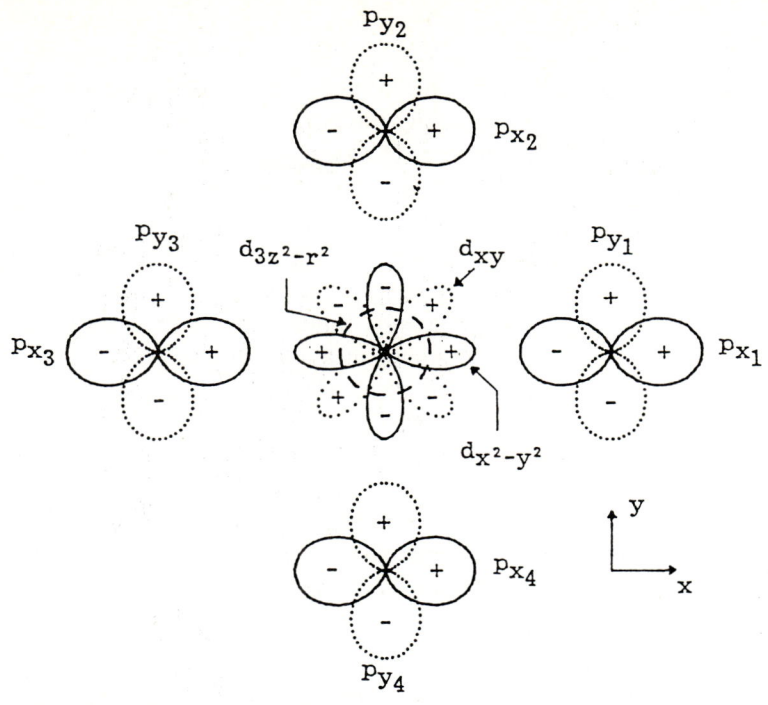

Fig. 8. Orbitals used in the (CuO$_4$)$^{6-}$ cluster calculation. The z direction points towards the reader. The O-2p_z and Cu-3d_{xz}, 3d_{yz} orbitals are not drawn

symmetry using the cluster approximation [28]. Consider the (CuO$_4$)$^{6-}$ cluster shown in Fig. 8. In the ionic limit we have Cu^{2+}(d^9) and O^{2-}($2p^6$) ions which can mix with a charge transfer state which we refer to as $d^{10}\underline{L}$. The d hole can be in b_1 ($x^2 - y^2$), a_1 ($3z^2 - r^2$), b_2 (xy) or e (xz, yz) symmetry and in each case it can hybridize with a ligand hole $\underline{L}$ of the same symmetry. These $\underline{L}$ states are linear combinations of $2p$ states on the surrounding 4 oxygens as given in Table 3.

Table 3. Irreducible representations (all gerade), one-hole basis functions and matrix elements for the (CuO$_4$)$^{6-}$ cluster in D_{4h} symmetry. The non-bonding ligand O-$2p$ orbitals are not included

Irreducible representation m	Cu $--3d$ basis d_m	O $--2p$ basis p_m
a_1	$d_{3z^2-r^2}$	$\frac{1}{\sqrt{4}}(p_{x_1} + p_{y_2} - p_{x_3} - p_{y_4})$
b_1	$d_{x^2-y^2}$	$\frac{1}{\sqrt{4}}(p_{x_1} - p_{y_2} - p_{x_3} + p_{y_4})$
b_2	d_{xy}	$\frac{1}{\sqrt{4}}(p_{y_1} + p_{x_2} - p_{y_3} - p_{x_4})$
e	d_{xz}, d_{yz}	$\frac{1}{\sqrt{2}}(p_{z_1} - p_{z_3}), \frac{1}{\sqrt{2}}(p_{z_2} - p_{z_4})$

The hybridization of each of these orbitals will be different, resulting in a different shift δ^n in Fig. 7, which in turn is the ligand field contribution to the crystal field splitting. The ligand field contribution to the splittings is a result of the different transfer integrals (T) mixing the transition metal d and anion (ligand) p orbitals. In octahedral symmetry (O_h) the transfer integrals for the t_{2g} orbitals (π bonding) are about 1/2 of those involving the e_g orbitals [35] $T(t_{2g}) \simeq T(e_g)/2$. In square planar symmetry the transfer integrals are $T(b_1) = \sqrt{3}T(a_1)$, $T(b_2) = \sqrt{2}T_e$ and as in O_h symmetry $T(b_2) \simeq 0.5T(b_1)$. Provided we can use perturbation theory the hybridization shift $\delta^n(i)$ is determined by $T^2(i)/\Delta$ where Δ is the charge transfer energy. Since $T(b_1)$ is the largest, the hole will be in a state with $x^2 - y^2$ symmetry in the ground state.

The resulting splitting is shown in Fig. 9 for parameters suitable for CuO and the high-T_c superconductors. Note that T_{pp} here represents the O-O hybridization, which in the solids evolves into the O-2p band width. From Fig. 9 we obtain the local d–d optical transitions at 1.3, 1.4 and 1.5 eV. These transitions are electric dipole forbidden and therefore weak (a factor of 10^3 lower than ordinary interband transitions), and can only be observed if they are in a band gap. Both intensities and energies are in disagreement with the assignments by *Geserich* et al. [36].

The procedure we have just described is called a configuration interaction approach, which is in general different from the usually used molecular orbital

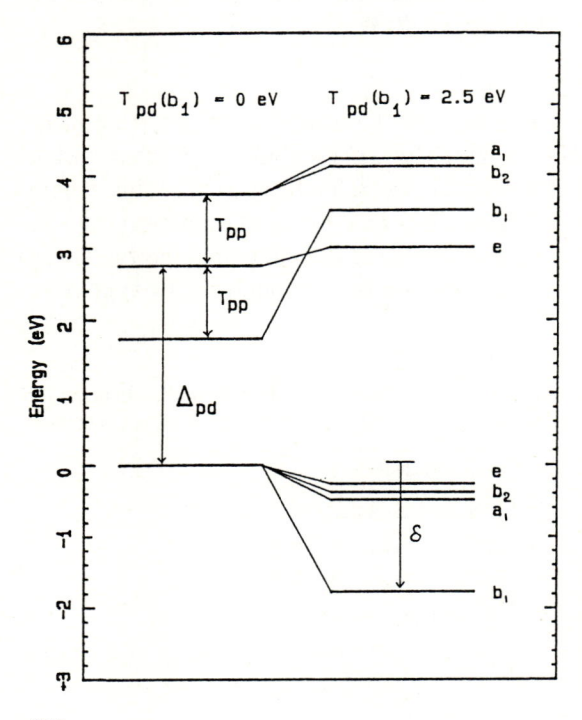

Fig. 9. The energy level scheme of the one-hole basis functions before and after Cu-3d/O-2p—ligand hybridization. The basis wave functions are defined in Table 3

approach to get the ligand field splitting. These differences become evident for states with more than one d hole, as in NiO and in the ionized states of CuO. In the molecular orbital approach, which is used in most textbooks, one basically assumes that the charge transfer energy (Δ) is much larger than the multiplet splitting, which is not in general the case for TM oxides. This is discussed in more detail in [34].

As can be seen from Fig. 7, hybridization will also strongly affect the ionization and electron addition states. These states determine the band gap, which by definition is given by (Fig. 7)

$$E_{gap} = E_I(n-1) + E_A(n+1) .$$

This is also the minimum energy required to create an electron and a hole (Fig. 3) that are sufficiently far apart for their interaction to be neglected.

7. Superexchange and Cu-O Exchange

This type of hybridization is also responsible for the Anderson [8] superexchange interactions between two TM ions via an intervening O ion. The possible paths for superexchange for the case of one hole in a $d_{x^2-y^2}$ orbital on each TM ion as for Cu^{2+} in an antiferromagnetic spin alignment are shown in Fig. 10. Also shown are the contributions to the superexchange interactions, which for this case are always antiferromagnetic. Notice here that the U for the $^1A_1$ d^8 state and not the Hund's rule $U(^3B_1)$ should be used. Also notice that even though $U \to \infty$ we still have a superexchange mechanism via dou-

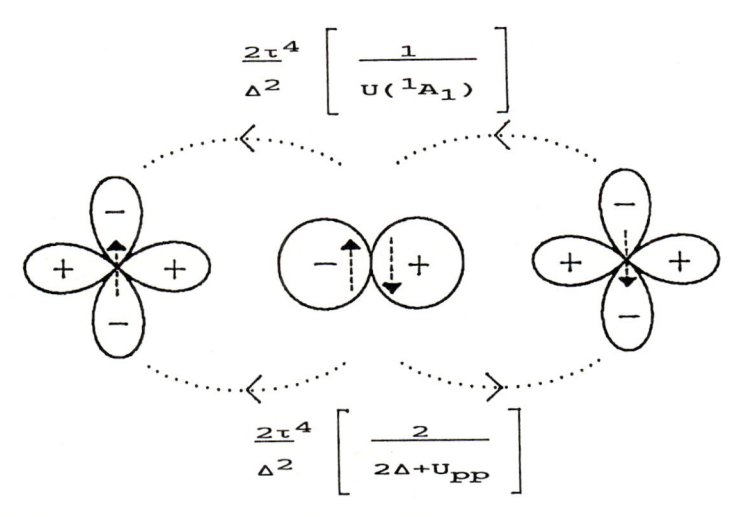

Fig. 10. An artist's concept of the two mechanisms contributing to the Anderson superexchange interaction

ble hole occupation of oxygen which is again only possible for the singlet, not triplet, state. The energy denominator now is $2/(2\Delta + U_{pp})$ where U_{pp} is the oxygen Coulomb interaction. More details can be found in [17], although there we neglected U_{pp} which is probably not a good approximation for the high T_c materials. The value one should use for the transfer integral τ here is $\tau_{sup} = (1/2)T(b_1)$ because only one oxygen is involved per Cu-Cu bond. The total superexchange interaction is then

$$J = \frac{T^4(b_1)}{4\Delta^2} \left(\frac{2}{2\Delta + U_{pp}} + \frac{1}{U(A_1)} \right) . \tag{7}$$

This is a modification of Anderson's relation, which is valid only for $U \ll \Delta$. Using this relation for the case of octahedral symmetry we have [34] shown that the Néel temperature trend for the TM monoxides is in excellent agreement with experiment whereas Anderson's relation gives a qualitatively incorrect trend as pointed out by *Oguchi* et al. [16]. This is demonstrated in Fig. 11. For details see [34].

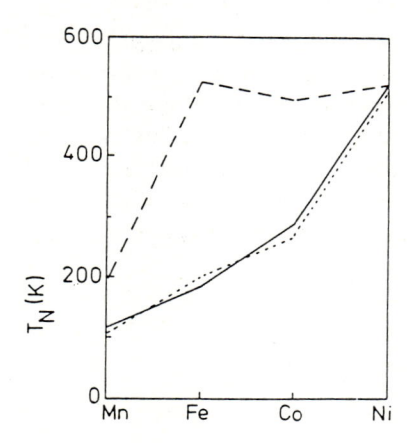

Fig. 11. Experimental Néel temperatures of the monoxides (solid line) compared with Anderson's results (dashed line) and our results (dotted line)

It is interesting to look at the dependence of the O-Cu exchange on the symmetry of the O hole state since various models for superconductivity assume various symmetries of O hole states. Consider a simple diatomic cluster compound of a Cu(d^9) ion with a hole in an $x^2 - y^2$ orbital and the O labeled 1 in Fig. 8. The O hole could be in a p_x, p_y or p_z orbital. The calculation of the exchange within perturbation theory involves the states $d_{x^2-y^2}p_i$, $d_{x^2-y^2}d_i$, $p_i p_j$ corresponding to $d^9\underline{L}$, d^8 and $d^{10}\underline{L}^2$. These configurations are coupled by a transfer integral which depends on symmetry and their diagonal energy depends on Δ, U_{pp} as well as $U_{dd}(x^2 - y^2, i)$. The exchange is then the energy difference between the singlet and triplet in each case. The Hamiltonian matrices are given in Table 4 for each of the above cases with the

TRIPLETS

p_x	$\|d_{x^2-y^2}p_x\rangle$	0	$T_{x^2-y^2}/\sqrt{3}$	
	$\|\bar{d}_{x^2-y^2}d_{3z^2-r^2}\rangle$	$T_{x^2-y^2}/\sqrt{3}$	$A-8B-\Delta$	
p_y	$\|d_{x^2-y^2}p_y\rangle$	0	T_{xy}	$T_{x^2-y^2}$
	$\|d_{x^2-y^2}d_{xy}\rangle$	T_{xy}	$A+4B-\Delta$	0
	$\|p_xp_y\rangle$	$T_{x^2-y^2}$	0	$\Delta+U_{pp}$
p_z	$\|d_{x^2-y^2}p_z\rangle$	0	T_{xz}	$T_{x^2-y^2}$
	$\|d_{x^2-y^2}d_{xz}\rangle$	T_{xz}	$A-5B-\Delta$	0
	$\|p_xp_z\rangle$	$T_{x^2-y^2}$	0	$\Delta+U_{pp}$

SINGLETS

$\|d_{x^2-y^2}p_x\rangle$	0	$T_{x^2-y^2}/\sqrt{3}$	$\sqrt{2}T_{x^2-y^2}$	$\sqrt{2}T_{x^2-y^2}$
$\|d_{x^2-y^2}d_{3z^2-r^2}\rangle$	$T_{x^2-y^2}/\sqrt{3}$	$A+2C-\Delta$	0	0
$\|d_{x^2-y^2}d_{x^2-y^2}\rangle$	$\sqrt{2}T_{x^2-y^2}$	0	$A+4B+3C-\Delta$	0
$\|p_xp_x\rangle$	$\sqrt{2}T_{x^2-y^2}$	0	0	$\Delta+U_{pp}$
$\|d_{x^2-y^2}p_y\rangle$	0	T_{xy}	$T_{x^2-y^2}$	
$\|d_{x^2-y^2}d_{xy}\rangle$	T_{xy}	$A+4B+2C-\Delta$	0	
$\|p_xp_y\rangle$	$T_{x^2-y^2}$	0	$\Delta+U_{pp}$	
$\|d_{x^2-y^2}p_z\rangle$	0	T_{xz}	$T_{x^2-y^2}$	
$\|d_{x^2-y^2}d_{xz}\rangle$	T_{xz}	$A+B+2C-\Delta$	0	
$\|p_xp_z\rangle$	$T_{x^2-y^2}$	0	$\Delta+U_{pp}$	

Table 4. Triplet and singlet two-hole states and their matrix elements for the oxygen hole in the p_x, p_y and p_z orbitals

$U_{dd}(x^2-y^2, i)$ taken from Table 2 and $T_{x^2-y^2} = T(b_1)/2$ and the π transfer integrals $T_{xy} = T_{xz} \simeq T_{x^2-y^2}/2$ as discussed above.

We see immediately a large difference between a p_x hole and p_y or p_z, which is a result of the $d_{x^2-y^2}$ orbital, which mixes only with p_x, being singly occupied. For the p_x hole only singlet d^8 states are involved so that the total hybridization contributes to the exchange whereas for a p_y or p_z hole both the singlet and triplet d^8 states can be involved so only the difference in the hybridizations contributes to the exchange. The exchange interaction is for these cases

$$J_{x,x^2-y^2} = E_S - E_T$$
$$\approx -2T_{x^2-y^2}^2 \left(\frac{1}{\Delta + U_{pp}} + \frac{1}{A + 4B + 3C - \Delta} \right), \tag{8}$$

$$J_{y,x^2-y^2} \approx +2T_{xy}^2 \frac{2C}{(A+4B-\Delta)(A+4B+2C-\Delta)}, \tag{9}$$

$$J_{z,x^2-y^2} \approx +2T_{xz}^2 \frac{2C+6B}{(A-5B-\Delta)(A+B+2C-\Delta)}, \tag{10}$$

showing that only for p_x do we get a large antiferromagnetic exchange. All the other exchange interactions are ferromagnetic and at least two orders of magnitude smaller than J_{x,x^2-y^2} and one order of magnitude smaller than the superexchange. It is interesting also to look at the exchange interaction between a ligand hole in a b_1 (see Table 3) symmetry orbital with a x^2-y^2 d-hole. This we will see later is the lowest energy hole state of the cluster. To get this we replace $T_{x^2-y^2}$ by $T(b_1)$ where $T(b_1) = 2T_{x^2-y^2}$ in the equation for J_{x,x^2-y^2} and get $J(b_1) = 4J_{x,x^2-y^2}$. This is an extremely large ($\approx 3-4\,\mathrm{eV}$) antiferromagnetic exchange. This is more than an order of magnitude larger than the superexchange.

8. Model Hamiltonian Parameters for the High T_c Superconductors

In order to understand and describe the physical properties of exotic materials like the high T_c superconductors [37] we usually resort to the study of intelligently chosen model Hamiltonians. These model Hamiltonians are usually too complicated to solve exactly so the next step is to find approximations whose validity is checked by comparison to experimental results. Since the physical properties are determined by low energy scale excitations we are in the end interested in the low energy scale parts of the problem. However, the model Hamiltonians, as for example the Hubbard model or the Anderson impurity or lattice model, are described with parameters which usually belong to a high energy scale and can be determined, often quite directly, using high energy experimental probes. These high energy scale parameters are not expected to be strongly dependent on the details of the crystal structure and therefore we are justified in choosing stable model compounds with well-defined chemical composition, and structure similar in well-chosen aspects to the actual materials we are interested in. This is of special importance for the high T_c superconductors, where a combination of surface sensitive electron spectroscopies with unknown or uncontrolled surface composition and structure has led to quite varying results.

It is generally accepted that the action is on the CuO_2 planes as far as the low energy scale electronic properties are concerned. This does not mean that the other ions and CuO chains are not important. In fact the ions in the other layers of these layered compounds are of utmost importance in setting up the Madelung potential and in setting the stage for the hole or electron doping of the CuO layers. Although it is possible that the out-of-plane apex oxygens could play a more direct active role we will neglect them to start with. It is also generally accepted that a generalized Hubbard-like Hamiltonian is a good starting point for describing the charge and spin degrees of freedom and the low energy scale physics. This is evidenced by the insulating antiferromagnetic character of the undoped "stoichiometric" materials with local moments on the divalent Cu ions. This type of behavior is typical of a Mott-Hubbard insulator. It is also generally accepted that the Hubbard Hamiltonian to be used must include both the 10-fold Cu-$3d$ states and the 6-fold O-$2p$ states, leaving us with a complicated multiband problem. It would be considerably more pleasant if the experimental parameters led to a simplification of the ultimate reduced Hamiltonian valid for the low energy scale physics.

There have been a large number of parameter estimates made from experimental studies like PES, core level XPS, Auger, X-ray absorption (for instance see [38–41]). These estimates have however very large ranges, especially in the charge transfer energy $0 \leq \Delta \leq 3$ and $5 \leq U_{dd} \leq 13$. The reason for this is that in most cases a detailed analysis was not done so that large shifts due to hybridization were incorrectly put in U and/or Δ. Also the multiplet structure of the d^8 states which ranges over 8 eV was not included, resulting in a rather arbitrary value of U. The most important problem however is that the valence band PES spectrum is sensitive to $U_{dd} - \Delta$ and insensitive to Δ itself. Also core level spectroscopies introduce (at least) one new parameter, the core hole potential (Q). In the Cu-$2p$ core XPS then for example one basically has two experimental numbers: the satellite position and relative intensity, to determine Q, Δ, and T_{pd}. U_{dd} cannot be obtained because the d^8 states are not involved. A third reason for the large spread is that experiments done on the high-T_c materials themselves are often of questionable usefulness because of the badly defined materials and surfaces. For example, in most XPS spectra of the $YBa_2Cu_3O_{7-\delta}$ system the O$-1s$ core line consists of (at least) two dominant components [42] separated by several eV in energy. It is now known that the higher binding energy one is almost certainly due to contamination, making previous analysis of O Auger data questionable.

Theoretically the situation is somewhat better. In the two detailed studies done by *MacMahan* et al. [43] and *Hybertsen* et al. [44] using density functional calculations and frozen configurations to calculate the U's, Δ, and T's, the results are quite consistent except again for $1 \leq \Delta \leq 4\,\text{eV}$, which seems to be difficult to pin down. Ab initio quantum chemical cluster calculations are in our opinion mostly unreliable in estimations of Δ and U because of the well-known problems of the long range screening effects due to polarizable ions and other embedding effects as discussed by *Janssen* et al. for NiO [45].

In principle all these parameters can be determined from a combination of photoelectron (PES) and inverse photoelectron spectroscopy (IPES). This is quite obvious since PES gives us the spectrum of all ionization states in the actual solid and IPES gives us all the electron affinity states of the solid. Unfortunately we have to solve the many-body system described by our model Hamiltonian in order to extract the parameters. However, as mentioned above, the high energy scale parameters are expected to be relatively weakly dependent on the details of the crystal structure and also on the translational symmetry. This suggests two approaches. First we can choose systems for which many-body effects of the type described by a Hubbard Hamiltonian are expected to be unimportant for the ground state, as well as for the first ionization states. Secondly we could replace the solid by a cluster of suitable choice so that the many-body problem has a small number of particles for which the model Hamiltonian can be solved and compared with experiment.

With regard to the first approach we could choose the compound Cu_2O, which has Cu linearly coordinated by oxygen ($d_{Cu-O} = 1.84$ Å) and oxygen tetrahedrally coordinated by Cu. It is a semiconductor with a band gap of 2.2 eV [46] which is also obtained from a combination of PES and IPES [17]. This is a closed shell system $Cu(3d^{10})O(2p^6)$ and therefore in the model Hamiltonian U_{pp} and U_{dd} are inoperative for the ground state as well as for the ionization (PES) and affinity (IPES) states, provided we are justified in not explicitly including the Cu $4s$ $4p$ and higher lying states in our model Hamiltonian.

The PES states will be the ligand field split Cu-$3d$ bands (d^9 like) and the O-$2p$ bands (p^5 or $d^{10}\underline{L}$) where $\underline{L}$ denotes a ligand (O-$2p$) hole. Since the PES spectrum at various photon energies is sensitive to different portions of the partial density of states (DOS) we can extract the d^9 like states with X-ray energies (XPS) and compare this with tight binding band structure and/or model Hamiltonian cluster calculations to extract quantities like the T_{pd}'s and even Δ as it is defined for CuO. This has been done in detail for Cu_2O. The parameters obtained from this are given in Table 5. Note that LDA band theory gives a very satisfactory description of the PES and IPES data [17].

Table 5. Nearest neighbor Cu-O distance, experimental band-gap energy, one-particle and two-particle energies for Cu_2O and CuO. The free ion optical values are used for the Racah B (0.15 eV), C (0.58 eV) and Slater F^2 (6.0 eV) parameters. The quantities $T_{pd}(b_1)/2$ and $T_{pp}/2$ are the commonly used hybridization parameters in, for instance, an extended Hubbard-like Hamiltonian

	Cu_2O	CuO	Units
d_{Cu-O}	1.84	1.95	Å
E_{gap}(experimental)	2.4	1.4	eV
Δ	2.50	2.75 – 3.50	eV
$T_{pd}(b_1)/2$	1.47	1.25	eV
$T_{pp}/2$	–	0.50 – 0.65	eV
$U_{dd}(^1A_{1g})$(averaged)	9.7	8.8	eV
$U_{pp}(^1D)$	5.7	–	eV
U_{pd}	< 1	–	eV

Since Cu_2O is a full band system for which one-electron theory describes the "one hole" states obtained with PES very well we can use the powerful Auger spectroscopy and related theory to obtain quite directly U_{dd}, U_{pp} and U_{pd}. This can be done because in Auger spectroscopy the final state now contains 2 holes for which exact solutions for the Hubbard-like Hamiltonian can be found [47]. The Auger spectrum can then be described in terms of the self-convolution of the one-hole DOS and the Coulomb interactions as described by *Cini* and *Sawatzky* [48]. Alternatively one can also solve the two-hole problem for the same finite clusters as those used to describe the photoelectron spectrum and compare it with experiment.

The $Cu(L_{23}M_{45}M_{45})$ Auger spectrum probes the local (because of the local core hole) projected Cu d^8 DOS involving U_{dd} directly and the O $KL_{23}L_{23}$ Auger spectrum probes the local O p^4 or $d^{10}\underline{L}^2$ DOS involving U_{pp}. These states are of course mixed because of T_{pd} provided they occur on nearest neighbors. T_{pd} mixes in states of $d^9\underline{L}$ character with a hole on a nearest neighbor ligand which involves U_{pd}, the nearest neighbor, O-Cu Coulomb interaction. The exact procedure used is described elsewhere [28] and we content ourselves here with showing the final result in Fig. 12, which shows the experimental O derived local two-hole DOS and the comparison to theory. The shaded region arises from mixing in of the d^9p^5 or $d^9\underline{L}$ states and the higher binding energy region from the atomic $d^{10}\underline{L}^2$ or p^4 states. From this we conclude that $U_{pd} < 1$ eV – screened down from the bare point charge value of 7.8 eV. The on-site U values obtained from this are also listed in Table 5.

A better comparison with the high-T_c superconductors is CuO, where the Cu-O distance is comparable (d_{Cu-O} = 1.95 Å as compared to 1.89 Å in La_2CuO_4 or 1.95 in $YBa_2Cu_3O_7$) and the Cu is in a nearly square planar coordination of oxygen. Also here Cu is divalent (d^9) and the superexchange interaction ($J = 0.1$ eV) is comparable to that of the high T_c materials. CuO is a semiconductor with a gap of 1.4 eV as found by PES/IPES [17].

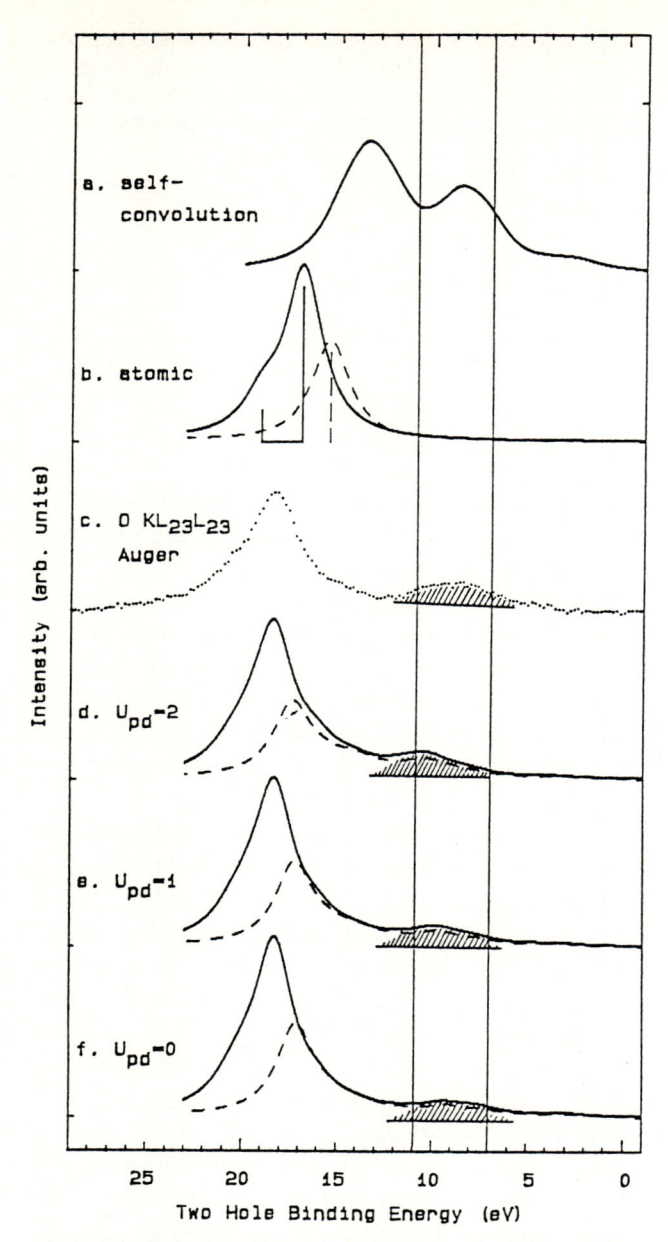

Fig. 12. O $KL_{23}L_{23}$ Auger spectrum and O $2p^4$ local DOS from Cu_2O. (**a**) Self-convolution of the O-2p partial DOS; (**b**) atomic O $2p^4$ DOS, calculated for a screened F^0 and atomic F^2 as listed in Table 5; (**c**) experimental O $KL_{23}L_{23}$ Auger spectrum, for which the kinetic energy scale has been transformed into a two-hole binding energy by subtracting the O $1s$ binding energy; (**d–f**) O $2p^4$ local DOS, calculated using a $(Cu_4O_5)^{6-}$ cluster with an O atom in the center, using $U_{pd} = 2$, 1 and 0eV and parameter values listed in Table 6. The solid lines in (**b, d–f**) are the singlet DOS, and the dashed lines the triplet DOS, reduced by a factor of 3

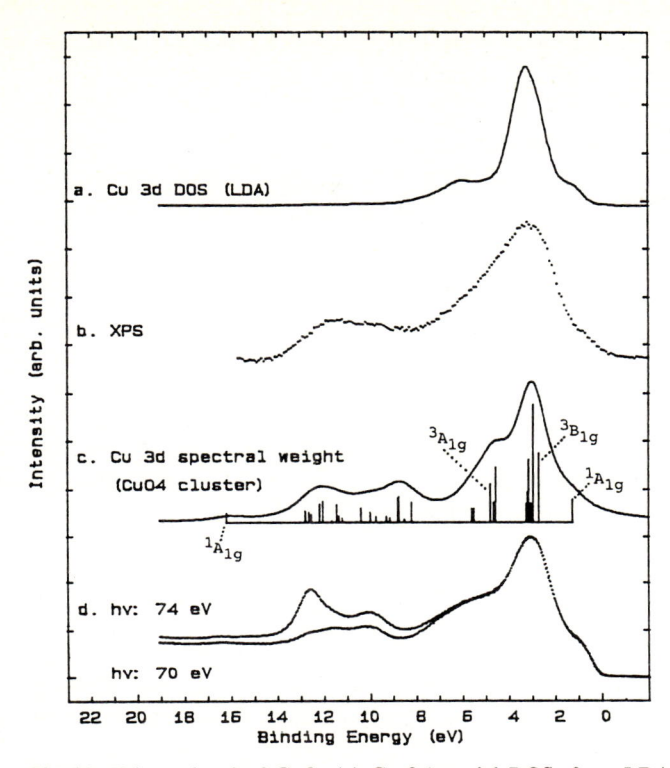

Fig. 13. Valence band of CuO. (a) Cu $3d$ partial DOS, from LDA band structure calculations [28]; (b) experimental valence band photoemission spectrum using the Al-$K\alpha$ (XPS) source; (c) Cu $3d$ spectral weight, calculated using a CuO$_4$ cluster with model parameters listed in Table 5; (d) experimental valence band photoemission spectrum at the Cu $3p$ resonance ($h\nu = 74$ eV) and out of resonance ($h\nu = 70$ eV).

Of special interest is the XPS valence band spectrum shown in Fig. 13 together with the LDA band structure calculation. Of special importance is the large energy spread of the d spectral weight as measured in XPS as compared to band theory. The states around 10–15 eV are due to d^8 final states which we get upon removal of *one* electron. These states occur at similar energies to the d^8 states in Cu$_2$O where they could only be reached by *two* electron removal. That they are indeed d^8 states is clear from their resonance behavior for photon energies at the Cu $3p$ edge [49, 50]. Also shown in Fig. 13 is a spectrum at resonance (74 eV) and out of resonance (70 eV) where we also see the large spread due to the Racah B and C parameters in the d^8 multiplets including a $^1A_{1g}$ state at 16 eV binding energy. Also shown in Fig. 13 is a simulation of the Cu $3d$ (XPS) spectral weight from a CuO$_4$ cluster calculation [28] using the model Hamiltonian with the parameters listed in Table 5. We make a special point of the calculated gap since the PES spectral shape is rather insensitive to the choice of Δ provided we keep $U_{dd} - \Delta$ constant. However, for $U_{dd} - \Delta > 0$ the gap is determined by the Δ and the oxygen band width or T_{pp}.

It is at this point that we notice that the translation symmetry is important because in the measurement of the gap with PES/IPES the top of the valence band and the bottom of the conduction band are used, but these include a dispersional width $W_v/2 + W_c/2$. Our cluster gap therefore should be taken larger than the measured gap by

$$E_{gap}(\text{cluster}) = E_{gap}(\exp) + \frac{W_v}{2} + \frac{W_c}{2}.$$

To obtain this we must know the extra broadening due to dispersion of the first ionization states and the first affinity states. This dispersional broadening on the other hand will depend on the choice of Δ and T_{pp} so we must determine these self-consistently. A reasonable estimate results in $\Delta = 3.5$ eV.

9. Influence of Doping or Non-stoichiometry

Very crudely speaking the influence of doping or non-stoichiometry is twofold. First, there is the impurity or defect potential which can locally strongly modify the electronic structure, and second, charge neutrality conditions may require the presence of holes or electrons. A good example is Li-substituted NiO, $Li_xNi_{1-x}O$, in which Ni^{2+} is replaced by Li^+, requiring an extra hole for charge neutrality. Neglecting first the impurity potential itself the hole could be either in a Ni $3d$ state, i.e. Ni^{3+} states, or in the O $2p$ band, depending on where we are in the phase diagram of Fig. 4. For $U > \Delta$ the holes will be in the O $2p$ band and for $U < \Delta$ as was previously assumed Ni^{3+} centers will exist. Recent O $1s$ X-ray absorption experiments [51] have shown that the holes are of primarily O $2p$ character, consistent with the PES and IPES conclusion that $U > \Delta$ (Table 1).

Why then is $Li_xNi_{1-x}O$ not a metal, or some may ask why is it not a high T_c superconductor? The answer is simple. In this case the impurity potential caused by Li^+ binds the O $2p$ hole, causing the hole state to be localized and in the band gap. A picture of this is given in Fig. 14, which shows the large gap material in Fig. 14a, a negligible impurity potential in Fig. 14b, resulting in a metallic state, and a strong impurity potential, resulting in an acceptor state in the gap, in Fig. 14c.

The high-T_c materials are quite different from $Li_xNi_{1-x}O$ in that the dopend is spatially well removed from the region doing the charge compensation. The impurity potential is therefore small and we have conditions leading to Fig. 14b. That this is the case is well established by electron energy loss spectroscopy [52] and X-ray absorption spectroscopy [53], which again show

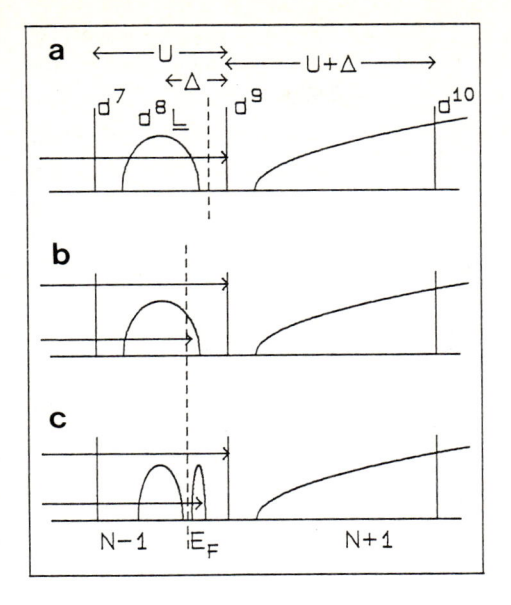

Fig. 14. (a) Sketch of the electron-removal and electron-addition states of NiO. Transitions in Li-doped NiO, (b) assuming a rigid-band model and (c) assuming localization of O $2p$ holes around lithium

the induced hole states in the high-T_c compounds to be of primarily O $2p$ character ($U > \Delta$) at the Fermi level.

10. Nature of the (Quasi)Particle States

We have just concluded from various investigations that the hole states induced by doping in the high-T_c materials are of primarily O $2p$ character. Since the added impurity potential is small at the CuO$_2$ planes these hole states are virtually identical with the first ionization states of the insulating materials, which as described above are states of primarily O $2p$ character for $U > \Delta$ as we found for CuO. The detailed character of these states is however considerably more interesting as can be concluded from Fig. 13. In Fig. 13c we have labelled the symmetry of the ionization states for the CuO$_4^{6-}$ cluster calculation and we see that the lowest ionization state is of $^1A_{1g}$ character. Upon looking at the wave function we discover that this state has one hole in a Cu $d_{x^2-y^2}$ orbital and the added hole in a combination of O $2p$ orbitals also of $x^2 - y^2$ symmetry. These holes are very strongly antiferromagnetically coupled to form a local singlet that looks much like a low spin Cu^{3+} (d^8) except that the charge distribution is quite different. From an Anderson-impurity-like calculation rather than a cluster model we have recently demonstrated [54] the stability of this local singlet state for suitable parameters and have studied the importance of the local square planar symmetry for its existence [28]. The square planar symmetry turns out to be very important since for octahedral symmetry the lowest energy state is $^3B_1$ (in D_{4h} representation) because the hybridization of $3d_{3z^2-r^2}$ with the O $2p$ states is then equal to that of the $3d_{x^2-y^2}$ orbital so that Hund's rule wins [55]. *Fujimori* [56] has investigated the role of the apex oxygens in this regard and has shown that if the O $2p$ orbital energy of the apex oxygens is 1.7 eV or more lower than that of the separate planar oxygen $2p$ orbitals the triplet state will be the lower state in spite of the larger Cu-apex oxygen distance. I do not think that such a large shift in the apex oxygen orbital energy is possible. We also have direct experimental evidence from the polarization dependent electron loss [57] and X-ray absorption [58] measurements that the oxygen hole states are in p_x and/or p_y orbitals, at least for the Bi compound.

That the lowest energy state is most likely a singlet has also been concluded by *McMahan* et al. [43] and is equivalent to the singlet state suggested by *Zhang* and *Rice* [59].

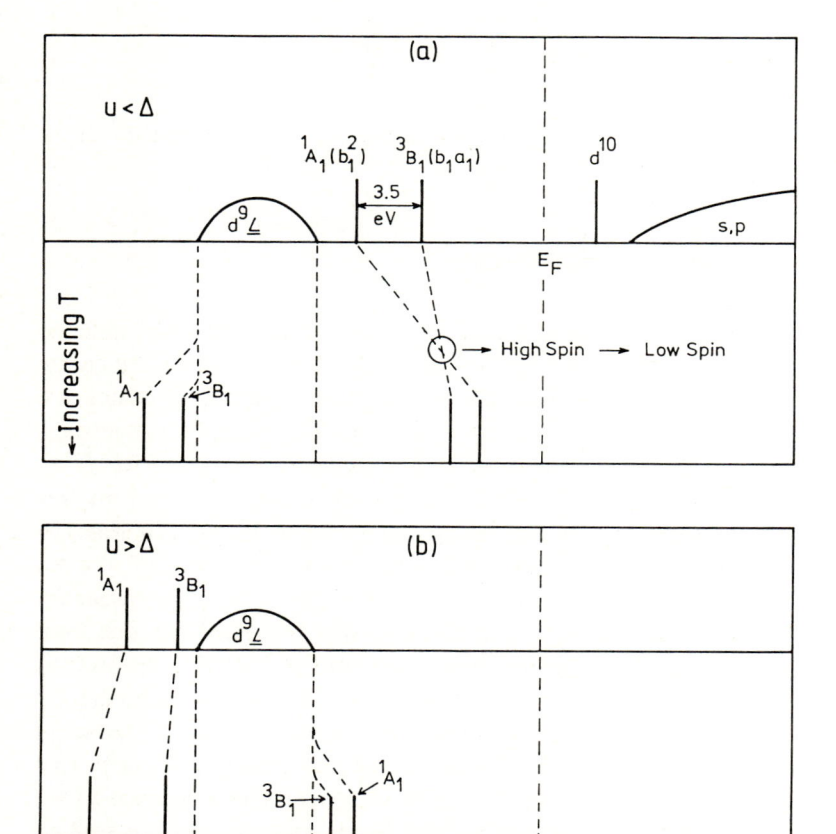

Fig. 15. An artist's concept of the shifts and changes occurring as we switch on the Cu $d - \mathrm{O}p$ transfer integral for the two situations $U < \Delta$ and $U > \Delta$

Because of the importance of this state, which if correct would form the quasiparticle states of the high T_c compounds, it is instructive to look again at the physics involved. In Fig. 15 we show the two extreme situations, $U < \Delta$ (top), which would be a Mott-Hubbard insulator, and $U < \Delta$ (below), which is a charge transfer insulator. In both cases we show two d^8 states, $^1A_1$ (both holes in $x^2 - y^2$ symmetry) and $^3B_1$ (one hole in $x^2 - y^2$ and the other in $3z^2 - y^2$) and we indicate what happens as we switch on the hybridization (T) of the d^8 states with the $d^9\underline{L}$ band states. As we have already seen in Fig. 6 for $U < \Delta$, very large T values are required before we get the crossing from high spin to low spin. For $U < \Delta$ however the first bound state pushed out of the band is a low spin singlet ($^1A_{1g}$) state for reasonable values of T. The reason for this is that the transfer integral for the $x^2 - y^2$ orbital $T(b_1) = \sqrt{3}T(a_1)$. This therefore happens only for low symmetry and as just demonstrated only for $U > \Delta$ or charge transfer gap systems. As pointed out by *McMahan* et al. [43] there is a further point which strengthens the case for such a singlet. It turns out that the O $2p$ ($\underline{L}$) states of $b_1(x^2 - y^2)$ symmetry are at the top of the oxygen band and therefore a bound state as described above costs little kinetic energy to form. This is because the ligand hole is already delocalized over 4 oxygens.

According to Fig. 13 the effective exchange interaction between such an extra ligand hole and an $x^2 - y^2$ symmetry orbital with a central Cu spin is enormously large ($^3A_{1g} - {}^1A_{1g}$ splitting is > 3 eV). This huge antiferromagnetic exchange is a good part of the energy stabilizing such a singlet state. Note that this is more than 20 times the superexchange interaction so that breaking 4 superexchange bonds in order to form such a singlet is no real problem energetically.

11. The Dispersion of the Quasiparticle State

Because of the large oxygen band width it turns out that the dispersion of these singlet states can be quite large. Recent calculations on larger clusters by *Kuramoto* et al. [60], *Fukuyama* et al. [61] and ourselves [62] have demonstrated that this dispersion is really quite large. We have recently shown from an exact model Hamiltonian calculation of the Cu_2O_7 cluster shown in Fig. 16 that the lowest energy states are very well represented by neglecting everything except the $3d_{x^2-y^2}$ and O-$2p_\sigma$ orbitals. The ground state or neutral state has 2 holes in a singlet or triplet with an energy difference due to superexchange as given in Table 6 for several parameter sets. The experimental value for La_2CuO_4 is $J = 0.12$ eV and for CuO somewhat smaller. The difference is primarily caused by the 146° band angle in CuO as compared to 180° in La_2CuO_4 together with the expected $\cos^2 \Theta$ dependence. The electron addition states are

Table 6. Effective singlet transfer integrals t (nearest neighbors) and t' (next nearest neigbors), and superexchange energies J (nearest neighbors) and J' (next nearest neighbors), calculated using parameter values listed in Table 5. All quantities in eV

	t	t'	J	J'
$U_{pd} = 0$	−0.44	0.18	0.24	0.024
$U_{pd} = 1$	−0.40	0.17	0.17	0.018

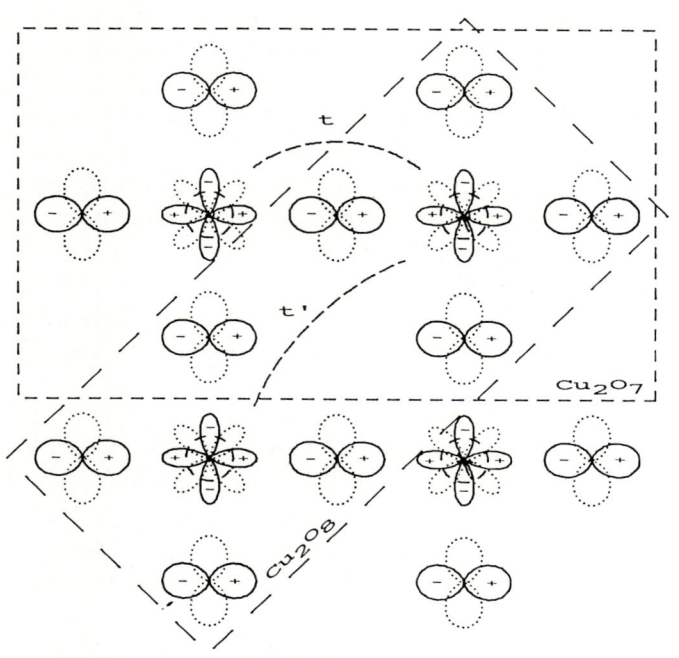

Fig. 16. The Cu_2O_7 and Cu_2O_8 clusters used in the calculations. t and t' are the effective singlet transfer integrals for nearest neighbors and next nearest neighbors respectively

one-hole states with a bonding–antibonding splitting of about 0.8 eV, resulting in a transfer integral of about $t \simeq 0.4$ eV.

The first electron removal state is a $^2A_{1g}$ state with an excited $^2B_{1u}$ state as its antibonding partner. The splitting is ≈ 0.9 eV and the states are very well represented by a bonding and antibonding combination of a state with two holes on the left in a $^1A_{1g}$ state and one hole on the right in an $x^2 - y^2$ state and the reverse with an effective transfer integral $t(^1A_{1g}) = 0.44$ eV as given in Table 6. This would cause a band width of the local singlet described by *Zhang* and *Rice* [59] and *Eskes* and *Sawatzky* [54] of $8t(^1A_{1g}) \approx 3.6$ eV.

It has been recently pointed out by *Lee* [63] that the next nearest neighbor hopping $t'(^1A_{1g})$ of the local singlet state could lead to new physics. Even for the localized (i.e. central Cu^{2+}) nearest neighbor, it is easy to see that $t'(^1A_{1g})$

can be quite large by looking at the Cu_2O_8 cluster shown also in Fig. 16. The large transfer integral is due to the T_{pp} integral connecting the four oxygens. We calculated an effective transfer integral $[t'(^1A_{1g}) \approx 0.2\,\text{eV}]$ showing that it is indeed not negligible. The fact that t and t' have opposite signs leads to a density of states which is strongly peaked at the top of the band.

Note that although the next nearest neighbor transfer integral (t') is quite large the next nearest neighbor superexchange interaction is very small although also antiferromagnetic.

12. A Model Hamiltonian

Perhaps the most important result reported here is that the lowest energy states for 3 holes in a Cu_2O_7 cluster can be well described by a bonding and antibonding combination of a state with a singlet on the left and a doublet on the right and vice versa. This demonstrates the stability of the local singlet and gives credence to identifying it as the quasiparticle state occupied upon doping. In the absence of doping we apparently have a charge transfer insulator with an extraordinarily large superexchange interaction. The Hamiltonian describing this situation is a so-called t, J-model [64], which is basically a single band Hubbard model but with extra freedom since in the Hubbard model $J = t^2/(U)$ whereas here it is an independent parameter. Within this model t is identified with our $t(^1A_{1g})$ and J with the superexchange as in the *Zhang* and *Rice* [59] description. We have found however that the next nearest neighbor transfer integral is also large so the t, t', J-model proposed by *Lee* [63] is even more realistic.

This single-band-like Hubbard model is based on the assumption that the dispersional broadening of the singlet state due to t and t' is not so large that the identity of the singlet quasiparticle cannot survive because of strong delocalization and mixing with other states. For example, if the singlet state band width is much larger than the energy by which the singlet in the impurity or cluster calculation is split off from the other band-like states one may question the validity of the impurity or cluster approach in describing the quasiparticle. Upon examining the parameters it seems that we are still in the regime in which a singlet quasiparticle and a t, t', J-model is applicable. This is evidenced by the survival of the singlet description in the Cu_2O_7 cluster as well as by the fact that the singlet in the CuO_4 cluster is about 1.5 eV below the other states whereas the dispersional half-width is also 1.5–2 eV. The top half of this quasiparticle band at least is well described by a t, t', J-model. Also as mentioned above t and t' have opposite signs and compete for the states at the top of the band, leading to narrower bands than expected at first glance. At too high dopings $x > 0.5$ this model will probably break down. The long range magnetic order or for that matter the short range order caused by J will further stabilize this singlet quasiparticle since any antiferromagnetic correlations will tend to reduce the singlet band width further. Numerical calculations by *Imada* [65] indicate that as long as $t/J < 10$ the identity of the singlet state survives and we can think of it as propagating in an antiferromagnetically coupled lattice of spins.

In conclusion then, the model Hamiltonian parameters obtained support a t, J or t, t', J single band model with $t = -0.4$ eV, $t' = 0.2$ eV and $J = 0.12$ eV. We all wonder what the low energy scale physics of this model Hamiltonian is and whether or not it will lead to high-T_c's be it via exotic flux phases or Fermi-liquid-like behavior or something resulting in a Fermi energy but a sort of smeared out Fermi surface in k space as the coupled holon spinon picture might give us.

At this point we still prefer a picture as presented earlier [55] in which the local singlet creates around it a large region (a RVB bag) which could be described by a singlet liquid because of the near degeneracy of the singlet liquid and Néel state for a 2D spin 1/2 Heisenberg system. This would lead to a new even lower energy scale Hamiltonian in which attractive coupling between the holes induced by doping is caused by the overlap of their singlet-liquid-like polarization clouds. However, at this point this is pure speculation and we will have to wait for the outcome of theoretical studies of the t, t', J-model.

Acknowledgements. This investigation was supported by the Netherlands Foundation for Chemical Research (SON) and the Foundation for Fundamental Research on Matter (FOM) with financial support from the Netherlands Organization for the Advancement of Pure Research (NWO).

References

1   N.F. Mott: *Metal Insulator Transitions* (Taylor and Francis, London 1974)
2   J.B. Goodenough: *Magnetism and the Chemical Bond* (Interscience-Wiley, New York 1963)
3   J.A. Wilson: In *The Metallic and Non-Metallic States of Matter*, ed. by P.P. Edwards, C.N.R. Rao (Taylor and Francis, London 1985), pp. 215–260
4   H.J. de Boer, E.J.W. Verwey: Proc. Phys. Soc. A **49**, 59 (1937)
5   F. Bloch: Z. Phys. **57**, 545 (1929)
      A.H. Wilson: Proc. R. Soc. London A **133**, 458 (1931)
6   N.F. Mott: Proc. Phys. Soc. A **62**, 416 (1949)
7   J. Hubbard: Proc. R. Soc. London A **277**, 237 (1964)
8   P.W. Anderson: Phys. Rev. **115**, 2 (1959); Solid State Phys. **14**, 99 (1963)
9   G. van der Laan, C. Westra, C. Haas, G.A. Sawatzky: Phys. Rev. B **23**, 4369 (1981)
10   A. Fujimori, F. Minami, S. Sugano: Phys. Rev. B **29**, 5225 (1984)
11   G.A. Sawatzky, J.W. Allen: Phys. Rev. Lett. **53**, 2239 (1984)
12   P. Hohenberg, W. Kohn: Phys. Rev. **136**, B864 (1964);
       W. Kohn, L.J. Sham: Phys. Rev. **140**, A1133 (1964)

13 A.R. Williams, U. von Barth: In *Theory of the Inhomogeneous Electron Gas*, ed. by S. Lundqvist, N.H. March (Plenum, New York 1983) p. 189
14 R. Coehoorn: Thesis, University of Groningen, The Netherlands (1985)
15 C.R. Ronda, G.J. Arends, C. Haas: Phys. Rev. B **35**, 4038 (1987)
16 K. Terakura, T. Oguchi, A.R. Williams, J. Kuebler: Phys. Rev. B **30**, 4734 (1984); O.K. Andersen, H.L. Skriver, H. Nohl, B. Johansson: Pure Appl. Chem. **52**, 93 (1979)
17 J. Ghijsen, L.H. Tjeng, J. van Elp, H. Eskes, J. Westerink, G.A. Sawatzky, M.T. Czyzyk: Phys. Rev. B **38**, 11322 (1988)
18 For a review see W.E. Pickett: Rev. Mod. Phys. **61**, 433 (1989)
19 J. Zaanen, G.A. Sawatzky, J.W. Allen: Phys. Rev. Lett. **55**, 65 (1985)
20 J. Zaanen, G.A. Sawatzky, J.W. Allen: J. Magn. Magn. Mater. **54–57**, 607 (1986)
21 J.S. Griffith: *The Theory of Transition Metal Ions* (Cambridge University Press, Cambridge 1961)
22 C. Herring: In *Magnetism*, Vol. IV, ed. by T. Rado, H. Suhl (Academic, New York 1964)
23 J.C.W. Folmer, F. Jellinek: J. Less-Common. Met. **76**, 153 (1980)
24 D.K.G. de Boer: Thesis, University of Groningen (1983) D.K.G. de Boer, C. Haas, G.A. Sawatzky: Phys. Rev. B **29**, 4401 (1984)
25 J. Zaanen: Thesis, University of Groningen (1986)
26 C.E. Moore: "Atomic Energy Levels" (NBS circular No. 467 U.S. GPO, Washington, DC, 1958 Vols. 1–3)
27 D. v.d. Marel, G.A. Sawatzky: Phys. Rev. B **37**, 10674 (1988) and references therein; D. v.d. Marel: Thesis, University of Groningen (1985)
28 H. Eskes, L.H. Tjeng, G.A. Sawatzky: To be published
29 J.C. Slater: *Quantum Theory of Atomic Structure* (McGraw-Hill, New York 1960)
30 Y. Tanabe, S. Sugano: J. Phys. Soc. Jpn. **9**, 753 (1954)
31 E. König, I. Kremer: *Ligand Field Energy Diagrams* (Plenum, New York 1977)
32 O. Gunnarsson, K. Schönhammer: Phys. Rev. B **28**, 4315 (1983); ibid B **31**, 4815 (1985)
33 J. Zaanen, G.A. Sawatzky: Phys. Rev. B, in press
34 J. Zaanen, G.A. Sawatzky: Can. J. Phys. **65**, 1262 (1987)
35 L.F. Mattheis: Phys. Rev. B **5**, 290 (1972)
36 H.P. Geserich, G. Scheiber, J. Geerk, H.C. Li, G. Linker, W. Assmus, W. Weber: Europhys. Lett. **6**, 277 (1988)
37 J.G. Bednorz, K.A. Müller: Z. Phys. B **64**, 189 (1986)
38 W.E. Pickett: Rev. Mod. Phys. **61**, 433 (1989)
39 K.C. Hass: In *Solid State Physics*, Vol. 42, ed. by H. Ehrenreich, D. Turnbull (Academic, New York 1989) p. 213
40 G. Wendin: J. de Phys. C9, 1157 (1987)
41 J.C. Fuggle, P.J.W. Weijs, R. Schoorl, G.A. Sawatzky, J. Fink, N. Nücker, P.J. Durham, W.M. Temmerman: Phys. Rev. B **37**, 123 (1988)
42 J.C. Fuggle: In Proc. Adriatico Research Conf. and Workshop on "Towards the Theoretical Understanding of High T_c Superconductors", ICTP Trieste (World Scientific, Singapore 1988)
43 A.K. McMahan, R.M. Martin, S. Satpathy: Phys. Rev. B **38**, 6650 (1988)
44 M.S. Hybertsen, M. Schlüter, N.E. Christensen: Phys. Rev. B **39**, 9028 (1989)
45 G.J.M. Janssen, W.C. Nieuwpoort: Phys. Rev. B **38**, 3449 (1988)
46 P.W. Baumeister: Phys. Rev. **121**, 359 (1961)
47 J. Kanamori: Prog. Theor. Phys. **30**, 275 (1963)
48 M. Cini: Solid State Commun. **24**, 681 (1971); G.A. Sawatzky: Phys. Rev. Lett. **39**, 504 (1977)
49 M.R. Thuler, R.L. Benbow, Z. Hurych: Phys. Rev. B **26**, 669 (1982)
50 J. Ghijsen et al.: Unpublished

51 P. Kuiper, G. Kruizinga, J. Ghijsen, G.A. Sawatzky, H. Verweij: Phys. Rev. Lett. **62**, 221 (1989)
52 N. Nücker, H. Romberg, X.X. Xi, J. Fink, P.S. Durham, W.M. Temmerman: Phys. Rev. B, in press
53 J.A. Yarmoff, D.R. Clarke, W. Drube, U.O. Karlsson, A. Talab-Ibrahimi, F.J. Himpsel: Phys. Rev. B **36**, 3967 (1987);
    P. Kuiper, G. Kruizinga, J. Ghijsen, M. Grioni, P.J.W. Weijs, F.M.F. de Groot, G.A. Sawatzky, H. Verweij, L.F. Feiner, H. Petersen: Phys. Rev. B **38**, 6483 (1988)
54 H. Eskes, G.A. Sawatzky: Phys. Rev. Lett. **61**, 1415 (1988)
55 H. Eskes, H. Tjeng, G.A. Sawatzky: In *Mechanisms of High Temperature Superconductivity*, ed. by H. Kamimura, A. Oshiyama (Springer, Berlin, Heidelberg 1989)p. 20
56 A. Fujimori: Phys. Rev. B, in press
57 N. Nücker, H. Romberg, X.X. Xi, J. Fink, B. Gegenheimer, Z.X. Zhao: Preprint
58 P. Kuiper et al.: To be published
59 F.C. Zhang, T.M. Rice: Phys. Rev. B **37**, 3759 (1988)
60 Y. Kuramoto et al.: In Proc. IBM Japan Int. Symp. on Strong Correlation and Superconductivity, 21–25 May 1989, Mt. Fuji, Japan
61 H. Fukuyama et al.: In Proc. IBM Japan Int. Symp. on Strong Correlation and Superconductivity, 21–25 May 1989, Mt. Fuji, Japan
62 H. Eskes, G.A. Sawatzky, L.F. Feiner: Preprint
63 P.A. Lee: In Proc. IBM Japan Int. Symp. on Strong Correlation and Superconductivity, 21–25 May 1989, Mt. Fuji, Japan
64 P.W. Anderson: Science **235**, 1196 (1987)
65 M. Imada: In Proc. IBM Japan Int. Symp. on Strong Correlation and Superconductivity, 21–25 May 1989, Mt. Fuji, Japan

The Electronic Structure of Previous and Present High-T_c Superconductors – Investigations with High-Energy Spectroscopies

J. Fink[1], *J. Pflüger*[1,*], *Th. Müller-Heinzerling*[1,**], *N. Nücker*[1], *B. Scheerer*[1], *H. Romberg*[1], *M. Alexander*[1], *R. Manzke*[2], *T. Buslaps*[2], *R. Claessen*[2], *and M. Skibowski*[2]

[1]Kernforschungszentrum Karlsruhe, Institut für Nukleare Festkörperphysik,
  Postfach 3640, D-7500 Karlsruhe, Fed. Rep. of Germany
[2]Institut für Experimentalphysik, Universität Kiel,
  Olshausenstr. 40–60, D-2300 Kiel, Fed. Rep. of Germany
* Present address: HASYLAB-DESY, Hamburg, Fed. Rep. of Germany
** Present address: Siemens AG, Karlsruhe, Fed. Rep. of Germany

Abstract. Electronic structure studies of previous and present high-T_c superconductors are reviewed. In particular, transition metal carbides and nitrides, A15 compounds and the cuprate superconductors are treated. We emphasize investigations performed by high-energy spectroscopies mainly by electron energy-loss spectroscopy, by photoemission spectroscopy, and inverse photoemission spectroscopy.

1. Introduction

Any microscopic understanding of the mechanism of superconductivity requires detailed knowledge of the electronic structure in the normal state. Within the conventional BCS theory for superconductivity, the superconducting transition temperature T_c is given by

$$kT_c = 1.14 \hbar \omega_c e^{-\frac{1}{\lambda}} \quad (1)$$

where λ is the electron-phonon coupling strength which is given by the product $\lambda = N(E_F) \cdot V$ of the single spin density-of-the states at the Fermi surface $N(E_F)$ with the pairing potential V. ω_C is the cut-off frequency of the phonon spectrum which is of the order of the Debye frequency. Within the Eliashberg theory which also explains the strong coupling superconductors, the electron phonon coupling constant is given by

$$\lambda = 2 \int_0^\infty \left(\alpha^2 \cdot F(\omega)/\omega \right) d\omega \quad (2)$$

where the Eliashberg function $\alpha^2 \cdot F(\omega)$ is given by

$$\alpha^2 \cdot F(\omega) = \sum_{kk'} \left| G_{kk'} \right|^2 \delta(E_k - E_F)\delta(E_{k'} - E_F)\delta(\omega - \omega_{k-k'})/N(E_F). \quad (3)$$

Here E_k is the energy of an electron of momentum k, $\omega_{k-k'}$ a phonon frequency belonging to momentum k-k', and $G_{kk'}$ the electron phonon coupling constant [1]. Since the Eliashberg function represents a Fermi-surface average over electron-phonon coupling, again the density-of-states at E_F comes into play. A crucial point for the validity of Eliashberg's theory is the applicability of Migdal's theorem, which implies a slowly varying electronic density-of-states around E_F on the scale of typical phonon frequencies $\hbar\omega_{Debye}$. So spectroscopic investigations of the electronic states around E_F may not only serve as an experimental check of the electronic data in $\alpha^2F(\omega)$, but also tell us whether $N(E)$ is indeed slowly varying around E_F, and thus whether Eliashberg's theory is applicable at all.

As mentioned above, the density-of-states at the Fermi level is an important parameter in electron-phonon coupling based theories for the superconducting transition temperatures. In Fig. 1 we show T_c of various superconductors versus the specific heat constant γ which is proportional

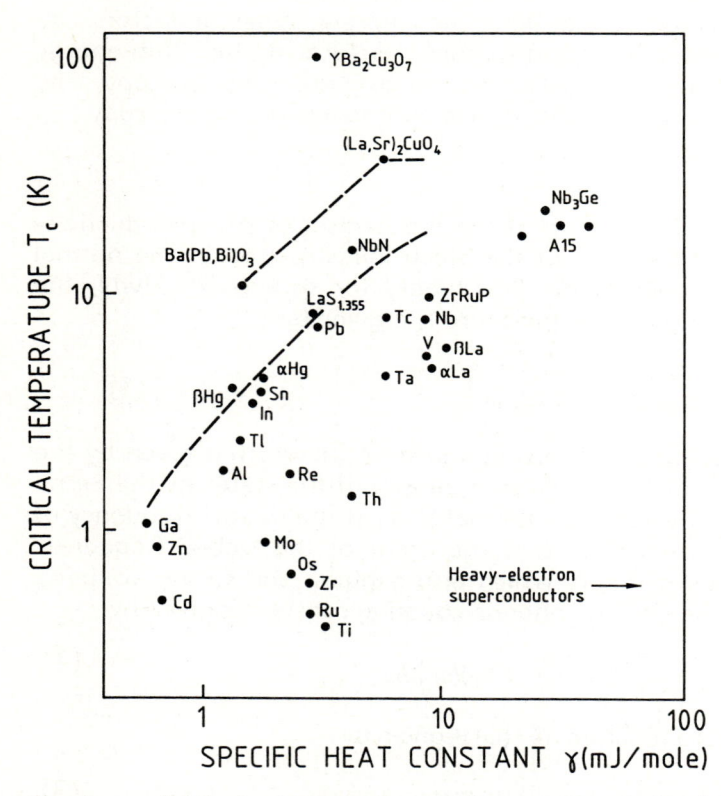

Fig. 1 Critical temperature for superconductivity versus the electronic specific heat constant γ, which is proportional to the density-of-states at the Fermi level.

to $N(E_F)$. According to these data, the various superconducting compounds may be classified as follows:
1. Normal superconductors where $N(E)$ is roughly constant over the energy range of phonon frequencies. Examples are Al, Pb, Nb, and NbN.
2. In the A15 compounds there is a particular high and strongly varying density-of-states at E_F leading to a strong electron-phonon coupling and to the highest T_c values of the previous high-T_c superconductors.
3. In some superconductors, such as V, $N(E_F)$ is so high that the Stoner condition for magnetism is almost satisfied. Then, T_c is depressed by spin fluctuations.
4. Heavy fermion superconductors have a very high specific heat constant but low critical temperatures for T_c. This is mediated by a Kondo-like weak hybridization of f electrons with spd conduction electrons.
5. The oxide superconductors, in particular the cuprates are well separated from the conventional superconductors. This suggests that the pairing interaction in the oxides is probably not mediated by phonons but by a different mechanism. On the other hand, we should realize that the CuO_2 planes responsible for superconductivity in the cuprates are rather diluted by ionic layers. Therefore, the experimental points should be shifted to higher γ values (about a factor of three). However, there still remains the clear separation between the previous and the new high-T_c superconductors.

In this contribution, we review electronic structure studies by high-energy spectroscopies of previous and present high-temperature superconductors. We first report on refractory materials such as transition-metal carbides and nitrides which can be described by a conventional Eliashberg theory. Then we report on A15 compounds, which were predicted to have a strongly peaked density-of-states close to the Fermi level. Finally we review recent studies of the electronic structure of the cuprate superconductors.

2. Refractory compounds

Refractory compounds such as the carbides and nitrides of the early 3d, 4d and 5d transition metals (TM), exhibit a number of exceptional physical properties. Among these are the high superconducting transition temperatures T_c which put them before 1986 into the class of high-T_c superconductors. The highest $T_c = 18$ K was observed for $NbC_{0.3}N_{0.7}$. The high T_c values, the high melting points and the great hardness are closely related to their electronic structure.

The formation of bands is shown in Fig. 2a. We start with a 3d level of the TM and a 2p level of the non-metal (NM). Due to the electric field gradient of the NM atoms, we have a crystal field splitting of the 3d level

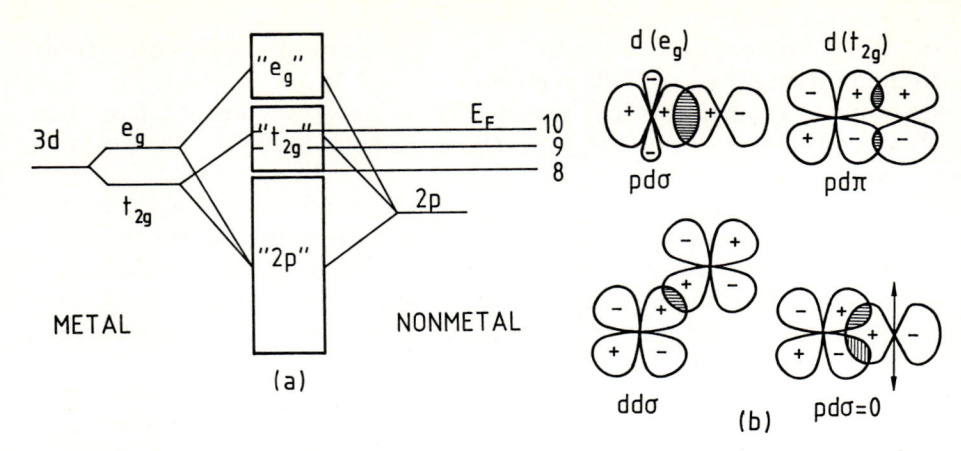

Fig. 2 (a): Schematic band structure of transition-metal carbides, nitrides and oxides. The position of the Fermi energy for 8, 9 and 10 valence electrons is indicated, corresponding to Ti(Zr)C, V(Nb)C and V(Nb)N. (b): Important bondings between transition metal 3d orbitals and non-metal 2p orbitals in refractory compounds (from W. Weber [4]).

into e_g and t_{2g} states. In Fig. 2b relevant bondings of the TM 3d orbitals and NM 2p orbitals are illustrated. The hardness and the high melting points of the refractory compounds can be understood by the strong hybridization between NM 2p and TM 3d orbitals. In the carbides, the energy difference between 3d levels and 2p levels is small. Due to the strong pdσ bonding, the antibonding "e_g" band is at the highest energy. The weaker antibonding pdπ interaction is partially compensated by the bonding ddσ coupling. Thus the "t_{2g}" band may be called a nonbonding band. We emphasize that there is no σ-type pd(t_{2g}) coupling possible (see Fig. 2b). The valence band ("2p" band) and the two conduction bands ("t_{2g}" band and "e_g" bands) have both 3d and 2p character. With increasing NM electronegativity, i.e., when going from the carbides via the nitrides to the oxides, the separation of the orbital energies E_d and E_p becomes larger and the p-d hybridization is less effective. Therefore, the valence band and the conduction band exhibits more and more pure NM 2p and TM 3d character, respectively [2]. In general, the TM 3d bands are well separated from the NM 2p bands. In the carbides and nitrides, there is an intermediate region of very low electronic density-of-states as the bands overlap to some extent. The compounds with 8 valence electrons (TiC,ZrC) have their Fermi energy in this low density region. This is probably the reason, why these compounds do not show superconductivity. In the compounds having 9 (TiN,VC,ZrN,NbC) and 10 (VN,NbN) valence electrons, the Fermi level is pushed more and more in

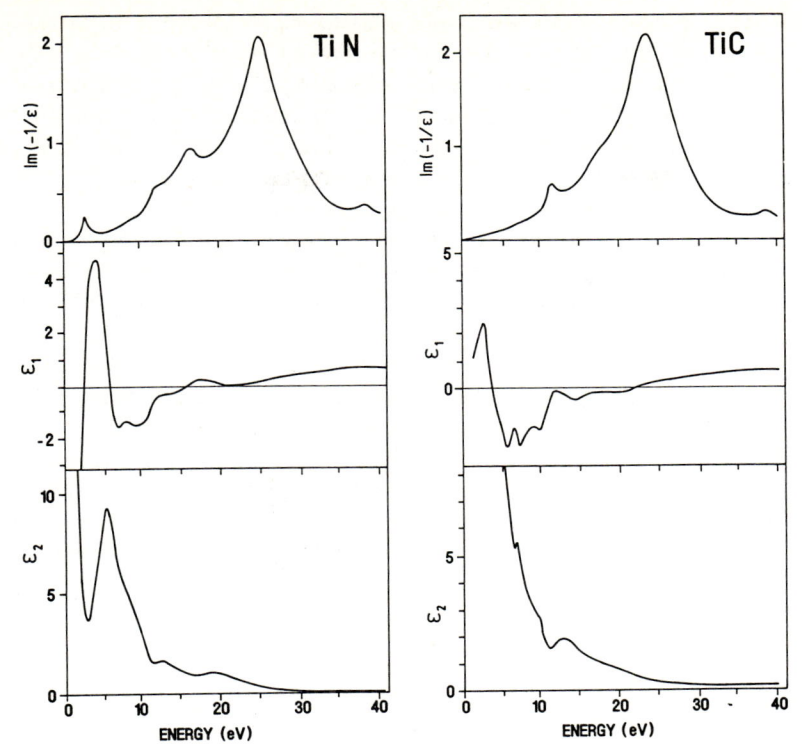

Fig. 3   Loss functions and Kramers-Kronig-derived real and imaginary parts of the dielectric functions ε_1 and ε_2 for the non-superconducting compound TiC and the superconducting compound TiN ($T_c = 5.5$ K).

the "t_{2g}" band having predominantly TM 3d character. Thus the density-of-states at E_F increases and there is a tendency of increasing T_c.

An extensive study of the electronic structure of these refractory compounds by electron energy-loss spectroscopy was performed by Pflüger et al. [3]. In Fig. 3 typical loss functions $Im(-1/\varepsilon(\omega,q))$ for small momentum transfer q compared to the extension of the Brillouin-zone are shown for TiN and TiC. The loss functions of both compounds are dominated by maxima at 23.5 (TiC) and 24.9 eV (TiN) due to a plasmon of all valence electrons. Since there is one more electron in TiN than in TiC, the plasmon energy for the former compound is higher. In the superconducting compound TiN there is a sharp maximum at 2.8 eV due to a collective excitation of the electrons in the partially filled "t_{2g}" band. This Drude part is also realized in the dielectric function, the real (ε_1) and the imaginary part (ε_2) of which are also shown in Fig. 3. The Drude-like decay of ε_2 is followed by interband transitions between the "2p" band and the partially filled "t_{2g}" band. In the non-superconducting compound TiC where E_F is situated in the very minimum between "2p" and "t_{2g}" band, the Drude part is small and there are strong $2p \rightarrow t_{2g}$ interband transitions which lead to a strong damping of the remaining Drude plasmon. Therefore, no maximum in the loss function is observed in this energy range. It should be remarked that a slightly different explanation of the results was given previously [3]. We emphasize that in all superconducting refractory compounds, the collective excitations of the charge carriers (predominantly TM 3d t_{2g} electrons) could be observed. In the 3d and 4d compounds of the third row there is, however, in some cases a considerable broadening of these plasmons.

Further information on the electronic structure of these transition metal compounds could be obtained by measuring transitions from core levels into unoccupied states. In the simplest approximation, the local density of unoccupied states can be studied. Since for small momentum transfer, dipole selection rules apply, the local p density of unoccupied states at the NM atoms is measured when exciting NM 1s electrons. In Fig. 4 we show typical 1s absorption edges of 3d carbides and nitrides

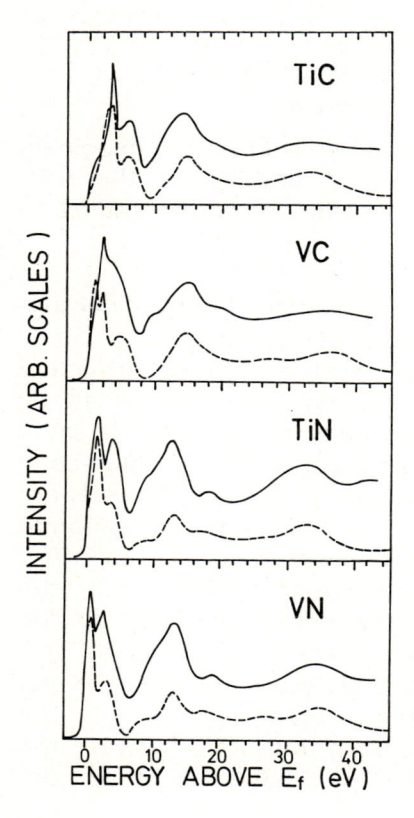

Fig. 4 Non-metal absorption edges of the K-shell of TiC, TiN, VC and VN (solid line). For comparison, calculations of unoccupied total density-of-states are shown (dashed line).

together with the calculated total unoccupied density-of-states. The calculated curves were broadened by the energy resolution and by an energy-dependent life-time broadening. There is a remarkable agreement in peak positions of the measured and the calculated curves. The intensities are almost correctly reproduced in the calculations, although matrix element effects providing the *local* density-of-states were neglected. As predicted by the band-structure calculations, the Fermi level corresponding to the absorption edges for TiC is in a region of low density-of-states while for TiN it is pushed into a region of high density-of-states in the "t_{2g}" band.

In all these spectra the first two peaks correspond to 3d-t_{2g} and 3d-e_g states hybridized with NM 2p states. Thus the crystal-field splitting of the 3d states can be clearly realized. The edges for VC and VN also indicate an almost rigid band like shift of E_F into the "t_{2g}" band with increasing number of valence electrons. For the superconducting properties it is important to note that in all superconducting carbides and nitrides we have a high density-of-states at the Fermi level. Within our experimental resolution of 0.2 eV no strongly varying density-of-states at the Fermi level could be detected. Thus the normal Eliashberg theory is applicable. The high density-of-states at E_F is the basis for a high superconducting transition temperature. Nevertheless, this fact alone does not lead to the high T_c observed e.g. for NbN. According to Weber [4] there is in addition a strong electron-phonon coupling due to a σ-type coupling of p and d(t_{2g}) orbitals which is at equilibrium position zero (see Fig. 2a). However, when the atoms are displaced from their equilibrium positions as indicated in Fig. 2b, the pdσ interaction leads to a strong electron-phonon interaction. To obtain high T_c it is therefore important to have the Fermi level in a band of t_{2g} character hybridized with NM p states. That this is the case can be directly seen in the NM 1s absorption edges in Fig. 4.

In Fig. 4 we see that the peak positions in absorption edges are remarkably well reproduced by the band structure calculations, although matrix elements are neglected. This is also true for peaks far above E_F and for the NM 1s absorption edges of the 4d compounds [3]. This indicates that core excitonic effects due to the remaining core hole are not important in these measurements. On the other hand, we show in Fig. 5 absorption edges from the 3d TM 2p level [5] which in principle probe the empty 3d states. It is well known [6] that for the 3d TM 2p absorption edges the core excitonic effects are strong. Therefore, the spectra no longer represent the local 3d unoccupied density-of-states. In principle, in all four spectra, the crystal field splitting of the 3d(e_g) and 3d(t_{2g}) states should be observed while the spectra only show the splitting in the carbides. In the nitrides, there is just one line for the 2$p_{3/2}$ (L_3) and the 2$p_{1/2}$ (L_2) excitations. A possible explanation for this

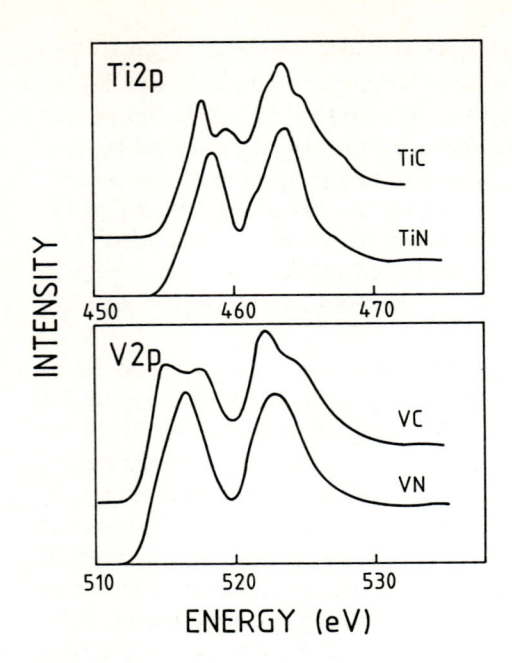

Fig. 5 $2p_{3/2}$ (L_3) and $2p_{1/2}$ (L_2) absorption edges of Ti and V in transition metal carbides and nitrides.

observation may be that in the nitrides, the TM-NM hybridization is weaker and therefore the screening of the core hole is weaker.

3. A15 compounds

By 1986, intermetallic compounds with A15 structure were the materials showing the highest superconducting transition temperatures (Nb$_3$Ge: $T_c = 23K$). Besides the high T_c values, other anomalies such as temperature-dependent Knight shifts and Pauli susceptibilities, softening of phonons, and structural phase transitions at low temperatures were observed (for reviews see Testardi [7] and Müller [8]). These unusual properties are again closely related to the electronic structure of these compounds. Already in 1961, Clogston and Jaccarino suggested that the anomalous properties of the A15 materials were caused by an extremely fine structure in the density-of-states near E_F. This feature was first explained on the basis of one-dimensional nonintersecting chains of A atoms along the three coordination axes in the A$_3$B structure. Later on, accurate first-principles band-structure calculations [9-11] indeed predicted narrow maxima in the density-of-states at E_F due to two flat bands near E_F which originate from a doubly degenerate Γ_{12} state at the Brillouin zone center. These two bands lead to a region of high density-of-states around E_F, approximately 0.5 eV wide. The two maxima itself have a width of about 50 meV. The Γ_{12} bands are made up predominantly of A atom orbitals (in particular

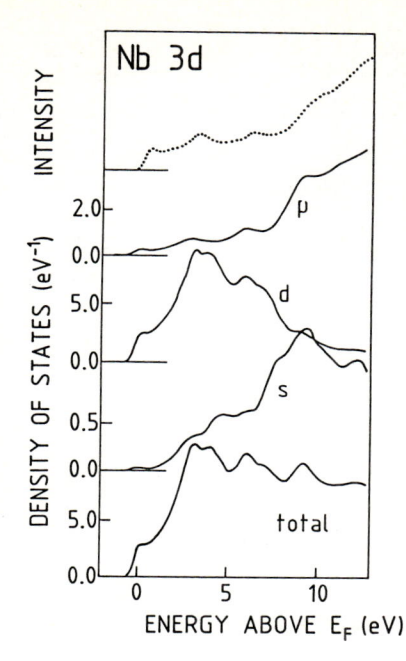

Fig. 6 Upper part: Nb 3d absorption edges of Nb$_3$Sn after background subtraction. Lower part: Calculated partial and total local density of unoccupied states at the Nb atoms broadened by the energy resolution of 0.6 eV.

d$_{3z^2-r^2}$) directed along the chains. Other orbitals such as d$_{xz,yz}$ and d$_{x^2-y^2}$ contribute to a lesser extent. The principal source of the Γ$_{12}$ band flattening is the *inter*chain interaction of d$_{3z^2-r^2}$ orbitals and d$_{xz,yz}$ orbitals. Thus the sharp structure in N(E) near E$_F$ arising from the flat Γ$_{12}$ bands is not at all an effect of one-dimensional energy bands (see Weber [4]).

The electronic structure of the unoccupied states close to the Fermi level has been studied by EELS using core level excitations [12,13]. Of particular importance were the Nb 3d absorption edges where the inherent broadening due to the lifetime of the core hole is only about 0.1 eV. Unfortunately the signal-to-background ratio is only 1 to 100, and therefore long counting times are needed to obtain good statistics. In Fig. 6 we show a representative spectrum taken on stoichiometric Nb$_3$Sn with an energy resolution of 0.6 eV. The spectrum is composed of the 3d$_{5/2}$ and 3d$_{3/2}$ absorption edges separated by a spin-orbit splitting of 3.06 eV. In the lower part, density of unoccupied states with different symmetries around the Nb atoms are plotted. These curves were obtained from tight-binding fits to self-consistent augmented-plane-wave calculations by adding a replica shifted by the spin-orbit splitting and weighted by the statistical degeneracy of the core level. After this, the curves were broadened by the experimental resolution and by lifetime broadening. Since these measurements were performed at low momentum transfer, dipole selection rules should apply and therefore the local p density-of-states should be probed. The 4f density of states at

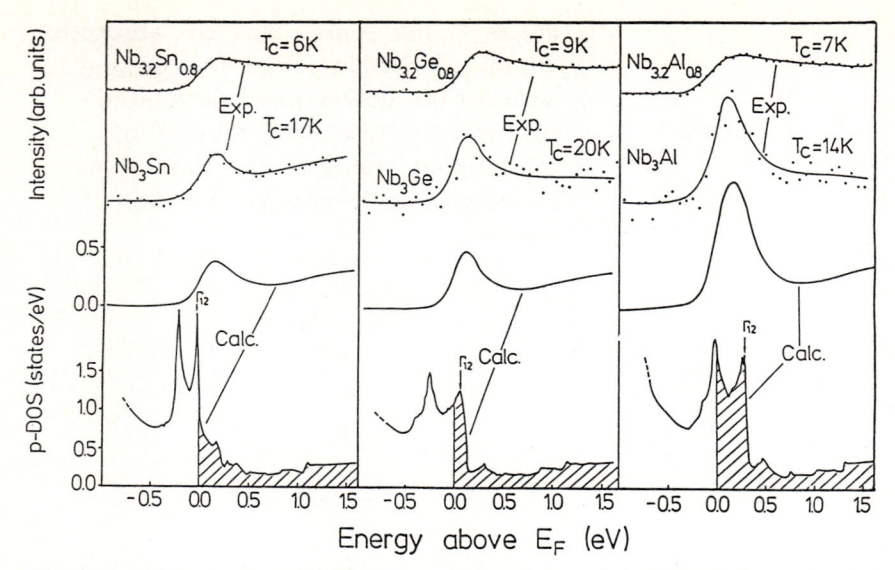

Fig. 7   Upper part: Nb $3d_{5/2}$ absorption edges in Nb_3Sn, Nb_3Ge, and Nb_3Al after background substraction measured with a resolution of 0.2 eV on off-stoichiometric low-T_c and near-stoichiometric high-T_c samples. The lines through the data are a guide to the eye. All curves were normalized to the spectral weight at 1 eV above E_F. Lower part: Calculated broadened density of unoccupied states and unbroadened density-of-states for the three compounds. Also indicated is the position of the Γ_{12} state relative to E_F.

the Fermi level should be small. Indeed, there is a rather good agreement of the measured absorption edges and the calculated spectrum with p symmetry.

The peak near E_F was measured with an improved spectrometer resolution of 0.2 eV for near-stoichiometric high-T_c and off-stoichiometric low-T_c samples of Nb_3Sn, Nb_3Ge and Nb_3Al. These spectra are shown in Fig. 7 together with calculations of the density-of-states and of absorption edges. The high-T_c samples indicate clearly a peaked density-of-state near E_F in agreement with the calculated spectra. Although our energy resolution is considerably weaker than the width, even significant differences between the three compounds can be realized. There is a trend of increasing peak heights when going from Nb_3Sn to Nb_3Ge and in particular to Nb_3Al which originates from a shift of the Fermi level from the top part of the high density-of-states region into the region between the two maxima. This leaves more and more of the high density-of-states region unoccupied and therefore contributes to the peak in the absorption edge. The off-stoichiometric low-T_c

samples exhibit much less pronounced peaks than the corresponding high-T_c samples. This is in agreement with the trend in the specific heat data. It seems as if the density-of-states peak is not just smeared out due to the enhanced impurity scattering, but as if part of the spectral weight has been shifted to impurity states away from E_F.

The high-resolution absorption edges of the A15 compounds verified the predicted high and strongly varying density-of-states at the Fermi level which is the basis for a strong electron-phonon coupling in these materials. These measurements also clearly showed that these superconductors can no more be treated in the framework of a normal Eliashberg theory, since the assumption of a constant density-of-states around E_F on a scale of optical phonon frequencies is no more fulfilled.

4. Cuprates

As shown in the Introduction the new high-T_c cuprate superconductors have a low density-of-states at the Fermi level as measured by the specific heat constant γ. At present there is a strong discussion whether these compounds are Fermi liquid systems at all. Therefore, an important prerequisite for the understanding of the mechanism of the superconductivity in these systems, namely the electronic structure, is rather unclear at present.

In the cuprate superconductors the important layers for superconductivitiy are probably the CuO_2 layers where a strong covalent bonding between Cu and O atoms is expected. The layers between CuO_2 sheets exhibit in most cases an ionic bonding, i.e., their electrons are energetically far away from the Fermi level. In Fig. 8 we illustrate the electronic structure of the two-dimensional CuO_2 layer for the example La_2CuO_4. We start with an ionic ansatz for Cu^{++} with a $3d^9$ configuration (one hole in the Cu 3d shell) and for O^{--} with a $2p^6$ configuration (completely filled O 2p shell). Due to the crystal field of a regular octahedron of O ions, the 3d level is split into e_g and t_{2g} states. In the crystal field of the distorted O octahedron there is a further splitting of the e_g states into $3d_{x^2-y^2}$ and $3d_{3z^2-r^2}$ states. The hybridization with the O 2p states is strongest for the $3d_{x^2-y^2}$ states in the CuO_2 plane. Therefore, a bonding and an antibonding $dp\sigma$ band is formed at the bottom and the top of all Cu-O bands, respectively. Filling the bands with electrons (La_2CuO_4 has an odd number of electrons), the $dp\sigma^*$ band is half filled, i.e. La_2CuO_4 should be a metal. A paramagnetic metal is also obtained by typical band-structure calculations [14] using the local density approximation (LDA) while in reality La_2CuO_4 is an antiferromagnetic insulator. The reason for this discrepancy is that the correlation effects on the Cu sites are not correctly taken into account by LDA band-structure calculations. It is well known that correlation effects lead to

$$La_2CuO_4$$

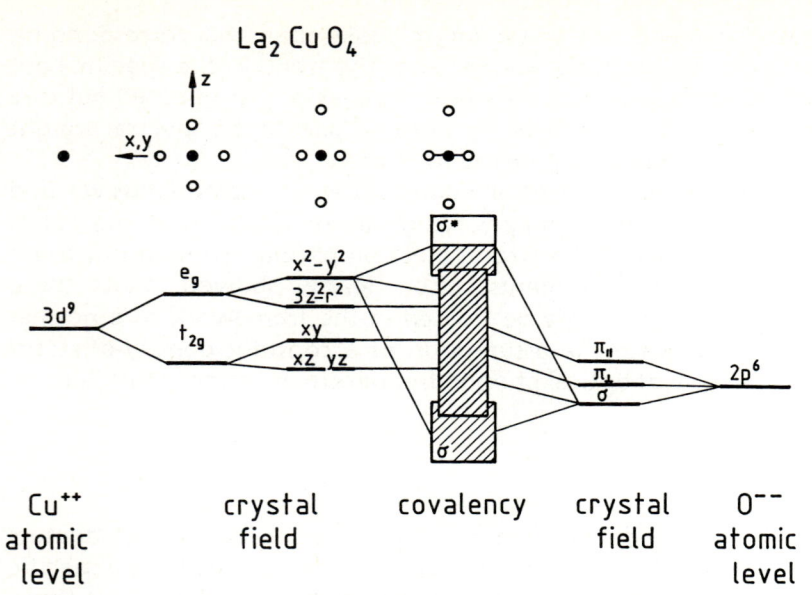

Fig. 8 The formation of electronic structure of CuO_2 layers in La_2CuO_4. Left side: Crystal field splitting of a Cu $3d^9$ level (Cu^{++}) in a regular and a distorted octahedron of O ions. Right side: Crystal field splitting of an almost filled O 2p state (O^{--}). Middle: Bands are formed by hybridization. In particular, the bonding and antibonding $dp\sigma$ bands are illustrated.

insulating properties of transition metal compounds even when they have an odd number of electrons per unit cell. A model for electronic structure of the transition metal compounds [15] was worked out before the discovery of the new high-T_c superconductors [16]. This model is based on two important parameters, namely the on-site Coulomb energy U on the Cu atoms and the charge-transfer energy Δ between Cu and O sites. For the late transition metal compounds it is generally assumed that U is larger than Δ and therefore these systems are charge-transfer insulators as illustrated in Fig. 9. The gap occurs between a valence band having predominantly O 2p character and a conduction band having predominantly Cu 3d character. The latter band is formally the Cu 3d upper Hubbard band. The lower Hubbard band should be well below the Fermi level and the O 2p band. The amount of hybridization of the O 2p and the Cu 3d band is at present controversial as is the exact value of U and in particular the value of Δ. Evaluation of high-energy spectroscopy data [17] yield $U \approx 6$-8 eV and $\Delta \approx 0.5$-3.5 eV. The CuO_2 planes can be p-type doped, i.e. electrons are removed from the CuO_2 planes, and holes will be formed having predominantly O 2p character. According to this model, upon n-type doping electrons will have predominantly Cu 3d

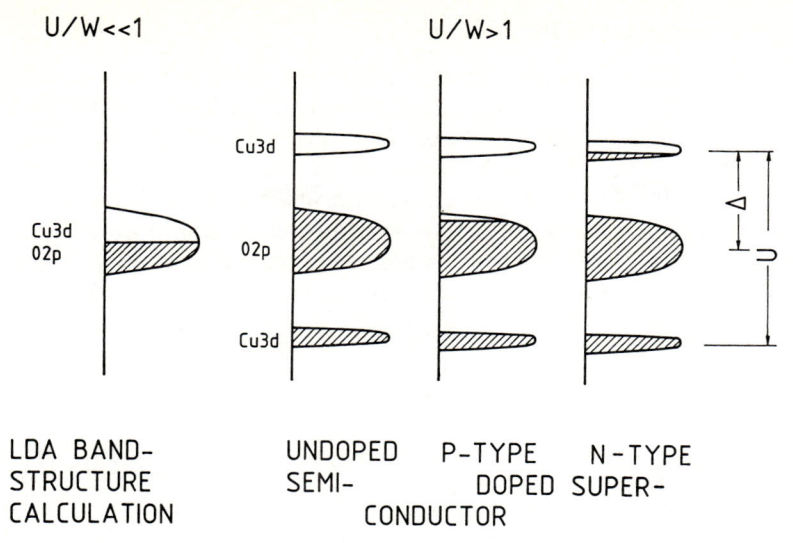

Fig. 9 Half filled Cu-O $3d_{x^2-y^2}$-$2p_{x,y}\sigma^*$ band derived from LDA band-structure calculations. When correlation effects on Cu sites are important, and the correlation energy U is larger than the charge-transfer energy Δ, a charge transfer insulator is formed. p-type and n-type doping leads to the formation of holes on O and electrons on Cu sites, respectively.

character. In a local picture there will be a formation of monovalent Cu. p-type doping is achieved in La_2CuO_4 by replacing in the LaO layers trivalent La by divalent Sr, in $YBa_2Cu_3O_7$ by CuO_3 chains, in $Bi_2Sr_2CaCu_2O_8$ by BiO layers, and in $Tl_2Ba_2Ca_2Cu_3O_{10}$ by TlO layers. n-type doping is achieved in Nd_2CuO_4 by replacing [18] trivalent Nd by tetravalent Ce or Th, or by replacing [19] divalent O by monovalent F. The model for the changes of the electronic structure upon doping as presented above is still strongly discussed. It is at present not clear whether in the doped system, a state is obtained which is again very close to that calculated in the LDA band-structure calculations or whether these compounds can no more be described [20] by Fermi liquids. Thus the importance of correlation effects in the doped systems is strongly discussed at present.

In the following we review investigations of the electronic structure of the cuprate superconductors by high-energy spectroscopies such as electron energy-loss spectroscopy in transmission, angular resolved photoemission and angular resolved inverse photoemission.

4.1 Plasmons and Interband Transitions

In Fig. 10a we show a typical loss function $\text{Im}(-1/\varepsilon(q,\omega))$ for a single crystal $Bi_2Sr_2CaCu_2O_8$ measured by EELS in transmission with electrons having a primary energy of 170 keV. The momentum transfer was $q = 0.1\ \text{Å}^{-1}$ which is small in comparison with the extension of the Brillouin zone and the direction was chosen parallel to the CuO_2 planes. The loss function, measuring collective excitations, is dominated by a broad feature around 20 eV which is caused by a plasmon of all valence electrons. Superimposed on this plasmon are low-lying core-level excitations and interband transitions. Details are described elsewhere

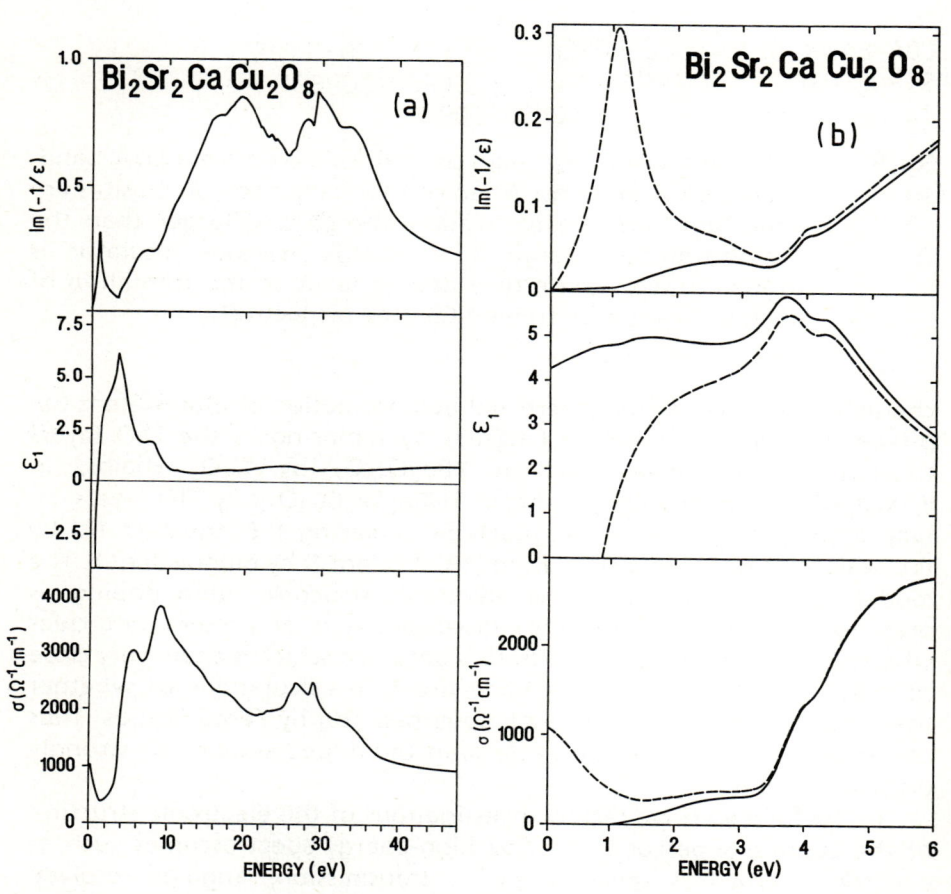

Fig. 10 (a) Loss function $\text{Im}(-1/\varepsilon)$, ε_1 and σ for $Bi_2Sr_2CaCu_2O_8$ measured with momentum transfer $q = 0.1\ \text{Å}^{-1}$ parallel to the CuO_2 plane. (b) Low energy part of (a) (dashed line). Solid line: "Drude" part removed.

[21]. At 1 eV, a plasmon of the charge carriers, probably holes on O, is observed. This plasmon is shown in more detail in Fig. 10b. It has a width of 0.7 eV which is probably predominantly caused by low-lying excitations and only partially by the finite relaxation time of the charge carriers. The low-lying excitations starting at about 50 meV were observed in optical experiments [22] but cannot be resolved in the present EELS experiment with an energy resolution of 0.15 eV. In Fig. 10 we also show the real part of the dielectric function ε_1 and the optical conductivity σ. In the latter function, a Drude-like decay is observed at energies below 1.5 eV. Removing the Drude-like part (low energy excitations are included), we obtain the solid line in Fig. 10b. In ε_1 a background dielectric function at the plasmon energy is derived to be $\varepsilon_\infty \approx 4.8$. A rise in the loss function and in the optical conductivity is observed at about 1.5 eV probably caused by charge-transfer excitations. A similar charge-transfer gap of about 1.7 eV was also observed [23] in the antiferromagnetic insulators La_2CuO_4 and $YBa_2Cu_3O_6$. Thus there is more and more evidence of a charge-transfer gap of about 1.7 eV in CuO_2 planes in various cuprates. This gives a lower limit for the charge-transfer energy Δ of about 1.7 eV.

In Fig. 11a we show measurements of the charge carrier plasmon as a function of momentum transfer q. While this plasmon is well pro-

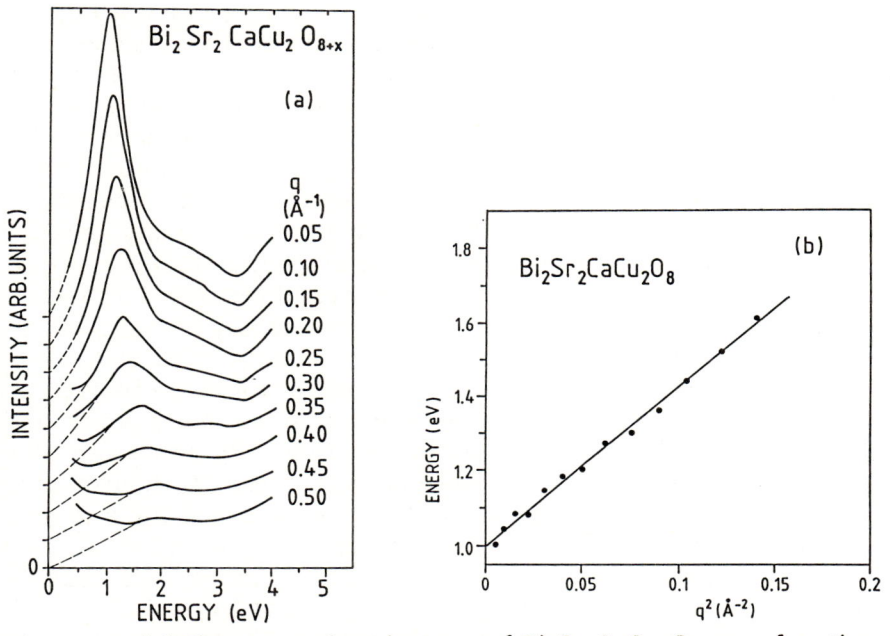

Fig. 11 (a) Charge carrier plasmon of $Bi_2Sr_2CaCu_2O_8$ as a function of momentum transfer q parallel to the CuO_2 planes. (b) Dispersion of the charge carrier plasmon.

nounced at lower q, at higher q it is strongly damped indicating that the plasmon line is dipping in a region of interband and intraband excitations. The plasmon energy as a function of squared momentum transfer is shown in Fig. 11b. An almost perfect linear relationship is observed. A plasmon dispersion $E_p(q) = E_p(0) + (\hbar^2/m)\alpha q^2$ is expected for a two-dimensional in-phase motion of charge carriers in parallel planes provided that the momentum transfer is parallel to the planes and smaller than $2\pi/a$ where a is the distance between the planes (long wavelength limit). The dispersion coefficient is then given [24] by $\alpha = 3/8[mv_F^2/E_p(0)]$. Compared to 3-dimensional plasmons the prefactor is 3/8 instead of 3/10. Using these relations the Fermi velocity can be calculated to be $v_F = 0.51 \cdot 10^8$ cm/sec. This value is slightly smaller than a value given in a previous publication [21] since in the latter the dispersion relation for a 3-dimensional plasmon was used. It is remarkable that the LDA band-structure calculations [25] yields a value $v_F(\|) = 0.49 \cdot 10^8$ cm/sec which is very close to the experimental value. It indicates that at least in this case the band-structure calculations reproduce quite well the average slope of the bands at the Fermi level.

4.2 Character and Symmetry of Holes in p-type Doped High-T_c Superconductors

In the following we report on electronic structure studies of high-T_c superconductors by core-electron excitations. In Fig. 12 we show O 1s absorption edges for $La_{2-x}Sr_xCuO_4$ measured by EELS [26]. In these spectra the local unoccupied density-of-states with p character at the O sites is measured. The binding energy of the O 1s level relative to the Fermi level has been determined using X-ray induced photoemission spectroscopy (XPS) to be $E_B = 528.5$ eV. Interpreting these spectra in terms of unoccupied local density-of-states, the binding energy corresponds to the Fermi level. The steep rise of spectral weight at about 2 eV above E_B is caused by La 5d and 4f states hybridized with O 2p states. In the undoped La_2CuO_4, there is almost no density-of-states at E_F expected for an antiferromagnetic insulator but not from LDA band structure calculations [14]. At about 2 eV a weak shoulder is observed. It may be caused by transitions into the upper Hubbard band which has predominantly Cu 3d character but also has some O 2p admixture. We emphasize that the intensity of this feature is rather low. With increasing dopant concentration a peak grows at the Fermi level and the intensity of this peak is roughly proportional to the dopant concentration [27]. In the model for the electronic structure described above, this peak is caused by holes in the band having predominantly O 2p character. Thus we see directly the density-of-states of the holes on O sites created upon doping. It is interesting to note that recent calculations [28] of inverse photoemission spectra using the method of exact diagonalization of

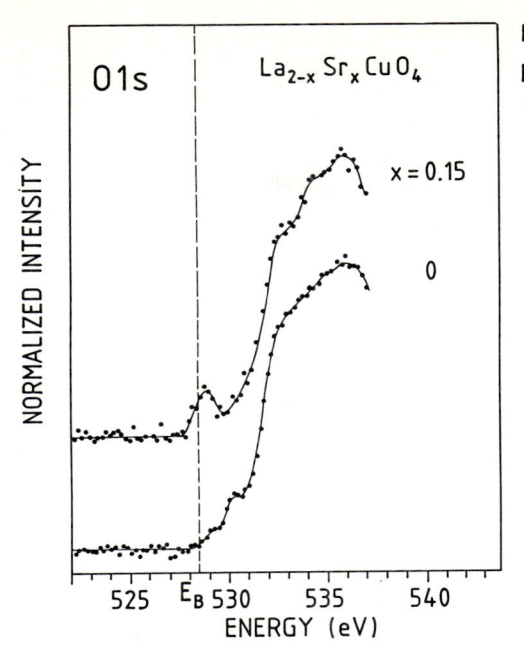

Fig. 12 O 1s absorption edges of polycrystalline $La_{2-x}Sr_xCuO_4$ for $x = 0$ and $x = 0.15$.

finite clusters also yield for the doped system the appearance of a peak with predominantly O 2p character. Because there was no change at E_F between undoped and doped samples in the XPS spectra [29] which probe mainly Cu 3d states, it was concluded that holes formed upon doping have almost completely O 2p character. Recent energy dependent photoemission measurements on well defined surfaces [30] indicate that the states at the Fermi level have about 1/3 Cu 3d and 2/3 O 2p character.

Similar O 1s absorption edges were obtained for the systems $YBa_2Cu_3O_{7-y}$ [26] (see Fig. 13) and $Bi_2Sr_2Ca_{1-x}Y_xCu_2O_8$ [31] (see Fig. 14). In the latter case the hole states have already disappeared for $x = 0.4$ where also T_c is close to zero. There are strong arguments that the prepeak in all these spectra is not caused by a core exciton, i.e., an interaction of the final states with the hole created on the core level: (i) The prepeak is observed in the metallic samples and not in insulating samples, contrary to what is expected for an excitonic line. (ii) From core-level excitation measurements on transition-metal carbides and nitrides as described in the previous section, we know that the interaction of the 2p final states with the 1s core-hole is smaller than the band width and that therefore the features in the measured absorption edges are close to the calculated total density-of-states. (iii) For $Bi_2Sr_2CaCu_2O_8$ there is a close similarity between the measured O 1s absorption edge and resonant inverse photoemission spectra [32]. In the latter experiment there is no core hole

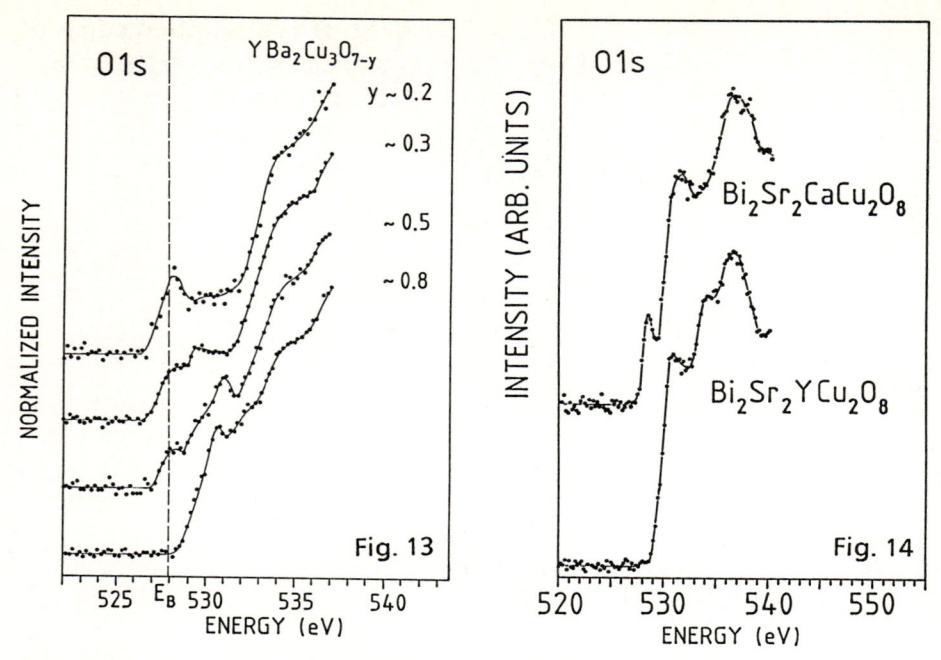

Fig. 13   O 1s absorption edges of polycrystalline samples of YBa$_2$Cu$_3$O$_{7-y}$.

Fig. 14   O 1s absorption edges of polycrystalline samples of Bi$_2$Sr$_2$Ca$_{1-x}$Y$_x$Cu$_2$O$_8$ for x = 0 and 1.

but an additional electron in the conduction band. (iv) The absorption edge was never found within error bars below the binding energy measured by XPS.

In all these systems the density of O 2p hole states just above the Fermi level is changed when changing the total number of valence electrons. This can be achieved by changing the O concentration or by replacing atoms by those having a different valency (La''' → Sr'' in La$_2$CuO$_4$, Ca^{++} → Y''' in Bi$_2$Sr$_2$CaCu$_2$O$_8$). When replacing atoms by other atoms with the same valency, e.g., Cu'' by Zn'' in YBa$_2$Cu$_3$O$_7$, the density of hole states is almost not changed [33]. A second example has recently been observed when replacing Y''' by Pr''' in YBa$_2$Cu$_3$O$_7$. Details will be published elsewhere. Although in both cases the density of hole states is almost not changed, superconductivity disappears and a semiconducting system is reached. This indicates that holes on O sites are necessary but not sufficient for superconductivity. Heavy doping leads probably to a localization of O 2p holes.

Orientation dependent measurements of core excitations on single crystals provide information on the symmetry of holes in high-T$_c$ superconductors [34]. Fig. 15a shows O 1s edges of a Bi$_2$Sr$_2$CaCu$_2$O$_8$

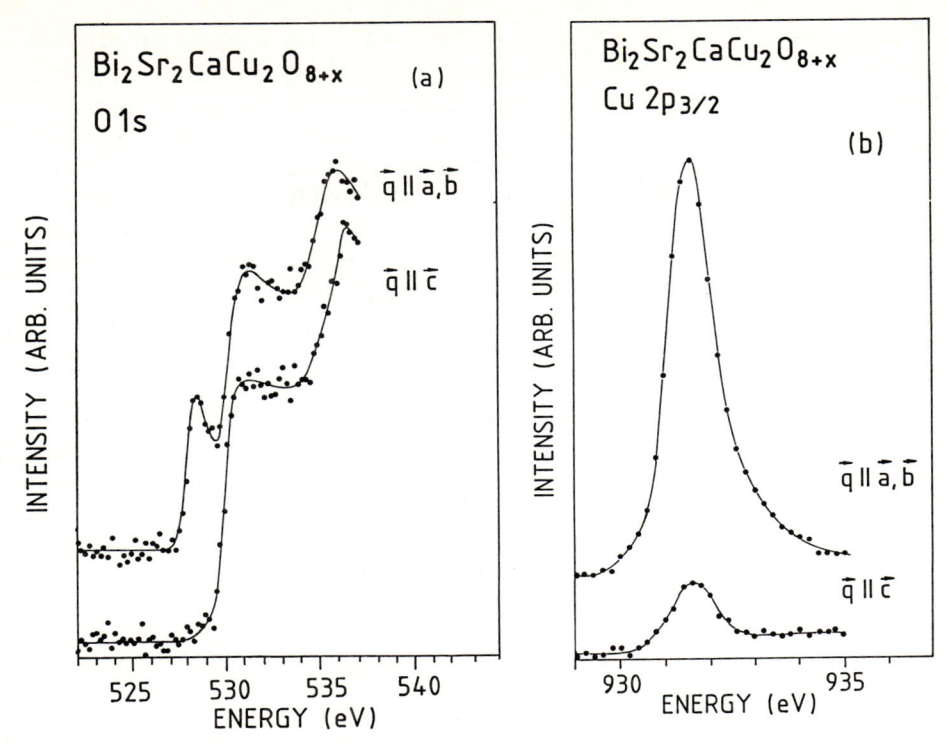

Fig. 15 Orientation dependent absorption edges of a $Bi_2Sr_2CaCu_2O_8$ single crystal. (a) O 1s absorption edges (b) Cu 2p absorption edges.

single crystal for momentum transfer q parallel and perpendicular to the CuO_2 (a,b) planes. For q∥a,b and for q∥c O $2p_{x,y}$ and $O2p_z$ orbitals, respectively, are reached by O 1s excitation. From the spectra it is immediately clear that there are only O $2p_{xy}$ states at the Fermi level and no O p_z states. There was a considerable discussion on the symmetry of O holes during the last two years. The data shown in Fig. 15a and similar results obtained by X-ray absorption spectroscopy (XAS) [35,36] clearly rule out all models for high-T_c superconductivity based on out-of-plane π-holes in the CuO_2 planes [37] and $2p_z$ holes on the apex O atoms [38,39]. However, these measurements cannot differentiate betwen O σ holes (as derived from LDA band-structure calculations, see above) or in-plane π-holes (as derived from cluster calculations [40]). Similar measurements [34] on $YBa_2Cu_3O_7$ single crystals show hole states with p_z symmetry which must be assigned to holes in the chains, either on O_4 or O_1 atoms. Since the p_z holes exhibit the lowest threshold energy and most of the band-structure calculations [41,42] predict the smallest O 1s binding energy for the O_4 atom, we ascribe the p_z holes to O_4 atoms. The measurements on $YBa_2Cu_3O_7$ together with the calculation of the chemical shift therefore strongly favour σ holes in agreement with recent NMR measurements [43,44]. In Fig. 15b we show orientation dependent measurements [34] of Cu $2p_{3/2}$ absorption edges of a single crystal of $Bi_2Sr_2CaCu_2O_8$ which probe the symmetry of unoccupied Cu 3d states, i.e., of the 3d states in the upper Hubbard band (see Fig. 9). As expected (see also Fig. 8) the spectra reveal that the symmetry of the unoccupied Cu states in predominantly $3d_{x^2-y^2}$. The intensity in the q∥c spectra was ascribed to a 10% admixture of probably $3d_{3z^2-r^2}$ states.

4.3 "n-type" Doped High-T_c Superconductors

Until the beginning of this year, all high-T_c superconductors were p-type doped, i.e., electrons were removed from the CuO_2 planes. More recently Tokura et al. [18] based on a chemical analysis and on measurements of a negative sign of the Hall effect claimed to have discovered the first n-type doped high-T_c superconductor $Nd_{2-x}Ce_xCuO_4$. This work was called a touchstone for theories of high-temperature superconductivity [45]. It is therefore of particular interest to investigate the electronic structure of these new materials by high-energy spectroscopies. In Fig. 16 we show O 1s absorption edges of the system $Nd_{2-x}Ce_xCuO_{4-\delta}$ together with Nd_2O_3 [46,29].

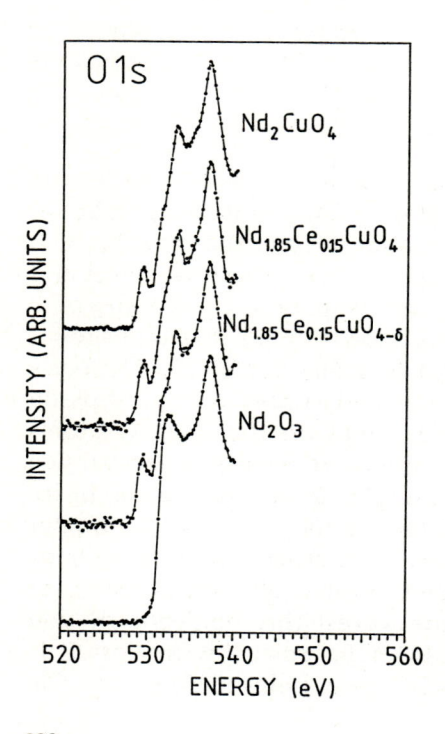

Fig. 16 O 1s absorption edges of Nd_2CuO_4, $Nd_{1.85}Ce_{0.15}CuO_4$, $Nd_{1.85}Ce_{0.15}CuO_{4-\delta}$ and Nd_2O_3.

Even the spectrum for the undoped Nd_2CuO_4 shows a prepeak with a threshold at 528.6 eV which is close to the O 1s binding energy measured by XPS [47]. The width of the prepeak is about 1.0 eV, considerably smaller than the width of the unoccupied O 2p density-of-states (~2 eV) calculated in a LDA band-structure calculation [48]. The intensity of the prepeak is slightly *increased* in the Ce-doped sample. Upon annealing the sample in a reducing atmosphere, the superconducting sample $Nd_{1.85}Ce_{0.15}CuO_{4-\delta}$ with $T_c = 19$ K and a Meissner fraction of 51% was obtained. Within error bars, there is no change in the O 1s absorption spectra for $\delta \sim 0$ and $\delta \neq 0$. In Nd_2O_3 no prepeak is observed. Recent orientation dependent measurements on an undoped Nd_2CuO_4 single crystal yield that the hole states are in orbitals parallel to the CuO_2 planes. Since the O atoms in the Nd_2O_2 planes have no planar symmetry, this result indicates that the prepeak is related to unoccupied O 2p states in the CuO_2 planes. Similar prepeaks have been observed recently in various similar n-type doped systems such as $Ln_{2-x}M_xCuO_{4-\delta}$ (Ln = Pr,Nd,Sm; M = Ce,Th) and $Nd_2CuO_{3.7}F_{0.3}$ which will be published elsewhere. The first explanation of the prepeak is that possibly the so-called electron superconductors are, due to non-stoichiometry, p-type doped superconductors and that the prepeak is then due to holes in a band having predominantly O 2p character. Recent electron diffraction data on these systems indicate that O migration leads to a disproportionation into two phases one having a superlattice and another having none [49]. The existence of hole states close to the Fermi level for semiconducting compounds (probably localized holes) was also observed in p-type doped systems (see above). The recent observation of antiferromagnetism in undoped Nd_2CuO_4 by neutron scattering [50] and μSR [51] would indicate that antiferromagnetism is not destroyed by localized holes. Another explanation of the prepeak is thinkable in terms of rare earth 4f states hybridized with O 2p states. This explanation can be clearly ruled out because the position of the prepeak does not vary when going from Pr to Sm cuprates. It is also not caused by Ce since it appears for undoped Th and F doped samples. Finally the prepeak may be caused by excitations into the upper Hubbard band which may have some O 2p admixture. However, it is very difficult to explain why the admixture of O 2p states should be so different for La_2CuO_4 and for Nd_2CuO_4. As pointed out above there is almost no indication of an upper Hubbard band in the O 1s spectrum of La_2CuO_4. Furthermore, since a gap of about 2 eV is observed in optical data, it is difficult to understand why the threshold of the prepeak is observed at the same energy as the binding energy of the O 1s level measured by XPS. At present we cannot rule out the second explanation of the prepeak (excitations into the upper Hubbard band), but there are many arguments against it. Finally we emphasize that the *increase* of unoccupied O 2p states upon Ce doping is very difficult to understand assuming n-type doping.

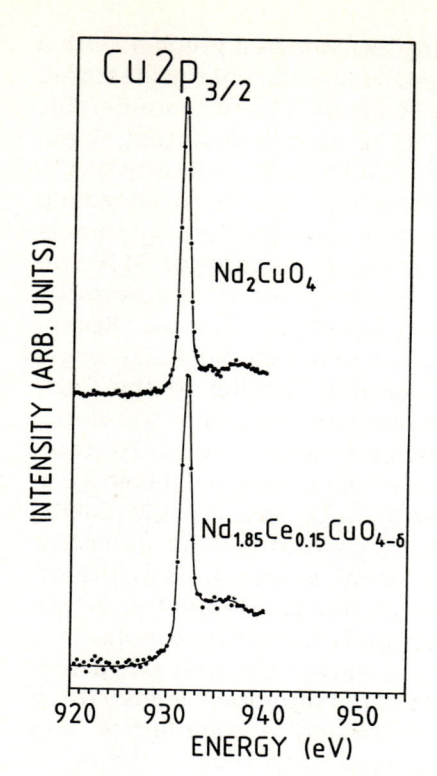

Fig. 17 Cu 2p absorption edges of Nd$_2$CuO$_4$ and the superconducting compound Nd$_{1.85}$Ce$_{0.15}$CuO$_{4-\delta}$.

In Fig. 17 we show Cu 2p absorption edges of Nd$_{2-x}$Ce$_x$CuO$_{4-\delta}$ which probe unoccupied 3d states. According to Fig. 9, a reduction of empty 3d states or the local formation of monovalent Cu is expected. Since there is a strong interaction of the core hole with 3d states, no detailed information on the shape and the energy position relative to the Fermi level is obtained. Rather a core excitonic line at about 931.5 eV is observed the intensity of which should be proportional to the number of unoccupied 3d states. Normalizing the spectra to the Nd 3d$_{5/2}$ line and taking into account the different Nd content, no change in intensity of the line at 931.5 eV was observed within error bars. Upon doping there is only a slight shift of the line to higher energies by about 0.2 eV. Furthermore, we have not observed the emerging of a shoulder between 932 and 934 eV typical for Cu' compounds [52,34]. We remark that upon doping there is a considerable increase in spectral weight above the core exciton line. We conclude that from these spectra we have no clear evidence for the formation of Cu' as derived, e g , from XAS spectra on the Cu 1s level [53].

Finally we mention that the measurements of the 3d$_{5/2}$ absorption edges of Nd and Ce in Nd$_{2-x}$Ce$_x$CuO$_{4-\delta}$ yield a valency of 3 for Nd and a valency close to that of CeO$_2$ (probably 3.5) for Ce. This would indicate an n-type doping for stoichiometric samples.

4.4 Electronic States at the Fermi Level and their Temperature Dependence

In this section we review recent high-resolution angle-resolved photoemission [54,55,56] and angle-resolved inverse photoemission measurements [57] on single crystalline $Bi_2Sr_2CaCu_2O_8$ which is today the prototype high-Tc superconductor for such experiments because chemically stable single-crystalline surfaces of high quality can be prepared from this material by in-situ cleavage under UHV conditions.

Typical angle-resolved photoemission and inverse photoemission spectra for various emission and incidence angles θ along the ΓX direction in the Brillouin zone are shown in Fig. 18. Similar photoemission data with a reduced energy resolution have been also reported by Takahashi et al. [58] and by Minami et al. [59]. The experimental peak positions are compared with band-structure calculations [60,61] in Fig. 19. In general, there is no good agreement between experiments and calculated band-structure. Concentrating on the energy region close to the Fermi level, the data indicate a crossing of a band along the ΓX direction. In the photoemission data for θ = 7.5° there is a more pronounced feature close to the Fermi level and an edge is observed which is almost completely determined by the energy resolution of $\Delta E_{1/2}$ = 60 meV and the Fermi-Dirac distribution function at room temperature. Moving to smaller angles (closer to the Γ-point) the peak disperses to higher binding energy and is considerably broadened. Close to the Γ-point there is only a very broad distribution. The unoccupied part of the band as seen in the inverse photoemission spectra with an energy resolution of $\Delta E_{1/2}$ = 0.6 eV shows up in a sudden increase of intensity close to the Fermi level for θ = 30-40°. The exact dispersion is difficult to evaluate since due to the reduced energy resolution there is only a broad intensity distribution above E_F. The data indicate that there is a Fermi edge on an energy scale of about 10 meV (taking data at lower temperatures and after deconvolution with the energy resolution) which is considerably smaller than magnetic exchange interactions (~100 meV). Since we know both from photoemission *and* inverse photoemission that the band crosses the Fermi level, the edge is not due to band-structure effects, e.g., a narrow band just below E_F, but due to a jump in the distribution function n(k). Probably this indicates that these systems are Fermi liquids (at least on an energy scale of about 10 meV).

The coherent part of the photoemission spectra can be represented (apart from one-electron matrix element effects) by the spectral function

$$A(\omega,\mathbf{k}) = \frac{1}{\pi} \frac{Im\,\Sigma(\omega,\mathbf{k})}{\left[\omega - \varepsilon_\mathbf{k} - Re\,\Sigma(\omega,\mathbf{k})\right]^2 + \left|Im\,\Sigma(\omega,\mathbf{k})\right|^2}$$

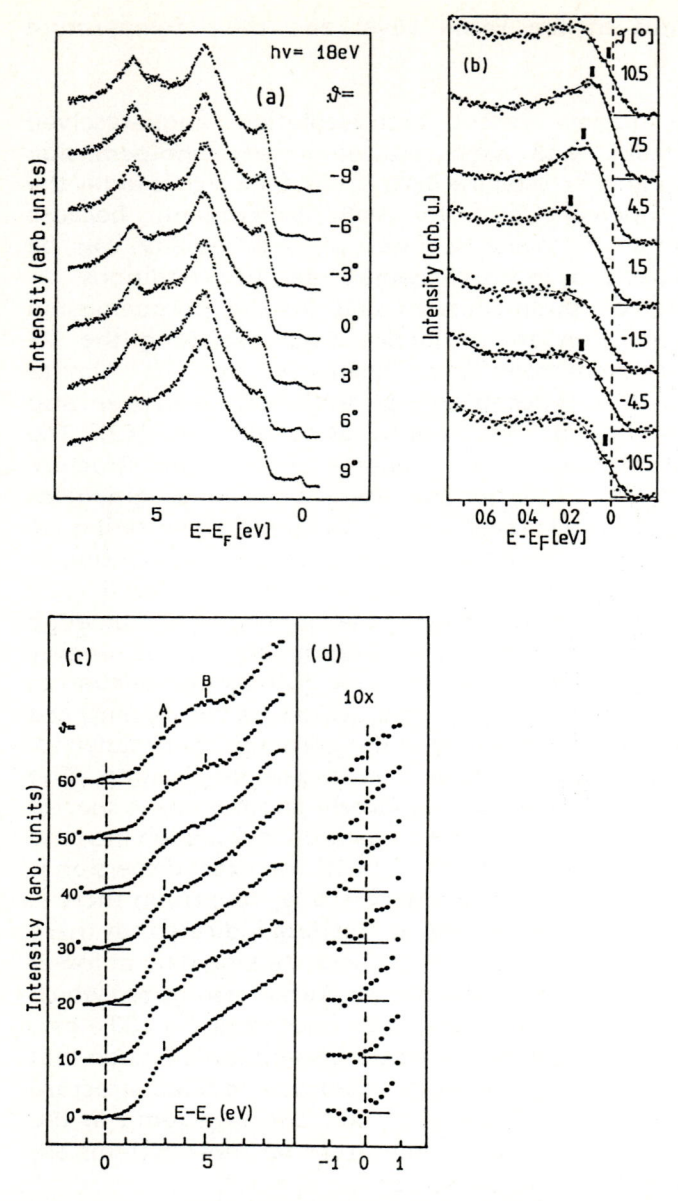

Fig. 18 (a) Angle-resolved photoemission spectra of a Bi$_2$Sr$_2$CaCu$_2$O$_8$ single crystal measured with 18 eV photon energy along the ΓX direction for various emission angles. (b) The same as (a) close to the Fermi level. (c) Angle-resolved inverse photoemission spectra along the ΓX direction. The spectra are normalized to the incident electron charge. (d) The same as (c) but close to the Fermi level with enlarged energy scale.

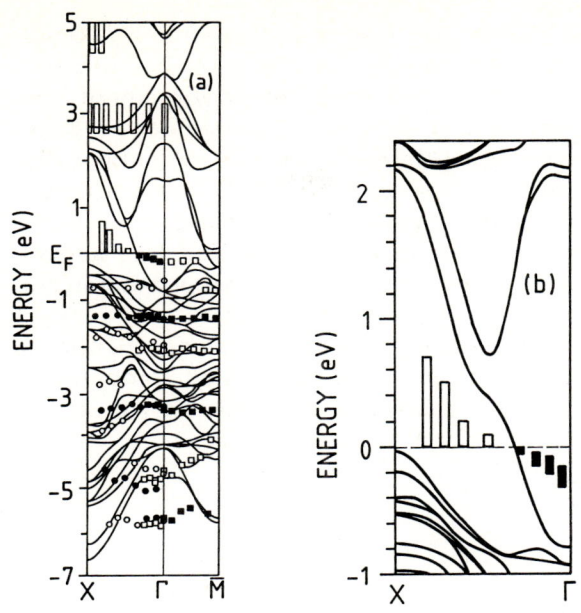

Fig. 19 (a) Band-structure calculations for $Bi_2Sr_2CaCu_2O_8$ by Massida et al. [60] and experimental inverse photoemission and photoemission results. Squares are taken with a primary photon energy of 18 eV, circles with 21.2 eV. Strong experimental features are marked by full symbols. (b) The same as (a) but close to the Fermi level along the ΓX direction. Band-structure from Krakauer and Pickett [61].

where ε_k are the band energies and $\Sigma(\omega,k)$ is the self energy function the imaginary part of which is zero at the Fermi level. In highly correlated systems, corrections of the band energies ε_k should be observed due to the self energy function, the real part of which should lead to an energy shift and the imaginary part to a broadening (a reduction of lifetime). This was observed in many systems such as alkali metals [62] or Ni metal [63]. For the CuO dpσ* band in cuprates, a very strong shift of about 500 meV compared to LDA band-structure calculations is observed (see Fig. 19b) and also the lifetime of the quasi-particles is strongly reduced when leaving the Fermi level. This broadening is so strong that at present it is unclear whether the spectral weight shown in Fig. 18b is only due to a coherent band-structure or whether there is also an incoherent background. Measurements of the lifetime as a function of the binding energy could provide a further criterion whether the metallic cuprates are Fermi liquids or not [64]. However, to obtain reliable data, a considerable improvement of energy resolution should be achieved. It is remarkable that the measured k value along the ΓX direction where the

CuO band crosses the Fermi level is well reproduced by the LDA band-structure calculation. Recent measurements along the ΓM direction where a BiO band crosses the Fermi level yield a similar agreement [65]. Therefore, the general statement can be given that, similar to other highly correlated systems (e.g., heavy fermion systems [66]) the Fermi surface is well reproduced by the present LDA band-structure calculations. While the calculated average slope of the bands at E_F is well reproduced by values derived from the plasmon dispersion (see above), the slope of the Cu-O band at E_F is not reproduced by the photoemission data. The origin of this discrepancy is at present not clear.

In order to study changes of the band-structure between the normal state and the superconducting state we have performed temperature dependent angle-resolved photoemission measurements. In Fig. 20 we

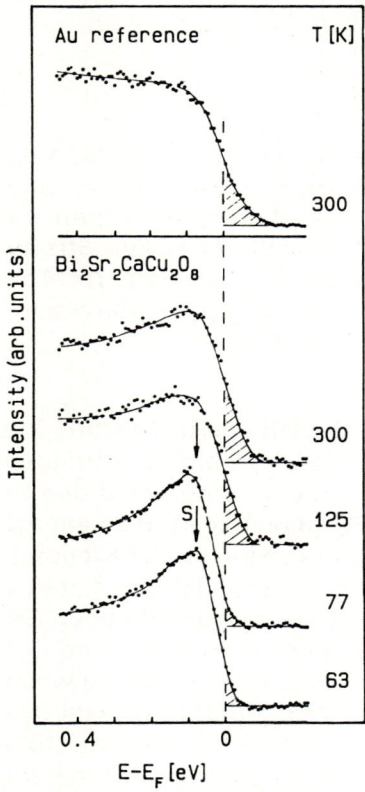

Fig. 20 Temperature dependence of the angle-resolved photoemission spectra of a $Bi_2Sr_2CaCu_2O_8$ single crystal taken for a k value along the ΓX direction where the band crosses the Fermi level. For comparison a Au reference spectrum is shown.

show such measurements on a $Bi_2Sr_2CaCu_2O_8$ single crystal taken with a $K_\parallel$ vector along the ΓX direction where the band crosses the Fermi level. For $T>T_c$ and within the given energy resolution of 60 meV, clear Fermi edges are measured. For $T<T_c$, spectral weight at and above the Fermi level is shifted to lower energy causing an enhancement of emission at about 80 meV below E_F (assigned by S). This causes a shift of the edge (the turning point of emission onset) by about 30 meV. We relate these experimental findings to the opening of a superconducting gap and derive from the energy shift a gap value of $\Delta = 30$ meV. The ratio $2\Delta(0)/k_BT_c \sim 8$ is about twice as large as the weak coupling value expected from BCS theory indicating that the new high-T_c cuprates are strong-coupling superconductors. Similar values for the gap were reported from angle-integrated photoemission measurements [67,68] and from recent angle-resolved measurements more along the ΓM direction [69]. The above mentioned measurements [54] have demonstrated for the first time that changes of the band structure between normal state and superconducting state can be measured by angle-resolved photoemission spectroscopy.

Acknowledgements

The authors thank W. Weber and H. Rietschel for many helpful discussion.

References

1. D.J. Scalapino in *Superconductivity*, edited by R.D. Parks (Marcel Dekker, New York, 1969) Vol. 1, p. 449
2. J.L. Calais, Adv. in Phys. **26**, 847 (1977); A. Neckel, Int. J. of Quantum Chem. **23**, 1317 (1983)
3. J. Pflüger, J. Fink, G. Crecelius, K.P. Bohnen, H. Winter, Solid State Commun. **44**, 489 (1982); J. Pflüger, J. Fink, W. Weber, K.-P. Bohnen, and G. Crecelius, Phys. Rev. B **30**, 1155 (1984); ibid **31**, 1244 (1985); J. Pflüger, J. Fink, and K. Schwarz, Solid State Commun. **55**, 675 (1985)
4. W. Weber, in *Electronic Structure of Complex Systems*, edited by P. Phariseau and W. Temmerman (Plenum Press, New York, 1984) p. 345
5. J. Pflüger, Kernforschungszentrum Karlsruhe (KfK) Report No. 3585 (1983)
6. J. Fink, Th. Müller-Heinzerling, B. Scheerer, W. Speier, F.U. Hillebrecht, J.C. Fuggle, J. Zaanen, and G.A. Sawatzky, Phys. Rev. B **32**, 4899 (1985)

7. L.R. Testardi, Rev. Mod. Phys. **47**, 637 (1975)
8. J. Müller, Rep. Progr. Phys. **43**, 641 (1980)
9. B.M. Klein, L.L. Boyer, D.A. Papaconstantopoulos, and L.F. Mattheiss, Phys. Rev. B **18**, 6411 (1978)
10. W.E. Pickett, K.M. Ho, and M.L. Cohen, Phys. Rev. B **19**, 1734 (1979)
11. L.F. Mattheiss and W. Weber, Phys. Rev. B **25**, 2248 (1982)
12. Th. Müller-Heinzerling, J. Fink, and W. Weber, Phys. Rev. B **32**, 1850 (1985)
13. Th. Müller-Heinzerling, Ph.D. thesis, University of Karlsruhe, 1985
14. L.F. Mattheiss, Phys. Rev. Lett. **58**, 1028 (1987)
15. A. Fujimori and F. Minima, Phys. Rev. B **30**, 957 (1984); J. Zaanen, G.A. Sawatzky, and J.W. Allen, Phys. Rev. Lett. **55**, 418 (1985); S. Hüfner, Z. Phys. B **61**, 135 (1985)
16. J.G. Bednorz and K.A. Müller, Z. Phys. B **64**, 189 (1986)
17. J.C. Fuggle, J. Fink, and N. Nücker, Int. J. Mod. Phys. B **2**, 1185 (1988); G.A. Sawatzky, this volume
18. Y. Tokura, H. Takagi, and S. Uchida, Nature **337**, 345 (1989)
19. A.C.W.P. James, S.M. Zakurak, and D.W. Murphy, Nature **338**, 240 (1989)
20. P.W. Anderson, Science **235**, 1196 (1987)
21. N. Nücker, H. Romberg, S. Nakai, B. Scheerer, J. Fink, Y.F. Yan, and Z.X. Zhao, Phys. Rev. B **39**, 12379 (1989)
22. Th. Timusk and D.B. Tanner, in *Physical Properties of High-Temperature Superconductors I*, edited by D.M. Ginsberg (World Scientific, Singapore, 1989)
23. J. Fink, N. Nücker, H. Romberg, and S. Nakai, in *Electronic Structure of High-T_c Superconductors*, edited by A. Bianconi (Plenum Press, in print)
24. J.D. Jackson, *Classical Electrodynamics* (Wiley, New York, 1962); A.L. Fetter, Ann. Phys. (NY) **81**, 367 (1973)
25. L.F. Mattheiss and D.R. Hamann, Phys. Rev. B **38**, 5012 (1988); L.F. Mattheiss, private communication
26. N. Nücker, J. Fink, J.C. Fuggle, P.J. Durham, and W.M. Temmerman, Phys. Rev. B **37**, 5158 (1988)
27. N. Nücker, J. Fink, B. Renker, D. Ewert, C. Politis, P.J.W. Weijs, and J.C. Fuggle, Z. Phys. B **67**, 9 (1987)
28. P. Horsak, W.H. Stephan, K. v. Szczepanski, M. Ziegler, and W. von der Linden, Physica C in print.
29. J.C. Fuggle, P.J.W. Weijs, R. Schoorl, G.A. Sawatzky, J. Fink, N. Nücker, P.J. Durham, and W.M. Temmerman, Phys. Rev. B **37**, 123 (1988)
30. R.S. List, A.J. Arko, R. Bartlett, C. Olson, A.-B. Yang, R. Liu, C. Gu, B.W. Veal, Y. Chang, P.Z. Jiang, K. Vandervoort, A.P. Paulikas, and J.C. Campazano, Physica C **159**, 439 (1989)

31. J. Fink, N. Nücker, H. Romberg, M. Alexander, S. Nakai, B. Scheerer, P. Adelmann, and D. Ewert, Physica C, in print
32. W. Drube, F.J. Himpsel, G.V. Chandrashekhar, and M.W. Shafer, Phys. Rev. B **39**, 7328 (1989)
33. M.L. denBoer, C.L. Chang, H. Petersen, M. Schaible, K. Reilly, and S. Horn, Phys. Rev. B **38**, 6588 (1988)
34. N. Nücker, H. Romberg, X.X. Xi, J. Fink, B. Gegenheimer, Z.X. Zhao, Phys. Rev. B **39**, 6619 (1989)
35. F.J. Himpsel, G.V. Chandrashekhar, A.B.McLean, and M.W. Shafer, Phys. Rev. B **38**, 11946 (1988)
36. P. Kuiper, M. Grioni, G.A. Sawatzky, D.B. Mitzi, A. Kapitulnik, A. Santaniello, P. de Padova, and P. Thiry, Physica C **157**, 260 (1989)
37. K.H. Johnson, M.E. McHenry, C. Counterman, A. Collins, M.M. Donovan, R.C. O'Handley, and G. Kalonji, Physica C **153-155**, 1165 (1988)
38. H. Kamimura, Jpn. J. Appl. Phys. **36**, 6627 (1987)
39. A. Fujimori, Phys. Rev. B **39**, 793 (1989)
40. Y. Guo, J.-M. Langlois, and W.A. Goddard III, Science **239**, 896 (1988)
41. W.E. Pickett, Modern Physics in print
42. M. Alouani and J. Zaanen, Europhys. Lett. in print
43. H. Alloul, T. Ohno, J.F. Marucco, and G. Collin, Physica C, in print
44. M. Takigawa, R.C. Hammel, R.H. Heffner, Z. Fisk, K.C. Ott, and J.D. Thompson, Phys. Rev. Lett., submitted
45. T.M. Rice, Nature **337**, 686 (1989)
46. N. Nücker, P. Adelmann, M. Alexander, H. Romberg, S. Nakai, J. Fink, H. Rietschel, G. Roth, H. Schmidt, and H. Spille, Z. Phys. B **75**, 421 (1989)
47. A. Fujimori, Y. Tokura, E. Eisaki, H. Takagi, S. Uchida, E. Takayama-Muromachi, Phys. Rev. B in print
48. S. Massidda, N. Hamada, J. Yu, and A.J. Freeman, Physica C **157**, 571 (1989)
49. C.H. Chen, D.J. Werder, A.C.W.P. James, D.W. Murphy, S. Zahurak, R.M. Felming, B. Batlogg, and L.F. Schneemeyer, preprint
50. R.J. Birgeneau, private communication
51. Y.J. Uemura, Physica C, in print
52. M. Grioni, J.B. Goedkoop, R. Schoorl, F.M.F. de Groot, J.C. Fuggle, F. Schäfers, E.E. Koch, G. Rossi, J.-M. Esteva, and R.C.Karnatak, Phys. Rev. B **39**, 1541 (1989)
53. J.M. Tranquada, S.M. Heald, A.R. Moodenbaugh, G. Liang, M. Croft, Nature **337**, 720 (1989)
54. R. Manzke, T. Buslaps, R. Claessen, and J. Fink, Europhys. Lett. **9**, 477 (1989)
55. R. Manzke, T. Buslaps, R. Claessen, M. Skibowski, and J. Fink, Physica C, in print

56. R. Manzke, T. Buslaps, R. Claessen, M. Skibowski, and J. Fink, Physica Scripta, in print
57. R. Claessen, R. Manzke, H. Carstensen, B. Burandt, T. Buslaps, M. Skibowski, and J. Fink, Phys. Rev. B **39**, 7316 (1989)
58. T. Takahashi, H. Matsuyama, H. Katayama-Yoshida, Y. Okabe, S. Hosoya, K. Seki, H. Fujimoto, M. Sato, and H. Inokuchi, Nature **334**, 691 (1988); Phys. Rev. B **39**, 6636 (1989)
59. F. Minami, T. Kimura, and S. Takekawa, Phys. Rev. B **39**, 4788 (1989)
60. S. Massidda, J. Yu, and A.J. Freeman, Physica C **152**, 251 (1988)
61. H. Krakauer and W.E. Pickett, Phys. Rev. Lett. **60**, 1665 (1988)
62. E. Jensen and E.W. Plummer, Phys. Rev. Lett. **55**, 1912 (1985)
63. A. Liebsch, Phys. Rev. Lett. **43**, 1431 (1979); D.E. Eastman, F.J. Himpsel, and J.A. Knapp, Phys. Rev. Lett. **40**, 1514 (1978)
64. M. Schlüter, private communication
65. see also C.G. Olson, Physica C, in print
66. G.G. Lonzarich, JMMM **76 & 77**, 1 (1988)
67. J.-M. Imer, F. Patthey, B. Dardel, W.-D. Schneider, Y. Baer, Y. Petroff, and A. Zettl, Phys. Rev. Lett. **62**, 336 (1989)
68. Y. Chang, Ming Tang, R. Zanoni, M. Onellion, R. Joynt, D.L. Huber, G. Margaritondo, P.A. Morris, W.A. Bonner, J.M. Tarascon, and N.G. Stoffel, Phys. Rev. B **39**, 4740 (1989)
69. C.G. Olson, R. Liu, A.-B. Yang, D.W. Lynch, A.J. Arko, R.S. List, B.W. Veal, Y.C. Chang, P.Z. Jiang, and A.P. Paulikas, Science in print

A Positive Experimental Test for Pairing Mechanisms Including $3d_{z^2}$ Hole Symmetry: Correlation of the Relative Weight of $3d_{z^2}$ vs $3d_{x^2-y^2}$ Hole States with the Critical Temperature

A. Bianconi[1,*], *P. Castrucci*[1], *A. Fabrizi*[1], *M. Pompa*[1], *A.M. Flank*[2], *P. Lagarde*[2], *H. Katayama-Yoshida*[3], *and G. Calestani*[4]

[1]INFM gruppo G4, Dipartimento di Fisica, Università degli Studi di Roma "La Sapienza", I-00185 Roma, Italy
[2]LURE, CNRS-CEA-MEN, Bâtiment 209 D, Université Paris Sud, F-91405 Orsay, France
[3]Department of Physics, Tohoku University, Sendai 980, Japan
[4]Istituto di Struttuiristica Chimica, Università di Parma, Viale delle Scienze, I-43100 Parma, Italy
*Also at: University of L'Aquila, Collemaggio, I-67100 L'Aquila, Italy

Abstract. We report experimental evidence for the increasing ratio between the weight of the Cu $3d_{z^2}$ holes and the $d_{x^2-y^2}$ holes with increasing T_c reaching a maximum for the 110K BiCaSrCuO superconductors. The energy splitting ΔE_z between the maxima of in-plane and out-of-plane polarized Cu L_3 XAS spectra decreases by increasing T_c by doping within each class of high T_c superconductors. The range of values of ΔE_z increases going from one class of high T_c superconductors to another class with higher T_c. The weight of the Cu $3d_{z^2}$ holes increases by about 5% going from the normal phase at 300K to the superconducting phase at 10K. These experimental results support the pairing mechanism for high T_c superconductors involving electronic excitations where the weight of the Cu $3d_{z^2}$ holes modulates T_c.

1. Introduction

In the BCS theory the superconductivity is a consequence of an instability of the Fermi sea to the formation of pairs of electrons. In classical low T_c (T_c<30K) superconductors the pairing mechanism is due to the attractive interaction due to exchange of phonons forming the Cooper pairs. The critical temperature T_c is proportional to $\omega_d \exp(-1/\lambda)$ where ω_d is the Debye frequency of the crystal, i.e. the highest frequency of the spectrum, and the coupling constant λ is proportional to $N(E_F)V/M$ where $N(E_F)$ is the density of states at the Fermi level, V is the strength of the electron-phonon coupling, and M is the mass of the ion. A classical experimental test for the BCS theory is the isotope effect which measures the decrease of T_c when increasing the mass of the ion M which enters in the coupling constant λ. There are indications that the maximum values of T_c that can be reached by phonon pairing are not higher than 40K. Therefore different mechanisms for high T_c superconductivity in cuprate perovskites based on electronic and magnetic excitations have been proposed. A key problem for these models is the determination of the electronic structure of these materials.

High T_c superconductors were discovered by Bednorz and Müller[1] by doping a divalent Cu compound La_2CuO_4, where the Cu ions were expected to have the $3d_{x^2-y^2}$ configuration stabilized by the long axial Cu-O distance. In the original idea of Bednorz and Müller guided by bi-polaron theories[2,3], the hole induced by Ba doping resulting in a metallic conductivity will go in a $3d_{z^2}$ orbital. In fact an itinerant d hole with $3d_{z^2}$ symmetry was expected to suppress the Jahn-Teller splitting and to induce an increase of T_c via a strong electron-coupling term λ.

The presence of a $3d^8$ configuration induced by doping was ruled out by joint analysis of x-ray absorption XAS and x-ray photoemission XPS data in the early days following the discovery of high T_c superconductivity above liquid nitrogen temperature in 1987 [4,5], and confirmed by many groups[6]. XAS has shown that the weight of the $3d^8$ configuration in the ground state is less than 5% [5]. The divalent Cu materials are antiferromagnetic systems with a stable $3d_{x^2-y^2}$ configuration with a large Cu oxygen hybridization with oxygen $2p_x$ and $2p_y$ in the CuO_2 plane giving a band formed by states of b_1 molecular symmetry of the CuO_4 square plane cluster. The large electronic correlation of the d holes was found to be determined by the d-d hole repulsion $U_{dd} \approx 8$ eV. Having ruled out the presence of the d^8 configuration the problem was the determination of the two hole states induced by doping.

The gap observed in optical excitation of divalent antiferromagnetic compounds is given in a cluster model by the energy to add a second hole to a divalent cluster. So the first ionization level in the undoped insulating compound is expected to be the singlet $\underline{d}(b_1)\underline{L}(b_1)$ two hole state. In the divalent systems other excited states are possible with similar excitation energies where the d hole is excited to a different orbital of different energy, such as to the $3d_{3z^2-r^2}$ (called also d_{z^2}) orbital with a_1 molecular symmetry separated with the b_1 orbital by the Jahn Teller energy Δ_{JT}. Electronic structure calculations show that the energy of the singlet excited states $\underline{d}(b_1)\underline{L}(b_1)$ and triplet $\underline{d}(a_1)\underline{L}(b_1)$ or $\underline{d}(b_1)\underline{L}(a_1)$ are nearly degenerate in La_2CuO_4[7].

The first experimental identification of the $(3d^9\underline{L})^*$ states induced by doping in the gap of insulating systems was given by the variation of the Cu L_3 XAS spectra of the YBaCuO system induced by oxygen doping and later also in the Sr doped La_2CuO_4 systems[4,5].

The nature of the states $\underline{d}^*\underline{L}^*$ is object of discussion. It was not possible to determine the symmetry of the $\underline{d}^*$ or $\underline{L}^*$ holes from unpolarized XAS data so the asterisk was used to indicate that both $\underline{d}^*$ and $\underline{L}^*$ holes can have different symmetries from the b_1 symmetry in the parent antiferromagnetic compound. The states induced by doping at the Fermi level $\underline{d}^*\underline{L}^*$ can be either $\underline{d}(b_1)^*\underline{L}(b_1)^*$ or $\underline{d}(a_1)^*\underline{L}(b_1)^*$, $\underline{d}(b_1)^*\underline{L}(a_1)^*$ and $\underline{d}(a_1)^*\underline{L}(a_1)^*$ states (to cite the most probable two hole configurations between all the possible ones involving also b_2 and e symmetries). The asterisk was also used to indicate that the energy to create a $\underline{d}^*$ or a $\underline{L}^*$ hole, parameters used to describe the electronic structure of superconductors, should be different from the energy to create a $\underline{d}$ or $\underline{L}$ hole in the antiferromagnetic insulators. This is confirmed by the fact that at low doping an impurity band of localized states is formed in the antiferromagnetic or charge transfer gap, by the fact that the changes of Cu site structure with doping, with superstructures and associated distortions, induce a change of the direct hopping between b_1 and a_1 or (b_1 and e) states, and finally by the fact that the presence of ligand holes $\underline{L}^*$, introduced by doping, induce a change in the energy to create a $\underline{d}^*$ hole (in comparison with the value to create a $\underline{d}$ hole in the divalent parent system) via the Coulomb interactions V between $\underline{d}$ and $\underline{L}$ holes.

For these reasons it is very difficult to calculate by band theories or cluster theories the nature of $(3d^9\underline{L})^*$ states at the Fermi level in real superconducting materials. Therefore only the experimental determination of the symmetry of the $(3d^9\underline{L})^*$ states at the Fermi level in real metallic and superconducting materials can solve the problem.

Several theories of pairing in high T_c superconductors have assumed that the carriers are singlet states $\underline{d}(b_1)\underline{L}(b_1)$ in the frame of a single symmetry b_1 band split into two bands by antiferromagnetic interaction and electronic correlation[8]. Other theories have assumed the hole formation induced by doping in a band of different symmetry from the antiferromagnetic parent material, such as the a_1 symmetry [9-18]: the three band model of Castellani et al. [9,10], the spin polaron pairing mechanism of Kamimura [11], the d-d excitation model [14-17], the Kanamori model [13], the interlayer pairing of Askenazi, et al. [18] require the presence of electronic states with two holes in two different orbitals. Other models require the presence of the oxygen hole in a non-bonding orbital[19].

Polarized Cu L_3 XAS (x-ray absorption spectroscopy) provides a unique experimental tool to determine the symmetry of the unoccupied Cu 3d states at the Cu site. In fact the matrix element for 2p->3d dipole transition which is about 40 times stronger than the 2p->4s transition amplifies the signal due to d unoccupied states giving the famous white lines at the absorption edge. The E//xy polarized Cu XAS of single crystals probes the $3d_{x^2-y^2}$, d_{xy}, and also d_{z^2} and d_{xz} orbitals while the E//z polarization only the d_{z^2} and d_{xz} and d_{yz} orbitals.

The study of YBaCuO system by polarized Cu L_3 XAS has given experimental evidence [20,21] that the jump of T_c from the 60K superconductors to the 90K superconductors is correlated with the formation of ligand holes $\underline{L}^*$ in the O4 site, not in the CuO$_2$ plane. In fact the Cu L_3 XAS spectra show the formation of a $\underline{c}3d^{10}\underline{L}^*$ final state Cu1($\underline{2p}$)O4($\underline{2p}$) with E//c polarization giving a peak at about 1.7 eV above the white line pushed up by the final state interaction $U_{\underline{cL}}$ between the core hole $\underline{c}$ and the ligand hole $\underline{L}$ as discussed by M. De Santis et al.[5]. These results have given experimental evidence for the formation of ligand holes in the second band formed by the chains in YBaCuO crossing the Fermi level in the high T_c phase of YBa$_2$Cu$_3$O$_{7-x}$ (x<0.2). This experimental result has been confirmed by Cu K-XANES studies[22]

The fact that the critical temperature of the YBaCuO system was correlated with the formation of oxygen holes out of the xy plane in the O4 site has addressed our interest to the coupling of the electronic states in the CuO$_2$ plane with the states out of plane. This coupling was assigned to a degree of freedom of the orbital momentum of the Cu 3d holes in the CuO$_2$ plane. The hypothesis assumed by many theories of high T_c superconductivity that the orbital momentum of the Cu 3d holes, L=2, was frozen in the $3d_{x^2-y^2}$ state, ml=2, was put under criticism and a finite probability for the Cu 3d hole in the CuO$_2$ plane to have some, ml=0, i.e. $3d_{z^2}$ component was pointed out. Evidence for a large intensity of the E//c Cu L_3 XAS white line and the energy splitting between in-plane and out-plane polarized spectra has been interpreted as showing the relevance of d holes with non-in-plane orbital component [21] however the unique interpretation of the data is limited by the presence of two different Cu sites in YBaCuO superconductors.

To take full advantage of the polarization dependence of the x-ray absorption spectra to detect the symmetry of the d hole in the $(3d^9\underline{L})^*$ or $\underline{d}^*\underline{L}^*$ configuration we have focussed our interest on superconductors with a single Cu site in the CuO_2 plane parallel to the xy direction. At the Trieste meeting "Towards understanding high T_c superconductivity"[23] and at the Rome symposium on the "Electronic structure of high T_c superconductors"[24] it was reported that the E//z component of the white line of polarized Cu L_3 XAS appears in the spectra of both LaSrCuO and BiCaSrCuO systems confirming the presence of relevant probability for the $\underline{d}^*$ hole in the superconductors to have also orbital components of a_1 or e symmetry. Therefore the presence of a component of the 3d hole in the z direction detected by E//z Cu L_3 XAS has been interpreted as giving experimental evidence for formation of both $\underline{d}(a_1)^*\underline{L}^*$ and $\underline{d}(b_1)^*\underline{L}^*$ states at the Fermi level in high T_c superconductors.

The investigation of the Cu L_3 XAS of $Bi_2CaSr_2Cu_2O_8$ has shown that the energy separation between the E//xy and the E//z components of the white line is 300-400 meV in an oriented film with 61K[23,24] which is smaller than the intrinsic band width estimated to be of 600-800 meV. The energy splitting ΔE_z between the maxima of the E//z and E//xy white lines was found to decrease with increasing T_c both in $La_{2-x}Sr_xCuO_4$ and in YBaCuO and at high doping near the highest values of T_c in each of the two classes was found to reach 40-50 meV. The energy splitting of final states in Cu L_3 XAS of doped systems with $\underline{d}^*\underline{L}^*$ configurations in the ground state should be associated with states a and b in the ground state of different symmetry and/or with different charge distribution such as the final states $\underline{c}3d^{10}\underline{L}(a)^*$ (E//z polarization) and $\underline{c}3d^{10}\underline{L}(b)^*$ (E//xy polarization) appear at different energies, as discussed in ref.5 for trivalent Cu systems, because of the difference between the maxima of the partial density of states of a and b ligand hole states corrected by the different final state interaction $V_{\underline{cL}(a)}$ and $V_{\underline{cL}(b)}$ between the core hole $\underline{c}$ and $\underline{L}(a)^*$ and $\underline{L}(b)^*$ hole respectively.

Electron energy loss allows measurement of bulk X-ray absorption spectra in transmission mode and the analysis of the q dependent EELS allows measurement of polarized spectra. This technique has confirmed our experimental results of polarized Cu L_3 XAS both for YBaCuO and BiCaSrCuO systems [25] on the intensity ratio between the E//z and E//xy components and on the line shape, confirming therefore the key physical point concerning the presence of $\underline{d}^*$ holes with non $d_{x^2-y^2}$ symmetry. A disagreement exists between the two techniques concerning the small energy splitting ΔE_z between E//z and E//xy white lines, which is of the order of 300 meV in $Bi_2CaSr_2Cu_2O_8$ and 50 meV in $La_{1.9}Sr_{0.1}CuO_4$, which was not measured by Fink et al. collecting experimental points with 200 meV energy interval.

Oxygen K-edge EELS of YBaCuO [25] show the appearing of a E//z peak at threshold going from 60K to 90K YBaCuO superconductors confirming that while in the low T_c phase the ligand holes have mostly E//xy components the jump from 60K to 90K phase is associated with formation of ligand holes with E//z component. Oxygen K-edge XAS and EELS of BiCaSrCuO has shown that the ligand holes $\underline{L}^*$ derived form the O(2p) orbitals have only a component in the CuO_2 plane, and the problem on the symmetry of these states in the plane is open, in fact these states can be $\underline{L}(b_1)$ or $\underline{L}(b_2)$ or $\underline{L}(a_1)$ in the plane. The presence of $\underline{L}(a_1)$ states can be expected because of mixing of $O(2p_x)$ and $O(2p_y)$ orbitals in the plane with the ring in the xy plane of the $3d_{3z^2-r^2}$ component of the $\underline{d}^*$ holes. The presence of these components can be due to dynamical or static distortions of the CuO_4 square plane and/or to a $3d_{3z^2-r^2}$ band close to Fermi level and giving a partial density of states of a_1 symmetry above the Fermi level.

To identify the electronic states relevant for superconductivity a long term systematic study on well characterized single phase samples with different T_c was carried out in order to establish the correlation between Tc and the changes of the electronic states. Here we report the results of a long term investigation of the BiCaSrCuO systems to identify the states that are correlated with the variation of T_c. We have studied the insulator $Bi_2Sr_2YCu_2O_8$, the single phase superconductors $Bi_2Sr_2Ca_{n-1}Cu_nO_{4+2n}$ (n=2) over a range of critical temperature (temperature for resistivity zero) from 61K to 82K and the single phase 3 layer superconductors $Bi_2Sr_2Ca_{n-1}Cu_nO_{4+2n}$ (n=3) with T_c in the range 95K - 105K.

2. Experimental

The L_3 x-ray absorption experiment has been carried out on the super-ACO storage ring of the synchrotron radiation facility LURE at Orsay. A double crystal $10\overline{1}0$ beryl monochromator with an energy resolution about 0.35 eV at 900 eV has been used. The absorption coefficient has been measured by total electron yield method. The thickness of the probed surface layer is about 100 or 200 Å. The crystals have been cleaved with the surface normal direction z in the direction perpendicular to the CuO_2 planes.

The polarized spectra of the single crystals with several incidence angles between the electric field of the photon beam E and the sample surface normal z (the z axis was taken parallel to the c axis) have been recorded by rotating the sample, using synchrotron radiation linearly polarized in the horizontal plane. The polarized XAS spectra have been measured at four different incidence angles ranging from 10° to 75° degrees and at the magic angle which gives the unpolarized spectra. The spectra at 10° and at 75° degrees have been found to be very close to the extrapolated (E//z) and (E⊥z) polarized spectra.

The line-shape of the white line shows a broadening going from (E⊥z) to (E//z) as measured by studying the changes of the white line of a powder sample of CuO with the incidence angle, but this effect is so small that it is really negligible in comparison with intrinsic changes due to the superconducting materials in the two polarizations. The energy resolution depends also on the working conditions of the electron beam in the storage ring, therefore variations of the width and of the line shape can be extracted only by comparing spectra taken in the same experimental conditions. For these reasons the spectrum of CuO reference sample has been recorded immediately after each spectrum.

The samples were analyzed by microwave absorption to determine the presence of a single T_c phase of the surface area under investigation. The infrared reflectivity from 10 meV to 1 eV was measured to check high conductivity giving reflectivity close to one at the lowest energies. X-ray diffraction was measured to identify the structure and orientation of the surface layer studied by XAS. We have investigated a large set of powder samples, crystals, oriented pellets, and films. The samples were grown in Sendai, Parma and Rome. Similar results were found on samples grown in different laboratories and were confirmed by many experimental runs over about 15 months. The single phase samples were selected with complete orientation of the surface layer under study, and were classified with the value of T_c for zero resistivity.

3. Results

The polarized spectra of a highly oriented pellet formed by large grains with dimension larger than 2000 Å of the antiferromagnetic insulating sample $Bi_2Sr_2YCu_2O_8$ crystal is shown in Fig.1. The spectra recorded at incidence angle 10^0 (considered as the $E \perp z$ spectrum) and at 75^0 (close to the $E // z$ polarized spectra) are plotted with the spectrum recorded at the magic angle. The unpolarized spectra have been normalized to 1, the very weak continuum due to 2p->εd transitions in the continuum has been subtracted and the sum of the polarized spectra

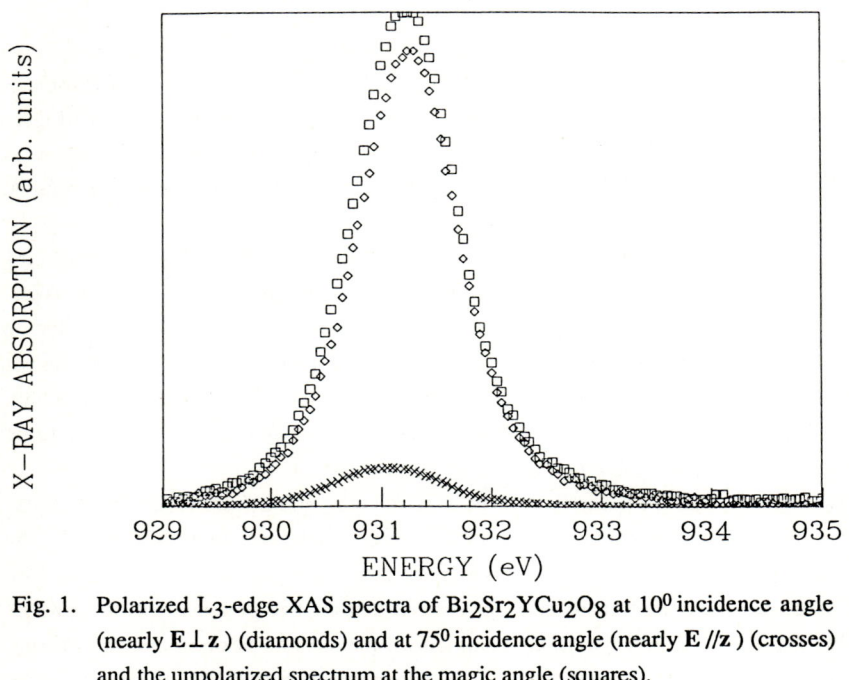

Fig. 1. Polarized L_3-edge XAS spectra of $Bi_2Sr_2YCu_2O_8$ at 10^0 incidence angle (nearly $E \perp z$) (diamonds) and at 75^0 incidence angle (nearly $E //z$) (crosses) and the unpolarized spectrum at the magic angle (squares).

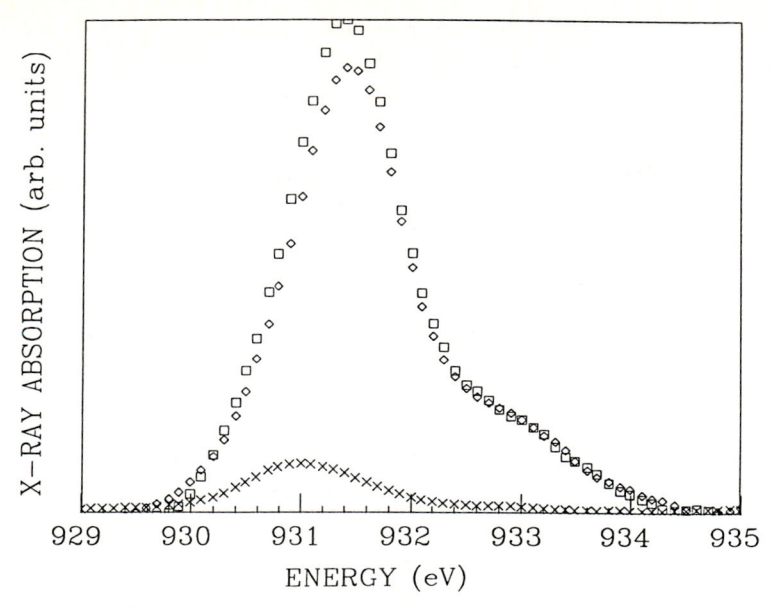

Fig. 2. Polarized L$_3$-XAS spectra of Bi$_2$Sr$_2$CaCu$_2$O$_8$ film T$_c$=61K (as in Fig.1).

gives the unpolarized spectrum. Fig. 2 shows polarized spectra of a well oriented 2000Å thick film 2212 (T$_c$=61K). Fig.3 shows the spectra of an oriented single phase pellet (T$_c$=70K).

Fig.4 shows the polarized spectra of a film with single 2223 phase T$_c$=95K. The samples were selected by measuring x-ray diffraction of the surface under study, in such a way that it was oriented with the c axis along the z direction and with less than 1% of unoriented microcrystals.

Finally Fig.5 shows the second derivative, multiplied by -1, of the unpolarized spectra of the insulator 2212-BiYSrCuO, of a 2212 sample with T$_c$=70K and of a third sample (T$_c$=110K) with 2223 structure. It is clear that the intensity of the low energy side of the white line, corresponding to the maximum of the white line of the E//z polarized spectrum increases with T$_c$.

The ratio of the integrated intensity of the white lines in the E//z and E//xy spectra of several samples with different T$_c$ is reported in Fig.6. In Fig.7 the energy separation ΔE_z between the E//z and E//xy white lines is reported for the three classes of systems including the LaSrCuO systems studied in ref. 24.

The results of Figs. 6 and 7 show the correlation between T$_c$ and 1) the ratio of the integrated intensities of the out-plane versus in-plane polarized spectra at room temperature and 2) the energy splitting between the in-plane and out-plane polarized spectra.

We have investigated the variation of the XAS by going from the normal phase at room temperature to the superconducting phase at 10K. The second derivative of unpolarized absorption spectra at room temperature and at 10K are shown in Figs. 8 and 9 for a 2212,

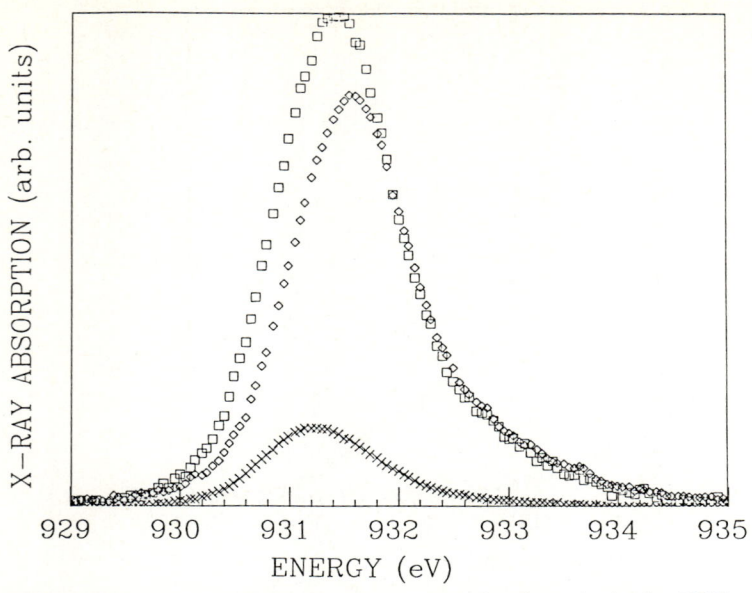

Fig. 3. Polarized L$_3$-edge XAS spectra of oriented single 2212 phase Bi$_2$Sr$_{1.5}$Ca$_{1.5}$Cu$_2$O$_8$ T$_c$=70K (notation as in Fig.1).

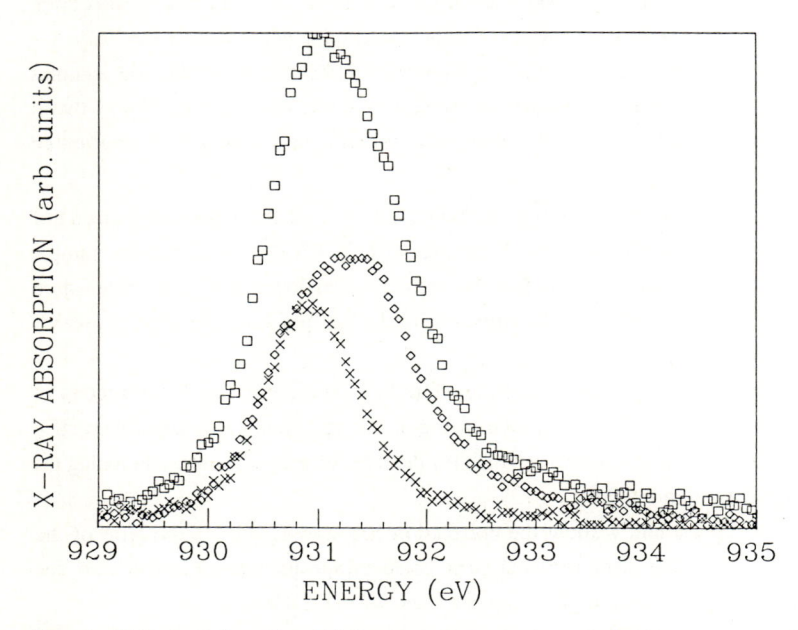

Fig. 4. Polarized L$_3$-edge XAS spectra of an oriented single phase BiSrCaCuO 2223 T$_c$=95K oriented film, nearly **E**⊥**z** (diamonds), nearly **E**//**z** (crosses) and the unpolarized spectrum at the magic angle (squares).

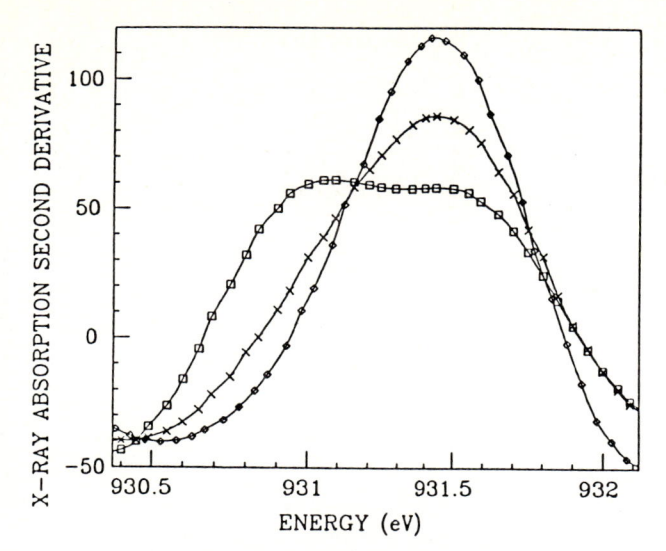

Fig. 5　Second derivative of the unpolarized spectra of insulating (diamonds) and of 70K (crosses) and 110K (squares) single phase superconductors.

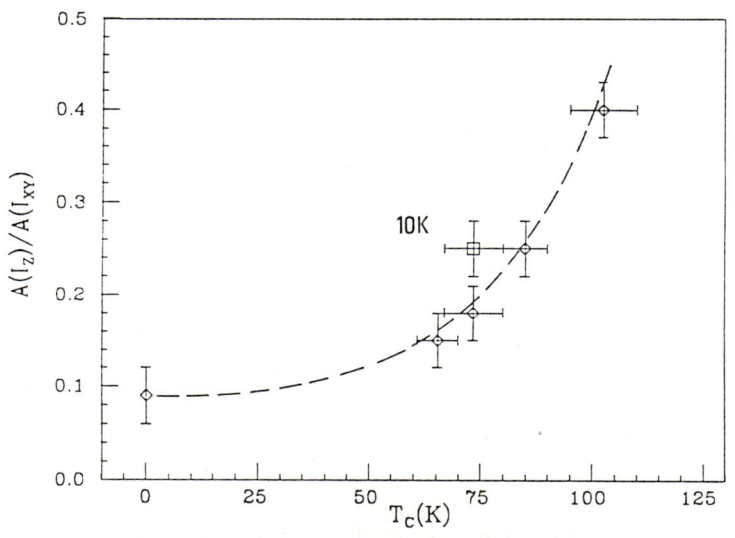

Fig. 6　Ratio between the integral of the polarized spectrum measured at 75^0 incidence angle (nearly $E//z$) and the integral of the polarized spectrum measured at 10^0 incidence angle (nearly $E//xy$) in samples with different T_c measured at room temperature and the ratio for the sample with $T_c=75K$ measured at 10K. The critical temperature of each sample is given as the middle ponit between T_c onset and the T_c, the error bars indicate the temperature range between T_c onset and the T_c.

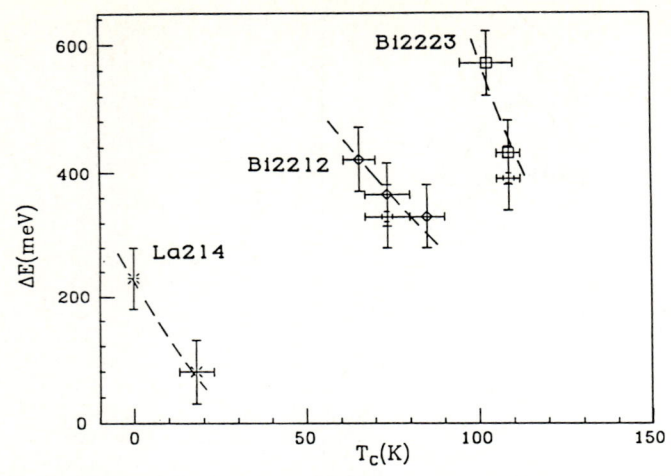

Fig. 7 Energy splitting ΔE_z between the peaks of polarized white lines in L$_3$-XAS ($E \perp c$ peak minus $E // c$ peak) as function of critical temperature (T_c) for the following classes of superconductors: La$_{2-x}$Sr$_x$CuO$_4$ (stars), BiSrCaCuO 2212 (diamond) and BiSrCaCuO 2223 (squares). The crosses with other symbols indicate the data collected at low temperature (10K).

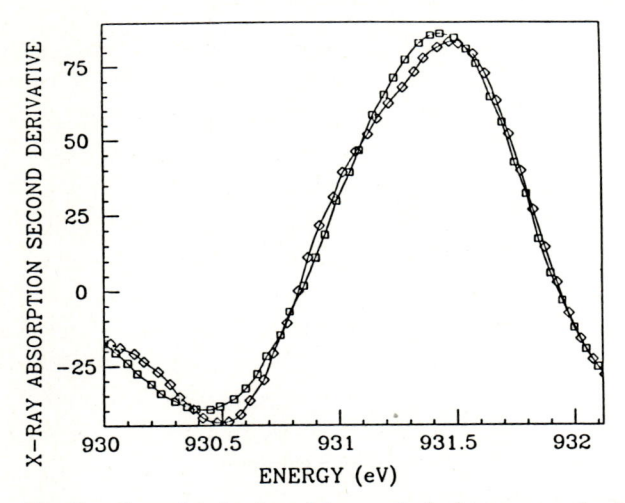

Fig. 8 Second derivative of the unpolarized spectrum of a single phase 2212 sample T_c=70K measured at room temperature (squares) and at 10K (diamonds).

T_c=70K, and a 2223, T_c=110K, sample respectively. The contribution of the E//z spectrum at lower energy increases at 10K as can be seen directly from Figs.8 and 9. The deconvolution with polarized spectra shows that for the 70K the E//z component increases by about 5% while the energy separation decreases by about 30 meV.

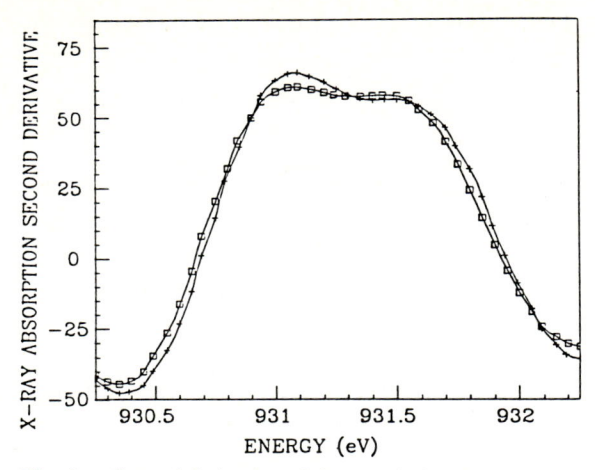

Fig. 9 Second derivative of the unpolarized spectrum of a single phase 2223 sample T_c=110K measured at room temperature (squares) and at 10K (crosses).

4. Discussion

a) On the integrated intensity of the E//xy and E//z white lines.

Cu L_3-XAS spectra probe the localized Cu 3d holes via the dipole matrix element. The spectra of divalent Cu^{2+} system, described by mixing of $3d^9$ and $3d^{10}\underline{L}$ in the ground state, are determined by a single final state $\underline{c}3d^{10}$. Neither final state multiplet splitting, there is only one core hole, nor final state hybridization with other possible final states will be present. Therefore the intensity of the single $\underline{c}3d^{10}$ white line will probe the probability of the d hole in the ground state. In hole doped Cu^{2+} systems, with $3d^9\underline{L}^*$ states, the possible final states are $\underline{c}3d^{10}\underline{L}^*$ and $\underline{c}3d^9$, as discussed by De Santis et al. [5]. The $\underline{c}3d^9$ final state, expected in the range of 938-945 eV because of the large Coulomb interaction between the core and the $\underline{d}$ hole, is not observed in the high T_c superconductors indicating the lack of probability for the $3d^8$ in the initial state. Under the condition that the $3d^8$ in the initial state and the $\underline{c}3d^9$ in the final state are not present it can be demonstrated that the ratio of the integrated oscillator strength of polarized E//z and E//xy XAS white lines, see Fig.6, is a measure of the ratio between the integrated partial density of states of $\underline{d}^*$ holes with E//z components of the angular moment and that of $\underline{d}^*$ holes with E//xy components of the angular moment in the ground state.

The present experimental findings show that the probability of the $\underline{d}^*$ holes with **non** $3d_{x^2-y^2}$ symmetry is in the range of 20-30% in high T_c superconductors, as probed by the E//z component of the Cu L_3 white line. The fact that the E//z components increases with T_c or with the doping indicates that the added ligand holes in the impurity band at the Fermi level induces a redistribution of the symmetry of the $\underline{d}^*$ states which were frozen in the b_1 symmetry, ml=2, in the insulator compound, and the d angular momentum can have also ml=0 (d_{z^2} or a_1) and/or ml=1 (d_{xz}, d_{yz} or e) components. The degrees of freedom for the d angular moment open the way to coupling of the electronic excitations with antiferrodistortive dynamical distortions.[26,27]

b) On the energy splitting of in-plane and out-of-plane polarized spectra.

The experimental data in Fig.7 show that the value of ΔE_z decreases on increasing T_c for the superconductors $La_{2-x}Sr_xCuO_4$, $Bi_2CaSr_2Cu_2O_{8-x}$ and $Bi_2Ca_2Sr_2Cu_3O_{10-x}$. For example the energy splitting ΔE_z for the 2212 superconductors was found to decrease from 420 ± 50 meV to 330 ± 50 meV, i.e. going from $T_c \sim 65K$ to $T_c = 75K$ with a rate of about 6 meV/K. For the LaSrCuO ΔE_z goes from 220 meV in La_2CuO_4 to 50 meV for a sample with $T_c = 20K$. It is clear that for high T_c materials in the limit of large doping the energy splitting goes toward zero and the bandwidths of E//xy and E//z become nearly the same indicating a larger mixing of ligands with different symmetries. It should be remarked that the threshold for the E//x and E//z spectra (the rising edge of the white line) are the same in all samples, and that the bandwidth of the E//xy is larger than that of the E//z line.

The interpretation of the energy splitting of the maxima of the final states in E//xy and E//z polarization depends on the description of the ground state.

1) Assuming that we have only a singlet state $\underline{d}(b_1)^* \underline{L}(b_1)^*$ we should have only one final state $\underline{cL}(b_1)^*$, this is in contradiction with the present experimental results.

2) Assuming that the ground state is formed by a singlet state $\underline{d}(b_1)^* \underline{L}(b_1)^*$ nearly degenerate with triplet state $\underline{d}(a_1)^* \underline{L}(b_1)^*$ and/or $\underline{d}(b_1)^* \underline{L}(a_1)^*$, according with the calculations of Annet et al. [7] for the lowest two hole states in La_2CuO_4, the final states in the XAS spectra will be two two final states $\underline{c}3d^{10}\underline{L}(a_1)^*$ and $\underline{c}3d^{10}\underline{L}(b_1)^*$ [5,28]. Therefore a possible intepretation of the present data is that the energy splitting of the two final states is probing the splitting between the $\underline{L}(a_1)^*$ and $\underline{L}(b_1)^*$ states. This energy splitting will be corrected by the energy difference between the final state interaction $V_{\underline{cL}(a1)}$ and $V_{\underline{cL}(b1)}$ between the core hole $\underline{c}$ and $\underline{L}(a_1)^*$ and $\underline{L}(b_1)^*$ hole respectively. Therefore, in this interpretation ΔE_z (the energy splitting) is a probe of the energy splitting between the $\underline{L}(a_1)^*$ and $\underline{L}(b_1)^*$ states which is related with the Jahn Teller splitting.

3) The angular resolved photoemission shows a band crossing the Fermi level with a dispersion of about 500 meV[29] and oxygen K-edge absorption shows the pinning of the Fermi level at the level of an impurity band[30]. Assuming that the ground state is described by a Fermi liquid, renormalized because of the large U_{dd}, with a single band crossing the Fermi level and assuming the one-electron approximation for core transitions, we will assign the low energy excitations, observed with both E//xy and E//z polarization, to the states **a** close to the Fermi level with a partial density of d states of both a_1 and b_1 symmetry and the final states at higher energy, observed with only E//xy polarization, to the upper part of the conduction band **b** with only d partial density of states of b_1 symmetry. The center of mass of the upper part of the band b is separated from the low energy states by ΔE_z. In this interpretation the final state interactions will change the line shape of the XAS giving for example the high energy tail which can be associated with $Cu(\underline{2p_x})\underline{L}(b_1)$ final states pushed up by V_{cL}.

Following this interpretation the XAS data show the presence of a band of a_1 symmetry, probably below the Fermi level, but mixed with the b_1 band in such a way to induce mixing between the two symmetries close to the Fermi level. This result is in qualitative agreement with most of the band structure calculations [6].

Finally the present results can be rationalized if the ground state is formed by a Fermi liquid with a mixture of a_1, b_1 and e molecular symmetries determined by possible crystalline distortions and mixing with an occupied $3d_{z^2}$ band close to the Fermi level.

c) The d-d excitation model.

These results point toward models for pairing mechanism where in the superconducting phase the d hole have **not only** the $3d_{x^2-y^2}$ symmetry.

In the d-d excitation model [14-17] the pairing mechanism is mediated by quadrupolar interaction between the oxygen hole and the Cu d states which induces virtual excitations of the $d_{x^2-y^2}$ hole into the d_{z^2} orbital. This proposed model for pairing in cuprate superconductors starts from the basic assumption that the symmetry of the d holes is not pure $d_{x^2-y^2}$ but has some admixture with d_{z^2} symmetry. This assumption is shown to be correct from our experimental data. In fact the spectral shape of the XAS spectra is modified by final state but the ratio of the integrated intensities of the polarized spectra is a measure of the ratio of integrated partial density of states of d holes having out-of-plane and in-plane orbital components in the ground state. The origin of this mixing can be due to a rhombic distortion of the Cu square plane in the CuO_2 plane given by the Bi induced reconstruction.

Following the prediction of the d-d virtual excitation theories for pairing both in the Weber[15] and Cox[17] model $k_B T_c$ is proportional to $\Delta_{JT} \exp(-1/\lambda)$, where Δ_{JT} is the energy splitting between the $d_{x^2-y^2}$ and d_{z^2} states while the coupling constant λ is related to the probability of the occupation of the d_{z^2} symmetry for the d holes and to $1/\Delta_{JT}$.

The Cu L_3 XAS provides the possibility to test experimentally the correlation between T_c and the probability of the $3d_{z^2}$ hole symmetry by measuring the ratio of the integrated intensities of the E//z versus E//xy Cu L_3 white line. The d-d excitation model predicts the increase of the critical temperature with increasing weight of the d_{z^2} component which enters in the coupling term λ in agreement with the present experiment, see Fig.6.

The d-d excitation model predicts the increasing of the critical temperature with decreasing splitting Δ_{JT}. The increasing value of T_c with decreasing Δ_{JT} is determined by the fact that Δ_{JT} enters also in the coupling constant. Assuming that the energy splitting ΔE_z between the in-plane and out-of-plane polarized XAS maxima is a probe of the $E(\underline{L}(a_1))-E(\underline{L}(b_1))$, which is related to Δ_{JT} then $\Delta E_z(T_c)$ will indicate a decrease of Δ_{JT} with increasing doping within each class of superconductors.

The variation of the XAS spectra from the normal phase to the superconducting phase shown in Figs. 8 and 9 indicate that the weight of the d_{z^2} component increases on going from the normal phase at 300K to the superconducting phase at 10K, both for the two layer 2212 systems and the three layer systems 2223. This effect can be associated with a structural phase transition at T_c, indicating a large electron-phonon coupling. In any case it shows that these structural changes go in the direction of increasing $3d_z^2$ character of the d holes.

5. Acknowledgements

We would like to thank M. Giura and his cryogenic group for characterization of the samples by microwave reflection, M. Capizzi and P. Calvani for measuring far-infrared reflectivity, C. Di Castro, C. Castellani, M. Grilli, and A. Kotani for useful discussions.

6. References

[1] J.G. Bednorz and K.A. Müller *Z. Phys. B-Condensed Matter* **64**, 189 (1986)

[2] B.K. Chakraverty *J.Phys.* **40**, L99 (1979)

[3] H. Thomas, these proceedings

[4] A. Bianconi, A. Congiu Castellano, M. De Santis, P. Rudolf, P. Lagarde, and A.M. Flank, *Solid State Commun.* **63**, 1009 (1987); A. Fujimori, E.Takayama-Muromachi, Y. Uchida and B. Okai, *Phys. Rev. B* **35**, 8814 (1987)

[5] A. Bianconi, A. Clozza, A. Congiu Castellano, S. Della Longa, M. De Santis, A. Di Cicco, K. Garg, P. Delogu, A. Gargano, R. Giorgi, P. Lagarde, A. M. Flank, and A. Marcelli *Jour. de Physique* **48** C9-1179 (1987); A. Bianconi, J. Budnick, A.M. Flank, A. Fontaine, P. Lagarde, A. Marcelli, H. Tolentino, B. Chamberland, C. Michel, B. Raveau, and G. Demazeau *Phys. Lett.* **127**, 285 (1988); M.DeSantis, A. Bianconi, A. Clozza, P. Castrucci, A. Di Cicco, M. De Simone, A.M. Flank, P. Lagarde, J. Budnick, P. Delogu, A. Gargano, R. Giorgi, T.D.Makris in *High T_c Superconductors: Electronic Sructure,* edited by A. Bianconi and A. Marcelli, Pergamon Press, Oxford, pag. 313 (1989).

[6] W.E. Pickett *Rev. Mod. Physics* **61**, 433 (1989); and *High T_c Superconductors: Electronic Structure* (proceedings of the international symposium on the electronic structure of high T_c superconductors, Roma 5-7 october 1989) edited by A.Bianconi and A. Marcelli, Pergamon Press, Oxford, 1989

[7] J.F. Annet. R.M. Martin, A.K. McMahan and S. Satpathy *Phys. Rev. B* 40, 2620 (1989); A.K. McMahan, R.M. Martin and S. Satpathy *Phys. Rev. B* **38**, 6650 (1988); A. Fujmori in *High T_c Superconductors: Electronic Sructure,* edited by A. Bianconi and A. Marcelli, Pergamon Press, Oxford pag.3 (1989); V.V.Flambaum and O.P.Sushkov *Physica C* **159**, 595 (1989).

[8] F.C. Zhang and T.M. Rice *Phys. Rev. B* **37**, 3759 (1988)

[9] C. Castellani, C. Di Castro and M. Grilli *Physica C* **153-155**, 1659 (1988).

[10] C. Castellani, C. Di Castro and M. Grilli *Int. Journal of Modern Physics* **1**, 659 (1988)

[11] H. Kamimura *Int. Journal of Modern Physics* **1**, 699 (1988).

[12] S. Kurihara *Physica C* **153-155**, 1247 (1988)

[13] T. Nishino, M. Kikuchi, and J. Kanamori *Solid State Commun.* **68**, 455 (1988)
[14] W. Weber, *Z. Phys. B- Condensed Matter* **70**, 323 (1988); W. Weber, *Advances in Solid State Physics* **28**, 141 (1988); and A.L. Shelankov, X. Zotos, and W. Weber *Physica C* **153-155**, 1307 (1988).
[15] Yu B. Gaididei and V.M. Loktev *Phys. Stat. Sol.* (b) **147**, 307 (1988)
[16] M. Jarrell, H.R. Krishnamurthy, and D.L. Cox *Phys. Rev. B* **38**, 4584 (1988)
[17] D.L. Cox, M. Jarrell, C. Jayaprakash, H.R. Krishna-murthy and J. Diez *Phys. Rev. Lett.* **62**, 2188 (1989)
[18] J. Askenazi and C.G. Kuper *Physica C* **153-155**, 1315 (1988); J. Askenazi, C.G. Kuper in *High T_C Superconductors: Electronic Sructure,* ref.6, pag.43 (1989)
[19] A. Aharony et al. *Phys. Rev. Lett.* **60**, 1330 (1988)
[20] A. Bianconi, M. De Santis, A.M. Flank, A. Fontaine, P. Lagarde, A. Marcelli, H. Katayama-Yoshida and A. Kotani. "*Physica C* **153-155**, 1760, (1988).
[21] A. Bianconi, M. De Santis, A. Di Cicco, A.M. Flank, A. Fontaine, P. Lagarde, H. Katayama-Yoshida, A. Kotani, and A. Marcelli *Phys. Rev. B* **38**, 7196 (1988).
[22] H.Tolentino, E. Dartyge A.Fontaine, G.Tourillon, T.Gourieux M.Maurer and M-F.Ravet in *High T_C Superconductors: Electronic Sructure,* ref.6, pag.245 (1989)
[23] A. Bianconi, P. Castrucci, M. De Santis, A. Di Cicco, A.M. Flank, P. Lagarde, H. Katayama-Yoshida, A. Marcelli, and Z.X. Zhao *Int. Jour. of Modern Physics* **1**,1151 (1988); A. Bianconi, P. Castrucci, M. De Santis, A. Di Cicco, A. Fabrizi, A.M. Flank, P. Lagarde, H. Katayama-Yoshida, A. Kotani, A. Marcelli, Z.X. Zhao and C. Politis *Modern Physics Letters B* **2**, 1313 (1988)
[24] A. Bianconi, P. Castrucci, A. Fabrizi, A.M. Flank, P. Lagarde, A. Marcelli, Y. Endoh, H. Katayama-Yoshida, and Z.X. Zhao, in *High T_C Superconductors: Electronic Sructure,* A. Bianconi and A. Marcelli eds, Pergamon Press,Oxford, pag. 281 (1989)
[25] J. Fink, N. Nücker, H. Romberg and S. Nakai in *High T_C Superconductors: Electronic Sructure,* edited by A. Bianconi and A. Marcelli, Pergamon Press, Oxford, pag. 293 (1989); *Phys. Rev. B* **39**, 6619 (1989); and in these proceedings
[26] K.I. Kugel and D.I. Khomskii *Sov. Phys. Usp.* **25** 231 (1982)
[27] J. Zaanen in *Int. Symposium on High Temperature Superconductivity*, Jaipur, July 6-8, 1988 edited by K.B.Garg, Oxford &IBH Publ. Co, New Delhi, pag. 31 (1989)
[28] K. Okada, and A. Kotani *J. Phys. Soc. Jpn* to be published,
[29] T. Takahashi, H. Matsuyama, H. Katayama-Yoshida, Y. Okabe, S. Hosoya, K.Seki, H. Fujimoto, M.Sato, and H. Inokuchi *Nature* **334**, 691 (1988); and *Phys. Rev.* **39**, 6636 (1989); R. Manzke, T. Buslaps, R. Claessen, M. Skibowski, and J. Fink *Physica Scripta* (1989) in press
[30] H. Matsuyama, T. Takahashi, N. Kosugi, H. Fujmoto, H. Katayama-Joshida, T. Kashiwakura, Y. Okabe, S. Sato, A. Yagishita, K. Tanaka, M. Sato, and H. Inokuchi *to be published*

Magnetic and Electronic Correlations in YBa$_2$Cu$_3$O$_{6+x}$

J.M. Tranquada

Physics Department, Brookhaven National Laboratory, Upton, NY 11973, USA

Abstract. YBa$_2$Cu$_3$O$_{6+x}$ is an antiferromagnetic insulator in its tetragonal phase at small x, and it becomes a superconducting metal in the large x orthorhombic phase. The transition between the two phases is controlled by the orientational ordering of CuO$_x$ chain segments and dielectric screening by the BaO layers. In the orthorhombic phase, holes from the chains are transferred to the CuO$_2$ planes, and the superconducting transition temperature scales with the density of O $2p$ holes. The antiferromagnetic ordering in the tetragonal phase is dominated by localized Cu moments in the CuO$_2$ planes. The two-dimensional (2D) coupling between those spins is unusually strong. The interaction between nearest-neighbor planes is sufficiently strong that spins in bilayers remain highly correlated well above the Néel temperature. Long-range order is destroyed when a significant density of O $2p$ holes is present in the planes; however, there is evidence that Cu moments survive and interact in the metallic phase.

1. Introduction

Before theorists can come to a consensus on the electron-pairing mechanism responsible for superconductivity in the layered copper-oxide compounds (or even agree on the Hamiltonian that contains the necessary basic physics of the problem), it is necessary to have a clear picture of the electronic and magnetic structure of these materials in the normal state. It is now generally agreed that the crucial structural element for the unusual superconductivity is the CuO$_2$ plane: a square lattice of copper atoms with an oxygen bridging each pair of nearest-neighbor coppers. The first superconducting cuprate discovered [1], La$_{2-x}$Ba$_x$CuO$_4$, contains only CuO$_2$ planes between layers of La(Ba)O. In the 90-K superconductor YBa$_2$Cu$_3$CuO$_7$ [2], however, each unit cell contains two CuO$_2$ planes plus a Cu layer in which oxygen bridges the Cu along just one direction, creating parallel CuO chains (see Fig. 1) [3]. The presence of the CuO chains causes the b lattice parameter (along the chain direction)

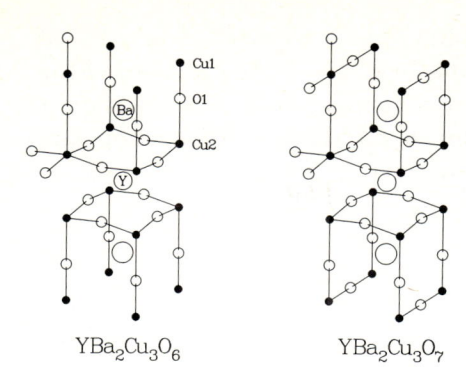

Fig. 1. Crystal structures of $YBa_2Cu_3O_6$, in which the Cu(1) sites are 2-fold coordinated, and $YBa_2Cu_3O_7$, in which the 4-fold coordinated Cu(1) sites form chains along the b axis.

to be longer than a, so that the structure is orthorhombic. When all of the oxygen is removed from the chains, as in $YBa_2Cu_3O_6$, the structure is tetragonal [4], and the material is an antiferromagnetic insulator [5, 6]. In the superconducting compound, the orthorhombic strain causes twinning: the chain direction can change by 90° at (110) boundaries between neighboring domains.

The presence of the chains and twin boundaries initially led to considerable confusion concerning the feature(s) responsible for the high superconducting transition temperature (T_c) in $YBa_2Cu_3O_7$. From a selective analysis of experimental results, it is now clear that it is the density of holes of predominantly O $2p$ character within the CuO_2 planes that determines T_c. The density of holes is controlled by the average occupancy x of the oxygen sites in the the CuO chain layer, and also the ordering of the oxygens on those sites. The distribution of holes between the chains and planes is stabilized electrostatically by displacements within the BaO layers. Furthermore, there is substantial evidence that the Cu atoms in the planes remain essentially 2+ (one $3d$ hole per Cu) in the presence of the O $2p$ holes, and that spin-spin interactions between neighboring Cu moments are significant in the metallic phase.

The experimentally determined phase diagram for $YBa_2Cu_3O_{6+x}$ is shown in Fig. 2. In the tetragonal phase the compound is an antiferromagnetic insulator, whereas it is a metallic superconductor when the lattice becomes orthorhombic. To understand the phase diagram it is necessary to consider the connections between the crystalline, electronic, and magnetic structures. Some important features of the oxygen dependence of the crystal structure will be reviewed first, followed by a short description of variations in the electronic structure with x. Magnetic order and correlations will then be discussed, and the paper concludes with a summary.

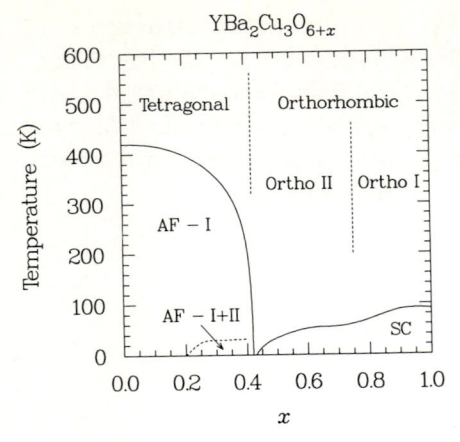

Fig. 2. Phase diagram of $YBa_2Cu_3O_{6+x}$. The positions of the phase boundaries with respect to x depend on sample preparation and treatment.

2. Oxygen Dependence of the Crystal Structure

The filling and ordering of the Cu(1)-O_x chains control the filling of valence band states, and hence they control the electronic and magnetic properties of $YBa_2Cu_3O_{6+x}$. Copper finds it energetically favorable to be either linearly 2-fold or planar 4-fold coordinated by oxygen. A 4-fold coordinated Cu can tolerate one or two more distant oxygen neighbors, such as the apical O(1) near the Cu(2) in the CuO_2 planes (see Fig. 1), but 3-fold coordination is uncommon. Accepting these observations as rules, it is clear that for arbitrary x the oxygen in the Cu(1) planes will cluster into chains so as to maximize the number of 2- and 4-fold coordinated Cu(1) sites. If the orientational order to the chains is coherent from layer to layer, then the crystal will be orthorhombic; however, at small x, where the chain segments become shorter and more separated, the orientation may vary between layers or within a layer, and the average symmetry becomes tetragonal [7].

Because the full or empty chains may be charged, like-chains will tend to repel each other. As a result, a set of full and empty chains will tend to order themselves into a regular pattern. The dominant type of ordering observed by electron [8] and x-ray [9] diffraction for intermediate x consists of alternating full and empty chains, with a resultant doubling of the unit cell along the a axis. Theoretical model calculations [10] which reproduce this ordering indicate that there are two distinct orthorhombic phases: Ortho I, consisting of full chains plus defects, and Ortho II, corresponding to the cell-doubled phase. The equilibrium width in x of the Ortho II phase depends on sample treatment (annealing temperature and oxygen partial pressure) [11, 12].

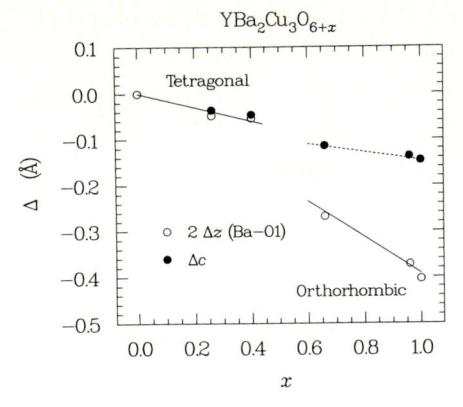

Fig. 3. Change in twice the Ba-O(1) intralayer spacing determined by neutron diffraction [3, 4] compared to the the change in the c lattice parameter as a function of x. Note that there are two BaO layers per unit cell.

As will be discussed in the next section, when oxygen atoms are added to an empty chain, they can obtain only half of the charge they require from their Cu(1) neighbors. Thus, either holes must be created in the chains, or electrons must be transferred from the planes to the chains, creating holes in the planes. Transferring electrons from the planes to the chains would change the electric field across the Ba-O(1) layer. The Ba and O(1) atoms are displaced from each other along the c-axis by an amount Δz(Ba-O1), and this separation contracts strongly with increasing x in the orthorhombic phase as shown in Fig. 3 [3, 4], consistent with plane-to-chain charge transfer [13]. Thus, the Ba-O(1) layer acts as a dielectric to screen the change in electric field due to the charge transfer. In the tetragonal phase the Ba-O(1) separation changes very little with x, suggesting that there is very little charge transfer. Thus, the density of holes in the planes should change in nearly a step-like way on going from the tetragonal to the orthorhombic phase.

3. Electronic Structure

It is now fairly well established that both the Cu(1) and Cu(2) atoms have a 2+ valence ($3d^9$ configuration) in the fully oxygenated $x = 1$ compound [14]. That the holes have a dominantly O $2p$ character is confirmed by x-ray absorption [15] and electron-energy-loss [16] measurements at the O K edge. The orientation dependence of O K edge measurements on single crystals indicate that the holes are shared between the chain and plane oxygens [17]. Evidence for O $2p$ holes in the planes and chains has also been obtained from nuclear magnetic resonance (NMR) studies probing ^{63}Cu [18], ^{17}O [19], and ^{89}Y [20] nuclei. Superconductivity occurs only when a significant density of holes is present in the planes. The hole density, and hence T_c, vary with x, and the details of that variation are controlled by the ordering of the oxygens in the Cu(1) layer.

If the oxygen atoms in the Cu(1) layer tend to cluster in chains, then there should be empty chain segments containing 2-fold coordinated Cu(1) for all $x < 1$. One expects these Cu(1) atoms to have a 1+ valence (i.e. a full $3d$ shell). A practical way to detect Cu(1)$^{1+}$ is through x-ray absorption measurements at the Cu K edge. As shown in Fig. 4, the $1s \rightarrow 4p_\pi$ transition for a Cu^{1+} ion, as in Cu$_2$O, is at a much lower energy than it is for Cu^{2+}, as in CuO. This low energy feature can be used as a fingerprint to indicate the presence of Cu^{1+}[21]. The lower part of Fig. 4 shows measurements on an oxygenated ($x = 1$) and an oxygen deficient ($x = 0.23$) sample, each of which had been uniaxially oriented in a magnetic field [22, 23]. Measurements with the x-ray polarization $\hat{\varepsilon}$ parallel to and perpendicular to the c-axis of each sample reveal that the distinctive low energy Cu^{1+} feature is present only for the oxygen deficient sample with the polarization perpendicular to the c-axis. Since the Cu^{1+} $4p_\pi$ final state must be oriented perpendicular to the ligand axis [24], it is clear that the observed Cu^{1+} feature is due to 2-fold coordinated Cu(1) sites. The presence of Cu^{1+} in oxygen-deficient YBa$_2$Cu$_3$O$_{6+x}$ has also been detected and characterized by optical spectroscopy [25] and by nuclear quadrupole resonance (NQR) [26] and nuclear magnetic resonance (NMR) [27] spectroscopy.

A measure of the amount of Cu^{1+} present is given by the area under the low energy peak obtained after subtracting off the $x = 1$ spectrum. It can be put on an absolute scale by normalizing to the Cu$_2$O spectrum. Such an analysis has been performed on a series of unoriented samples having a range of x values [23], and the results are shown in Fig. 5. If all of the Cu(1) atoms are in 2- or 4-fold sites, then the number of 2-fold sites is equal to $1 - x$, corresponding to the solid line in Fig. 5. The data are consistent with essentially all of the Cu(1) sites being in full or empty chains, with the atoms in empty chains having a valence of 1+.

It seems likely that the oxygen atoms cluster into chain segments in the tetragonal phase; however, it is interesting to consider another limit in which evergy oxygen is isolated and sits between two 3-fold coordinated Cu(1) atoms. If one assumes that the 3-fold sites have a 2+ valence, then the density of Cu^{1+} would be $1 - 2x$, corresponding to the dashed line in Fig. 5. Taking into account a possible systematic error in the normalization, the data appear more consistent with chain segments in the tetragonal phase.

From simple valence counting, the result that the density of Cu(1)$^{1+}$ atoms is equal to $1 - x$ implies that the density of holes n_h is equal to x.

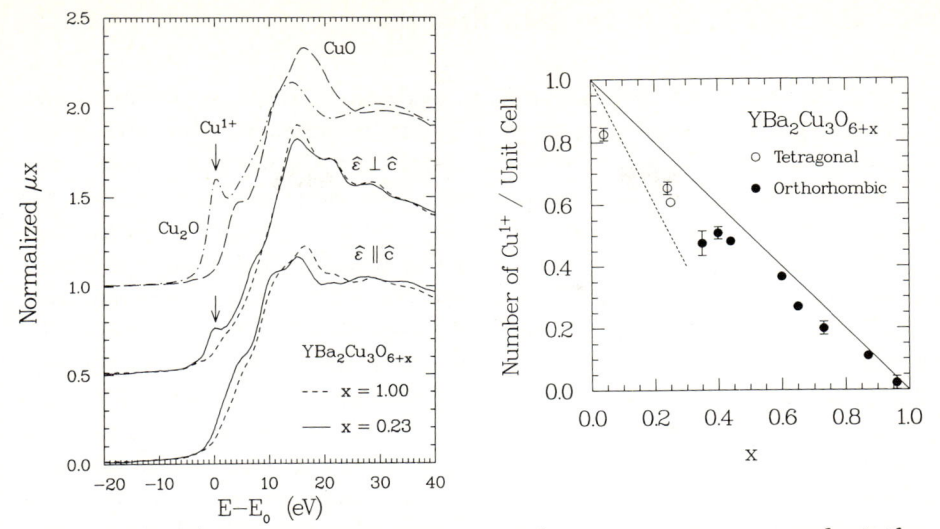

Fig. 4. (Left) X-ray absorption near edge structure measured at the Cu K edge, with $E_0 = 8980$ eV. At the top are data for unoriented powders of Cu_2O and CuO. Below are spectra, for uniaxially aligned samples of $YBa_2Cu_3O_{6+x}$ with $x = 0.23$ and 1.00, measured with the x-ray polarization vector $\hat{\varepsilon}$ perpendicular to and parallel to the c-axis direction $\hat{c}$. From Ref. [23].

Fig. 5. (Right) Number of Cu^{1+} ions per unit cell measured in $YBa_2Cu_3O_{6+x}$ plotted as a function of x. The solid line represents $n_{Cu^{1+}} = 1 - x$, while the dashed line indicates $n_{Cu^{1+}} = 1 - 2x$. From Ref. [23].

For the quenched samples used in the Cu K edge study, it was observed that $T_c \sim x$ [23]. It follows that $T_c \sim n_h$. A similar result was obtained in a muon spin relaxation study [28]. By studying the relaxation rate for superconducting $YBa_2Cu_3O_{6+x}$ in an applied magnetic field, it was established that, at least for $x \lesssim 0.9$, $T_c \sim n_h/m^*$, where m^* is the effective mass of the holes. For annealed samples, a wide plateau develops at ~ 60 K in the T_c vs. x curve [11], corresponding to the Ortho II phase. It has been argued that the plateau structure is due to percolation effects resulting from the inhomogeneity of the chain ordering on the length scale of the superconducting coherence length (20–30 Å) [29].

4. Magnetic Structure and Excitations

In the fully reduced compound $YBa_2Cu_3O_6$ the Cu(2) sites in the planes have magnetic moments by virtue of their 2+ valence, while the 2-fold coordinated Cu(1) sites are nonmagnetic. The Cu moments within the CuO_2 planes order antiferromagnetically because of nearest-neighbor superexchange interactions, and the planes couple together antiferromagnetically along the c axis, as shown in Fig. 6(a) [6]. The magnetic Bragg peak intensity is proportional to the square of the staggered magnetization, which in turn is proportional to the average spin $\langle S \rangle$. Experimetally, a value of $M = g\mu_B \langle S \rangle \approx 0.65\mu_B$ was observed at low temperature [30, 31] where μ_B is the Bohr magneton. Assuming a typical value of $g \sim 2.2$ for a spin-$\frac{1}{2}$ Cu^{2+} ion, one obtains $\langle S \rangle \approx 0.30$. As will be discussed below, the magnetic interactions within the CuO_2 planes are much stronger than those between the planes, and it is commonly believed that the spin dynamics should be very similar to those of a 2D Heisenberg system. Because of the very small size of the spin, quantum corrections become important, and a large zero-point spin fluctuation is expected. Spin-wave analysis applied to the spin-$\frac{1}{2}$, 2D Heisenberg model yields $\langle S \rangle = 0.30$ [32], in very good agreement with the experimental result. Thus, the neutron diffraction results are consistent with localized magnetic moments of approximately 1 μ_B and large fluctuations in spin orientation even at low temperatures.

The temperature dependence of the magnetic ordering has been studied as a function of oxygen content x by muon spin rotation [33], neutron diffraction [30, 31], and Cu NQR [34], and the results are summarized by the magnetic phase boundaries shown in Fig. 2. With increasing x,

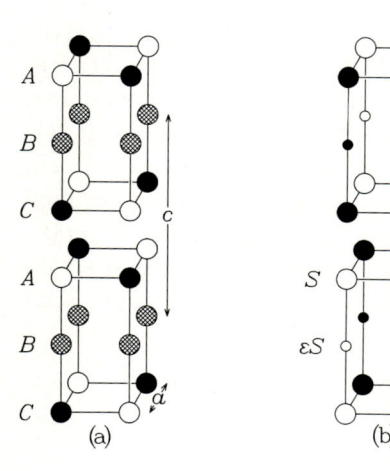

Fig. 6. (a) Magnetic spin structure for $YBa_2Cu_3O_{6+x}$ with x near zero. Only copper atoms are shown for clarity; cross-hatched circles represent nonmagnetic Cu^{1+} ions, while black and white circles indicate antiparallel spins at Cu^{2+} sites. Solid lines connect pairs of sites bridged by oxygen atoms. (b) Second type of spin structure expected for larger x. The average spin at a B-layer site is a fraction ε of the spin on a CuO_2 layer.

oxygen enters the Cu(1) layer and creates some Cu(1)$^{2+}$. Clustering of oxygen into chains should also create some holes, but very few of these are transferred to the CuO$_2$ planes, as discussed in the previous two sections. This conclusion is confirmed by the observation that T_N stays fairly high over a large range of x, in contrast to La$_{2-x}$Sr$_x$CuO$_{4-y}$ in which a very small amount of doping destroys the long-range order [35]. In YBa$_2$Cu$_3$O$_{6+x}$ the magnetic order is not destroyed by mobile holes until the tetragonal-orthorhombic phase boundary is reached. The magnetic Cu(1)$^{2+}$ ions are frustrated in the Type II structure of Fig. 6(a). By Hund's rules, one expects the Cu(1) to couple ferromagnetically to its Cu(2) neighbors [36, 37], causing the Cu(2) spins in next-nearest-neighbor planes to be ferromagnetically aligned. When a sufficient number of moments are present on Cu(1) sites, one would expect the Type II structure of Fig. 6(b) to be favorable [38]. The average Cu(1) spin should be much smaller than the average Cu(2) spin.

Superlattice peaks corresponding to the Type II structure were observed to coexist with Type I peaks below 40 K in one single crystal with $x \sim 0.35$ studied by neutron diffraction [39]. As the temperature decreased below 40 K, the intensities of the Type I peaks decreased while the Type II peaks grew. A more common observation in single crystal diffraction studies is that, for $x \gtrsim 0.2$, the Type I magnetic Bragg peaks begin to decrease as the temperature is lowered below 30–50 K [see Fig. 7(a)], but instead of new superlattice peaks appearing, 2D diffuse scattering is found [31, 40]. Furthermore, there is a low energy component to the inelastic scattering which is in excess of that expected for the temperature dependence of spin-waves from the CuO$_2$ layers, as shown in Fig. 7(b) and (c). The intensity of the low-energy scattering reaches a maximum in the temperature range where the magnetic Bragg intensity begins to decrease, suggesting that it is due to critical fluctuations of Cu(1)$^{2+}$ moments whose spin direction freezes at low temperature. Similarly, a recent magnetic susceptibility study [41] has revealed a Curie-like contribution that increases in magnitude with x up to the tetragonal-orthorhombic phase boundary, but which disappears below 20–30 K. Evidence for a low-temperature transition to a spin-glass-like phase comes from Cu NQR measurements. For $x \geq 0.2$, the spin-echo decay rate for the NQR for Cu(1)$^{1+}$ diverges at 20 K [42]. A single broad peak is observed at low temperature. If the low-temperature magnetic structure were of Type II, then all Cu(1) sites should see the same constant hyperfine field, and the Cu(1)$^{1+}$ peak should be split in two. In the only case where such a splitting was observed, the sample was found to contain a small amount of Fe impurity [43], which tends to substitute on the Cu(1) site. Substituting magnetic Co ions on Cu(1) sites also induces Type II

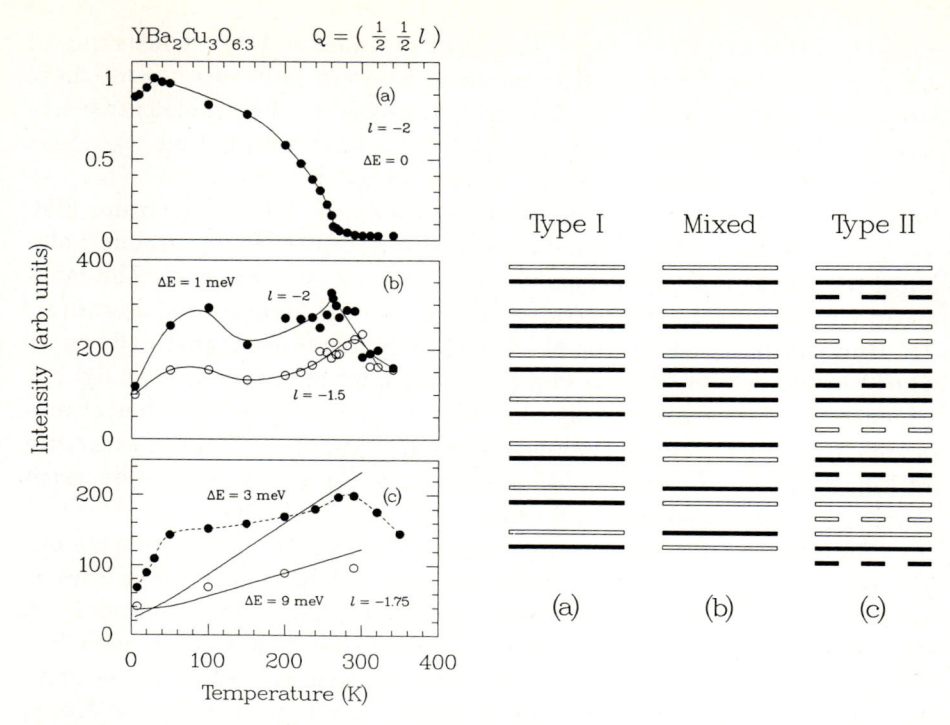

Fig. 7. (Left) Temperature dependence of elastic and inelastic magnetic scattering in $YBa_2Cu_3O_{6.3}$. (a) Elastic Bragg peak intensity at $(\frac{1}{2},\frac{1}{2},-2)$. (b) Inelastic intensity for $\Delta E = 1$ meV at $(\frac{1}{2},\frac{1}{2},-2)$ and $(\frac{1}{2},\frac{1}{2},-1.5)$. (c) Background corrected inelastic intensity at $(\frac{1}{2},\frac{1}{2},-1.75)$ for $\Delta E = 3$ and 9 meV. The solid lines represent the calculated temperature dependence normalized to the 9 meV point at 200 K, taking the spectrometer resolution function into account; the dashed line is a guide to the eye. From Ref. [40].

Fig. 8. (Right) Schematic diagrams of different magnetic structures, as discussed in the text. Each bar represents an antiferromagnetic CuO_2 plane in which the Cu spins have a simple Néel order. Black and white bars indicate layers with antiparallel spins. Broken bars represent magnetically ordered Cu(1) layers. From Ref. [40].

order [44]. The Type II order observed in $Nd_{1+y}Ba_{1-y}Cu_3O_{6+x}$ [45] is probably caused by the presence of excess Nd^{3+} on Ba^{2+} sites.

The evidence indicates that the $Cu(1)^{2+}$ moments do order in some way at low temperature, but that they do not exhibit the coherent order of the Type II structure. If random Cu(1) layers couple to the planes with local Type II structure, then a mixture of Type I and Type II structures could form, as illustrated in Fig. 8. Isolated Type II layers would contribute diffuse rather than Bragg scattering, and their presence would reduce the number of CuO$_2$ planes contributing to the Type I superlattice peaks. Most Cu(1)$^{1+}$ sites would see no hyperfine field, consistent with the Cu NQR results.

A classical antiferromagnet would have perfect Néel order at zero temperature. At a finite temperature the spins will fluctuate around their zero temperature orientations, and these fluctuations can be analyzed in terms of harmonic spin waves. In a quantum mechanical antiferromagnet spin waves are present even at $T = 0$, and correspond to the zero-point fluctuations of the spins. In a nearest-neighbor Heisenberg model, the Hamiltonian for spin-spin interactions has the form

$$H = J \sum_{\langle i,j \rangle} \mathbf{S}_i \cdot \mathbf{S}_j,$$

where the sum is over nearest-neighbor pairs and J is called the exchange energy. Diagonalization of the Hamiltonian leads to a dispersion relation for the spin-wave energy $\hbar\omega$ as a function of wavevector $\mathbf{q}$, measured relative to a magnetic Bragg point. For small q the dispersion is linear, with a velocity $c = \omega/q$ proportional to the exchange energy J. Hence, by measuring the slope of the spin-wave dispersion with inelastic neutron scattering one can determine the strength of coupling between the spins.

Anderson [46] was the first to point out that the superexchange interaction between nearest-neighbor coppers within a CuO$_2$ plane should be unusually strong, and that the magnetism should be essentially two-dimensional in character. These speculations were initially confirmed by neutron scattering studies of La$_2$CuO$_4$ [47]. For a 2D system, the scattering cross section does not depend on the momentum transfer perpendicular to the plane, so that in reciprocal space Bragg points are replaced by Bragg rods. Figure 9 shows the (hhl) zone in reciprocal space for YBa$_2$Cu$_3$O$_{6+x}$, with the rod for antiferromagnetic scattering from a CuO$_2$ layer indicated by the hatched line. Below T_N, magnetic Bragg peaks are observed at the points indicated by the squares. Assuming that the spin dynamics are dominated by the exchange energy $J_\parallel$ between nearest-neighbors within a CuO$_2$ plane, one expects to observe spin waves rising steeply in energy as $\mathbf{q}$ is scanned in the $[hh0]$ direction away from the 2D rod, as indicated by Scan A in Fig. 9. In an inelastic neutron scattering measurement, if one fixes the energy transfer $\Delta E = \hbar\omega$ and performs Scan A, spin wave peaks should be observed at $q_\parallel = \pm\omega/c$, where $\mathbf{q}_\parallel$ is the component of $\mathbf{q}$ perpendicular to the rod. Such scans, with $\Delta E = 3$, 9, and 15 meV and measured on a single crystal with $x \sim 0.3$, are shown in Fig. 10. Because of the very large spin-wave veloc-

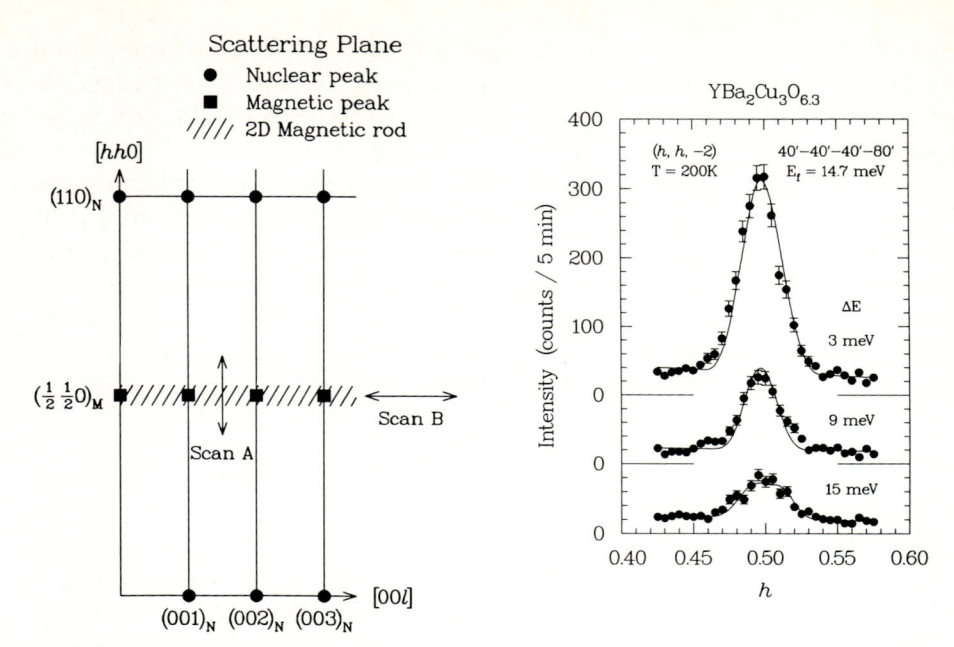

Fig. 9. (Left) Reciprocal space (hhl) zone. The hatched line along $(\frac{1}{2},\frac{1}{2},l)$ is the magnetic ridge for two-dimensional scattering; A and B indicate scans across and along the ridge, respectively.

Fig. 10. (Right) Several constant ΔE scans of type A (see Fig. 9) across the 2D magnetic ridge in a $YBa_2Cu_3O_{6.3}$ single crystal at 200 K. The intensities have been adjusted slightly to correct for the presence of $\lambda/2$ contamination seen by the incident beam monitor. The solid lines are fits to the data assuming linear spin-wave dispersion and taking into account the spectrometer resolution function. From Ref. [40].

ity, the two spin-wave peaks cannot be resolved. By taking into account the spectrometer resolution function, one can fit the unresolved peaks as indicated by the solid lines, yielding the crude estimate $J_{\|} = 80^{+60}_{-30}$ meV. Inelastic light-scattering measurements of spin-pair excitations indicate that $J_{\|} \approx 120$ meV in a single crystal with $x \approx 0.0$ [48].

If the CuO_2 layers were truly uncoupled, then there would be no spin-wave dispersion along the $[00l]$ direction indicated by Scan B in Fig. 9. In reality, the Cu spins are coupled weakly along the c axis, and because there are two Cu(2) atoms per unit cell, the spin-wave modes are split into acoustic and optical branches, in analogy with phonon modes in a non-Bravais lattice. Two different c-axis couplings must be considered: $J_{\perp 1}$, between nearest-neighbor (nn) CuO_2 planes, and $J_{\perp 2}$, between next-nn

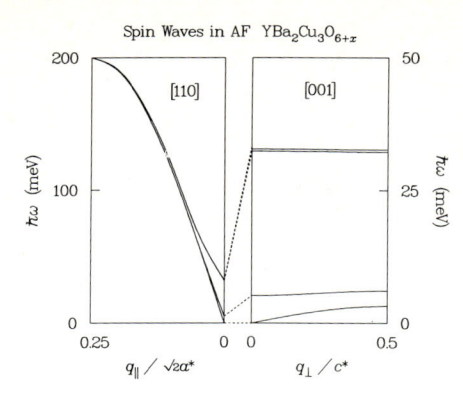

Fig. 11. Schematic diagram of spin-wave dispersion in $YBa_2Cu_3O_{6+x}$. Note that the energy scales for the two panels differ by a factor of four. The 2D nature of the magnetic interactions makes the dispersion very large along $q_\parallel$ but extremely weak along $q_\perp$. From Ref. [40]

layers through the Cu(1) sites. It turns out that the direct exchange interaction corresponding to $J_{\perp 1}$ is much stronger than $J_{\perp 2}$. As a result, $J_{\perp 1}$ determines the zone-center splitting of the acoustic and optical frequencies, and $J_{\perp 2}$ is responsible for the dispersion along $[00l]$. There is also a weak XY-like anisotropy in the CuO_2 planes due to spin-orbit coupling, which splits the acoustic and optical branches into in-plane and out-of-plane modes. A schematic diagram of the spin-wave dispersion is shown in Fig. 11 [40].

The scattering cross section for each branch is modulated by the square of a magnetic structure factor $g(\mathbf{q})$, where

$$g(\mathbf{q}) = \begin{cases} \sin(\pi z l), & \text{for acoustic modes;} \\ \cos(\pi z l), & \text{for optical modes;} \end{cases}$$

and where $z = 0.28$ is the relative separation between nn Cu(2) sites along the c axis [40, 49]. Figure 12 shows scans along the $(\frac{1}{2}\frac{1}{2}l)$ direction at $\Delta E = 6$ meV for several different temperatures [40]. The solid line is a fit using spin-wave theory and taking the spectrometer resolution function into account. The phase and amplitude of the modulation make it clear that only acoustic modes contribute to the scattering. Quite remarkably, the structure factor modulation is still present well above T_N, indicating that CuO_2 bilayers remain strongly correlated in the absence of three-dimensional order. The optical modes are not observed up to at least 30 meV, providing the limit $J_{\perp 1} \gtrsim 2$ meV [40]. $J_{\perp 2}$ and the exchange anisotropy are much smaller, being on the order of $10^{-4} \times J_\parallel$.

What happens to the magnetic moments in the metallic orthorhombic phase of $YBa_2Cu_3O_{6+x}$? As discussed in the previous section, high-energy spectroscopies indicate that the Cu(2) atoms retain their 2+ valence for all x, suggesting that they should remain magnetic. Temperature-dependent magnetic susceptibility measured for varying

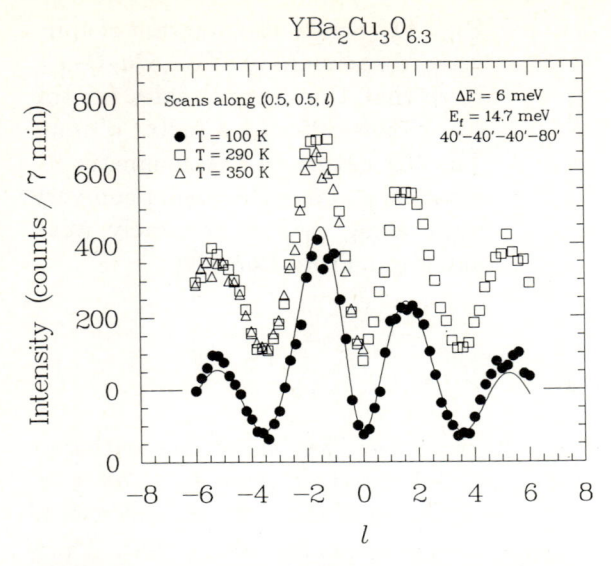

Fig. 12. Constant ΔE scan of type B (see Fig. 9) along the 2D rod with $\Delta E = 6$ meV for YBa$_2$Cu$_3$O$_{6.3}$. The modulation is due to the inelastic structure. The enhancement near $l = -1.5$ is due to a "focussing" effect of the spectrometer resolution function. The solid line is a fit which is discussed in the text. From Ref. [40].

oxygen content and corrected for a small Curie-like contribution is shown in Fig. 13 [30, 50]. At low temperatures the susceptibility has the shape expected for a 2D antiferromagnet [51], and it appears to evolve continuously as the boundary between the antiferromagnetic and the metallic phases is crossed. A much more drastic change would be expected if the magnetic moments and/or the correlations suddenly disappeared in the orthorhombic phase.

The results of neutron scattering searches for low energy magnetic scattering have been mixed so far. Mezei et al. [52] observed a magnetic cross section at very low energies (≤ 1.5 meV) and low temperature (< 75 K) in a powder sample with $x = 0.59$ and $T_c = 47$ K. Fairly strong spin-wave-like scattering was recently observed in a large orthorhombic crystal with $x \sim 0.45$ and $T_c \approx 15$ K [53]; however, initial measurements on several other crystals with larger oxygen concentrations were negative [40]. Brückel et al. [54] found negligible magnetic scattering for $|\hbar\omega| < 25$ meV in an $x = 1$ powder sample, and they concluded that most of the Cu atoms are nonmagnetic. However, that conclusion is based on the assumption that most of the magnetic scattering should

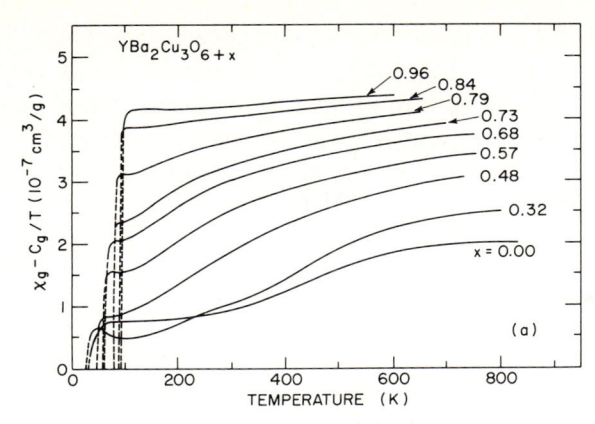

Fig. 13. Magnetic susceptibility, corrected for a Curie-like contribution, plotted vs. temperature for a range of oxygen concentrations in $YBa_2Cu_3O_{6+x}$. From Ref. [30].

occur at low energies, whereas spin-wave theory indicates that for the antiferromagnetic tetragonal material most of the spectral weight occurs at $\hbar\omega > J_{\parallel}$. A short magnetic correlation length or a shift in spectral weight to higher energies could explain the negative observations at low energies. Cu NMR and NQR studies appear to give positive evidence for the presence of interacting Cu^{2+} spins in $YBa_2Cu_3O_7$ [55].

5. Summary

The magnetic and electronic correlations in $YBa_2Cu_3O_{6+x}$ are controlled by the oxygen-filling and ordering of the $Cu(1)$-O_x layers. Oxygen atoms tend to cluster in CuO chains in order to minimize the number 3-fold coordinated coppers. In the orthorhombic phase, ordering of the oxygens in chains creates O $2p$ holes, a large fraction of which are transferred to the CuO_2 planes. Each BaO layer acts as a dielectric which distorts to help screen the charge transfer. The reduction in the number of holes on removal of oxygen is partially compensated by the conversion of $Cu(1)^{2+}$ to $Cu(1)^{1+}$. T_c tends to scale with the density of mobile O $2p$ holes. In the tetragonal phase the chain segments become short and lack orientational order; very few holes are transferred to the planes.

Antiferromagnetism occurs throughout the insulating tetragonal phase. CuO_2 bilayers are coupled antiferromagnetically, and they remain correlated in the absence of long-range order. $Cu(1)^{2+}$ atoms in

Cu(1)-O chain segments tend to couple ferromagnetically to the CuO_2 layers at low temperature. Holes in the CuO_2 planes reduce the magnetic correlation length and kill long-range order. Nevertheless, there is evidence that Cu^{2+} moments survive and interact in the metallic state. More work is required to understand the interactions between O $2p$ and Cu $3d$ holes in the metallic phase.

Acknowledgements

I am grateful for interactions and discussions with many colleagues and collaborators, especially including R. J. Birgeneau, V. J. Emery, S. M. Heald, D. C. Johnston, B. Keimer, A. R. Moodenbaugh, A. H. Moudden, M. Sato, G. Shirane, S. K. Sinha, and Y. J. Uemura. This work was supported in part by the U.S.-Japan Cooperative Neutron Scattering Program. Research at Brookhaven is supported by the Division of Materials Sciences, U.S. Department of Energy under contract No. DE-AC02-76CH00016.

References

1. J. G. Bednorz and K. A. Müller, *Z. Phys.* B **64**, 189 (1986).
2. M. K. Wu, J. R. Ashburn, C. J. Torng, P. H. Hor, R. L. Meng, L. Gao, Z. J. Huang, Y. Q. Wang, and C. W. Chu, *Phys. Rev. Lett.* **58**, 908 (1987).
3. M. A. Beno, L. Soderholm, D. W. Capone II, D. G. Hinks, J. D. Jorgensen, J. D. Grace, I. K. Schuller, C. U. Segre, and K. Zhang, *Appl. Phys. Lett.* **51**, 57 (1987); D. E. Cox, A. R. Moodenbaugh, J. J. Hurst, and R. H. Jones, *J. Phys. Chem. Solids* **49**, 47 (1988).
4. D. C. Johnston, A. J. Jacobson, J. M. Newsam, J. T. Lewandowski, D. P. Goshorn, D. Xie, and W. B. Yelon, in *Chemistry of High-Temperature Superconductors*, edited by D. L. Nelson, M. S. Whittingham, and T. F. George, ACS Symposium Series No. 351 (American Chemical Society, Washington, DC, 1987); A. Renault, G. J. McIntyre, G. Collin, J. P. Pouget, and R. Comès, *J. Phys. (Paris)* **48**, 1407 (1987).
5. N. Nishida *et al.*, *Jpn. J. Appl. Phys.* **26**, L1856 (1987); *J. Phys. Soc. Jpn.* **57**, 722 (1988).
6. J. M. Tranquada, D. E. Cox, W. Kunnmann, H. Moudden, G. Shirane, M. Suenaga, P. Zolliker, D. Vaknin, S. K. Sinha, M. S. Alvarez,

A. J. Jacobson, and D. C. Johnston, *Phys. Rev. Lett.* **60**, 1567 (1988); J. Rossat-Mignod, P. Burlet, M. J. G. M. Jurgens, J. Y. Henry, and C. Henry, *Physica C* **152**, 19 (1988); W.-H. Li, J. W. Lynn, H. A. Mook, B. C. Sales, and Z. Fisk, *Phys. Rev. B* **37**, 9844 (1988); D. Petitgrand and G. Collin, *Physica C* **153**, 192 (1988).

7. J. L. Hodeau, C. Chaillout, J. J. Capponi, and M. Marezio, *Solid State Commun.* **64**, 1349 (1987).

8. G. Van Tendeloo, H. W. Zandbergen, and S. Amelinckx, *Solid State Commun.* **63**, 603 (1987); T. Takabatake, M. Ishikawa, Y. Nakazawa, and K. Koga, *Physica C* **152**, 424 (1988); D. J. Werder, C. H. Chen, R. J. Cava, and B. Batlogg, *Phys. Rev. B* **37**, 2317 (1988); **38**, 5130 (1988).

9. R. M. Fleming, L. F. Schneemeyer, P. K. Gallagher, B. Batlogg, L. W. Rupp, and J. V. Waszczak, *Phys. Rev. B* **37**, 7920 (1988).

10. L. T. Wille, A. Berera, and D. de Fontaine, *Phys. Rev. Lett.* **60**, 1065 (1988); A. Berera and D. de Fontaine, *Phys. Rev. B* **39**, 6726 (1989); R. Kikuchi and J.-S. Choi (to be published).

11. R. J. Cava, B. Batlogg, C. H. Chen, E. A. Rietman, S. M. Zahurak, and D. Werder, *Phys. Rev. B* **36**, 5719 (1987).

12. Y. Nakazawa and M. Ishikawa, *Physica C* **158**, 381 (1989).

13. R. J. Cava, B. Batlogg, K. M. Rabe, E. A. Rietman, P. K. Gallagher, and L. W. Rupp Jr, *Physica C* **156**, 523 (1988); W. K. Kwok, G. W. Crabtree, A. Umezawa, B. W. Veal, J. D. Jorgensen, S. K. Malik, L. J. Nowicki, A. P. Paulikas, and L. Nunez, *Phys. Rev. B* **37**, 106 (1988); P. F. Miceli, J. M. Tarascon, L. H. Greene, P. Barboux, F. J. Rotella, and J. D. Jorgensen, *Phys. Rev. B* **37**, 5932 (1988).

14. For example, see the review by K. C. Hass, *Solid State Phys.* **42**, (1989).

15. J. A. Yarmoff, D. R. Clarke, W. Drube, U. O. Karlsson, A. Taleb-Ibrahimi, and F. J. Himpsel, *Phys. Rev. B* **36**, 3967 (1987); P. Kuiper, G. Kruizinga, J. Ghijsen, M. Grioni, P. J. W. Weijs, F. M. F. de Groot, G. A. Sawatzky, H. Verweij, L. F. Feiner, and H. Petersen, *Phys. Rev. B* **38**, 6483 (1988).

16. N. Nücker, J. Fink, J. C. Fuggle, P. J. Durham, and W. M. Temmerman, *Phys. Rev. B* **37**, 5158 (1988).

17. N. Nücker, H. Romberg, X. X. Xi, J. Fink, B. Gegenheimer, and Z. X. Zhao, *Phys. Rev. B* **39**, 6619 (1989).

18. M. Takigawa, P. C. Hammel, R. H. Heffner, and Z. Fisk, *Phys. Rev. B* **39**, 7371 (1989).

19. M. Horvatić, Y. Berthier, P. Butaud, Y. Kitaoka, P. Ségransan, C. Berthier, H. Katayama-Yoshida, Y. Okabe, and T. Takahashi (to be published).

20. H. Alloul, P. Mendels, G. Collin, and P. Monod, *Phys. Rev. Lett.* **61**, 746 (1988).
21. L.-S. Kau, D. J. Spira-Solomon, J. E. Penner-Hahn, K. O. Hodgson, and E. I. Solomon, *J. Am. Chem. Soc.* **109**, 6433 (1987).
22. S. M. Heald, J. M. Tranquada, A. R. Moodenbaugh, and Y. Xu, *Phys. Rev. B* **38**, 761 (1988).
23. J. M. Tranquada, S. M. Heald, A. R. Moodenbaugh, and Y. Xu, *Phys. Rev. B* **38**, 8893 (1988).
24. T. A. Smith, J. E. Penner-Hahn, K. O. Hodgson, M. A. Berding, and S. Doniach, in *EXAFS and Near Edge Structure III*, edited by K. O. Hodgson, B. Hedman, and J. E. Penner-Hahn (Springer-Verlag, Berlin, 1984), p. 58.
25. M. K. Kelly, P. Barboux, J.-M. Tarascon, D. E. Aspnes, W. A. Bonner, and P. A. Morris, *Phys. Rev. B* **38**, 870 (1988).
26. W. W. Warren, Jr., R. E. Walstedt, G. F. Brennert, R. J. Cava, B. Batlogg, and L. W. Rupp, *Phys. Rev. B* **39**, 831 (1989).
27. M. Horvatić, P. Ségransan, C. Berthier, Y. Berthier, P. Butaud, J. Y. Henry, M. Couach, and J. P. Chaminade, *Phys. Rev. B* **39**, 7332 (1989).
28. Y. J. Uemura et al., *Phys. Rev. Lett.* **62**, 2317 (1989).
29. Y. Kubo and H. Igarashi, *Phys. Rev. B* **39**, 725 (1989).
30. J. M. Tranquada, A. H. Moudden, A. I. Goldman, P. Zolliker, D. E. Cox, G. Shirane, S. K. Sinha, D. Vaknin, D. C. Johnston, M. S. Alvarez, A. J. Jacobson, J. T. Lewandowski, and J. M. Newsam, *Phys. Rev. B* **38**, 2477 (1988).
31. J. Rossat-Mignod, P. Burlet, M. J. Jurgens, C. Vettier, L. P. Regnault, J. Y. Henry, C. Ayache, L. Forro, H. Noel, M. Potel, P. Gougeon, and J. C. Levet, *J. Phys. Colloq.* **49**, C8-2119 (1988).
32. P. W. Anderson, *Phys. Rev.* **86**, 694 (1952); M. E. Lines, *J. Phys. Chem. Solids* **31**, 101 (1970).
33. J. H. Brewer et al., *Phys. Rev. Lett.* **60**, 1073 (1988).
34. M. Matsumura, H. Yamagata, Y. Yamada, K. Ishida, Y. Kitaoka, K. Asayama, H. Takagi, H. Iwabuchi, and S. Uchida, *J. Phys. Soc. Jpn.* **57**, 3297 (1988); P. Mendels and H. Alloul, *Physica C* **156**, 355 (1988).
35. R. J. Birgeneau and G. Shirane, in *Physical Properties of High Temperature Superconductors I*, edited by D. M. Ginsberg (World Scientific, Singapore, 1989).
36. Y. Guo, J.-M. Langlois, and W. A. Goddard III, *Science* **239**, 896 (1988).

37. A. H. Moudden, G. Shirane, J. M. Tranquada, R. J. Birgeneau, Y. Endoh, K. Yamada, Y. Hidaka, and T. Murakami, *Phys. Rev. B* **38**, 8720 (1988).
38. Y. Lu and B. Patton (to be published).
39. H. Kadowaki, M. Nishi, Y. Yamada, H. Takeya, H. Takei, S. M. Shapiro, and G. Shirane, *Phys. Rev. B* **137**, 7932 (1988).
40. J. M. Tranquada, G. Shirane, B. Keimer, S. Shamoto, and M. Sato, *Phys. Rev. B* (to be published).
41. W. E. Farneth, R. S. McLean, E. M. McCarron, III, F. Zuo, Y. Lu, B. R. Patton, and A. J. Epstein, *Phys. Rev. B* **39**, 6594 (1989).
42. M. Matsumura, H. Yamagata, Y. Yamada, K. Ishida, Y. Kitaoka, K. Asayama, H. Takagi, H. Iwabuchi, and S. Uchida, *J. Phys. Soc. Jpn.* **58**, 805 (1989).
43. H. Lütgemeier and B. Rupp, *J. Phys. Colloq.* **49**, C8-2147 (1988).
44. P. F. Miceli, J. M. Tarascon, L. H. Greene, P. Barboux, M. Giroud, D. A. Neumann, J. J. Rhyne, L. F. Schneemeyer, and J. V. Waszczak, *Phys. Rev. B* **38**, 9209 (1988); P. F. Miceli, J. M. Tarascon, P. Barboux, L. H. Greene, B. G. Bagley, G. W. Hull, M. Giroud, J. J. Rhyne, and D. A. Neumann, *Phys. Rev. B* **39**, 12375 (1989); P. Zolliker, D. E. Cox, J. M. Tranquada, and G. Shirane, *Phys. Rev. B* **38**, 6575 (1988).
45. J. W. Lynn, W.-H. Li, H. A. Mook, B. C. Sales, and Z. Fisk, *Phys. Rev. Lett.* **60**, 2781 (1988); A. H. Moudden, G. Shirane, J. M. Tranquada, R. J. Birgeneau, Y. Endoh, K. Yamada, Y. Hidaka, and T. Murakami, *Phys. Rev. B* **38**, 8720 (1988).
46. P. W. Anderson, *Science* **235**, 1196 (1987).
47. G. Shirane, Y. Endoh, R. J. Birgeneau, M. A. Kastner, Y. Hidaka, M. Oda, M. Suzuki, and T. Murakami, *Phys. Rev. Lett.* **59**, 1613 (1987); Y. Endoh et al., *Phys. Rev. B* **37**, 7443 (1988).
48. K. B. Lyons, P. A. Fleury, L. F. Schneemeyer, and J. V. Waszczak, *Phys. Rev. Lett.* **60**, 732 (1988).
49. M. Sato, S. Shamoto, J. M. Tranquada, G. Shirane, and B. Keimer, *Phys. Rev. Lett.* **61**, 1317 (1988).
50. D. C. Johnston, S. K. Sinha, A. J. Jacobson, and J. M. Newsam, *Physica C* **153-155**, 572 (1988).
51. See, for example, L. J. de Jongh and A. R. Miedema, *Adv. Phys.* **23**, 1 (1974).
52. F. Mezei, B. Faragó, C. Pappas, Gy. Hutiray, L. Rosta, and L. Mihály, *Physica C* **153-155**, 1669 (1988).

53. J. M. Tranquada, H. Chou, G. Shirane, S. Shamoto, and M. Sato (unpublished).
54. T. Brückel, H. Capellmann, W. Just, O. Schärpf, S. Kemmler-Sack, R. Kiemel, and W. Schaefer, *Europhys. Lett.* **4**, 1189 (1987).
55. See, for example, F. Mila and T. M. Rice, *Physica C* **157**, 516 (1989).

Magnetic Correlations and Spin Dynamics in $La_{2-x}Sr_xCuO_4$ from NQR Relaxation

A. Rigamonti[1], *F. Borsa*[1], *M. Corti*[1], *T. Rega*[1], *J. Ziolo*[1], *and F. Waldner*[2]

[1]Department of Physics "A. Volta", Unita' GNSM-CISM and
 Sezione INFN-Pavia, Italy
[2]Physik-Institut der Universität Zürich, CH-8001 Zürich, Switzerland

Abstract. La and Cu NQR relaxation measurements in $La_{2-x}Sr_xCuO_4$ for $0 \leq x \leq 0.3$ in the temperature range 1.6–450 K are analyzed in terms of Cu^{2+} magnetic correlations and spin dynamics. It is described how the magnetic correlation that would result from the Cu-Cu exchange is reduced by mobile charge defects related to the x-doping, with a liquid-like thermal bath of magnetic excitations. A comprehensive picture is given for the x and T dependence of the correlation time, of the correlation length and of the Néel temperature $T_N(x)$, as well as for the electrical resistivity and the magnetic susceptibility. It is discussed how, in the superconducting samples, the mobile defects also cause the decrease, for $T \rightarrow T_c^+$, of the hyperfine Cu electron–nucleus effective interaction, yielding a coexistence of quasi-localized reduced magnetic moments from $3d$ Cu electrons and of a Fermi liquid from the O p-hole carriers. The temperature behavior of the effective hyperfine field around the superconducting transition is derived.

1. Introduction

In the CuO_2-based superconductors a crucial role is played by the magnetic correlations and the spin dynamics of the Cu^{2+} ions and by the effects of Sr doping and/or oxygen depletion on these properties via the related charge defects. Successful results from theoretical models and a number of experimental investigations support this point of view [1–3].

In the thermodynamical-statistical description, the magnetic correlation length ξ controls the site-dependent order parameter $M(r)$; ξ is the distance over which the magnetization changes significantly, causing a substantial decrease of a local fluctuation of $M(r)$ [4]. In terms of microscopic spin variable S, the correlation length controls the spatial part of the space-time spin correlation function $G_{\alpha\alpha}(r,t) = \langle S_\alpha(r,t)S_\alpha(0,0)\rangle/[S(S+1)/3]$ (where α denotes the spin component and the brackets an equilibrium ensemble average) or of its quantum mechanical equivalent $\langle S_i^*(0)S_j(t)\rangle = \text{Tr}\{\varrho_0 S_i^*(0)S_j(t)\}$ (ϱ_0 is the equilibrium density matrix).

In fact, by referring, for experimental convenience, to the Fourier transform of $G(r,t)$, namely the dynamic structure factor (and omitting α for simplicity of notation),

$$S(q,\omega) = \int e^{i(\omega t - q \cdot r)} G(r,t) dr\, dt = \langle |S_q|^2 \rangle J_q(\omega)\,, \tag{1.1}$$

the mean square amplitude of the collective fluctuations at wave vector q, $\langle |S_q|^2 \rangle = (1/2\pi) \int S(q,\omega) d\omega$ can be written [4]

$$\langle |S_q|^2 \rangle = \xi^2 f(q\xi)\,, \tag{1.2}$$

where ξ is in lattice units and $f(q\xi)$ is a homogeneous function of the product $q\xi$, which, through the usual expansion or in mean field approximations, has the form $f \simeq [(1+(q\xi)^2]^{-1}$.

Also the time-dependent part of the correlation function, and then $J_q(\omega)$ in (1.1), can be related to the correlation length ξ. In fact, in strongly correlated systems, where $\xi \gg 1$, dynamical scaling [5] allows one to write

$$J_q(\omega) = \frac{2\pi}{\Gamma_q} F(\omega/\Gamma_q)\,, \tag{1.3}$$

where Γ_q is the decay rate of the fluctuations and $F(\omega/\Gamma_q)$ a well-behaved function depending on ω only through the ratio ω/Γ_q. For $\omega \to 0$ (namely fluctuations probed at times much longer than the microscopic time scale) one has $F(\omega/\Gamma_q) \simeq 1$, while

$$\Gamma_q \propto \frac{1}{\xi^z} g(q\xi) \tag{1.4}$$

with $g \simeq f^{-1}$ and z a characteristic dynamical exponent. By considering that in a paramagnet well above the ordering temperature, where $\xi \simeq 1$, one has $\Gamma_q \simeq \omega_e$, ω_e being the Heisenberg exchange frequency

$$\omega_e = \frac{1}{\hbar} \left(\frac{8}{3} J^2 n S(S+1) \right)^{1/2}\,, \tag{1.5}$$

where J is the exchange interaction and n the number of nearest-neighbor magnetic ions, we can describe the cooperative spin dynamics through a decay rate of the fluctuations of the form

$$\Gamma_q \simeq \frac{\omega_e}{\xi^z} f^{-1}\,, \tag{1.6}$$

which emphasizes how the spin fluctuations slow down when the correlation length increases. This occurs, typically, on approaching the transition from the paramagnetic state to an ordered one.

One of the successful experimental tools for the study of magnetic correlations and spin dynamics in paramagnetic systems undergoing phase transitions and/or of restricted lattice dimensionality has proved to be nuclear relaxation [6]. In NMR-NQR the nucleus is coupled to the spins of the magnetic ions via the magnetic dipolar and transferred hyperfine interactions, which can be cast in the form

$$\mathcal{H}_{ne} = -\gamma\hbar \left(I_z h_z(t) + \frac{1}{2}I_+ h_-(t) + \frac{1}{2}I_- h_+(t) \right),$$

where $\boldsymbol{h}(t)$ is a fictitious fluctuating magnetic field

$$\boldsymbol{h}(t) = -\gamma_e \hbar \sum_i \left(\frac{\boldsymbol{S}_i}{r_i^3} - \frac{3(\boldsymbol{r}_i \cdot \boldsymbol{S}_i)\boldsymbol{r}_i}{r_i^5} - \frac{A_i}{\gamma}\boldsymbol{r}_i \cdot \boldsymbol{S}_i \right) \qquad (1.7)$$

[the z-axis is along the external magnetic field in NMR while in NQR z is the direction of the greatest principal component of the electric field gradient (EFG) tensor V_{zz}]. By considering the time-dependent perturbation $\mathcal{H}_{ne}$ on the Zeeman or on the quadrupole NQR levels, one can relate the nuclear relaxation rate to the spectral densities of the field $\boldsymbol{h}(t)$. For example, for $I = 1/2$ (first-order process and classical magnetic lattice [6]) one gets

$$T_1^{-1} = 2W_M = \frac{\gamma^2}{4}[J_+(\omega_R) + J_-(\omega_R)] \qquad (1.8)$$

where

$$J_\pm = \int e^{-i\omega_R t} \langle h_\pm(0) h_\pm^*(t) \rangle dt \qquad (1.9)$$

are the spectral densities at the resonance frequency ω_R of the correlation functions for the field components.

In Sect. 2 of this work we briefly discuss the behavior of the nuclear relaxation rate driven by electronic spin dynamics in two-dimensional (2D) paramagnets (PA) undergoing the transition to the 3D antiferromagnetic (AF) state, the transition-metal phosphorus trichalcogenides. On one hand, the analysis of the ^{31}P spin–lattice relaxation in these systems illustrates how the correlation effects in the spin dynamics are reflected in the NMR relaxation for "conventional" 2D paramagnets. On the other hand, this analysis will provide a comparative basis to show how the [139]La NQR relaxation in $La_{2-x}Sr_xCuO_{4-y}$ displays unconventional features.

In Sect. 3, some results of [139]La NQR spin–lattice relaxation measurements are presented. It is pointed out that over most of the x and T range the dominant relaxation mechanism is actually associated with the Cu^{2+} magnetic correlation and it is emphasized how, in the case of classical slowing-down of paramagnons as well as in the possible presence of quantum spin-liquid excitations, it would be very difficult to explain some features of the NQR relaxation in the framework of the conventional approach.

In Sect. 4 an interpretative model is discussed. In this model mobile charge defects cause a dynamical quenching of the correlation that would result from the strength of the Cu-Cu exchange coupling. Thus one has a classical liquid-like thermal bath of magnetic excitations, with a concentration-dependent and temperature-activated correlation time. The NQR data allow one to derive the x dependence of the activation energy.

On the basis of these findings a comprehensive justification of the x and T behavior of the results is obtained. Moreover, since the relaxation rates turn out to be directly related to the defect-controlled correlation length, an evaluation of ξ from T_1 is possible without adjustable parameters.

In Sect. 5 of the paper we discuss the conclusions that can be drawn from the NQR T_1 data and compare them with the experimental results obtained with other techniques. For example, we show that at $T \simeq 77\,\mathrm{K}$ the x-dependence of ξ is in good agreement with the results obtained by neutron scattering through the q-dependence of the static structure factor. The x-dependence of the transition temperature T_N from the PA to the 3D AF state is also

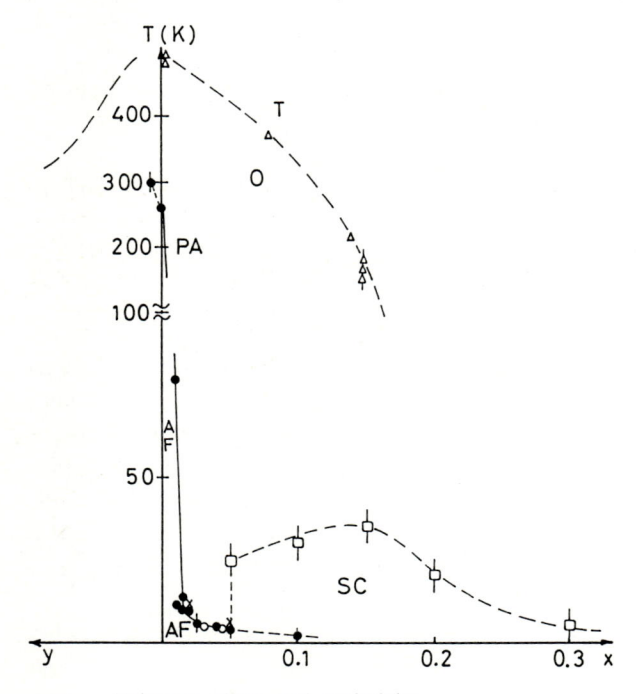

Fig. 1. Phase diagram for $La_{2-x}Sr_xCuO_{4-y}$ (figure adapted from [7]). For a more detailed picture of the x-dependence of the transition temperature $T_N(x)$ from the PA phase to the 3D-ordered AF phase and the appropriate references, see Fig. 10

□ Meissner effect and resistivity
● NQR T_1 and T_2
× μSR
△ Neutron diffraction
○ Specific heat

discussed. The behavior of other quantities, like magnetic susceptibility and electrical resistivity, that are expected to be controlled by the mobile charge defects is examined. In particular, it is shown that the resistivity data in the low temperature range for x smaller than 0.05 support the x-dependence of the activation energy derived from NQR T_1. The section ends with concluding remarks about the properties of moving extended magnetic excitations.

Section 6 is devoted to the ^{63}Cu NQR results obtained in samples with $0.1 \leq x \leq 0.3$ around the transition temperature T_c to the superconducting phase. In particular, it is emphasized how the NQR spin–lattice data lead to a picture of quasi-localized reduced $3d$ Cu magnetic moments (with the decrease of the effective hyperfine electron–nucleus coupling constant starting well above T_c) and of a Fermi liquid of p-holes. The temperature behavior of the effective hyperfine field around T_c is derived.

For convenience of the discussion we shown in Fig. 1 the phase diagram for $La_{2-x}Sr_xCuO_{4-y}$ as derived for our samples from various measurements (Meissner effect, resistivity, T_1 and T_2 NQR, specific heat), also including a few data (μSR, neutron scattering) by other authors in nominally correspondent samples.

2. Nuclear Relaxation in the 2D Paramagnets FePS$_3$ and MnPS$_3$ Around the Antiferromagnetic Ordering

Transition-metal phosphorus trichalcogenides MPX_3 have a layered structure whereby the magnetic ion M (Fe, Mn or Ni) occupies a honeycomb lattice while the MX_6 octahedra (X =S or Se) share the edges. The center of the M^{2+} hexagons are occupied by a P-P pair, each of the P atoms being shifted out of the plane by a displacement $\delta = \pm 1.1$ Å along the octahedral threefold axis (Fig. 2).

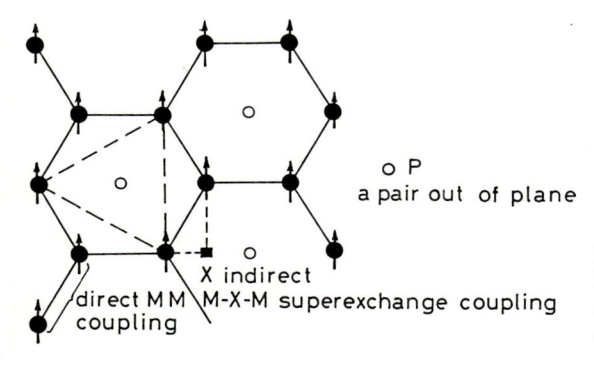

Fig. 2. MM coupling: AF for Mn, FE for Fe, zero for Ni. $M - X - M$ coupling: AF or FE (depending on details). See [8]

In the neighborhood of the PA-AF transition the ^{31}P NMR relaxation is mostly due to the correlated spin dynamics of the six nearest neighbor (n.n.) magnetic ions of the hexagon nearby. Then, starting from (1.7), by introducing the collective spin coordinate $S_q(t) = (1/\sqrt{N})\sum_i S_i(t)\exp(i\mathbf{q}\cdot\mathbf{r}_i)$, one can write

$$\langle h_+(0)h_-(t)\rangle = \frac{1}{N}\sum_q \left(h'_q\langle S_q^\pm(0)S_q^\pm(t)\rangle + h''_q\langle S_q^z(0)S_q^z(t)\rangle\right), \quad (2.1)$$

where the factors h'_q and h''_q are the Fourier transform of the lattice functions that in (1.7) couple $\mathbf{h}(t)$ to the spin operators $S_i(t)$. These factors are very sensitive to the symmetry of the critical fluctuation, namely of the spin fluctuation characterized by the wave vector $\mathbf{q}_c$ which describes the staggered order that sets in below T_N.

The factors h_q can often be expanded around the values $h_q = h_{q_c}$. Then, from (1.8), by considering in (1.9) $\omega_R \to 0$, by taking into account (1.2), (1.3) and (1.6) in (2.1) and by writing

$$\frac{1}{N}\sum_q \frac{\langle |S_q|^2\rangle}{\Gamma_q} \to \int dq\, q^{d-1}\xi^2 \frac{1}{\omega_e}\xi^z [f(q\xi)]^2 \quad (2.2)$$

(d is the lattice dimensionality), one can obtain the temperature behavior of the relaxation rates for $T \to T_N$ [9]. Here we stress only the following:

i) MnPS$_3$ - The critical dynamics induces an enhancement of the q_c-correlation, yielding zero effective field at the P site (namely $h'_{q_c} = h''_{q_c} = 0$), thus causing a kind of anti-divergence of the relaxation rate for $T \to T_N^+$; the terms at $q \neq q_c$ lead to a contribution of the form

$$T_1^{-1} \propto (\xi)^{z-d}\ln\xi, \quad (2.3)$$

which for a dynamical exponent z, see (1.4), around 1.5–2 [5] becomes relevant only for T close to T_N, where ξ tends to diverge.

ii) FePS$_3$ - The symmetry of the critical dynamics is such that $h''_{q_c} \neq 0$; therefore, the progressive condensation of the spin dynamics, for $T \to T_N^+$, on the critical mode is experienced by the P nucleus and the related slowing down yields

$$T_1^{-1} = C\frac{1}{\omega_e}(\xi)^{2-d+z}, \quad (2.4)$$

where the constant C is of the order of $(\gamma h_{\text{eff}})^2$, h_{eff} being the static field measured at the P site in the antiferromagnetic phase.

Thus for a second-order or a quasi-second-order PA $\to$ AF transition, where ξ goes as $(T - T_0)^{-\nu}$ [4] (ν is a critical exponent, $T_0 \simeq T_N$), for

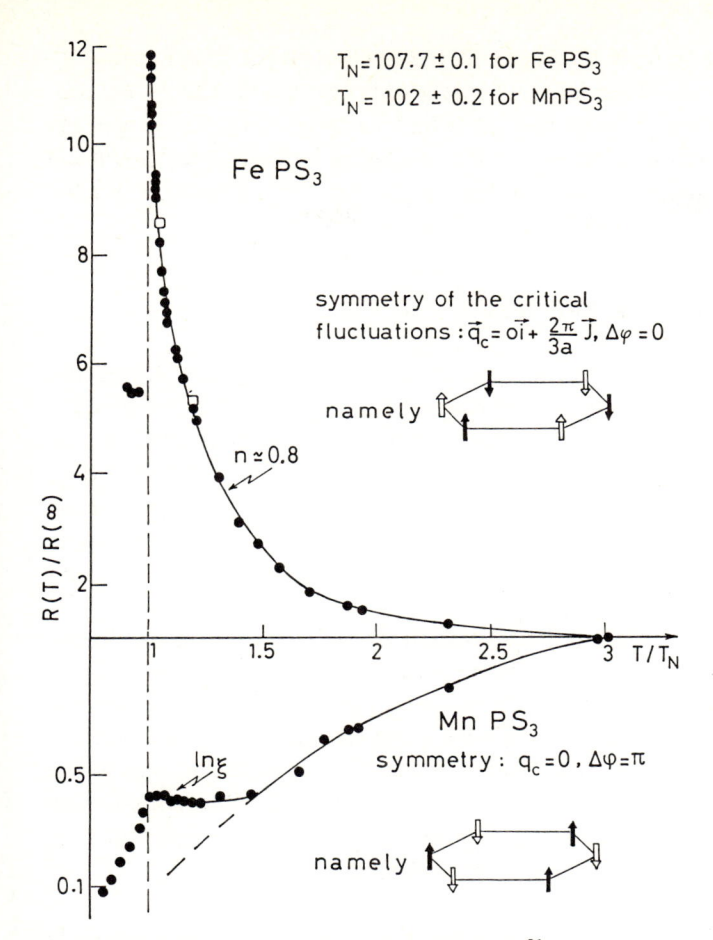

Fig. 3. Comparison of the experimental data for the ^{31}P nuclear relaxation rates in the 2D paramagnets FePS$_3$ and MnPS$_3$ on approaching the PA-AF transition with the behavior expected from a conventional picture of critical spin dynamics (data from [9]). The symmetries of the critical fluctuations [given through the wave vectors q_c and the phase factors $\Delta\varphi(q_c, q_c')$ for the fluctuations in the two triangular sublattices, see Fig. 2], correspond to the spin configurations of the magnetic ions around the hexagon as sketched in the figure

$d = 2$, eq. (2.4) predicts a divergence of the relaxation rate with an exponent $n \simeq \nu z$. In Fig. 3 the temperature behavior of the relaxation rates $R = T_1^{-1}$ (normalized to the infinite temperature limit), predicted for FePS$_3$ and for MnPS$_3$ on the basis of the theoretical picture outlined above, are compared with the experimental findings.

We conclude this section on the behavior of the nuclear relaxation around conventional 2D PA-AF transitions with a comment on the effects of introducing some disorder to the interactions. In samples of the type Fe$_{1-x}$Zn$_x$PS$_3$, preliminary ^{31}P NMR results [10] support the general belief that in the high concentration regime of disordered magnets only moderate effects are observed on the transition temperature $T_N(x)$ and on other magnetic properties [11]. Therefore the critical behavior of the relaxation rates driven by the 2D spin-dynamics also does not show drastic effects for $x \ll 1$. This is not the case, as we will see, for the x-doping in $La_{2-x}Sr_xCuO_4$.

3. NQR Relaxation in $La_{2-x}Sr_xCuO_4$

The ^{139}La NQR spin–lattice relaxation times are obtained by monitoring the growth of the population difference between the $\pm 5/2 \leftrightarrow \pm 7/2$ levels, after a proper sequence of rf pulses causing saturation. An analogous process was carried out for the ^{63}Cu nucleus in terms of the $\pm 1/2 \leftrightarrow \pm 3/2$ NQR levels. The recovery process, in principle, can be driven by two mechanisms: i) time dependence of the EFG components induced by lattice vibrations (quadrupole mechanism); ii) fluctuations of the field $h(t)$, see (1.7), induced by the spin dynamics $S_i(t)$ (magnetic mechanism).

The dominant contribution to the relaxation rate induced by the quadrupolar mechanism is expected to come from optical modes, which usually cause modulations in the EFGs larger than the acoustic modes, particularly from the ones at lower frequencies. An order of magnitude estimate of the quadrupolar contribution can be obtained by considering the EFG modulation due to the Jahn-Teller optical mode in which La vibrates against the five neighboring oxygen ions and by using data from neutron scattering for the frequency ω_0 and the dispersion width of the branch. One derives [12]

$$T_1^{-1} \simeq 5 \times 10^{-4} T^2 \tag{3.1}$$

for $T \gtrsim \hbar\omega_0/k$, while for lower T the relaxation rate should decrease faster than T^2. A comparison of this estimate with the experimental data allows one to conclude that over most of the x and T range that we will take into consideration the quadrupolar relaxation mechanism can be disregarded [13]. Thus the experimental results for T_1 will be analyzed in terms of the correlated dynamics of the nine Cu^{2+} ions around a given La ion (Fig. 4). Furthermore, it appears from the data reported in the figure, in the light of (1.7), that the dominant contribution to the magnetic relaxation mechanism is from the four n.n. Cu^{2+} magnetic ions, within a few percent.

Therefore, starting from (1.8) and (1.9), by considering the structure of the NQR La ($I = 7/2$) levels and again observing that the resonance frequency ($\omega_R \simeq 18\,\text{MHz}$) can be set equal to zero, one has

$$T_1^{-1} \simeq 23 W_M = \frac{23}{2}\gamma^2 \int \langle h_+(0) h_-(t)\rangle dt \,. \tag{3.2}$$

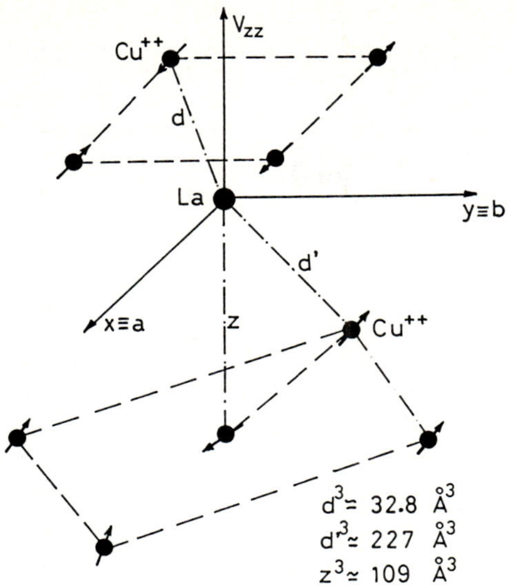

Fig. 4. Magnetic environment of Cu^{2+} ions around a given La ion in La_2CuO_4

$d^3 \simeq 32.8 \ \text{Å}^3$
$d'^3 \simeq 227 \ \text{Å}^3$
$z^3 \simeq 109 \ \text{Å}^3$

Then, from (2.1), by considering that the h'_q and h''_q factors are different from zero for $q = q_{AF}$ and weakly q-dependent, by following (2.2), in the light of (1.2), (1.4) and (1.6) we can write for the relaxation rate

$$T_1^{-1} \simeq 11.5 (\gamma h_{\text{eff}})^2 \frac{1}{\omega_e} \xi^{2-d+z} \quad \text{for} \quad {}^{139}\text{La} \ . \tag{3.3}$$

One should stress that h_{eff} can be measured from the splitting of the La NQR doublet in the AF phase. One has [14] $h_{\text{eff}} \simeq 100$ G. An analogous description holds for the NQR relaxation of ^{63}Cu. Here the nucleus probes the hyperfine field from the same ion and thus T_1^{-1} is sensitive to the autocorrelation function $\langle S(0)S(t)\rangle$ only. Therefore the h_q factors are q-independent. For the $\pm 1/2 \leftrightarrow \pm 3/2$ transition one has $T_1^{-1} = 6W_M$ and thus the magnetic contribution to the ^{63}Cu relaxation rate can be written, from (3.3),

$$(T_1^{-1})_{\text{Cu}} = (T_1^{-1})_{\text{La}} \frac{3}{11.5} \left| \frac{\gamma_{\text{Cu}} h_{\text{eff,Cu}}}{\gamma_{\text{La}} h_{\text{eff,La}}} \right|^2 . \tag{3.4}$$

The effective field at the Cu site can be estimated as around 78×10^3 G from the observation, at $T = 1.3$ K, of the Cu NMR in antiferromagnetic La_2CuO_4 [15].

In Fig. 5 the temperature behavior of the relaxation rate vs T is shown for different x doping. In Fig. 6 the temperature behavior of T_1^{-1} as a function of T in La_2CuO_{4-y} is shown for a group of samples nominally at $y = 0$ and for one sample kept for about 50 h at $T \simeq 500$ where slight depletion of oxygen

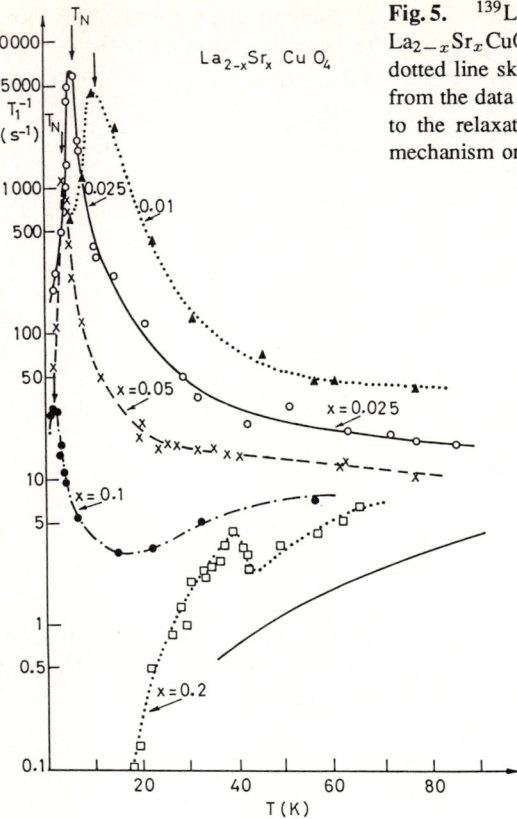

Fig. 5. ^{139}La spin–lattice relaxation rates in $La_{2-x}Sr_xCuO_4$ for $x = 0.01, 0.025, 0.05$ and 0.1. The dotted line sketches the behavior deduced for $x = 0.2$ from the data in [16]. The solid line is the contribution to the relaxation rate expected from the quadrupolar mechanism on the basis of (3.1)

($y > 0$) occurred. The analysis of the results shown in Fig. 5 and 6 in the light of (3.3) for a conventional slowing-down of the correlation allows one to note the following:

i) From (1.5) and $J \gtrsim 500\,\text{K}$ [17] one has for ω_e at least the value $\omega_e = 10^{14}$ rad s^{-1}, yielding, in the limit of no correlation, $(T_1^{-1})_\infty \leq 2\,\text{s}^{-1}$. Thus, referring to the values of the relaxation rates around T_N (Fig. 6), from (3.3) one derives $\xi(T_N) \simeq (3000)^{1/z}$, namely from about 55 to about 200 lattice units, for z from 1.5 to 2. This value of the in-plane correlation length is not unrealistic in La_2CuO_4 in view of the strong Cu^{2+} exchange coupling. For example, in the classical Heisenberg model [4] $\xi = \exp[2JS(S+1)/kT]$, yielding $\xi \simeq 2.6 \times 10^3$ for $T \simeq 300\,\text{K}$. Thus one could infer that the degree of correlation is limited by defects and/or quantum fluctuations. However, the most relevant observation that one can draw from the data in pure La_2CuO_4 is the lack, for $T > T_N$, of critical effects analogous to the ones observed in classical 2D paramagnets (Sect. 2): the relaxation rate does not decrease on increasing the temperature above T_N as one would expect for ξ going as $(T - T_N)^{-\nu}$.

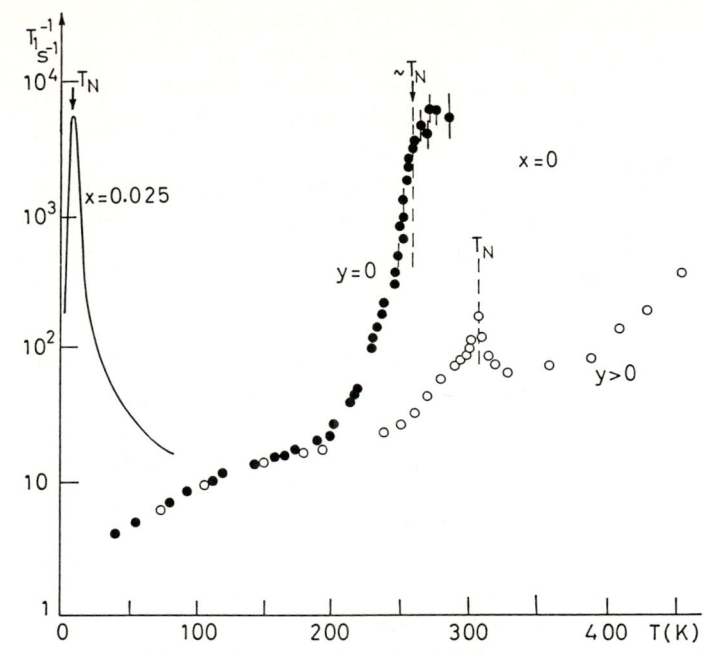

Fig. 6. Temperature dependence of the ^{139}La relaxation rate in nominally pure La_2CuO_4 and for a sample with slight oxygen depletion ($y > 0$; (o)). For comparison the behavior of the sample at $y \simeq 0$ and $x = 0.025$ is sketched (solid line – data in Fig. 5)

ii) The data for $T \to T_N^+$ in the samples at $x \neq 0$ (Fig. 5) are also difficult to explain in the framework of a conventional correlation and slowing-down description. In particular, if one forces the T behavior of T_1^{-1} into a power law, the exponent $n \simeq \nu z$ [see discussion after (2.4)] turns out to be strongly x-dependent.

iii) There is extreme sensitivity to x and y doping. A few percent of Sr^{2+} in place of La^{3+} causes dramatic changes in the spin-dynamics driving the nuclear relaxation process, with a marked drop in the PA $\to$ AF transition temperature (Fig. 5). Although only qualitatively, this effect can be noticed also for oxygen depletion (Fig. 6). In this case the increase in T_N (see Fig. 1) is probably due to the effect of the interplane coupling (note that T_N is deduced not only from the peak in the relaxation rate but also from the onset of a doublet for the NQR line due to the static internal magnetic field [14]). In a mean field description [4], one has

$$T_N \simeq J'\xi^2(T_N) \,. \tag{3.5}$$

iv) At $T \simeq T_N(x)$ one has a value of the relaxation rate which is almost independent of x, for $y = 0$. In view of (3.3) this seems to suggest that the transition to the 3D AF state occurs when ξ has reached a given value ξ_{2D}^*, regardless of the x and T values. Unless one assumes that the interplane coupling J' is strongly x-dependent, this implies that the mean field argument [see (3.5)] by which the 3D transition can occur at a lower temperature with a shorter $\xi(T_N)$ does not hold.

In the presence of several effects which evidence the inapplicability of a conventional description one should take into account the possibility that the "lattice" of the electronic spins should not be viewed as a classical one (as implicit in our approach) but rather as a quantum spin liquid (QSL). The RVB model [18] predicts excitations which should involve the nuclei via the magnetic hyperfine interaction, in a way somewhat analogous to the magnons in ordered magnetic systems. A crude evaluation of the role of QSL excitations in causing nuclear relaxation through a scattering process has been given [7, 12]. One derives a relaxation rate which depends on the details of the dispersion relation for the RVB excitations (which is a rather controversial matter). A plausible feature is a large enhancement of $(T_1)^{-1}$ with respect to the value $(T_1^{-1})_\infty$ for uncorrelated spins, with a weak, if any, temperature dependence above T_N. Thus, the occurrence of the QSL excitations could possibly explain, but only in a qualitative way, the experimental findings in nominally pure La_2CuO_4. We are still left with the need to explain most of the above points i)–iv), and in particular to include in the picture the role of the electronic holes due to x-doping (or of the extra electrons associated with missing oxygen) in controlling the relaxation rates through damping effects on the spin correlations.

4. Effects of Charge Defects on the Correlation Length and Spin Fluctuations

One can assume that Sr-doping with $x \lesssim 0.05$ or O-depletion, while leaving the Cu magnetic moment unaffected (Sect. 6), introduces mobile charge defects, like holons or charged solitons, which destroy the long-range spin correlation and cause a dynamical, locally disordered 2D magnetic state. Mobile defects have been observed as magnetic solitons in low-dimensional insulating paramagnets [19, 20] and their effect on the correlation length has been studied by a resonance technique [21]. In particular, it has been pointed out that the critical broadening of the EPR line in 2D AF could be explained by thermally activated mobile 2D solitons and such excitations, with density $n \sim \exp(-E/k_B T)$, could be the main cause of correlation length limitation, since $(n)^{1/d} \sim \xi^{-1}$ [21].

With regard to the nuclear spin–lattice relaxation process, the effect of mobile defects on the correlation can be heuristically taken into account by writing for the spin-correlation function [see (1.9) and (2.1)]

$$\sum_q \langle S_q^z(0) S_{-q}^z(t) \rangle = e^{-(\omega_e \xi^{-z})t} e^{-t/\tau_d} . \tag{4.1}$$

In (4.1) $\omega_e \xi^{-z}$ is the effective critical frequency marking the slowing-down of the correlations, see (3.3), while τ_d is an effective correlation time describing the loss of correlation in the local field due to the mobile defects. τ_d is related to the concentration of the mobile defects: $\tau_d = \tau_0 n^{-1}$, τ_0^{-1} being of the order of the exchange frequency ω_e; τ_d can be interpreted as the mean time interval between consecutive "collisions" of a defect with a nucleus. A similar argument has been pointed out by *Ramakrishnan* [22]: the time for the "annealing" of a local spin configuration by a mobile hole has been written $\tau_h = \Delta_h^{-1} n_h^{-1}$, where n_h is the hole density and Δ_h the hole band width.

From (4.1), in the light of (2.1) and (1.9), one sees that (3.3) (with $d = 2$) is now modified as

$$\begin{aligned} T_1^{-1} &= 11.5(\gamma h_{\text{eff}})^2 \tau_d / (\omega_e \xi^{-z} \tau_d + 1) \\ &\simeq 11.5(\gamma h_{\text{eff}})^2 \tau_0 n^{-1} = 11.5(\gamma h_{\text{eff}})^2 \tau_0 \xi^2 , \end{aligned} \tag{4.2}$$

where we have made the assumption $\omega_e \xi^{-z} \tau_d \ll 1$ in order to have the degree of correlation limited by the defects rather than by thermal fluctuations, on the basis of the experimental results. Then the temperature behavior of the relaxation rates for $T \to T_N^+$ (see Fig. 5), for $x \lesssim 0.05$ (condensation on the superconducting state would cause a decrease of the magnetic fluctuations, see Sect. 6), can be analyzed in terms of a correlation length which is controlled by the mobile defects and which increases when they are progressively frozen on lowering the temperature.

In Fig. 7 it is shown that the behavior of T_1^{-1} in the paramagnetic phase is rather well fitted by a thermally activated concentration $n(x,T)$ of mobile defects, see (4.2): $n(x,T) = n(x,\infty) \exp(-E/T) = x \exp(-E/T)$. This expression, based on the experimental findings (see later on), should be valid for low temperatures. Furthermore one could have a multiplicative factor which takes into account an effective extension of the defects. The activation energy E is inversely proportional to x-doping, indicating a cooperative character of the motions. From the fit of the data for E vs x (see also Fig. 12) one deduces $E \simeq (0.85/x)$ K. The intercept for $T \to \infty$ shows a qualitatively correct dependence and τ_0 turns out to be $\tau_0 \simeq 2 \times 10^{-15}$ s, of the order of ω_e^{-1} as expected. Thus (4.2) can be rewritten

$$T_1^{-1} = a \frac{1}{x} e^{b/xT} , \tag{4.3}$$

where $b \simeq 0.85$ K, while a is $(a)_{\text{La}} \simeq 0.32$ s^{-1} for ^{139}La and $(a)_{\text{Cu}} \simeq 1.7 \times 10^3$ s^{-1} for ^{63}Cu, see (3.4).

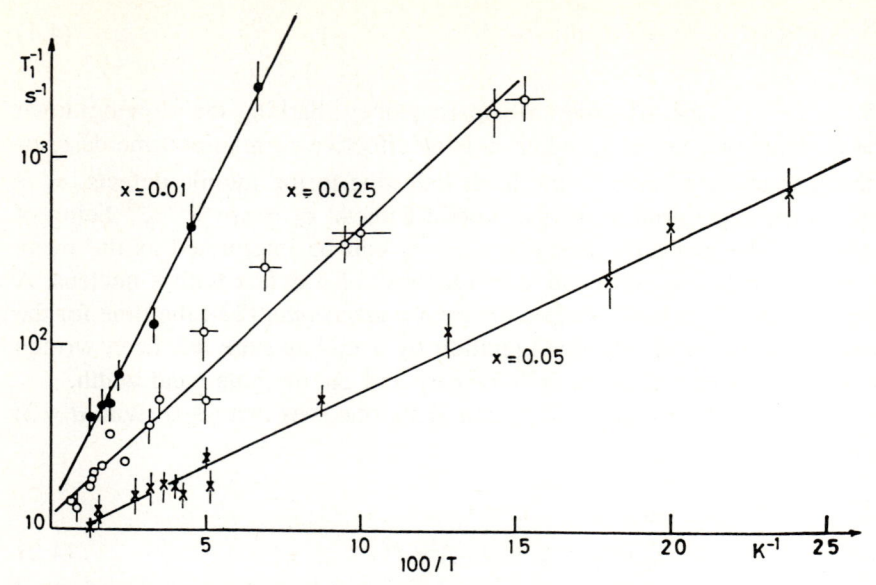

Fig. 7. Semilogarithmic plot of the data for T_1^{-1} in Fig. 5 as a function of $100/T$, for $x = 0.01$, 0.025 and 0.5 (from [12]). The results for $x = 0.1$ have not been plotted since the sample shows superconductivity. However, the increase of T_1^{-1} is in substantial agreement with an activated temperature behavior and yields $E \simeq 8\,\mathrm{K}$ (see Fig. 12)

It is interesting to note that, according to (4.3), if one rescales the NQR T_1 data by the factor aT, then $y = (T_1 T a)^{-1}$ should fall on a universal curve of the variable $z = xT$ given by

$$y = z^{-1} e^{0.85 z^{-1}} \qquad (4.4)$$

regardless of the x and T values, as well as for different nuclei (provided that they probe the magnetic fluctuations of the Cu^{2+} ions). The data for ^{139}La do appear to gather around the universal function (4.4) [12]. The data for ^{63}Cu NQR T_1 in $La_{2-x}Sr_xCuO_4$ at $T = 77$ K appear consistent with (4.4) within the experimental errors and by considering some uncertainty in the constant a (Table 1). (For lower temperatures, see Sect. 6.)

Table 1. Values of $y = (T_1 T a)^{-1}$ for ^{63}Cu in $La_{2-x}Sr_xCuO_4$, at $T = 77$ K

x	$y[K^{-1}]$	
	Experimental value	Theoretical value (4.4)
0.3	3.2×10^{-2}	4.5×10^{-2}
0.2	5.5×10^{-2}	6.8×10^{-2}
0.15	5×10^{-2}	9.3×10^{-2}
0.1	6×10^{-2}	14×10^{-2}

Finally, it is worth mentioning that a value for $\tau_0 \simeq 2.3 \times 10^{-15}$ s, in agreement with our estimate in $La_{2-x}Sr_xCuO_4$, has been deduced also in $YBa_2Cu_3O_{7-\delta}$ from the high-temperature determination of the dynamic Cu NMR tensor [23, 24].

5. General Conclusions and Comparisons with the Findings from Other Techniques

The model of mobile defects in the 2D Heisenberg lattice of the Cu^{2+} ions causing a liquid-like bath of magnetic excitations appears to explain quantitatively the NQR T_1 in $La_{2-x}Sr_xCuO_4$. Now we want to derive from the model other general properties for these crystals, comparing them with the experimental results obtained by means of other techniques.

From (4.2), since h_{eff} has been measured from the spectra [14, 15] and τ_0 is given by the extrapolation of the lines in Fig. 7, the magnetic correlation length ξ can be estimated, without adjustable parameters, from the measurements of the NQR T_1:

$$\xi \simeq \begin{cases} 1.8/\sqrt{(T_1)_{La}} \\ 2.4 \times 10^{-2}/\sqrt{(T_1)_{Cu}} \end{cases} \tag{5.1}$$

where we have used $h_{eff} \simeq 1000$ G for La and 7.8×10^3 G for Cu and for τ_0 the value 2×10^{-15} s.

In Fig. 8 the values for ξ estimated at $T \simeq 77$ K from (5.1) are reported. One observes that, within the experimental errors, the estimates from NQR data are in agreement with those from neutron scattering. One should remark that, while neutron scattering yields the correlation length by means of a static-type measure (inelastic neutron scattering cannot be used to resolve the low-frequency spin dynamics, which is probed by NQR relaxation [26]), the estimate from T_1 is based on the spin dynamics. Thus, in the light of (4.2) and (2.4), one has a kind of dynamical scaling whereby the effective characteristic frequency of the spin fluctuations goes as ω_e/ξ^2. The correlation length can be thought of as the instantaneous average distance between defects and τ_0 as the time needed to cover the distance in a random walk, so that $\tau_d = \tau_0 \xi^2$.

As regards the temperature dependence of ξ, direct comparison with neutron scattering is not possible. In the temperature interval 10–300 K the in-plane correlation length has been deduced [25] to be of the order of 18 ± 6 Å, for $x \simeq 0.11$; unfortunately the range of the error is of the order of the effect which is expected from the temperature variation if one uses for ξ the expression

$$\xi \simeq \frac{1}{\sqrt{x}} e^{0.85/2xT} \tag{5.2}$$

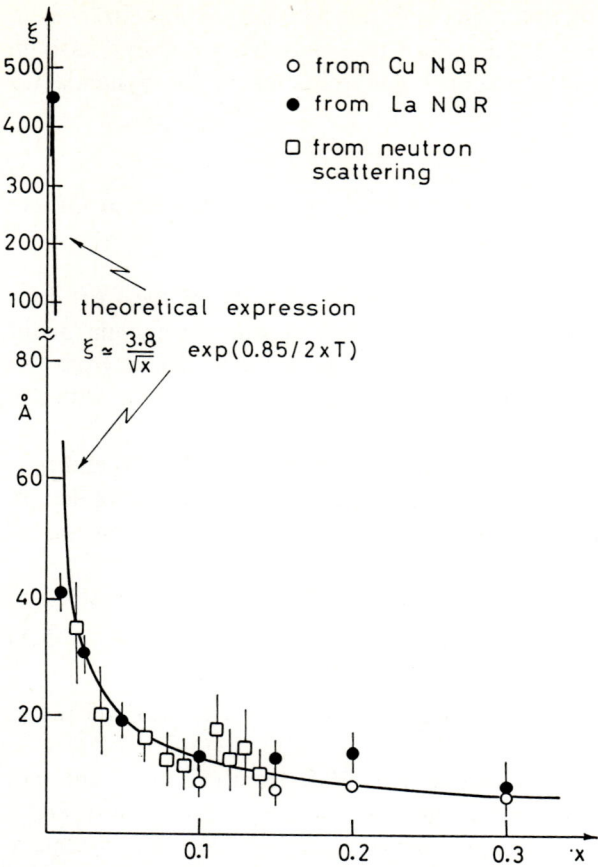

Fig. 8. In-plane magnetic correlation in La$_{2-x}$Sr$_x$CuO$_4$ at $T = 77$ K, in Å, as derived from NQR T_1 for ^{139}La and ^{63}Cu according to (5.1) (the value for nominally pure La$_2$CuO$_4$ at $T \gtrsim T_N$ is also reported), and comparison with the results obtained through neutron scattering [2.5]. For illustration, the theoretical function for ξ implicit in the interpretative model described in Sect. 4 is also shown (solid line)

deduced from the NQR T_1's (Sect. 4). Neutron scattering measurements in samples at $x \lesssim 0.05$ could possibly detect the more consistent growth of the in-plane correlation in the low-temperature range.

The temperature dependence of ξ for nominally pure La$_2$CuO$_4$ [27] is shown in Fig. 9. One should stress that in the case of a very small concentration of defects, intrinsic effects of thermal or quantum fluctuations [28] limiting the degree of correlation cannot be neglected. An attempt to fit the data according to (5.2) (see Fig. 9) yielded a concentration of defects around 0.4×10^{-3}. This observation has a limited meaning also because in the nominally pure La$_2$CuO$_4$ the defects should primarily be due to oxygen depletion ($y > 0$).

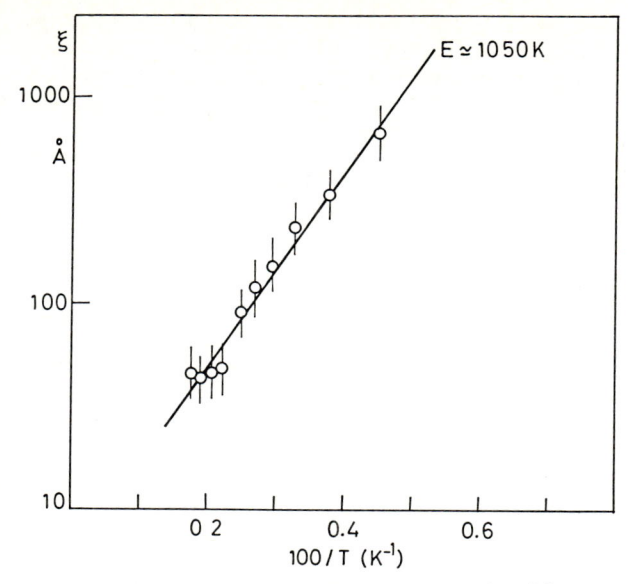

Fig. 9. Values for ξ in nominally pure La_2CuO_4 deduced from neutron scattering data [27] reported in a semilogarithmic plot vs $100/T$

In the light of observation iv) in Sect. 3, one can try to derive the x-dependence of T_N by stating that ξ_{2D} is around 100 lattice units. From (5.2) one has

$$T_N(x) = 0.85/2x \ln 100\sqrt{x} \ . \tag{5.3}$$

In Fig. 10 we compare the behavior expected for $T_N(x)$ from (5.3) with the experimental results obtained by different techniques.

One should note that from NQR T_1 it is not possible to derive the character of the 3D ordered phase, namely 3D AF or spin-glass. However, anomalies in the specific heats observed [32] around T_N favor a 3D AF state. A 3D character of ξ has recently been argued from neutron scattering [25]. Finally one can observe from Fig. 10 that for the sample at $x = 0.1$ coexistence of superconductivity and magnetic order, or heterogeneity in the properties, occurs [37].

Since the mobile defects that in our model limit the magnetic correlation are charged, they should contribute to the electrical conductivity. Thermally activated resistivities are well-known features in the semiconducting samples. The interest here is in a possible x-dependence of the activation energy and in the comparison with the NQR results. In Fig. 11 we report some representative results for the resistivity, crudely extracted from the data published by other authors for $La_{2-x}Sr_xCuO_4$ samples in pellets. The data seem to fit an activated

457

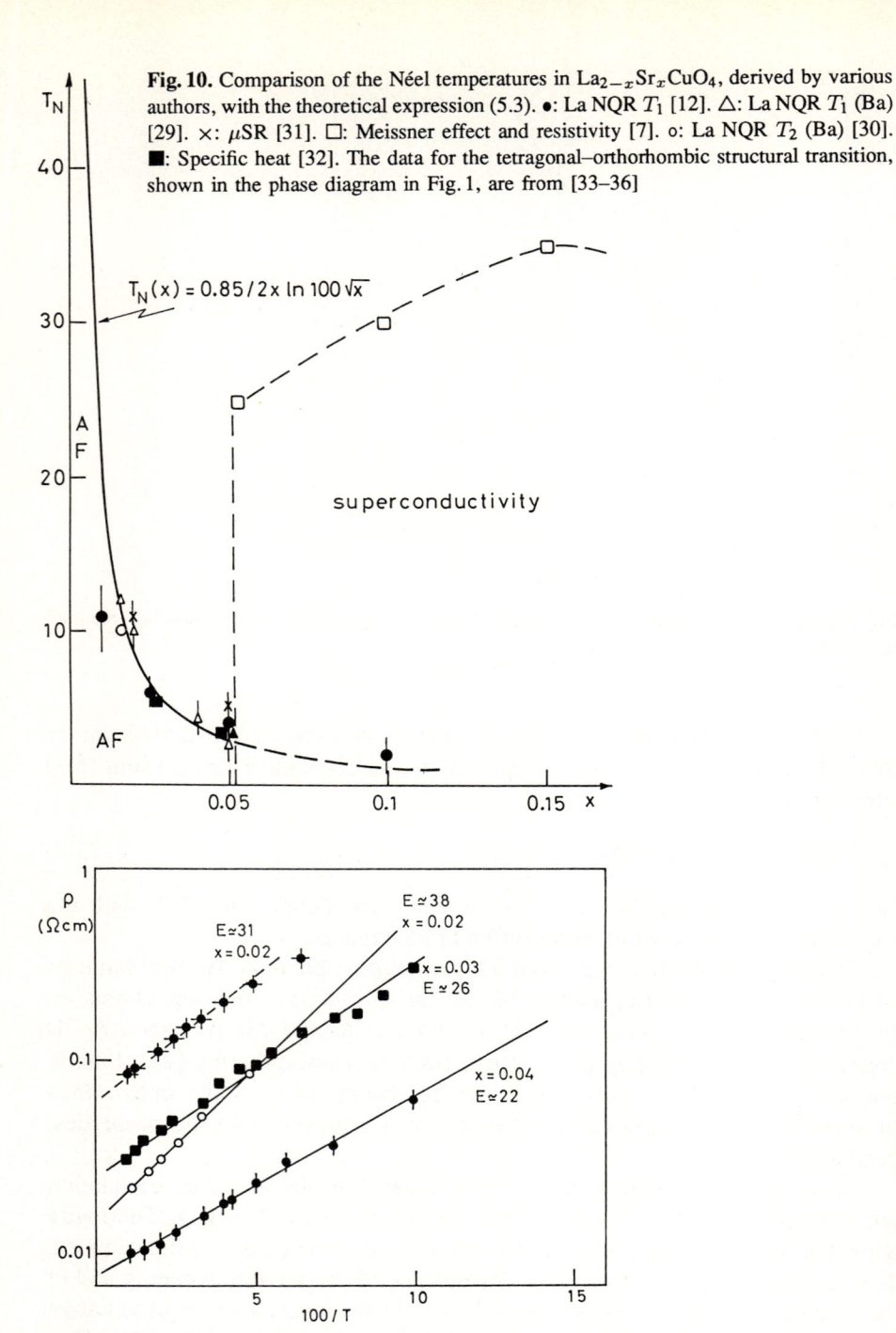

Fig. 10. Comparison of the Néel temperatures in $La_{2-x}Sr_xCuO_4$, derived by various authors, with the theoretical expression (5.3). •: La NQR T_1 [12]. △: La NQR T_1 (Ba) [29]. ×: μSR [31]. □: Meissner effect and resistivity [7]. ○: La NQR T_2 (Ba) [30]. ■: Specific heat [32]. The data for the tetragonal–orthorhombic structural transition, shown in the phase diagram in Fig. 1, are from [33–36]

$T_N(x) = 0.85/2x \ln 100\sqrt{x}$

Fig. 11. Representative experimental results for the resistivity, in a semilogarithmic plots vs $100/T$, for $La_{2-x}Sr_xCuO_4$ (• and ○ [38]; ■ [39]; + [40])

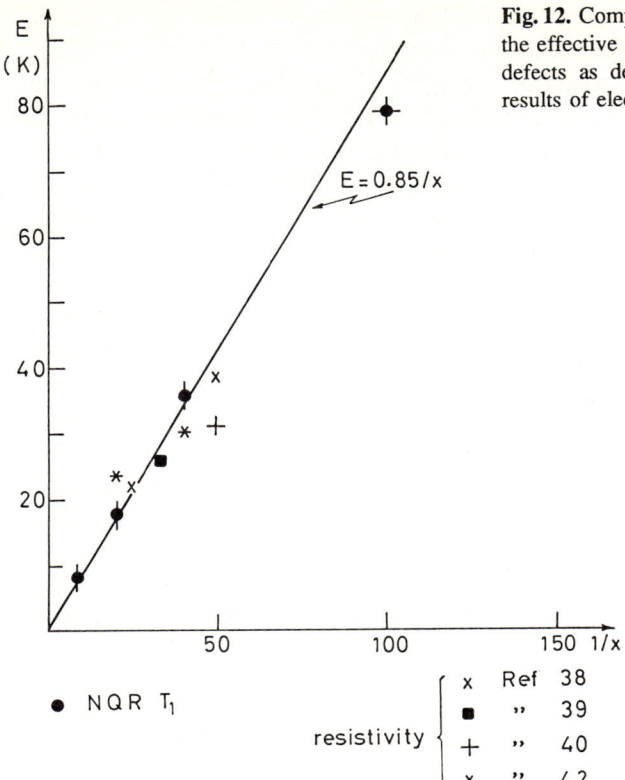

Fig. 12. Comparison for the x-dependence of the effective activation energy for the mobile defects as derived from NQR T_1 with the results of electric resistivity (data in Fig. 11)

law, at least in the temperature range which has been considered in view of comparison with our NQR results [41]. The relevant fact, however, is the x-dependence of the effective activation energy. As shown in Fig. 12, the values for E extracted from the resistivity measurements are in substantial agreement with the law $E = 0.85/x$ derived from NQR T_1.

As regards the magnetic susceptibility, it has recently been pointed out [43] that the contribution $\chi_{2D}(x,T)$ from the 2D Heisenberg Cu spin sublattice diplays a maximum χ_{max} at a $T = T_{max}$ which drops drastically with x. However χ_{max} is rather independent of x. These observations are in qualitative agreement with the model of Sect. 4. In fact, the susceptibility is related to the magnetic correlation length and the maximum in χ_{2D}, for $T \gg T_N$ (so that 3D correlation effects are still negligible), should correspond to a given value of ξ_{2D}. [Note that $\chi_q \propto S(q) = \xi^2 f(q\xi)$ according to (1.2).] Thus Sr-doping causes the drop of T_{max} through the drastic reduction induced in ξ. When, on lowering T, the correlation length increases again because of the partial freezing of the defects, then about the same χ_{max} is restored at a lower temperature. See also Sect. 6.

459

Finally, thermally activated mobile defects should be detectable in the temperature behavior of specific heats, similarly to the soliton contribution. Qualitative considerations, based on the form of $n(x,T)$ suggested from the interpretation of the NQR T_1, would indicate the possible occurrence of a broad maximum in the specific heat vs temperature. Careful analysis of the data [32] is under way in order to single out this effect from the other contributions.

We conclude this section with some remarks about the nature of the mobile defects. The NQR relaxation was assumed to be caused mainly by slowly moving extended magnetic excitations. For approximately round excitations, the spin correlation length is related to the density n of these mobile excitations by $\xi^{-2} \sim n$. As a simple approximation, n might follow an Arrhenius law with excitation energy E.

Such a model gave an alternative interpretation of electron spin resonance line broadening in various undoped layered AFMs [21]. The experimental excitation energy E was in accord with the prediction [44] of metastable states of nearly reversed spins in 2D Heisenberg magnets with the shape described by Skyrme [45] (skyrmions). Similarly, this model would also be consistent with the neutron diffraction data, yielding roughly the energy $E = 4\pi|J|S^2$ of the skyrmions. This prediction [44] is also consistent with the classical renormalization group result [46] $\xi \sim \exp(2\pi|J|S^2)$ used as the classical basis for the "quantum spin liquid" (QSL) [28, 47]. The weakness of the magnetic anisotropy in La_2CuO_4 is essential, since the size of the skyrmions is reduced by increased deviation of the isotropic exchange interaction [21].

The same form $\xi^{-2} \sim n \sim \exp(-E/T)$ also describes the NQR data in doped $La_{2-x}Sr_xCuO_4$ samples in the wide range of doping above the Néel temperature T_N for the onset of 3D order. However, and this is a surprising result, the apparently low activation energies E resulting from these fits turn out to be inversely proportional to the doping x. In addition, the simple assumption that 3D order is triggered by ordered regions about 100 lattice units wide could fit the observed concentration dependence of T_N. Moreover, published electrical conduction data seem to follow this simple approach in the temperature region explored by the NQR data.

At present, it is not easy to extract the simple relation $E \sim 1/x$ from existing theories. Too many important factors are involved, such as the nature of the defects caused by the doping, the charge distribution, the spin arrangement, the influence on the copper spins, and, most important, the collective effects produced by defects with average separations r_{av} of the order of ten lattice units.

The experimental fact that reducing the average separation r_{av} reduces the activation energy $E \sim 1/x \sim r_{av}^2$ seems to imply, in addition to Coulomb repulsion, the presence of a strong attractive effect between charged defects [2]. This attractive force between charges seems to be measurable by NQR even for x values too low to bring about superconductivity in $La_{2-x}Sr_xCuO_4$.

6. ^{63}Cu NQR Relaxation Around the Superconducting Transition

Let us briefly discuss the ^{63}Cu relaxation rates in the x and T range around the onset of the superconducting phase (see Fig. 1). The experimental results are reported in Fig. 13. Comparison with the theoretical estimates obtained by using (4.3) shows that the perturbative approach based only on the reduction of the degree of correlation of the Cu magnetic moments cannot hold for large x and $T \lesssim 70$ K: for example, a meaningless value $\xi \ll 1$ would be derived. This is not surprising: a large number of extra holes must affect the lattice of the correlated Cu^{2+} spins by modifying the magnetic properties of the ion itself, particularly when delocalization of the $3d$ electrons occurs. In Fig. 13

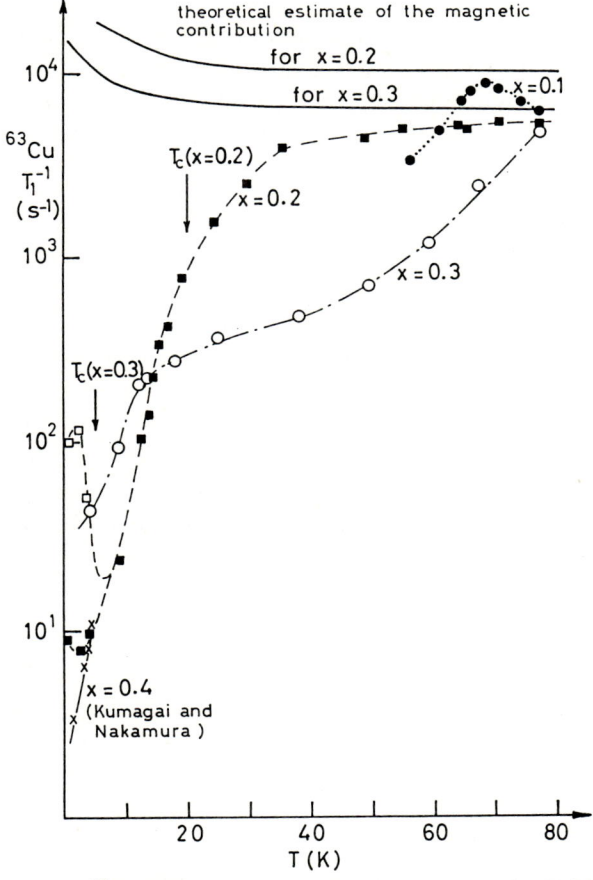

Fig. 13. ^{63}Cu NQR relaxation rates vs T in La$_{2-x}$Sr$_x$CuO$_4$ for $x = 0.1(\bullet)$, $x = 0.2(\blacksquare)$ and $x = 0.3(\circ)$. The solid lines at the top are the theoretical behavior estimated from the extension of (4.3). The data for $x = 0.4$ (metallic, nonsuperconducting state) are from *Kumagai* and *Nakamura* [37]. In the sample at $x = 0.1$ a very short T_2, around 5 μs, prevents measurements below 45 K

we also note the following features (which in part have been observed also in YBCO compounds by other authors [48–51].

i) The $T_1 T$ = const law (Korringa relation), characteristic of the relaxation process driven by the hyperfine interaction with delocalized carriers, is not obeyed (while for $x = 0.4$, namely the normal nonsuperconducting state, the Korringa relation does hold).

ii) No hump is detected in T_1^{-1} just below T_c, although there is one in conventional BCS superconductors [52].

iii) A drastic decrease in the relaxation rate occurs for $T < T_c$, which is not fitted by the law $T_1^{-1} \sim \exp(-\Delta/T)$ characteristic of the process of Cooper-pair breaking with increasing temperature.

iv) At $T \lesssim 10\,\text{K}$ for $x = 0.2$, clear evidence of coexistence of superconductivity and magnetic effects is provided by the recovery law. One detects, in fact, two exponentials, one with a long relaxation time and the second one yielding a divergence in T_1^{-1} similar to the ones observed on approaching the PA $\to$ AF transition (Fig. 5).

A remarkable fact pointed out by the data in Fig. 13 is the drastic reduction of the relaxation rate which appears to start above T_c. Analogous "precursor" effects of superconductivity have recently been observed in the range 90–110 K in $YBa_2Cu_3O_{6.7}$ ($T_c \simeq 61$ K) and interpreted in terms of pair formation and energy gap formation beginning in the CuO_2 plane above the 3D superconductivity [49].

A simple explanation for the suppression, on cooling, of the relaxation rate due to magnetic fluctuations can be given in the framework of our model by abandoning the perturbative approach and assuming instead, for $x \gtrsim 0.1$, a reduction of the effective nucleus-electron hyperfine interaction in the Cu ion. In this case, from (4.2) and (4.3) [see also (3.4)] one can write for the Cu relaxation rate

$$T_1^{-1} = 3\gamma^2 |h_{\text{eff}}(x,T)|^2 \frac{1}{x} \tau_0 e^{0.85/xT} \ . \tag{6.1}$$

Then the x and T dependence of the effective field $h_{\text{eff}}(x,T)$ at the Cu nucleus can be obtained from a comparison of the experimental results with the theoretical estimates derived in the assumption of charge defects which destroy the magnetic correlation while h_{eff} = const. In this way one obtains the effective hyperfine filed reported in Fig. 14.

One should note that for $T \lesssim 4.2\,\text{K}$ the value for h_{eff} derived on the basis of (6.1) is around 1 kG, in close agreement with the static hyperfine field measured from Cu NQR spectra (*Kumagai* and *Nakamura* [37]) and corresponding to a reduction of the Cu magnetic moment of around 0.01 μ_B.

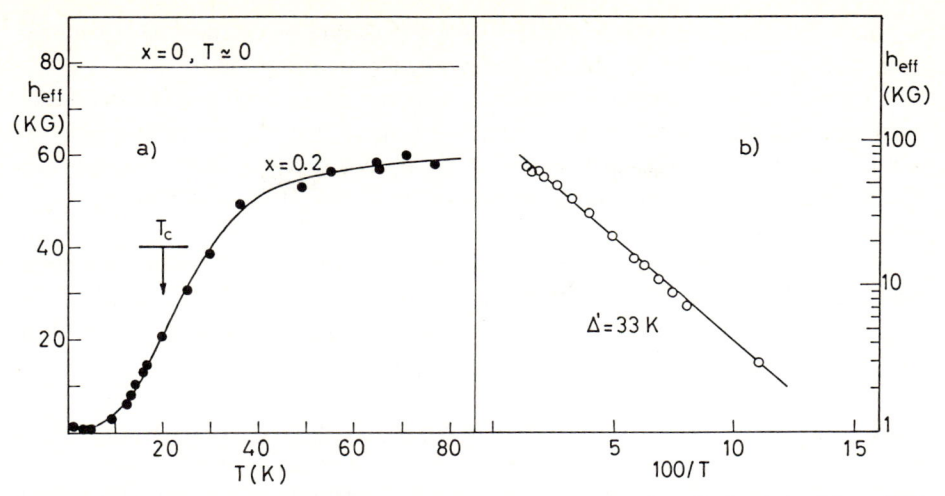

Fig. 14. (a) Effective hyperfine field at the Cu nucleus vs T, for $La_{1.8}Sr_{0.2}CuO_4$, derived from (6.1) and the experimental data in Fig. 13; (b) h_{eff} in a semilog plot vs $100/T$; the value $h_{eff} \simeq 78.8$ kG is from the Cu NMR observation in antiferromagnetic La_2CuO_4 at $T = 1.3$ K [15]

One can observe that the temperature dependence of h_{eff} seems to be of the form $h_{eff} \propto \exp(-\Delta'/T)$. The activation energy $\Delta' \simeq 33$ K could be related to the pairing energy, yielding $S = 0$ and thus causing indirectly the decrease in the Cu electron–nucleus hyperfine interaction. It should be remarked that no discontinuity or change of slope is observed in h_{eff} at T_c.

The drop of the effective intralayer Cu–Cu supercharge coupling constant and of the effective moment per Cu ion suggested from magnetic suspectibility measurements for $x \gtrsim 0.1$ [43] is the counterpart of the decrease of the effective hyperfine coupling constant. Therefore, in the superconducting samples one should take into account two nuclear relaxation mechanisms, both associated with magnetic interactions: one with the nearly localized Cu $3d$ electrons (localization is supposed to be related to a strong on-site repulsive interaction) with an x and T dependence of the effective magnetic moment, and a second one from the Fermi liquid of the p-holes induced at the oxygen sites by Sr doping. This separation is obviously an idealization and a correct picture should take into account the correlation effects among itinerant electrons [53].

Strong support of the picture of the simultaneous presence of correlated fluctuations of almost localized Cu spins and of p-hole carriers behaving like a Fermi liquid is given by the ^{17}O relaxation rate in enriched $YBa_2Cu_3O_7$ [54]. Because of symmetry considerations analogous to the ones discussed in Sect. 2, the ^{17}O nucleus is not sensitive to the AF correlated fluctuations of the n.n. Cu spins ($h_{qAF} = 0$). Then almost the only relaxation process detected is that driven by the p-hole carriers and a T behavior of the ^{17}O relaxation rate similar to the one for BCS superconductors is observed.

Two kinds of magnetic relaxation mechanisms have been indicated since the early detection of different temperature behaviors of the Cu relaxation rates for Cu (I) and Cu (II) in YBCO [48]. A more recent confirmation comes from Cu NMR spectra and spin–lattice relaxation in $YBa_2CU_3O_7$ and $YBa_2Cu_3O_{6.75}$ single crystals [55].

Finally it is worth mentioning that direct confirmation of the Sr-concentration dependence of the hyperfine field at the Cu nuclei in $La_{2-x}Sr_xCuO_4$ comes from recent nuclear heat capacity measurements, with a sharp reduction for $x \gtrsim 0.05$ of the nuclear Schottky coefficient [56].

Acknowledgements. The study of the La_2CuO_4-based superconductors is being carried out in Pavia through a collaboration involving F. Borsa, M. Corti, A. Rigamonti and T. Rega (Dottorato in Fisica, University of Pavia) and J. Ziolo (University of Katowice, ICTP Program for Training and Research in Italian Laboratories, Trieste). The samples were synthesized by G. Flor and the conductivity measurements were by G. Chiodelli and G. Flor (Department of Physical Chemistry, University of Pavia). The specific heat measurements were performed by S. Aldrovandi and A. Lascialfari (Department of Physics A. Volta, University of Pavia). Useful discussion with G. D'Ariano and G. Senatore are gratefully acknowledged. Thanks are due to P. Orlandi for his valuable technical assistance and for the graphics. The contribution to the realization of the manuscript by Viviana Minassi is herewith much appreciated.

References

1 L.J. De Jongh: Physica C **152**, 171 (1988); In Proc. NATO Adv. Res. Workshop on Condensed Systems of Low Dimensionality, to appear in J. Chim. Phys.; Proc. 1st Int. Symposium on Superconductivity, Nagoja (August 1988), to be published by Springer
2 R.J. Birgeneau, M.A. Kastner, A.Aharony: Z. Phys. B **71**, 57 (1988) and references therein
3 E.Y. Loh, T. Martin, P. Prelovsek, D.K. Campbell: Phys. Rev. B **38**, 2494 (1988)
4 H.E. Stanley: *Introduction to Phase Transitions and Critical Phenomena* (Oxford University Press, Oxford 1971)
5 P.C. Hohenberg, B.I. Halperin: Rev. Mod. Phys. **49**, 435 (1977)
6 See various contributions in *Local Properties at Phase Transitions*, ed. by K.A. Müller, A. Rigamonti (North-Holland, Amsterdam 1976); see also F. Borsa, A. Rigamonti: In *Magnetic Resonance at Phase Transition*, ed. by F.J. Owens, C.P. Poole, H.A. Farach (Academic, New York 1979) and A. Rigamonti: NMR-NQR studies at structural transitions. Adv. Phys. **33**, 115 (1984), where the general aspects of NMR and NQR and a reformulation of the relaxation process applied to phase transitions and critical phenomena can be found
7 A. Rigamonti, F. Borsa, M. Corti, T. Rega, J. Ziolo, G. Flor: Proc. of the XXIV Ampère Meeting, Poznan 1988 (North-Holland, Amsterdam 1989)
8 G. Le Flem, R. Brec, G. Ouvard, A. Louisy, P. Segransan: J. Phys. Chem. Solids **43**, 455 (1982)
9 J. Ziolo, S. Torre, A. Rigamonti, F. Borsa: J. Appl. Phys. **63**, 3095 (1988); for further details, including the effect of the external magnetic field on the spin dynamics and on the transition temperature, see S. Torre, J. Ziolo: Phys. Rev. B, May (1989)
10 A. Rigamonti, F. Tabak, S. Torre: To be published
11 T. Enoki, I. Tsujikawa: J. Phys. Soc. Jpn. **39**, 317 (1975); for susceptibility measurements in $Fe_{1-x}Zn_xPS_3$ see P. Dolile, J.J. Steger, A. Wold: Inorg. Chem. **14**, 2400 (1975)

12 F. Borsa, M. Corti, T. Rega, A. Rigamonti: Nuovo Cimento D **11**, 1785 (1989)
13 The conclusion on the negligibility of the quadrupolar relaxation mechanism in the ^{139}La NQR relaxation for $T < 80\,\text{K}$ and $x < 0.2$, as well as for $T \simeq T_N \simeq 300\,\text{K}$ in pure La_2CuO_4, is also supported by an analysis of the experimental recovery plots for the various NQR lines and/or for different initial conditions of the relaxation process. A method has been devised that allows the separation of the magnetic and quadrupolar contributions from the tangents at the origin of recovery plots obtained under different conditions. See T. Rega, to be published
14 J. Ziolo, F. Borsa, M. Corti, A. Rigamonti: Physica C **153**, 725 (1988) and references therein
15 T. Tsuda, T. Shimizu, H. Yasuoka, K. Kishio, K. Kitazawa: J. Phys. Soc. Jpn. **57**, 2908 (1988)
16 H. Seidel, F. Hentsch, M. Mehring, J.G. Bednorz, K.A. Müller: Europhys. Lett. **5**, 647 (1988)
17 Y. Endoh et al.: Phys. Rev. B **37**, 7443 (1988) and references therein; K.B. Lyons et al.: Phys. Rev. Lett. **60**, 732 (1988); M. Sato et al.: Phys. Rev. Lett. **61**, 1317 (1988)
18 P.W. Anderson: Science **235**, 1196 (1987); Phys. Rev. Lett. **59**, 2497 (1987)
19 See various contributions in *Magnetic Excitations and Fluctuations*, ed. by V.L. Lovesey, U. Balucani, F. Borsa, V. Tognetti, Springer Ser. Solid-State Sci., Vol. 54 (Springer, Berlin, Heidelberg 1984)
20 F. Borsa, J.P. Boucher, J. Villain: J. Appl. Phys. **49**, 1326 (1978); J.P. Boucher: Solid State Commun. **33**, 1025 (1980)
21 F. Waldner: J. Magn. Magn. Mater. **54**, 873 (1986); ibid **31**, 1203 (1983)
22 T.V. Ramakrishnan: Physica C **153**, 155 (1988)
23 C.H. Pennington, D.J. Durand, C.P. Slichter, J.P. Rice, E.D. Bukowski, D.M. Ginsberg: Phys. Rev. B **39**, 2902 (1989)
24 In a recent Cu NMR study in YBCO [M. Horvatic et al.: Phys. Rev. B **39**, 7332 (1989)] the relationship of Cu T_1 to the correlation length of the 2D AF correlations in the CuO plane has been qualitatively discussed along arguments similar to the ones developed in Sects. 4 and 5 of the present work
25 R.J. Birgeneau et al.: Phys. Rev. B **38**, 6614 (1988); R.J. Birgeneau, Y. Endoh, K. Kakurai, Y. Hidaka, T. Murakami, M.A. Kastner, T.R. Thurston, G. Shirane, K. Yamata: Phys. Rev. B **39**, 2868 (1989)
26 G. Aeppli et al.: Phys. Rev. Lett. **62**, 2052 (1989). In this paper a frequency resolution limit around $10^{13}\,\text{rad}\,\text{s}^{-1}$ is indicated; it is also shown that in La_2CuO_4 conventional spin waves describe the magnetic dynamics for momentum and energy transfers below $0.15\,\text{Å}^{-1}$ and $0.1\,\text{eV}$ respectively. Furthermore it is deduced how doping changes in the magnetic dynamics through a combination of softening and damping effects, qualitative support for our picture of low-frequency diffusive modes due to the mobile defects
27 D. Vaknin et al.: Phys. Rev. Lett. **58**, 2802 (1987); G. Shirane et al.: Phys. Rev. Lett. **59**, 1613 (1987); Y. Endoh et al.: Phys. Rev. B **37**, 7443 (1988); see also E. Manousakis, R. Savador: Phys. Rev. Lett. **62**, 1310 (1989), where the neutron scattering data taken in La_2CuO_4 have been analyzed in terms of the 2D $S = 1/2$ Heisenberg model versus a nonlinear σ model
28 S. Chakravarty, B.I. Halperin, D.R. Nelson: Phys. Rev. Lett. **60**, 1057 (1988); see also E. Manousakis, R. Salvador: [27]
29 Y. Kitaoka et al.: J. Phys. Soc. Jpn. **57**, 734 (1988)
30 I. Watanabe et al.: J. Phys. Soc. Jpn. **56**, 3028 (1988)
31 J.I. Budnick et al.: Europhys. Lett. **5**, 651 (1988)
32 S. Aldrovandi, F. Borsa, A. Lascialfari: To be published
33 M. Francois, K. Yvon, P. Fischer, M. Decroux: Solid State Commun. **63**, 35 (1987)
34 D.M. Paul, G. Balakrishnan, N.R. Bernhoeft, W.I.F. David, W.T.A. Harrison: Phys. Rev. Lett. **58**, 1976 (1987)
35 M.J. Rosseinsky, K. Prassides, P. Day, A.J. Dianoux: Phys. Rev. B **37**, 2231 (1988)

36 R.J. Birgeneau et al.: Phys. Rev. Lett. **59**, 1329 (1987);
T.R. Thurston, R.J. Birgeneau et al.: Phys. Rev. B **39**, 4327 (1989)
37 In YBa$_2$Cu$_3$O$_{6.55}$ coexistence of superconductivity and antiferromagnetism has been claimed: Petitgrand et al: J. de Phys. **49**, 1815 (1988);
A. Weidinger et al.: Phys. Rev. Lett. **62**, 102 (1988).
A recent Cu-NQR study in La$_{2-x}$Sr$_x$CuO$_4$ for $0.12 < x < 0.4$ [K. Kumagai, Y. Nakamura: Physica C **157**, 307 (1989)] has indicated the coexistence of superconductivity and magnetic order of Cu d-electrons, with a tiny magnetic moment
38 W. Kang et al.: J. de Phys. **48**, 1181 (1987)
39 J.G. Bednorz, K.A. Müller: Rev. Mod. Phys. **60**, 590 (1988)
40 D.R. Harsham et al.: Phys. Rev. B **38**, 852 (1988)
41 The resistivity of crystals grown under different conditions and, in particular, of one crystal at $x = 0.02$ grown in Li flux [M.A. Kastner et al.: Phys. Rev. B **37**, 111 (1988)] has been found to display, over a wide temperature range (600–4 K) a temperature behavior of the form $\varrho(T) \sim \exp(T_0/T)^\nu$ with $\nu \sim 0.25$, in agreement with the Mott variable-range hopping law. [See also M. Oda et al.: Solid State Commun. **67**, 257 (1988)]. The crucial point of a possible x-dependence of T_0 does not appear to have been discussed. Finally we mention that a coupling between magnetic and transport properties has recently been studied in crystalline La$_2$CuO$_{4-y}$ [S.W. Cheong et al.: Phys. Rev. B **29**, 6567 (1989)]
42 G. Chiodelli, G. Campari-Vigano', V. Massarotti, G. Flor: Z. Naturforsch. (1989) and private communication
43 D.C. Johnston: Phys. Rev. Lett. **63**, 957 (1989)
44 A.A. Belavin, M.A. Polyakov: Pis'ma Zh. Eksp. Theor. Fiz. **22**, 503 (1975); JETP Lett. **22**, 245 (1975)
45 T.H.R. Skyrme: Proc. R. Soc. London **262**, 237 (1961)
46 P. Young: In *Ordering in Strongly Fluctuating Condensed Matter Systems*, ed. by T. Riste (Plenum, New York 1979)
47 A. Auerbach, D.P. Arovas: Phys. Rev. Lett. **61**, 617 (1988)
48 M. Mali, D. Brinkmann, L. Pauli, J. Roos, H. Zimmermann, J. Hullinger: Phys. Lett. A **124**, 112 (1987);
D. Brinkmann: Physica C **153**, 75 (1988) and references therein
49 W.W. Warren Jr., R.E. Walstedt, G.F. Brennert, R.J. Cava, R. Tycko, R.F. Bell, G. Dabbagh: Phys. Rev. Lett. **62**, 1193 (1989) and references therein
50 Y. Kitaoka, S. Hiramatsu, T. Konda, K. Asayama: J. Phys. Soc. Jpn. **57**, 31 (1988);
K. Ishida, Y. Kitaoka, K. Asayama: J. Phys. Soc. Jpn., to be published
51 T. Imai, T. Shiizu, T. Tsuda, H. Yasuoka, T. Takabatake, Y. Nakazawa, M. Ishikawa: J. Phys. Soc. Jpn. **57**, 1771, 2280 (1988)
52 L.C. Hebel, C.P. Slichter: Phys. Rev. **113**, 1504 (1957)
53 T. Kayama and M. Tachiki [Phys. Rev. B **39**, 2279 (1989)] have recently given a theoretical treatment of the relaxation process below T_c along this line, in terms of a dynamical spin susceptibility for correlated electrons. A sharp decrease in T_1^{-1} below T_c is obtained, with the assumption of a T-independent hyperfine coupling constant. No precursor effects above T_c are considered
54 K. Ishida, Y. Kitaoka, K. Asayama, H. Katayama-Yoshida, Y. Okabe, T. Takahashi: J. Phys. Soc. Jpn. **57**, 2897 (1988);
Y. Kitaoka is gratefully thanked for providing preprints with experimental data prior to publication
55 See M. Horvatic, P. Ségransan, C. Berthier, Y. Berthier, P. Butaud, J.Y. Henry, M. Couach, J.P. Chaminade: [24]
56 N. Wada, H. Muro-oka, Y. Nakamura, K. Kumagai: Physica C **157**, 453 (1989)

A Credo for a Spy: Nuclear Spin Interactions in Superconductors

M. Mehring

2. Physikalisches Institut, Universität Stuttgart,
Pfaffenwaldring 57, D-7000 Stuttgart 80, Fed. Rep. of Germany

Abstract. The interaction of nuclear spins with magnetic and electric fields caused by conduction electrons and orbital motion is discussed. The difference between chemical and Knight shift tensors is treated in detail. A discussion on nuclear spin relaxation processes due to conduction electrons in the normal and superconducting states is also included. I will demonstrate that the formation of the vortex lattice in high-temperature superconductors is clearly visible in the field distribution of NMR spectra.

1. Introduction

Nuclear spin resonance techniques such as NMR (Nuclear Magnetic Resonance) and NQR (Nuclear Quadruple Resonance) have proved extremely useful for the investigation of classical superconductors [1–5]. The cornerstones of success in this area include gap opening determined by nuclear spin relaxation [1, 2, 6, 7], Knight shift collapse due to Cooper pairing [3, 8, 9], field distribution in the vortex lattice [10–12], effective charge and oxidation state determination [13, 14], and competition between magnetic ordering and superconductivity [15–17], to name just a few. The NMR techniques are nondestructive and do not introduce "foreign" species into the sample. They rely on the natural "spies" in the system, namely those atoms which possess a nuclear spin. There are plenty. Some have spin $I = 1/2$, like [207]Pb, [85]Y, [203,205]Tl, and therefore no quadrupolar interaction, others have $I > 1/2$, such as [63,65]Cu ($I = 3/2$), [17]O ($I = 5/2$), [23]Al ($I = 5/2$), [139]La ($I = 7/2$), with usually strong quadrupolar interaction. Moreover, the new high-temperature superconductors usually contain a number of different nuclei at different positions of the unit cell, i.e. spies all over the place as members of the natural unit cell community. They keep everything very secret, i.e. the solid-state phenomena are unaffected by the "spies" since all nuclear spin interactions with conduction electrons are negligibly small compared with electron and phonon energies.

What is there to spy on? Well, among other things there are electric field gradients at particular nuclear sites leading to effective charges of the atoms, chemical shifts leading to information on the oxidation stage, Knight shifts giving us the paramagnetic susceptibility in the normal state and demonstrating singlet pairing in the superconducting state, the spectral distribution in the vortex state, and spin lattice relaxation containing a wealth of information on electron spin dynamics.

Space does not permit me to treat everything in detail, instead I will dwell on special aspects which seem to me most interesting. In particular the whole area of quadrupolar interactions, which has been studied in detail in high-temperature superconductors, will not be covered here.

2. Static and Dynamic Hyperfine Fields: Chemical Shifts and Knight Shifts

There are two basically different types of electron-nuclear spin interactions, namly (i) those caused by the orbital motion of all the electrons and (ii) those due to the electron spin. Both types of interactions derive from fast electron motion, which creates an average local field at the nuclear site, leading to a shift δ of the resonance line. Among metal physicists all the contributions to the lineshift δ are called the "Knight shift", whereas in other areas of chemistry and physics the first part (i) is called the "chemical shift", because it is caused by the orbitals surrounding the nucleus, and the second part (ii) is called the "paramagnetic" shift or, in metals, the Knight shift. Since in Knight's original investigation [18a] on copper this paramagnetic shift due to the Pauli susceptibility of the conduction electrons was the dominant part, I will make a distinction in the following between (i) chemical shift δ_c and (ii) (paramagnetic) Knight shift K_s, where the index s stands for the (electron) spin contribution [18b]:

$$\tilde{\delta} = \tilde{\delta}_c + \tilde{K}_s . \tag{1}$$

Note that both the chemical shift and Knight shift are second rank tensors with individual principal axes, which may not coincide. The totally observed shift tensor $\tilde{\delta}$ may therefore have a principal axis system which coincides with neither $\tilde{\delta}_s$ nor $\tilde{K}_s$. We are therefore dealing in general with three principal values for each tensor, i.e. δ_j, δ_{cj}, K_{sj} ($j = 1, 2, 3$) and three possibly different principal axes. The center of gravity of the line in a powder sample appears at the isotropic value $\delta_{iso} = (\delta_1 + \delta_2 + \delta_3)/3$. Both δ_c and K_s are very complex and contain a number of different contributions, like diamagnetic and paramagnetic orbital (Van Vleck) contributions in δ_c, and Fermi contact, core polarization, dipolar contributions, superexchange, etc., in K_s [18b]. The disentanglement of all these contributions is by no means trivial and requires a great deal of experience and comparison with "reference compounds" [18b].

The Knight shift tensor for ^{63}Cu was indeed determined in high-temperature superconductors [8, 18c]. There is a nice way of separating δ_c and K_s in the superconducting state (singlet pairing). At very low temperatures K_s should vanish because all electron spins are frozen. Indeed this has been observed in classical as well as high-temperature superconductors [3, 8, 9]. An example will be discussed in the next section.

3. Thallium-Containing High-Temperature Superconductors: A Case Study

Three different thallium-containing superconductors, namely $Tl_2Ba_2CuO_6$ ($n = 1$; $T_c = 80$ K), $Tl_2Ca_1Ba_2Cu_2O_8$ ($n = 2$; $T_c = 108$ K) and $Tl_2Ca_2BaCu_3O_{10}$ ($n = 3$; $T_c = 125$ K), will be discussed in the following. The number of CuO_2 planes is labelled by $n = 1, 2, 3$. There are in addition two TlO planes separating the CuO_2 layers. We have performed NMR experiments on ^{205}Tl in a 4 T field [14a]. Figure 1 shows a comparison of ^{205}Tl NMR spectra for the three compounds in the normal state [19]. The dominant line is due to ^{205}Tl in the TlO layers. The small line to the left of the dominant line was found by us to be caused by a defect in which Ca is replaced by ^{205}Tl (defect line) [14a]. Note that the defect line is absent in the $n = 1$ compound, which does not contain a Ca position in the unit cell. The "defect line" was also observed by others [21] but assigned to an oxygen-deficient TlO layer.

The shift δ of the dominant line is represented by (1). There is an appreciable distribution of shifts caused by the tensorial character, leading to characteristic lineshapes with singularities [20], most obviously seen for the $n = 1$ compound. This is well known for non-conducting materials [20]. We have argued that the shifts of the "dominant line" are due to the chemical shift tensor $\tilde{\delta}_c$ [14a]. Comparison with reference compounds containing Tl^+ and Tl^{3+} ions, respectively, shows that the oxidation state of Tl in the TlO layers is Tl^{3+} (Fig. 1). The defect line, see Fig. 1, turned out to be even more important than the dominant line. Note that the defect line is shifted strongly down from the Tl^+ and Tl^{3+} reference positions. The Ca, which is partially replaced by Tl, is located between two CuO_2 planes, i.e. in the immediate vicinity of the conducting layers. This is an ideal position for a nuclear spy who wants to observe the electron dynamical action going on the CuO_2 layers. Correspondingly the defect line shows a dramatic negative Knight shift of $K_s = -0.3\%$ [19]. Figure 2 displays its temperature dependence, which freezes out in the superconducting state [19]. Three conclusions can be drawn from these observations: (i) Singlet pairing takes place below T_c; (ii) at $T = 0$ with $K_s = 0$ only the chemical shift δ_c is left, which corresponds in this case to Tl^+; and (iii) the negative sign of the Knight shift hints at an indirect core polarization of the Tl ion. A similar effect has been observed on ^{85}Y in $YBa_2Cu_3O_7$ by *Alloul* and co-workers [22]. The importance of this finding can be appreciated from Fig. 3. Two different possible bonding mechanisms in the CuO_2 plane are sketched, namely σ-bonding (left) and π-bonding (right). The

Fig. 1. ^{205}Tl NMR spectra for $T > T_c$ of three different Tl-containing high-temperature superconductors ($n = 1, 2, 3$) according to [19]. Chemical shift references for Tl$^+$ and Tl^{3+} ions are indicated by dashed lines. Note the "defect line" below the Tl$^+$ reference position

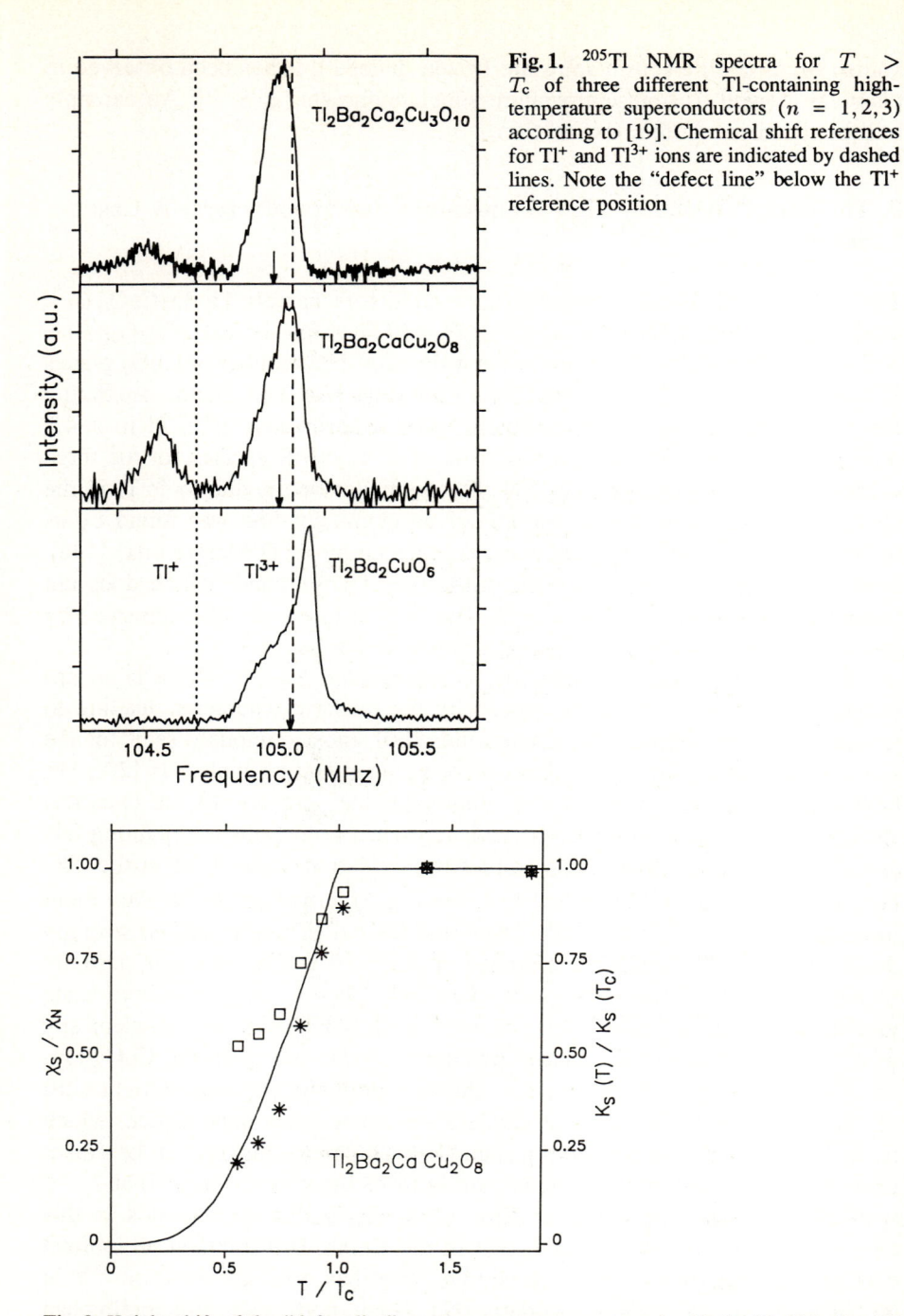

Fig. 2. Knight shift of the "defect line" versus temperature compared with the Yoshida line [3] for the electron spin susceptibility according to [19]. The stars correspond to the assumption that the defect is Tl$^+$, whereas the squares correspond to Tl^{3+}

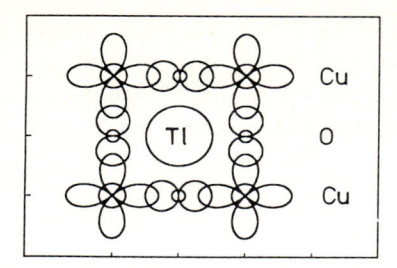

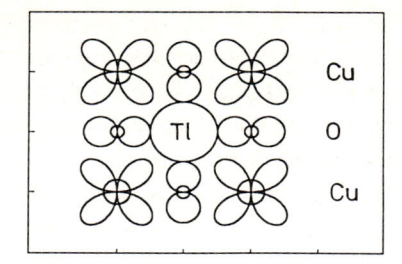

Fig. 3. Sketch of the orbital scenario in the CuO_2 planes for σ-bonding (left) and π-bonding (right). The negative Knight shift of the Tl ion favours the σ-bonding configuration (see text)

Tl ion is located in the middle between two of these CuO_2 layers. If π-bonding (right) were dominant, appreciable overlap with the Tl orbitals would occur, resulting in a large positive Knight shift. Instead, a negative Knight shift is observed experimentally. This is expected for the σ-bonding situation shown in the left part of Fig. 3.

4. Nuclear Spin Relaxation

The nuclear spin lattice relaxation rate T_1^{-1} probes the dynamic nature of conduction electrons and different sorts of magnetic fluctuations. Quadrupole interactions are not discussed here. It proved to be crucial for an understanding of the superconducting gap opening [1, 2] and was intensively studied in high-temperature superconductors [6, 7, 9, 14a, 23–27]. The mechanisms leading to nuclear spin lattice relaxation T_1 in the high-temperature superconductors are still under discussion. The anisotropy of the gap parameter $\Delta(k)$ in k-space complicates the situation and a simple activated behavior such as [1]

$$T_1^{-1} \sim \exp(-\Delta/k_B T) \quad \text{for} \quad T \ll T_c \tag{2}$$

expected for isotropic BCS-type superconductors is hardly observed. If the data are interpreted in this way, usually a rather large value of $2\Delta/k_B T_c \simeq 6$–8 is obtained [14a]. An analysis based on an anisotropic gap parameter by Japanese groups [23, 26] leads to

$$T_1^{-1} \sim T^\alpha \tag{3}$$

behavior in a certain temperature range, with $\alpha = 3.0$–4.5.

Even in the normal state T_1^{-1} does not behave "normally", i.e. a Korringa law [$T_1^{-1} \sim T$ or $\alpha = 1$ in (3)] is not usually observed. It seems that the messages from the nuclear spy are cryptically encoded, at least as far as spin lattice relaxation is concerned.

It is evident that some "intelligence" is required in order to interpret the relaxation data correctly. The simple laws (2) and (3) are obviously too simple minded to deal with the realistic situation. There seems to be highly correlated motion of the electrons and the electron spins in this type of material. Correspondingly, these correlations show up in the electron dynamics picked up by the nuclear spins. In order to deal with this situation one has to start from the general expression [28]

$$T_1^{-1} = 2k_B T \sum_q A_q A_{-q} \chi''_\perp(q, \omega_n)/\omega_n \tag{4}$$

where $\omega_n = \gamma_n H_0$ is the Larmor frequency of the nucleus with gyromagnetic ratio γ_n in the magnetic field H_0; A_q (in rad s^{-1}) is the q-dependent hyperfine interaction, where q is the wave vector transferred in the electron scattering process. The electron dynamics contributes through the imaginary part $\chi''_\perp(q, \omega_n)$ of the transverse susceptibility. It is evident from (4) that it makes sense to plot $(T_1 T)^{-1}$ versus temperature in order to learn more about the temperature dependence of the dynamic susceptibility $\chi(q, \omega)$. The relaxation data are presented in this form in Fig. 4. Note the pronounced cusp in the relaxation rate near T_c. A similar presentation of the data was proposed in [23, 25].

In an ordinary metal (no electron–electron correlations) the dynamic susceptibility $\chi(q, \omega)$ is temperature independent (Pauli susceptibility and its dynamic counterpart) and the Korringa law [$(T_1 T)^{-1}$ = const.] holds. This is so

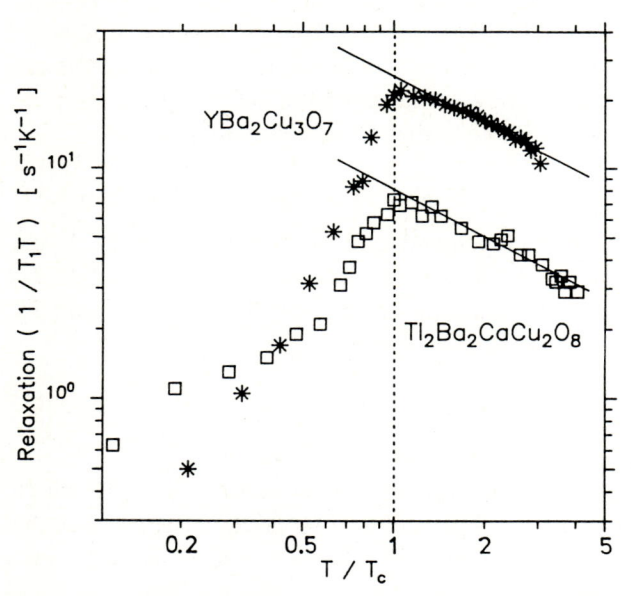

Fig. 4. Spin lattice relaxation rate $(T_1 T)^{-1}$ versus temperature comparing the ^{63}Cu relaxation (stars) in the 1-2-3 compound (taken from [23]) with ^{205}Tl relaxation (squares) in the $n = 2$ compound [14a, 27]

even when the susceptibility is enhanced by correlations. For strong correlation effects $\chi(q,\omega)$ becomes temperature dependent, as is observed in Fig. 4 in the normal state $(T > T_c)$. In the fast fluctuating regime (in conductors) the Larmor frequency is negligible compared with the correlation rate τ_c^{-1} of the electrons and we are in the $\omega_n \to 0$ regime. If we replace the sum over the q-space in (4) by averaged parameters we arrive at the simplified relation

$$(T_1 T)^{-1} = 2k_B \langle |A|^2 \rangle \langle \chi_q \rangle \quad \text{with} \quad \langle \chi_q \rangle = \sum_q \chi''_\perp(q,\omega_n)/\omega_n , \qquad (5)$$

which separates the nuclear spin interaction $\langle |A|^2 \rangle$ from the electron dynamics summarized in $\langle \chi_q \rangle$.

Note, that $\langle \chi_q \rangle \sim T^\alpha$ with $\alpha = 2-3$ for $T < T_c$ and $\alpha = -0.7$ for $T > T_c$ as demonstrated in Fig. 4. Dynamic scaling arguments might be applied in order to understand these exponents [29]. A detailed theoretical investigation seems to be required in order to throw more light on this behavior. Rigamonti will perhaps deal with these questions in his lecture.

The hump in the relaxation rate near T_c that has been reported in some investigations [14a, 25, 26] is most likely not a Hebel-Slichter [1] phenomenon. In fact the increase in relaxation rate starts already at T_c^*, well above T_c, as was pointed out in [14a]. It is most likely due to fluctuations in the superconducting order parameter around T_c as was worked out by *Maniv* and *Alexander* [30a], and which can be expressed as [30b]

$$(T_1^{-1})/(T_{1\text{Korringa}}^{-1}) = \left(\frac{\pi}{2k_F\xi}\right)^2 I \ln\left\{\left[r + A\left(\frac{\omega_n}{T}\right)^2\right]^{-1}\right\} \qquad (6)$$

where ξ is the coherence length, $r = \ln(T/T_c) \simeq (T - T_c)T_c$, $A \simeq 10^{-2}$ and $I \simeq 10^2$ is a constant which depends on the anisotropy of the band structure. A significant enhancement of T_1^{-1} at T_c of this sort has been observed in organic superconductors [30b].

5. The Vortex Lattice and Its field Distribution

In the following I will briefly discuss the vortex lattice in superconductors of the second kind and its consequence for the NMR signal. It is well established that the applied magnetic field B_a penetrates superconductors of the second kind in the range $B_{c1} < B_a < B_{c2}$. The average magnetic field

$$\bar{B} = B_a + (1 - N)\mu_0 M \qquad (7)$$

is reduced from the applied field by the magnetization $M < 0$ (diamagnetic shielding), which is modified by the demagnetizing factor N. High-

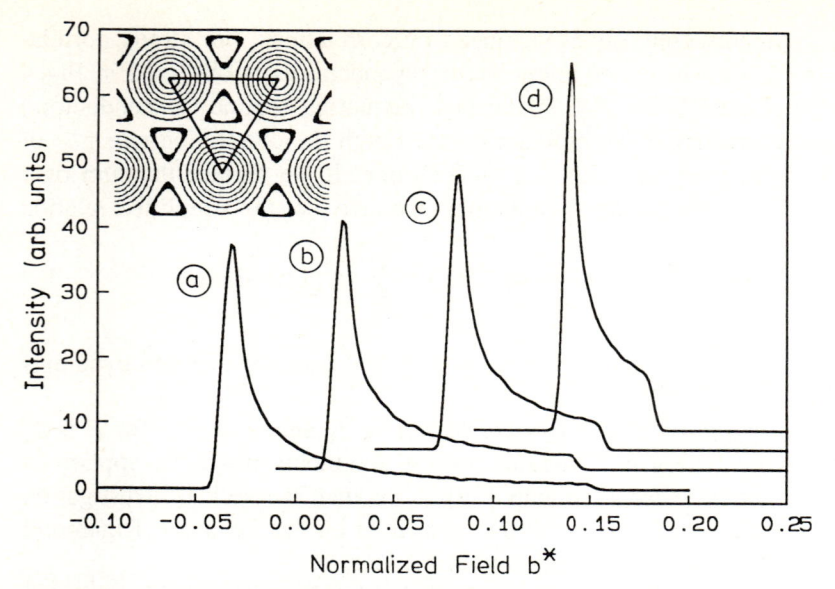

Fig. 5. Calculated field distribution spectra due to a triangular vortex lattice (see text). Spectra a–d correspond to different coherence lengths ξ but the same penetration depth λ. The following values for ξ^* were used in the calculation (see text): $\xi^* = 0.1$ (a), 0.15 (b), 0.2 (c) and 0.25 (d). The centre of gravity of the spectra was shifted to the right in 0.05 increments in order to separate them

temperature superconductors are well described by the London approximation since the Ginzburg-Landau parameter $\kappa = \lambda/\xi > 50$ is large. λ is the magnetic field penetration depth and ξ is the coherence length. A section of a triangular vortex lattice is shown in the inset of Fig. 5. The separation d of the vortices is determined by the average magnetic field as [5, 10–12]

$$\bar{B} = \phi_0/(\sqrt{3}d^2/2), \tag{8}$$

where $\phi_0 = 2.0678 \times 10^{-15}$ T m^2 is the flux quantum. For an applied field of, e.g., 4 T the separation d equals 245 Å. With the applied field $B_0 \| z$ the field distribution in the x, y-plane can be expressed in the London limit as [12]

$$B(r) = \bar{B} + \bar{B} \sum_{G \neq 0} b(G) \exp(G \cdot r) \quad \text{with}$$

$$b(G) = \frac{\exp(-0.5\xi^2 G^2)}{1 + \lambda^2 G^2}, \tag{9}$$

where $r = (x, y)$ and the sum is taken over all reciprocal lattice vectors G except $G = 0$.

By choosing the appropriate basis vectors for r (along the triangular sides) and the appropriate reciprocal lattice basis one can derive the following normalized field expression to be discussed below:

$$b^*(n_1, n_2) = \sum_{k_1, k_2 \neq 0,0} b^*(k_1, k_2) \exp\left[i\frac{2\pi}{N}(n_1 k_1 + n_2 k_2)\right] \quad (10a)$$

with

$$b^*(k_1, k_2) = \frac{(\lambda^*)^2 \exp\left[-0.5(\xi^*)^2 (2/\sqrt{3})(2\pi)^2(k_1^2 + k_2^2 - k_1 k_2)\right]}{1 + (\lambda^*)^2 (2/\sqrt{3})(2\pi)^2(k_1^2 + k_2^2 - k_1 k_2)} \quad (10b)$$

where every point of the vortex unit cell can be reached by

$$\boldsymbol{r} = d\left[n_1 \hat{x} + n_2(\tfrac{1}{2}\hat{x} + \tfrac{\sqrt{3}}{2}\hat{y})\right]/N \quad (11)$$

with $n_1, n_2 = 0, 1, 2, \ldots, N - 1$. The sum is taken over $k_1, k_2 = 0, 1, 2, \ldots$ except $k_1 = k_2 = 0$. The field $b^*(n_1, n_2)$ is expressed in (10) in a form suitable for the application of Fast Fourier Transform (FFT) algorithms. The parameters λ^* and ξ^* are parametrized as

$$\lambda^* = \frac{\lambda}{A^{1/2}} = \left(\frac{\lambda^2 \bar{B}}{\phi_0}\right)^{1/2} \; ; \quad \xi^* = \frac{\xi}{A^{1/2}} = \left(\frac{\xi^2 \bar{B}}{\phi_0}\right)^{1/2}, \quad (12)$$

where $A = \sqrt{3}d^2/2$ is the area of the triangular lattice unit cell.

From an experimental point of view the dimensionless normalized field

$$b(\boldsymbol{r}) = [B(\boldsymbol{r}) - \bar{B}]/\bar{B} = (\lambda^*)^{-2} b^*(\boldsymbol{r}) \quad (13)$$

is a useful quantity because it can be determined from the anisotropic NMR line.

The importance of formulating $b^*(r)$ according to (10) is the fact that $b^*(r)$ is rather "insensitive" to variations in λ^* for $\lambda^* > 1$. This is the situation in typical NMR fields ($1\mathrm{T} \leq \bar{B} \leq 10\,\mathrm{T}$). The *characteristic shape* of the magnetic field distribution

$$f(B) = \iint \delta(B(x,y) - B) dx\, dy \quad (14)$$

is independent of λ^* in this range and only the scaling of the B axis is affected by λ^*. It is therefore preferable to define the dimensionless field distribution

$$f(b^*) = \iint \delta(b^*(x,y) - b^*) dx\, dy \,. \quad (15)$$

Several field distribution functions calculated according to (10) and (15) are shown in Fig. 5. The characteristic shape is dominated by a singularity at the

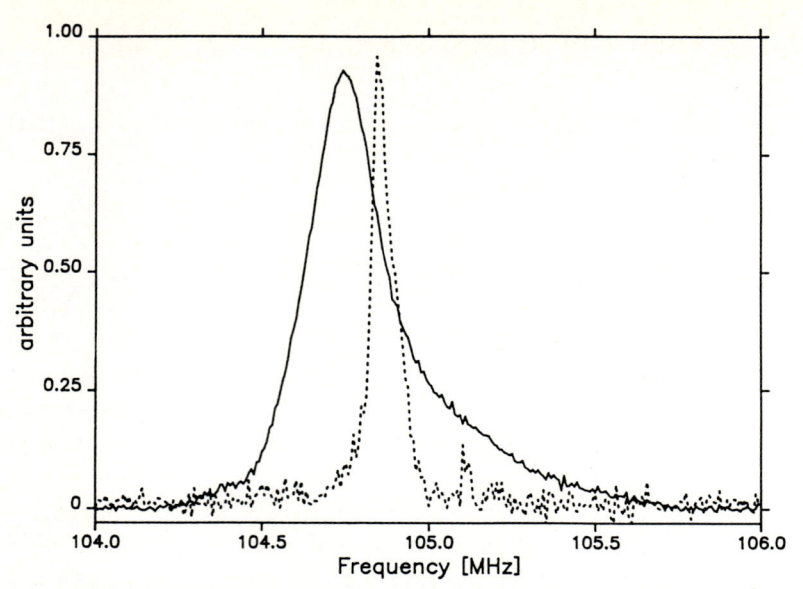

Fig. 6. ^{205}Tl spectrum of Tl$_2$Ba$_2$CuO$_6$ at 5 K ($T \ll T_c$) obtained by zero field cooling compared with the $T > T_c$ spectrum (dotted line) for an oriented powder sample with B_a parallel to the c-axis, according to [31]. Note the asymmetric broadening due to the vortex lattice

saddle point (b_s^*) below the average field ($b^* = 0$) in the field distribution and a monotonic decrease for larger b^* values. For short coherence length $\xi^* \leq 0.1$ no "step" is observed at the maximum field value b_{max}^*. For larger coherence lengths (see e.g. Fig. 5d) a step in the field distribution at b_{max}^* is clearly observed and the width of the distribution is appreciably reduced, i.e. even for constant λ^* the width of the field distribution $f(B)$ depends on the coherence length. The experimentally determined field distribution observed at 5 K for ^{205}Tl of an oriented ($B_a \parallel c$) powder sample of Tl$_2$Ba$_2$CuO$_6$ in a 4 T field is presented in Fig. 6 [31]. It shows the typical characteristic field distribution shown in Fig. 5a,b for short coherence length.

The penetration depth λ can readily be obtained by fitting the experimentally observed spectra with the field distribution according to (10–15). This analysis has been performed for Tl$_2$Ba$_2$CuO$_6$ with $B_a \parallel c$ and will be discussed elsewhere [31].

There are a few useful quantities which can be determined experimentally. These are the

(a) second moment

$$M_2 = \langle (B(r) - B)^2 \rangle = \phi_0^2 \lambda^{-4} M_2^* \quad \text{with}$$
$$M_2^* = \sum_{k_1, k_2 \neq 0, 0} [b^*(k_1, k_2)]^2 \; ; \tag{16}$$

(b) maximum field

$$B_{max} = \bar{B} + \phi_0 \lambda^{-2} b^*_{max} \quad \text{with} \quad b^*_{max} = \sum_{k_1, k_2 \neq 0,0} b^*(k_1, k_2) ; \quad (17)$$

(c) saddle field

$$B_s = \bar{B} + \phi_0 \lambda^{-2} b^*_s \quad \text{with} \quad b^*_s = \sum_{k_1, k_2 \neq 0,0} (-1)^{k_2} b^*(k_1, k_2) , \quad (18)$$

where all parameters have been defined before and $b^*(k_1, k_2)$ is given by (10). Besides fitting the experimentally observed field distribution with the calculated one by varying the penetration depth λ, it might be useful to extract first a numerical quantity which allows one to get a good estimate of ξ^*. Such a quantity which does not depend on λ^* is

$$\tilde{m}_2 = (\bar{B} - B_s)/M_2^{1/2} = -b^*_s/M_2^{*1/2} . \quad (19a)$$

An alternative measure of ξ^* would be the relation

$$\tilde{m}_1 = (\bar{B} - B_s)/(B_{max} - \bar{B}) = -b^*_s/b^*_{max} . \quad (19b)$$

Table 1. Calculated parameters for different ξ^* values

ξ^*	0.0	0.05	0.1	0.15	0.2	0.25	0.3
$\tilde{m}_1$	0.034	0.14	0.21	0.27	0.31	0.327	0.33
$-10^2 b^*_s$	3.67	3.6	3.2	2.6	1.8	1.1	0.57
$\tilde{m}_2$	0.60	0.65	0.72	0.79	0.819	0.82	0.82
$10^3 M_2^*$	3.7	3.0	1.9	1.0	0.47	0.17	0.048

A few values of $\tilde{m}_1, \tilde{m}_2$ for different ξ^* values are listed in Table 1. A note of warning seems to be appropriate. The second moment M_2 must be determined from the field distribution spectra. It has been the custom in μSR analysis to extract the second moment from the decay function in the time domain (inverse Fourier transform of the field distribution). However, the field distribution is not a Gaussian, as is evident from Figs. 5 and 6, and the decay function is not an exponential. Both assumptions have been employed in the μSR analysis in the past [32]. It is, however, recommended that one performs the more complete analysis involving the $\tilde{m}_j$ ($j = 1, 2$) parameters and lineshape fittings as discussed here.

In summary, it was the discovery of high-T_c superconductors by *Bednorz* and *Müller* [33] which led many researchers to "secret" and "not-so-secret" investigations about the mechanisms of this novel class of materials. NMR has made significant contributions, although not all pertinent questions could be answered.

Acknowledgments: I would like to thank my co-workers F. Hentsch, N. Winzek and U. Grosshans for discussions and for performing the ^{205}Tl experiments. I am indebted to A. Simon and his co-workers for the preparation of the Tl superconductors and to E.H. Brandt for introducing me to the vortex lattice. Last but not least I would like to thank K.A. Müller and J.G. Bednorz for the encouragement to pursue this work. The Bundesministerium für Forschung und Technologie (Project-Nr. A3N55774) has given financial support.

References

1   L.G. Hebel, C.P. Slichter: Phys. Rev. **113**, 1504 (1957)
2   Y. Masuda, A.G. Redfield: Phys. Rev. **125**, 159 (1962)
3   K. Yoshida: Phys. Rev. **110**, 769 (1958)
4   G.M. Androes, W.D. Knight: Phys. Rev. **121**, 779 (1961)
5   D.E. MacLaughlin: Solid State Phys. **31**, 1 (1976)
6   J.T. Markert, T.W. Noh, S.E. Russek, R.M. Cotts: Solid State Commun. **63**, 847 (1987)
7   H. Seidel, F. Hentsch, M. Mehring, J.G. Bednorz, K.A. Müller: Europhys. Lett. **5**, 647 (1988)
8   M. Takigawa, P.C. Hammel, R.H. Heffner, Z. Fisk: Phys. Rev. B **39**, 7371 (1989)
9   Y. Kitaoka, Sh. Hiramatsu, T. Kondo, K. Asayama: J. Phys. Soc. Jpn. **57**, 30 (1988)
10   (a) A.G. Redfield: Phys. Rev. **162**, 367 (1967)
       (b) W. Fite, A.G. Redfield: Phys. Rev. Lett. **17**, 381 (1966)
11   (a) J.-M. Delrieu, J.-M. Winter: Solid State Commun. **4**, 545 (1966)
       (b) J.-M. Delrieu: J. Low Temp. Phys. **6**, 197 (1972)
12   (a) E.H. Brandt, A. Seeger: Adv. Phys. **35**, 189 (1986)
       (b) E.H. Brandt: J. Low Temp. Phys. **73**, 355 (1988)
13   (a) M. Mali, D. Brinkmann, L. Pauli, J. Roos, H. Zimmermann, J. Hullinger: Phys. Lett. A **124**, 112 (1987)
       (b) M. Mali, J. Roos, D. Brinkmann: Physica C **153–155**, 737 (1988)
       (c) H. Riesemeier, Ch. Grabow, E.W. Scheidt, V. Müller, K. Lüders: Solid State Commun. **64**, 309 (1987)
14   (a) F. Hentsch, N. Winzek, M. Mehring, H. Mattausch, A. Simon: Physica C **158**, 137 (1989)
       (b) T. Imai, H. Yasuoka, T. Shimizu, Y. Ueda, K. Kosuge: Tech. Report ISST No. 1989, July 1988
15   (a) Y. Kitaoka, S. Hiramatsu, K. Ishida, T. Kohara, K. Asayama: J. Phys. Soc. Jpn. **56**, 3024 (1987)
       (b) Y. Kitaoka, S. Hiramatsu, K. Ishida, K. Asayama, H. Takagi, H. Iwabuchi, S. Uchida, S. Tanaka: J. Phys. Soc. Jpn. **57**, 737 (1988)
16   I. Watanabe, K. Kumagai, Y. Nakumura, T. Kimura, Y. Nakamichi, H. Nakajima: J. Phys. Soc. Jpn. **56**, 3028 (1987)
17   H. Lütgemeier: Physica C **153–155**, 95 (1988)
18   (a) W.D. Knight: Solid State Phys. **2**, 93 (1956)
       (b) D. Köngeter, M. Mehring: Phys. Rev. B **39**, 6361 (1989)
       (c) C.H. Pennington, D.J. Durand, D.B. Zax, C.P. Slichter, J.P. Rice, D.M. Ginsberg: Phys. Rev. B **37**, 7944 (1988)
19   N. Winzek, F. Hentsch, M. Mehring, H. Mattausch, A. Simon, J.B. Kremer: Physica C (submitted)

20 M. Mehring: *Principles of High Resolution NMR in Solids*, 2nd ed. (Springer, Berlin, Heidelberg 1983)
21 K. Fujiwara, Y. Kitaoka, K. Asayama, H. Katayama-Yoshida, Y. Okabe, T. Takahashi: J. Phys. Soc. Jpn. **57**, 2893 (1988)
22 H. Alloul, T. Ohno, P. Mendels: Submitted to Phys. Rev. Lett.
23 T. Imai, T. Shimizu, T. Tsuda, H. Yasuoka, T. Takabatake, Y. Nakazawa, M. Ishikawa: J. Phys. Soc. Jpn. **57**, 1771 (1988)
24 M. Horvatic̀, P. Segransan, C. Berthier, Y. Berthier, P. Butaud, J.Y. Henry, M. Couach, J.P. Chaminade: To be published
25 P. Wzietek, D. Köngeter, A. Auban, D. Jerome, J.M. Bassat, J.P. Coutures, B. Dubois, Ph. Odier: Europhys. Lett. **8**, 363 (1989)
26 Y. Kitaoka, K. Ishida, K. Fujiwara, T. Kondo, K. Asayama, H. Katayama-Yoshida, Y. Okaba, T. Takahashi: J. Magn. Magn. Mater. **76–77**, 527 (1988)
27 F. Hentsch, N. Winzek, M. Mehring, H. Mattausch, A. Simon, J.B. Kremer: Physica C (in press)
28 T. Moriya: J. Phys. Soc. Jpn. **18**, 516 (1961)
29 (a) F. Borsa, A. Rigamonti: In *Magnetic Resonance of Phase Transitions*, ed. by F.J. Owens et al. (Academic, New York 1979)
    (b) F. Borsa, M. Corti, T. Rega, A. Rigamonti: To be published
    (c) C. Bourbonnais: In *Low Dimensional Conductors and Superconductors*, ed. by D. Jerome, L.G. Caron, NATO ASI Series B 155, 155 (1987)
30 (a) T. Maniv, S. Alexander: Solid State Commun. **18**, 1198 (1976)
    (b) F. Creuzet, C. Bourbonnais, D. Jerome, D. Schweitzer, H.J. Keller: Europhys. Lett. **1**, 467 (1986)
31 M. Mehring, F. Hentsch, H. Mattausch, A. Simon: Z. Phys. B **77**, 355 (1989)
32 (a) G. Aeppli, R.J. Cava, E.J. Ansaldo, J.H. Brewer, S.R. Kreitzman, G.M. Luke, D.R. Noakes, R.F. Kiefl: Phys. Rev. B **35**, 7129 (1987)
    (b) F.N. Gygax, B. Hitti, E. Lippelt, A. Schenk, D. Cattani, J. Cors, M. Decroux, O. Fischer, S. Barth: Europhys. Lett. **4**, 473 (1987)
33 J.G. Bednorz, K.A. Müller: Z. Phys. B **64**, 189 (1986)

Part IV

Theoretical Models

Phonons and Charge-Transfer Excitations in High-Temperature Superconductors

A.R. Bishop

Theoretical Division, Los Alamos National Laboratory,
Los Alamos, NM 87545, USA

Abstract. Some of the experimental and theoretical evidence implicating phonons and charge-transfer excitations in HT superconductors is reviewed. It is suggested that superconductivity may be driven by a synergistic interplay of (anharmonic) phonons and electronic degrees of freedom (e.g. charge fluctuations, excitons).

I. Introduction

The wealth of experimental and theoretical studies of the new oxide superconductors begins to suggest more focused directions regarding mechanisms. While contributions from exotic and spin-based pairing mechanisms remain active possibilities, we believe that charge-transfer (CT) (excitonic) mechanisms synergistically assisted by electron-phonon coupling (and possibly accompanied by spin fluctuations) are indicated by a number of results and by the materials themselves. The high polarizability of O^{2-} and the familiar strong phonon modes and affinity for structural instabilities which characterize perovskite-like materials are ideal conditions to synergistically enhance low-lying metal-oxygen charge transfer channels. This is true of the simple layered 2-1-4 structures such as $La_{1-x}Sr_xCuO_4$ but is further augmented by the "sandwich" structures characterizing 1-2-3 $YBa_2Cu_3O_{7-\delta}$, Bi (4-3-3-4) and Tl (2-1-2-2), etc. In these latter cases we have proposed that the dynamic polarizability of the environment surrounding CuO_2 planes plays an important role in enhancing T_c.

Elsewhere [1] we have described how a large body of experimental information can be rationalized within such a theoretical scenario. Here we mention a few additional experimental facts which have become available more recently.

Raman scattering indicates certain strong phonon modes and the optical conductivity [2] has now converged on a picture of electrons scattering from "some" high frequency excitation in a phonon-like regime (500-700K): the implied high phonon frequency and strong interaction are certainly natural if oxygen is centrally involved, consistent with our discussion in section 2. The mid-IR absorption observed in the frequency dependent conductivity for $YBa_2Cu_3O_{7-\delta}$ thin films has now been resolved into a series of absorptions, probably of electronic (e.g. charge transfer (excitonic)) origin [3]. In an untwinned single crystal of $EuBa_2Cu_3O_{7-\delta}$ a strong mid IR feature has been observed [4] to be both narrow, polarized in the chain direction, and changing its frequency with δ, all consistent [5] with our theory of low-lying excitons in the Cu-O chains (see section 2(i)).

Phonons are also implicated by inelastic neutron scattering (suggesting, e.g., "breathing" modes in CuO_2 planes [6]), by (small) phonon anomalies which track T_c [7], and by tunneling data, particularly for $BiCaSrCu_2O_y$ [8]. Elastic constant anomalies appear as precursors of superconductivity in single crystal 2-1-4 samples. Strong electron-phonon coupling is almost certainly the driving mechanism in $Ba_xK_{1-x}BiO_3$ [9]. Strongly anharmonic phonons [6] may be suggested by recent neutron and Raman scattering evidence for more phonon branches than expected on symmetry grounds, as well as by apparent doping-dependent isotope effects. In the 1-2-3 and high temperature Bi and Tl sandwich structures, recent copper K-edge XAS data suggest [10] that the O(4) "bridging" oxygen is characterized by a double well anharmonicity which is very sensitive in the superconducting transition region. This is highly suggestive of a synergistic coupling between the axial phonon dynamics and electronic degrees of freedom involved in the superconductivity. Such a picture is further supported by the anomalous oscillator strengths of axial O(4) phonon modes in both IR [7] and Raman [11] spectra. In section 2(ii) we interpret this enhancement in terms of a coupling between phonons and CT electronic modes -- which we have elsewhere (see section 2(i)) also suggested as driving the superconductivity. The scenario of strongly nonlinear (e.g., Jahn-Teller-, buckling-, tilting-, or breathing-mode) phonons, perhaps coupled to CT excitons or spin fluctuations, may also occur in the CuO_2 planes of the 2-1-4 materials, but data is incomplete so far [6], although evidence for anomalous Raman activity of specific plane modes is beginning to appear. The ubiquitous large linear resistivity temperature regime could also be ascribed to the presence of two dominant phonons/excitons: electron interactions with a low frequency mode ($\hbar\omega \lesssim k_B T_c$) could drive T_c, while scattering from high frequency modes ($\hbar\omega \gg k_B T_c$) would dominate resistivity.

Some recent tunneling [8] and photoemission [12] data have suggested strong coupling BCS-like behavior with $2\Delta(o)/k_B Tc \sim 7$. This is somewhat high for conventional strong coupling Eliashberg theory with linear phonons and Coulomb effects controlled by μ^*. However, a synergistic coupling of e.g. nonlinear phonons, or phonons and CT excitons (as suggested above) has yet to be investigated systematically. Again, strong Coulomb interactions, strong coupling between phonons and superconductivity, or anisotropic pairing, will take us beyond the conventional Eliashberg framework. We will reemphasize these issues in section 3.

Our primary intention here will be to summarize some of the model Hamiltonians being employed to address aspects of the scenario outlined above.

2. Model Hamiltonians and Charge Transfer

Deciding on an appropriate microscopic model hamiltonian to describe the complex structures of the new materials is not straightforward and not a settled issue: 1-band or multiband; π- or σ- or other orbitals?; 2-dimensional or (anisotropic) 3-dimensional? Interpretation of experimental probes lags behind the quality of the measurements themselves in some cases, and sample quality continues to be a limiting issue for characterizing intrinsic properties. Interpreting photoemission so as to extract model parameters is still controversial, although band structure and various degrees of ab initio quantum chemistry are now converging with respect to parameter assignments for Hubbard models. However, since we believe that Hubbard models will eventually have to be seriously extended in these materials (via anharmonic electron-phonon coupling [13], Hubbard V,W [14,15], multiple-bands [15], etc.), it remains premature to attempt to fix model parameter values. We therefore prefer to explore parameter space widely in simple models and allow experiment to guide the most important augmentation of those models. With this philosophy in mind, we describe here (i) a U-V Hubbard model; (ii) a local charge-transfer-phonon cluster model; (iii) a "-ve U" model of BCS superconductivity; and (iv) a 1-dimensional analog system. Each model allows us to address different aspects of the puzzle.

2(i) The Extended Hubbard Model

We first consider a single CuO_2 plane. One of the simplest Hamiltonians that takes into account both Cu and O sites can be written

$$H = \varepsilon_d \sum_i d_i^\dagger d_i + \varepsilon_p \sum_j p_j^\dagger p_j - \sum_{\langle i,j \rangle} t_{i,j} d_i^\dagger p_j + h.c$$

$$+ U_d \sum_i n_{d\uparrow i} n_{d\downarrow i} + U_p \sum_j n_{p\uparrow j} n_{p\downarrow j} + V \sum_{\langle i,j \rangle} n_{d,i} n_{p,j}, \qquad (1)$$

where $t_{i,j}$ is the hopping matrix element between Cu $d_{x^2-y^2}$ and O $p_{x,y}$ orbitals, ε_d and ε_p are respective local energy levels, $U_d(U_p)$ is the on-site Coulomb repulsion and V is the nearest-neighbor Coulomb repulsion between Cu and O sites. Since *both* spin and charge fluctuations can be soft, Hamiltonian (1) differs qualitatively from the Hubbard Hamiltonian, where charge fluctuations are suppressed. If we ignore U_d, U_p and V the resulting band structure consists of bonding (B) and antibonding (AB) Cu-O bands along with a flat nonbonding (NB) oxygen band [16,17]. At stoichiometry in La_2CuO_4 and $YBa_2Cu_3O_6$ the 2D AB band is $\frac{1}{2}$-filled and these materials should be metals. However, experimentally both exhibit antiferromagnetism [18,19], indicating the importance of correlations. To understand the effect of correlations in (1) we will first use weak coupling theory which enables explicit evaluation of relevant quantities. We find that the physics in the superconducting regime is dominated by *low lying charge transfer* excitations. Since oxide superconductors appear to be more in the intermediate coupling regime, our result should be interpreted in a qualitative sense. However, quantum chemistry calculations for finite clusters (below) also demonstrate the importance of charge transfer excitations. We therefore expect that overall features of our theory persist in the intermediate coupling regime, and this has indeed been supported by a variety of calculations in the strong coupling regime using, e.g. slaved-boson analytic techniques [20], exact numerical diagonalizations or quantum Monte Carlo [21].

Consider the case when the AB band in the CuO_2 plane is $\frac{1}{2}$-filled. After projecting the Hamiltonian (1) onto the AB band we obtain the following effective Hamiltonian:

$$H_{AB}^{eff} = \sum_{\vec{k},\sigma} \xi_{\vec{k}} c^{\dagger}_{\vec{k},\sigma} c_{\vec{k},\sigma}$$

$$+ \sum_{\vec{k},\vec{k}',\vec{q}} \sum_{\sigma,\sigma'} [\bar{U} + 2\bar{V}cos(q_x a) X_{\vec{k}+\vec{q}} X_{\vec{k}} + 2\bar{V}cos(q_y a) Y_{\vec{k}+\vec{q}} Y_{\vec{k}}] c^{\dagger}_{\vec{k}-\vec{q},\sigma'} c^{\dagger}_{\vec{k}'+\vec{q},\sigma} c_{\vec{k},\sigma} c_{\vec{k}',\sigma'} \,, \quad (2)$$

where a is the Cu-O separation, $\xi_{\vec{k}} = \frac{1}{2}(\varepsilon_d + \varepsilon_p) - \mu + \frac{1}{2}[(\varepsilon_d - \varepsilon_p)^2 + 16t^2(\sin^2 k_x a + \sin^2 k_y a)]^{\frac{1}{2}}$, $\bar{U} \equiv U_d \cos^4 \theta + U_p \sin^4 \theta$, $\bar{V} \equiv V \cos^2 \theta \sin^2 \theta$, with $\cos^2 \theta = \frac{1}{2} + \frac{1}{2}[1 + \frac{16\tau^2}{(\varepsilon_d - \varepsilon_p)^2}]^{-\frac{1}{2}}$, τ being the average of the Cu-O hybridization over the Fermi surface. $X_{\vec{k}}$ and $Y_{\vec{k}}$ are coherence factors.

In weak coupling, we can study collective modes of the system by considering the equation for the response function $\chi_{\sigma,\sigma'}(\vec{q},\omega)$ arising from the sum of ring and ladder diagrams:

$$\chi_{\sigma,\sigma'}(\vec{q},\omega) = \chi_0(\vec{q},\omega) + \delta_{\sigma,\sigma'} \chi_0(\vec{q},\omega) I(\vec{q}) Tr \chi(\vec{q},\omega) - \chi_0(\vec{q},\omega) J(\vec{q}) \chi_{\sigma,\sigma'}(\vec{q},\omega) \,, \quad (3)$$

where $\chi_0(\vec{q},\omega)$ is the Lindhard function in the AB band and we have restricted our attention to the interband response. The direct interaction $I(\vec{q})$ and exchange interaction $J(\vec{q})$ follow from (2):

$$I(\vec{q}) = \bar{U} + 2\bar{V}cos(q_x a)\alpha_x(\vec{q}) + 2\bar{V}cos(q_y a)\alpha_y(\vec{q}) \,, \quad (4)$$

$$J(\vec{q}) = \bar{U} + 2\bar{V} \sum_{i=1}^{2} [|\beta_x^i(\vec{q})|^2 + |\beta_y^i(\vec{q})|^2] \,. \quad (5)$$

In the above equation we have used the definitions

$$\alpha_x(\vec{q}) \equiv \chi^{-1}(\vec{q},\omega=0) \sum_{\vec{k}} K_0(\vec{k},\vec{q}) X_{\vec{k}+\vec{q}} X_{\vec{k}} \,,$$

$$\beta_x^i(\vec{q}) \equiv \chi^{-1}(\vec{q},\omega=0) \sum_{\vec{q}} K_0(\vec{k},\vec{q}) X_{\vec{k}+\vec{q}} \gamma_x^i(\vec{k}) \,,$$

where $\gamma_x^1 = \cos k_x a$, $\gamma_x^2 = \sin k_x a$. $K_0(\vec{k},\vec{q})$ is defined by $\chi_0(\vec{q},\omega=0) \equiv \sum_{\vec{k}} K_0(\vec{k},\vec{q})$. The definitions of quantities with subscript y are obvious.

At half-filling, the Fermi surface has perfect nesting and there will be a tendency to open a gap on the Fermi surface. An instability can occur in both the charge and spin density channels and the resulting CDW and SDW will have periodicity determined by the nesting vectors $\vec{Q}_0 = (\frac{\pi}{2a}, \pm\frac{\pi}{2a})$. One can show by solving (3) that the relative stability of a CDW or SDW depends on the sign of $I(\vec{q})$: First note that $\chi_0(\vec{Q}_0, \omega = 0) \propto -\frac{1}{D}ln^2(\frac{D}{T})$, where $D \sim 2t$. From (3) follows $ln^2(\frac{D}{T_{SDW(CDW)}}) = \lambda_{SDW(CDW)}^{-1}$, where $\lambda_{SDW} \propto J(\vec{Q}_0) > 0$ and $\lambda_{CDW} \propto J(\vec{Q}_0) - 2I(\vec{Q}_0)$, $T_{SDW(CDW)}$ being the transition temperature of the spin (charge) density wave. Obviously, for positive $I(\vec{Q}_0)$, $T_{SDW} > T_{CDW}$, while for negative $I(\vec{Q}_0)$ the situation is reversed. Since $I(\vec{Q}_0) = \bar{U} > 0$ the ground state in weak coupling is SDW, *irrespective* of the relative size of $\bar{U}$ and $\bar{V}$. This SDW reflects the qualitative features of the observed antiferromagnetic state. Thus, the *intraband* particle-hole excitations arising in (1) account for the antiferromagnetism of these compounds.

However, away from half-filling, when holes are added to the system either by doping or by varying the oxygen content, the situation may change qualitatively. This can be appreciated by considering $\chi_0(\vec{q}, \omega = 0)$. As x holes are added to the AB band, the maximum of $\chi_0(\vec{q}, \omega = 0)$ moves from $\vec{Q}_0$ to some $\vec{Q}(x)$, such that $|\vec{Q}(x)| < |\vec{Q}_0|$ and $2\bar{V}\alpha_x(\vec{Q}(x))\cos Q_x(x)a + 2\bar{V}\alpha_y(\vec{Q}(x))\cos Q_y(x)a$ becomes progressively more negative. (Note that $\alpha_{x,y}(Q(x)) < 0$). The maximum occurs for $\vec{Q}(x) = (\pm Q, \pm Q)$ where $Q \simeq \frac{\pi}{2a} - \frac{\mu'}{t}$, for $\varepsilon \ll t$ and $\mu' \equiv 2t - \mu$. The relation between x and μ is $x \simeq (\frac{\mu'}{2t})ln(\frac{2t}{\mu'})$. For $\bar{V}$ large enough it is possible to attain a situation where

$$I(Q_y(x)) = \bar{U}(x) + 2\bar{V}(x)\alpha_x(\vec{Q}(x))cosQ_x(x)a + 2\bar{V}(x)\alpha_y(\vec{Q}(x))cosQ_y(x)a < 0 ,\quad (6)$$

for some concentration of holes $x > x_c$, x_c being defined by $I(\vec{Q}(x_c)) = 0$. This signals the preference for CDW formation. The real, static CDW deformation does not occur, however, since away from half-filling the nesting features are diminished. We then have *low-lying*, finite wavevector $(\vec{Q}(x))$, temperature independent, collective charge excitations in the CuO$_2$ plane, which will strongly influence the effective electron interactions. For some region of doping $0 < x_c \leq x \leq x_{c_1} < 1$, which can be calculated from (3), the frequency of these short wavelength charge fluctuations $\omega_{CF}(\vec{Q}(x))$ will be lower than the corresponding spin fluctuation frequency $\omega_{SF}(\vec{Q}(x))$, and they can be utilized to build an attractive retarded interaction which can lead to superconductivity. Such a region will occur only for $\varepsilon \equiv \varepsilon_d - \varepsilon_p \lesssim t$ and for $\bar{V} \gtrsim \frac{1}{4}\bar{U}$. Explicit results for the frequencies of collective modes illustrating this crossover are given in [22]. The generic behavior is shown in Fig. 1(a). The upturn in ω_{CF} is caused by a decrease of $\bar{V}$ as one empties the AB band.

Note that our *intraband* charge excitation is distinct from the local charge transfer exciton proposed in [23], although the underlying theme, relying on the low energy cost for Cu ↔ O charge transfer (see below), is similar. Softening of *interband* charge fluctuations is driven by $\bar{V}$ alone [22,23] but are importantly enhanced by local field effects [23].

We now discuss superconductivity. Following the Eliashberg formalism outlined in [24] we define the matrix element of the pairing interaction,

$$\lambda_{CF(SF)}(\vec{k}, \vec{k}') = -2\alpha_{CF(SF)}^2 \int_0^\infty d\omega \frac{F(\vec{k}, \vec{k}'; \omega)}{\omega} ,\quad (7)$$

where $\alpha_{CF(SF)}$ is the electron-charge fluctuation (spin fluctuation) vertex, and $F(\vec{k}, \vec{k}'; \omega)$ is given by $-\frac{1}{\pi}\text{Im}\chi^{CF(SF)}(\vec{k}, \vec{k}'; \omega)$, where $\chi^{CF(SF)}(\vec{k}, \vec{k}'; \omega)$ is obtained from (3). In (7) $\vec{k}$ and $\vec{k}'$ are constrained to lie on the Fermi surface. We have used the solution of Eq. (3) to evaluate (7) in various orbital states as a function of model parameters. In Figure 1(b) we show the typical results for the ground state of the CuO$_2$ plane as a function

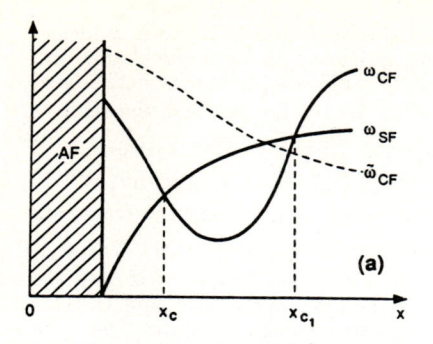

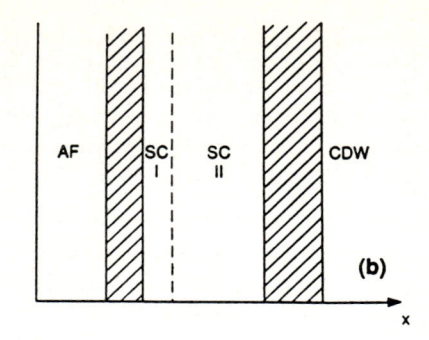

Figure 1. a) The generic x-dependence of ω_{SF} and ω_{CF} for the CuO_2 plane. The region of $\omega_{CF} < \omega_{SF}$ occurs only for sufficiently large $\bar{V}/\bar{U}$ as explained in the text. $\tilde{\omega}_{CF}$ refers to the interband charge excitations which may lead to the CDW (charge disproportionation) distortion at high doping. b) The qualitative ground state phase diagram as a function of x. Different regions are discussed in text. A CDW phase is most likely for $\varepsilon = 0$.

of x. We have set $\varepsilon = 0$. SC I is a d-wave superconductivity, produced by low-lying spin fluctuations. In this region T_c is expected to be *low*, due to self-energy corrections and dynamic pair-breaking. For $x > x_c$, the superconductivity crosses over to the s-wave type (SC II), where charge fluctuations dominate and T_c may be higher since the above limitations are absent for an isotropic superconducting state. High T_c's are still problematical in pure C-T pairing because of competing trends of coupling constants and μ^*, however it is very important to note (see below) that *phonons* may enhance T_c in the SC II phase (they basically have no effect in SC I). Since T_c calculations are notoriously unreliable one should interpret our results as representing the qualitative trend produced by doping the AB band. Finally, for large doping, the AB-NB *interband* charge fluctuations can lead to a CDW phase, a charge disproportionated state with a "frozen-in" Cu ↔ O charge transfer and lattice distortion. In the shaded regions there is a competition between different ground state symmetries and it is possible that the system is a metal at T = 0. This phase diagram is very suggestive of a similarity between high T_c oxides and the "old" Ba-Pb-Bi-O superconductor [25,26]. The newly discovered K-Ba-Bi-O superconductor increasingly clearly belongs to the same class.[9] This behavior is also in qualitative agreement with recent data on the 2-1-4 material which indicate that T_c saturates at 40K for x between 0.15 and 0.25 and then drops quickly to zero at $x \sim 0.30$ [27]; data on T_c in 1-2-3 materials as a function of hole density in the plane now appear to be similar [28].

Why is T_c much higher in 1-2-3 and the new Bi and Tl systems than in 2-1-4? We conjecture [1] that the dynamic polarizability of the environment of CuO_2 planes is crucial for elevating T_c. As an *illustration*, we consider $YBa_2Cu_3O_{6.9}$ specifically. Consider a single Cu-O chain along the y-axis. (We emphasize that chains are not crucial to our general concept.) We can use the same Hamiltonian (1) where now the relevant atomic orbitals are Cu $d_{y^2-z^2}$ and O $p_{y,z}$. In 1D there is perfect nesting at any band filling. The ensuing instability is driven by deformation of wavevector $Q_y = 2k_F^c$. Just as in the plane, as holes are added to the AB band, Q_y decreases from $\frac{\pi}{2a}$ ($\frac{1}{2}$-filled) to 0 (AB band empty), and $I(Q_y)$ can become negative, leading to strong charge fluctuations. The actual CDW instability is prevented by hopping, $t_\perp$, between chains and planes, by quantum fluctuations, and by disorder. The reduced dimensionality, however, is likely to make chains more polarizable than planes. In this way we use the high polarizability of a 1D system at wavevector $2k_F^c$ to further enhance an intrinsic polarizability of the Cu-O bond (see below). The high polarizability of the chains will affect the electron-electron interactions on neighboring planes, via an effective chain-plane Coulomb coupling, W. This effective interaction for electrons on planes is given by $n_c W^2 \text{Tr}\chi^{chain}(q_y,\omega)$, where

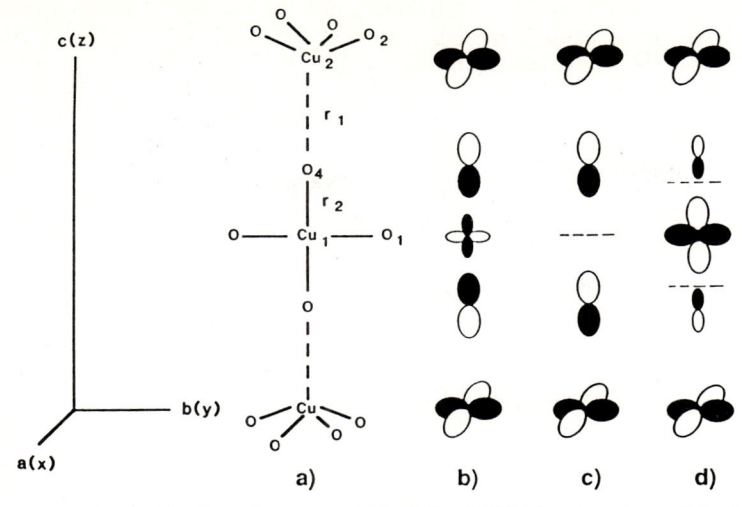

Figure 2. a) Cu_3O_{12} cluster model of the PCP-like structure. b), c) and d) denote the B, NB and AB combinations, respectively. The phase of an atomic orbital is denoted by shade. The Cu $d_{x^2-y^2}$ orbitals in the planes are pictured for the reader's orientation; they do not interact with the O p_z orbitals.

n_c is the linear density of chains. Note that, this interaction is the *same* for two electrons on adjacent CuO_2 planes or on a single plane.

In reality the situation in $YBa_2Cu_3O_{6.9}$ is more complicated. In addition to the linear Cu-O chain, there are bridging oxygens (O(4)), which control the plane-chain coupling, W. The presence of these oxygens induces additional transverse polarization of the chains, favoring intraplane pairing. This mode is a part of an *interband* response, not included in (3). Consider the cluster in Figure 2a. The upper and lower Cu have the coordination of the planes (Cu_p), while the central atom is a unit of the chain (Cu_c). The orbitals of particular interest are the $d_{x^2-y^2}$ orbitals on the Cu_p, the $d_{y^2-z^2}$ orbital on Cu_c, and the p_z oribtals on the bridging O(4) oxygens. Note that the bridging O p_z orbitals have the wrong symmetry to interact directly with $d_{x^2-y^2}$, but can interact with the $d_{y^2-z^2}$ orbital. These three atomic orbitals on the chain combine to give B, NB, and AB molecular orbitals. The NB orbital will develop into a narrow O p_π band along the chain axis involving O p_z orbitals, while the AB orbital develops into the Cu $d_{y^2-z^2}$ band considered above. Current *ab initio* quantum chemistry calculations [29] on the cluster in Figure 2 suggest that the interband charge transfer mode, in which an electron moves between the NB O p_π orbital and the AB Cu $d_{y^2-z^2}$ band, occurs at low energy, ω_t, of the order of a few 0.1 eV, and that the ground state of the Cu_3O_{12} cluster model for $YBa_2Cu_3O_{6.9}$ involves an "empty" Cu $d_{y^2-z^2}$ orbital and a pair of electrons in the O p_π NB orbital ($Cu^{3+}-O^{2-}$). The state in which the electron is excited from the NB orbital to $d_{y^2-z^2}$ ($Cu^{2+}-O^{1-}$) lies a few 0.1 eV above the ground state. Which of these two states has lower energy depends on the nature of the correlations included in the configuration interaction. The common denominator in all our calculations, however, is the presence of low-lying charge transfer excitations. By contrast, in cluster models of $YBa_2Cu_3O_{6.5}$ the lowest charge transfer excitations occur at energy $\sim$ several eV.

If the above clusters are ordered in chains, there will be strong plane-chain Coulomb coupling, since both the intraband and interband excitations lead to rearrangement of the charge distribution on the bridging oxygens. The strength of the pairing interaction should scale approximately as $\frac{1}{r_1^2}$, where r_1 is the Cu(2)-O(4) distance. With both types of charge transfer excitations included, the retarded attraction will be somewhat stronger with the electrons on the same plane. Nevertheless, the beauty of the $YBa_2Cu_3O_{6.9}$ sandwich structure is that the bridging oxygens can couple two distinct planes, and we might even expect that the system will still take advantage of interplane pairing, thereby defeating in large part the direct Coulomb repulsion. The most favorable situation for interplane pairing would occur in orthorhombic $YBa_2Cu_3O_{6.9}$ with highly ordered chains, where the longitudinal intraband exciton could be dominant. Estimates of T_c and gap function symmetry have been given in [1], including both longitudinal and transverse excitons, and intra- as well as inter-plane pairing. Alternative scenarios for the role of off-plane polarizations enhancing or driving in-plane superconductivity are outlined in section 3.

We emphasize that many related chain complexes (see section 3(iv)) have strong CDW (charge disproportionation) tendencies, which are further strengthened by coupling to the lattice. Therefore, it is indeed plausible to expect that the physics of Cu-O chains in $YBa_2Cu_3O_{6.9}$, with electron-phonon interactions and out-of-chain (plane) O polarizations included, is dominated by their proximity to a CDW.

2(ii) Charge-transfer-phonon coupling:

Here we present a simple three level cluster model [30] which addresses the possibility that charge transfer fluctuations are responsible for the anomalously large intensity of the 155 cm^{-1} Ba mode observed by Genzel et al.[7] The model also relates to the accompanying Raman signature and the c-axis anomaly observed by Cava et al., [31] and suggests that these features should be related to a change in the hole density in the plane.

Anomalously large oscillator strengths have been reported for the IR phonon mode observed in $YBa_2Cu_3O_{7-\delta}$ at 155 cm^{-1} for $\delta \sim 0$. This mode has been labeled as the "Barium" mode and it can be quite accurately described as a rigid motion of the $[CuO_3]$ cluster (chain+two $O(4)$) in one direction and Ba^{+2} ions in the opposite direction, (Fig. 3). The oscillatory strength of this mode was found to be ~15 times larger than the

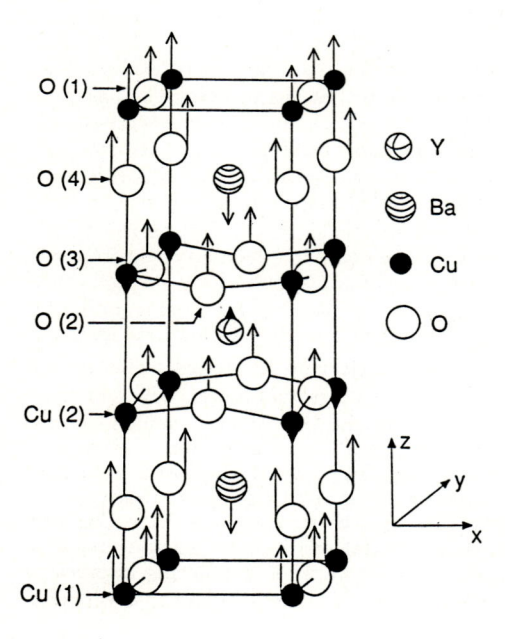

Figure 3. Unit cell of $YBa_2Cu_3O_7$ with the relative ionic displacements of the "Barium mode" at 155 cm^{-1}. (After ref. [7].)

prediction of a pure lattice dynamics calculation.[7] This is of course consistent with our earlier discussion here of the importance of the c-axis activity.

We assume that the charge of a CuO_3 cluster in the $YBa_2Cu_3O_7$ system is -4. There are several reasons for such a thesis. If the charge of the cluster is equal to -3, then the charge of the conducting planes CuO_2 should be -2, which implies half filling of the conducting band, and the appearance of the antiferromagnetism. This is in contradiction with the observation of superconductivity in $YBa_2Cu_3O_7$. However, if the charge of the cluster is -4, or more realistically the average is between -3 and -4, then the charge of the planes is less than -2. The finite density of holes in the conducting band may then lead to superconductivity as in doped La_2CuO_4. In oxygen-deficient samples, a part of the electrons from the oxygens lacking in the $CuO_{3-\delta}$ cluster is presumably transferred to the plane. This transfer will suppress the superconductivity, but will also result in a structural transformation along the c-axis, as has been indeed observed by Cava et al.[31] Additional experimental support for our assignment are the X-ray absorption measurements of Tranquada et al.[32] They have found that the "doping" (=reduction of oxygen content) increases the number of Cu^{+1} ions per unit cell from zero to 1 on account of the reduction in the number of Cu^{+2} ions; They have attributed Cu^{+1} ions to the chains ($Cu1$). We also note that Monien and Zawadowski [33] have recently given a complementary discussion of Fano-like Raman scattering enhancement by electronic degrees of freedom in a 2-band CT model and coupling to a specific (buckling) oxygen plane mode at $\simeq 330$ cm^{-1}.

Assuming therefore that the total charge of the CuO_3 cluster is -4, we have only four formal charge states of the cluster with the charge configurations,

$$
\begin{array}{llll}
O(4)^{-1} & O(4)^{-2} & O(4)^{-2} & O4^{-2} \\
| & | & | & | \\
Cu(1)^{+1} - O1^{-2} & Cu(1)^{+2} - O1^{-2} & Cu(1)^{+1} - O1^{-2} & Cu(1)^{+1} - O1^{-1} \\
| & | & | & | \\
O(4)^{-2} & O(4)^{-2} & O(4)^{-1} & O(4)^{-2}
\end{array}
$$

Note that the first and third states are asymmetric in charge distribution, and hence possess a dipole moment along the c-axis. The fourth state also has a dipole moment but along the chain. For simplicity we shall neglect this state in our qualitative model since it does not contribute directly to IR activity of the 155-mode polarized along the c-direction.

In the $YBa_2Cu_3O_6$ system we have quite a different situation. The oxygens $O1$ are removed from the chains, and one part of their electrons fills the d-shells of $Cu1$ ions, while the remainder is transferred to the conducting sheets CuO_2. The total charge of our cluster is then reduced to -3. Consequently both the p-shells on $O4$ oxygens and the d-shell on the $Cu1$ copper ion in the cluster are completely filled and no low energy charge transfer fluctuation is possible. There exists only one formal charge state of the cluster: $O4^{-2}$, Cu^{1+}, $O4^{-2}$.

The fluctuations between formal charge states in the $YBa_2Cu_3O_7$ cluster, can be described approximately by the following Hamiltonian,

$$H_{el} \begin{Bmatrix} e_2 & t & 0 \\ t & e_1 & t \\ 0 & t & e_2 \end{Bmatrix}. \qquad (8)$$

Here, the diagonal matrix elements e_1, e_2 correspond to the energies of the microscopic states, while the off-diagonal term t is the transition amplitude from one state to another. The Hamiltonian (8) can also be associated with a "single particle" Hamiltonian describing the motion of a hole among the $Cu1$ and two $O4$ ions. In this sense e_1 and e_2 are on-site energies of the hole on the $Cu1$ and $O4$ ions. The direct hopping between two $O4$ oxygens is neglected. Note that the microscopic state of the $YBa_2Cu_3O_6$ cluster is the situation when the hole is removed from the system. We shall frequently refer to Hamiltonian (8) as an "electronic" part of the total Hamiltonian, since the charge transfer fluctuations are nothing but electron motion within the cluster. The effects of including electron correlation effects within the cluster (e.g. Hubbard U and V) will be reported elsewhere. Here we only note that their major effects are to renormalize the parameters in Eq.(8): it is important to appreciate that the levels used are implicitly fully dressed many-body ones.

The matrix elements of H_{el} quite generally depend on the lattice deformation of the unit cell of $YBa_2Cu_3O_{7-\delta}$. We are particularly interested in the coupling of the charge transfer fluctuations to the 155-mode. The 155-mode is associated with almost rigid motion of the whole CuO_3 cluster between the CuO_2 sheets. This motion will primarily change the energies e_1, e_2, rather than the hopping amplitude t. The counterpart of the 155-mode is the Raman active mode corresponding to the symmetric oscillation of the $O4$ ions within the cluster. The frequency assigned to this Raman mode is 515 cm^{-1}. Both modes can be described approximately by the asymmetric and symmetric oscillations, respectively, of the $O4$ oxygens in the c-direction. Let Δ_1 and Δ_2 denote shifts of the $O4$ oxygen ions from their equilibrium positions. Keeping only terms linear in $\Delta_{1,2}$ in a Taylor expansion of e_1, e_2, we obtain the interaction Hamiltonian between the phonon and charge transfer fluctuations,

$$H_{el-ph} = -\lambda \begin{pmatrix} \Delta_1 & 0 & 0 \\ 0 & -\Delta_1 + \Delta_2 & 0 \\ 0 & 0 & -\Delta_2 \end{pmatrix}, \quad (9)$$

where λ is the coupling constant. Other forms of electron-phonon couplings can also be included (e.g. through t) and lead to similar results. Note also that eq. (9) is oversimplified: there is in fact a (small) asymmetry to the coupling constants for IR and Raman modes.

We assume that the "phonon part" of the Hamiltonian is given by the familiar expression

$$H_{ph} = \frac{1}{2} K(\Delta_1^2 + \Delta_2^2) + \frac{1}{2} M(\dot{\Delta}_1^2 + \dot{\Delta}_2^2), \quad (10)$$

where M and K are effective mass and elastic constant, respectively.

The Hamiltonians (9) and (10) can be rewritten in terms of coordinates corresponding to the asymmetric (IR) and symmetric (Raman) modes:

$$\begin{matrix} u \\ v \end{matrix} = \frac{\Delta_1 \pm \Delta_2}{\sqrt{2}}, \quad (11)$$

The model defined by the Eqs. (8-10) is quite general. It can alternatively be associated with the charge transfer fluctuations between the CuO_2 planes and CuO_3 cluster. In this case we assign the following formal charge states

$$\begin{array}{ccc} [CuO_2]^{-1} & [CuO_2]^{-2} & [CuO_2]^{-2} \\ | & | & | \\ [CuO_3]^{-4} & [CuO_3]^{-3} & [CuO_3]^{-4} \\ | & | & | \\ [CuO_2]^{-2} & [CuO_2]^{-2} & [CuO_2]^{-1} \end{array},$$

described again by Hamiltonian (1). The phonon modes Δ_1 and Δ_2 now correspond to the motion of the sheets CuO_2 with respect to the chain unit CuO_3, which is also part of displacements involved in the 155-mode. Conradson and Raistrick [10] have emphasized the importance of this charge transfer channel from an analysis of their EXAFS/XANES data. Presumably this charge transfer fluctuation also contributes to the IR activity of the 155-mode. The most important charge transfer has yet to be unambiguously assigned but ab initio quantum chemistry is in progress [34] and finds a strong enhancement of the dipole derivative for the rigid c-axis motion of a CuO_4 cluster. This enhancement is a combination of the inherent oxygen polarization and charge transfer excitations within the cluster. Note that from a solid state model perspective oxygen polarizability may also be included via anharmonic electron-phonon coupling, in the spirit of shell models.[13]

Our model is a generalization of a two level system arising in many physics contexts. In fact a similar two level system has already been applied for the anomalous IR activity of the 240-mode in La_2CuO_4.[35] The scenario presented here can apply in a plane cluster through coupling to an appropriate phonon, and evidence for in-plane (e.g. Jahn-Teller, breathing, buckling) phonon modes has been reported [6].

The interaction Hamiltonian (9) will result in a renormalization of both IR and Raman phonon frequencies. It can also cause a static deformation of the cluster, which may be important for understanding of the structural transformations observed by Cava et al.[31] The renormalized frequency can be found within the usual RPA treatment of the interaction (9), Fig. 2a. In the case of the IR mode the renormalized frequency is given by

$$\omega_{IR}^2 = \omega_o^2 - \frac{\lambda^2}{M_{eff}} \Pi_{MM}(\omega) , \qquad (12)$$

where ω_o is the bare phonon frequency, $\sqrt{\frac{K}{M_{eff}}}$, while Π_{AB} is the electron polarization,

$$\Pi_{AB} = \iota \int \frac{d\varepsilon}{2\pi} Tr(G(\omega + \varepsilon) A\, G(\varepsilon)\, B) . \qquad (13)$$

In the above expression $G(\omega)$ is the electron Green function equal to

$$G(\omega) = [\omega - H_{el}]^{-1} . \qquad (14)$$

The pole of $G(\omega)$ with the lowest real part is shifted into the upper part of the complex ω plane, while the other poles are shifted into the lower part of the ω plane. The matrix M in expression (12) is the part of the interaction (9) corresponding to the IR mode, viz.

$$M = \frac{1}{\sqrt{2}} \begin{pmatrix} 1 & 0 & 0 \\ 0 & 0 & 0 \\ 0 & 0 & -1 \end{pmatrix} . \qquad (15)$$

From expression (13) we can obtain the real part of the electron polarization and hence the <u>change</u> of frequency of the IR mode due to the electron-phonon interaction (9):

$$\frac{\delta(\omega^2)}{\omega^2} = \frac{\lambda^2}{M_{eff}\omega_o^2} \Pi_{MM}(0) . \qquad (16)$$

In $YBa_2Cu_3O_6$ there are no relevant charge transfer fluctuations, and consequently the frequency of the IR mode (and the frequency of any other mode) remains unchanged, i.e. ω_o. The above expression is therefore the relative difference between the frequencies of the 155-mode for $YBa_2Cu_3O_7$ and $YBa_2Cu_3O_6$ systems. It has been found experimentally that this difference is $\simeq 10\%$ (for $YBa_2Cu_3O_7$ $\omega_{IR} \simeq 155\ cm^{-1}$, while for $YBa_2Cu_3O_6$ $\omega_{IR} \simeq 168\ cm^{-1}$).[7] The right-hand side of Eq. (16) is proportional to a conventional definition of the dimensionless electron-phonon coupling constant,

$$\tilde{\lambda} = \frac{\lambda^2}{M_{eff}\omega_o^2 W} ,$$

where W is the width of the excitation spectrum of the charge transfer fluctuations

$$2\sqrt{2t^2 + (\frac{e_2 - e_1}{2})^2} .$$

In order to find the polarizability we need to determine the operator $\hat{P}$ of the dipole momentum of the system. This contains several terms with different origins,

$$\hat{P} = \sqrt{2}(Q\, u - p_o\, M) . \qquad (17)$$

The first term is the dipole momentum induced by the asymmetric motion of $O4$ ions.

Here Q is the oxygen charge and u is the displacement of the IR mode defined in Eq. (11). The second electronic term depends on the microscopic state of the cluster. The quantity p_o is the dipole momentum of the first (third) microscopic state; it is of the order of the electron charge multiplied by the distance between $Cu1$ and $O4$ ions. M is a matrix defined by Eq. (15). In the case of the plane-chain-plane cluster, Q is the averaged charge of the CuO_2 planes per unit cell, while p_o is of the order of the electron charge multiplied by the distance between $Cu1$ and $Cu2$ ions.

The polarizability α of the system is given by the correlation function of the dipole momentums. The phonon part of $\hat{P}$ leads to the familiar lattice dynamics term,

$$\alpha_{ph}(\omega) = 2 \frac{Q^2}{M_{eff}} \frac{1}{\omega_{IR}^2 - \omega^2} . \tag{18}$$

The electron part of the polarizability is proportional to the polarization Π_{MM},

$$\alpha_{el}(\omega) = 2p_o^2 \, \Pi_{MM}(\omega) . \tag{19}$$

This is the zeroth order term in the electron-phonon interaction. The RPA treatment of the interaction (9) results in an additional term with a pole at the phonon frequency ω_{IR},

$$\delta\alpha_{el}(\omega) = 2 \frac{(p_o \lambda \Pi_{MM})^2}{M_{eff}} \frac{1}{\omega_{IR}^2 - \omega^2} . \tag{20}$$

In $YBa_2Cu_3O_6$ this additional term is absent because the charge transfer fluctuations are suppressed. The combination of the Eqs. (18) and (20) gives the polarizability with an effective oscillator strength equal to

$$f = 2 \frac{Q^2}{M_{eff}\omega_o^2} \left(1 + \left(\frac{p_o \, \lambda \, \Pi_{MM}(0)}{Q} \right)^2 \right) \tag{21}$$

for the 155-mode. The difference δf between the oscillator strengths of this phonon mode in $YBa_2Cu_3O_7$ and $YBa_2Cu_3O_6$ systems is therefore given by

$$\frac{\delta f}{f} = (\frac{p_o \, \lambda \, \Pi_{MM}(0)}{Q})^2 . \tag{22}$$

According to the measurements of Genzel et al. [7] the quantity (22) is equal to 15.

Equations (16) and (22) can be used for independent estimates of the coupling constant λ for $YBa_2Cu_3O_{7-\delta}$. If we assume that the polarization $\Pi_{MM}(\omega=0)$ is given by the inverse of the typical energy of the charge transfer fluctuations, i.e. of the order of $(0.1 \text{ eV})^{-1}$, then, according to the Eq. (16) the coupling constant is

$$\lambda \simeq 0.12 \sqrt{\frac{M_{eff}}{M_O}} \frac{eV}{\overset{\circ}{A}} ,$$

while from Eq. (22) it follows that

$$\lambda \simeq 0.19 \frac{eV}{\overset{\circ}{A}} .$$

Here, M_O stands for the oxygen mass. In order to improve agreement between these two estimates it is necessary to assume that the effective mass M_{eff} of the 155-mode is of the order of four oxygen masses: this effective mass requirement increases if the energy of the charge transfer fluctuations increases (i.e. $\Pi_{MM}^{-1} > 0.1$ eV). The estimated values of λ are relatively small in comparison to coupling constants in some other materials, e.g. polyacetylene. Interestingly, other estimates of λ based on the superconducting transition temperatures lead to similar values.[36]

The static deformation of the corresponding symmetric (Raman) mode is given by the averaged value of its interaction with CT fluctuations,

$$v_o = \frac{\lambda}{\sqrt{2}K} \langle \begin{pmatrix} 1 & 0 & 0 \\ 0 & -2 & 0 \\ 0 & 0 & 1 \end{pmatrix} \rangle_o . \qquad (23)$$

For $YBa_2Cu_3O_7$, v_o exists for any finite value of the coupling constant λ, and is equal to

$$v_o = -\frac{\lambda}{2\sqrt{2}M_{eff}\omega_o^2} \left(1 + \frac{3(e_2 - e_1)}{2\sqrt{2t^2 + (\frac{e_2-e_1}{2})^2}}\right) . \qquad (24)$$

However, this deformation is equal to zero in $YBa_2Cu_3O_6$. The expression (24) corresponds therefore to the difference of the $Cu1$-$O4$ distances in $YBa_2Cu_3O_7$ and $YBa_2Cu_3O_6$ systems, which is found experimentally to be equal to 0.04 Å, or $\approx 2\%$. The static deformation of the Raman mode will also renormalize the on-site energies, so that the energies e_1 and e_2 are shifted by $2\lambda v_o/\sqrt{2}$ and $-\lambda v_o/\sqrt{2}$ respectively. For self consistency we must take into account these corrections in all previous expressions. Alternatively, for the plane-chain-plane cluster the static deformation (24) is measured to be equal to 0.13 Å, or $\approx 3\%$.

A static deformation of the asymmetric (IR) mode is possible only for sufficiently large coupling constants λ. The critical value of λ can be estimated from the expression (5) by requiring that the renormalized frequency ω_{IR} is equal to zero. Since the change of the frequency in the IR mode is proportional to the dimensionless coupling constant, we find that the necessary condition for this transition is $\tilde{\lambda} \geq \omega_{CT}^2/t^2 \sim 1$. Note that such a static deformation of the IR mode will induce a permanent dipole momentum in the unit cell, and it is therefore to be associated with an (*anti-*)ferroelectric transition.[37] In this region of parameter space, the dynamics of the IR phonon mode is essentially nonlinear. The corresponding equation of motion of this mode can be found in an adiabatic approximation valid for $t^2 \gg \lambda \dot{u} \approx \lambda u \omega_{IR}$. For small phonon displacements, we obtain an equation with a dominant cubic nonlinearity:

$$M\ddot{u} = -K\,u + K\,\tilde{\lambda}\,\frac{2t^2}{\omega_{CT}^2}\,u\left(1 - \left(\frac{u}{u_o}\right)^2\right) , \qquad (25)$$

where the coefficient of the nonlinear term is

$$u_o^{-2} = \frac{\lambda^2}{W^2\omega_{CT}^2}(2t^2 + W(e_1 - e_2)) .$$

Clearly, the three level model defined by Eqs. (8-10) can be used only for some special composition of $YBa_2Cu_3O_{7-\delta}$. Namely, the number of the holes within the cluster can be changed only discontinuously, in our case from the one hole to the case with no holes. As we have already discussed, these two cases correspond to the compositions with $\delta = 0$ and $\delta = 1$ respectively. However, in a case with an intermediate composition $0 < \delta < 1$, the crystal of $YBa_2Cu_3O_{7-\delta}$ consists of clusters with one hole and clusters without holes. The simplest approach to such situation is the *virtual crystal approximation*. (Extension to a coherent potential approximation is also straightforward.) According to this approximation the overall electron Green function is given by the linear interpolation of the Green functions of the two limiting known cases,

$$G(1 \geq \delta \geq 0) = x\,G(\delta = 1) + (1 - x)\,G(\delta = 0) .$$

Here parameter x is the fraction of the clusters with no holes. According to the measurements of Tranquada et al.,[32] this parameter is nothing but the number of Cu^{+1} ions per unit cell, and it is equal approximately to δ for $\delta < 0.6$, while for $\delta > 0.6$ it is $\approx 2\delta - 1$. By using the formula (13) for the electron polarization, we find that Π_{MM} is proportional to $(1-x)$,

$$\Pi_{MM}(1 \geq \delta \geq 0) = (1-x)\, \Pi_{MM}(\delta = 0) \ .$$

The composition dependence of the frequency of the IR mode can be obtained from Eq. (12). It is given by a linear interpolation of the frequencies for $\delta = 0$ and $\delta = 1$,

$$\omega_{IR}(1 \geq \delta \geq 0) = x\, \omega_{IR}(\delta = 1) + (1-x)\, \omega_{IR}(\delta = 0) \ .$$

A nonlinear behavior in δ of the frequency ω_{IR} is expected only for δ close to 0.6, when an orthorhombic-tetragonal transition occurs. The electron contribution to the oscillatory strength of the IR mode is quadratic in the electron polarization Π_{MM}, and consequently it will show a more dramatic δ-behavior. It decreases linearly for $\delta \geq 0$, and eventually for $\delta \leq 1$ it disappears as $(\delta - 1)^2$. The virtual crystal approximation also predicts the linear x-behavior of the static deformation of the Raman mode. Associated with such a static deformation there is also a self-consistent change in the hole density in the planes/chains. This is presumably reflected in the rapid change in T_c at $\delta \sim 0.5$.

Finally, we note that more detailed microscopic modeling of phonon-CT coupling requires studies of clusters explicitly including, e.g., Hubbard U and V and electron-phonon interactions. Such studies are in progress.

2(iii) Pairing-Bag excitations

Here we focus on one aspect of *strong* coupling (small coherence length) superconductors, namely the possibility of polaron or bag-like deformation of the condensate, which may have distinct experimental consequences -- e.g. for observations of the tunneling density of states and assignment of a SC "gap."

In the pairing theory of superconductivity, the ground-state energy gap Δ is uniform in space for a translationally invariant system.[38] When a single quasiparticle is added to the system, the gap is assumed to remain uniform so that the excitation is in a plane-wave state k. As in Koopmans's theorem for the Hartree-Fock approximation to extended sytems, Δ is unaltered by the presence of the excitation as the volume of the system tends to infinity.

For nonzero temperature, the gap decreases because of the finite density of quasiparticles, with Δ vanishing at the transition temperature T_c, where the quasiparticle density is of order $(\xi/a)^2$ per coherence volume ξ^3, with a the mean electron spacing and ξ the zero-temperature coherence length. Since $\xi/a \approx 10^3$ for conventional superconductors, the depression of the gap by the addition of *one* quasiparticle in a coherence volume is extremely small.

However, for the new layered high-temperature superconductors the corresponding ξ/a is of order 1-10. Therefore, when a quasiparticle is excited in a coherence volume the gap is locally substantially reduced even at zero temperature. This reduction forms a baglike potential which, if sufficiently strong, self-consistently traps the quasiparticle, as in a self-trapped polaron. Here the pairing field plays the role of the phonon field of the polaron.

We study the structure of such localization pairing-bag solutions to the Bogoliubov-de Gennes equations on a two-dimensional square lattice in the context of the "negative-U" Hubbard model.[39] As in the case of spin bags,[40] for a half-filled band we find self-consistent solutions for Δ which have "cigar" or "star" shapes depending on the symmetry of the orbital in which the quasiparticle is initially placed, as well as the initial spatial form of Δ. While these solutions break translational symmetry, this can be restored by forming linear combinations of such localized configurations suitably phased to create a momentum eigenstate.

We consider the negative-U Hubbard model on a 2D square lattice to model the CuO_2 planes in the oxide superconductors. The Hamiltonian is

$$\tilde{H} = -\sum_{nms} t_{nm} c^\dagger_{ms} c_{ns} - \sum_{nm} V_{nm} c^\dagger_{m\uparrow} c^\dagger_{n\downarrow} c_{n\downarrow} c_{m\uparrow} \ . \tag{26}$$

In the mean-field pairing approximation for the superconducting phase one has

$$H = -\sum_{nms} t_{nm} c_{ms}^{\dagger} c_{ns} - \sum_{nm} [\Delta_{nm}^* c_{n\downarrow} c_{m\uparrow}$$
$$+ h.c. - |\Delta_{nm}|^2/V_{nm}] \,, \tag{27}$$

with the self-consistency condition

$$\Delta_{nm}^* \equiv V_{nm} \langle \phi | c_{m\uparrow}^{\dagger} c_{n\downarrow}^{\dagger} | \phi \rangle \,, \tag{28}$$

where V_{nm} is positive for an attractive potential and $|\phi\rangle$ is a quasiparticle occupation number state. H can be diagonalized by making the Bogoliubov-Valatin transformation:

$$c_{ns}^{\dagger} = \sum_i [u_{ni}^* \gamma_{is}^{\dagger} + \nu_{nis} \gamma_{i,-s}] \,. \tag{29}$$

Here $\gamma_{is}^{\dagger}$ are the quasiparticle creation operators,

$$[H, \gamma_{is}^{\dagger}] = E_{is} \gamma_{is}^{\dagger}, \quad [H, \gamma_{is}] = -E_{is} \gamma_{is}, \quad E_{is} \geq 0 \,. \tag{30}$$

Without loss of generality we have chosen phases so that u_{nis} is spin independent. The u and ν amplitudes can be determined by our taking matrix elements of the equation of motion of the bare operators c between an initial state $|\phi\rangle$ and the state having one more quasiparticle in orbital i. As usual, we work within the grand canonical ensemble. Using $\nu_{nis} = -s\nu_{ni}$ one finds

$$E_i u_{ni} = \sum_m [-t_{nm} u_{mi} + \Delta_{nm} \nu_{mi}] \,, \tag{31}$$

$$E_i \nu_{ni} = \sum_m [t_{nm} \nu_{mi} + \Delta_{nm}^* u_{mi}] \,, \tag{32}$$

and the self-consistency condition becomes

$$\Delta_{nm} = V_{nm} \sum_i [u_{ni} \nu_{mi}^* (1 - N_{i\downarrow}) - u_{mi} \nu_{ni}^* N_{i\uparrow}] \,, \tag{33}$$

where N_{is} is the quasiparticle occupation number. Evaluating the expectation value of H in the state $|\phi\rangle$, we find the total energy given by

$$E_\phi = -\sum_i E_i \left(1 - \sum_s N_{is}\right) + \sum_{nm} \frac{|\Delta_{nm}|^2}{V_{nm}} \,. \tag{34}$$

For an N×N lattice, we note that there are $2N^2$ amplitudes u_n and ν_n. While this leads to $2N^2$ eigenvalues E_i, only the positive values of E_i correspond to the energy required to create a quasiparticle, whether it be a quasielectron above the Fermi surface, or a quasihole below; all physical excitation energies are necessarily positive. The negative eigenvalues correspond to the energy released when a quasiparticle (either electronlike or holelike) is destroyed. Therefore, we are only interested in the positive-energy solutions, and sum only over $E_i > 0$ in (34).

We have solved the equations numerically on periodic $N \times N$ square lattices with N = 4 to 16. Extremal solutions to the coupled equations (31) and (32) were sought by iteration in $\{u_{ni}, \nu_{ni}\}$ until self-consistency (33) was obtained [39] for a given quasiparticle occupation, with various initial Δ_{nm} profiles as "seeds." The eigenvalue distribution E_i and symmetries of the associated eigenvectors $\{u_{ni}, \nu_{ni}\}$ were studied. To illustrate bag states we show here results for a *local* "negative U''", $V_{nm} = V_0 \delta_{nm}$, and pure *near-neighbor*, isotropic hopping of strength t. In Fig. 4(a) we show the result of placing one additional quasiparticle in an orbital near the flat region of the square Fermi surface. Localization occurs, mirroring the symmetry of this orbital and producing a cigar shape. If orbitals

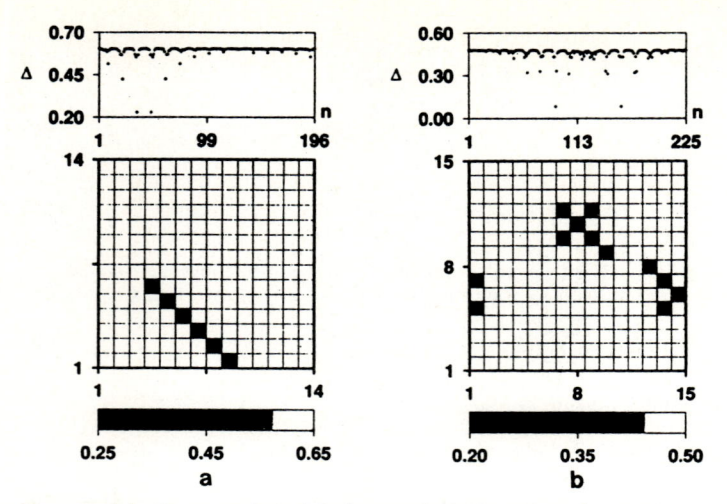

Figure 4(a). One quasiparticle has nucleated a cigar-shaped bag on a 14×14 lattice with periodic boundary conditions and $V_0 = 2.5t$. The top part of the figure shows Δ vs the site index n in a linear array (row by row). The lower part shows Δ on the square lattice as a gray-scale plot. The bag was seeded at the sites where it nucleated. The energy level of the quasiparticle is pulled into the gap. Its value is $0.430t$ and the band of higher levels starts at $0.565t$.

Figure 4(b). Two quasiparticles occupying the two lowest energy levels have formed two star-shaped bags on a 15×15 lattice for $V_0 = 2.25t$. The energy levels are again pulled into the gap and are almost degenerate, with values $0.363t$ and $0.366t$. The band of higher energy levels starts at $0.464t$. Note that the orientation of the stars was seeded to be slightly off diagonal in this case.

near corners of the square Fermi surface are occupied, the localized state assumes their symmetry -- the crossed cigar or star shown in Fig. 4(b). The local deformations of δ_{nm} are accompanied (see figure captions) by an eigenlevel being drawn into the uniform gap present in the absence of added quasiparticles: the greater the localization, the deeper the gap state. The remaining continuum states suffer energy-level shifts which are limited to near-gap states as the bag becomes more delocalized. The binding energy E_b of a bag also increases with V_0/t, where the binding energy is defined relative to an added quasiparticle with uniform Δ. For example, on a 10×10 lattice and for star bags in the same initial Fermi-surface orbital, we find $(V_0/t, \Delta/t, E_b/t)$ with values $(2.5, 0.615, 0.088)$ and $(3.0, 0.854, 0.147)$. In general the star bags seem to have a few percent greater binding energy than the corresponding cigar bags.

Inclusion of quantum fluctuations around the bag states described above is in progress. As in the strong-coupling polaron, the true quantum mechanical states must be constructed out of the present broken-symmetry states by a phased, translationally invariant sum, to yield a *band* of bag states. Note that the hole and order-parameter field must move together at a large velocity ($\approx v_f$). This motion is expected to deform the bag, determine limiting velocities, and even prevent the condensate from following the hole in some cases. The influence of a chemical potential will also be reported elsewhere -- the large density of states at half-filling is most favorable for superconductivity, but varying electron density changes the Fermi-surface structure and therefore bag shapes.

The physical consequences of such pairing-bag states are under investigation, including electromagnetic absorption, quasiparticle and Josephson tunneling, thermodynamics, fluctuation effects, quasiparticle recombination, etc. We have also developed a time-dependent extension of the present theory, which will allow study of dynamics of bag-formation and their interactions, localized bag oscillations, and decay channels -- for instance of quasiparticle excited states to the uniform ground state by phonon and microwave emission.

2(iv) Halogen-bridged transition metal linear chain crystals

These materials have been of interest to chemists for several decades [41]. However, they have only recently begun to receive systematic attention and detailed consideration in the physics community [42]. While they are important in their own right, it should be stressed that they represent a controlled class of materials in which to probe many of the issues raised by the new oxide superconductors -- in terms of theoretical techniques, collective ground state mechanisms, doping- and photo-induced defects, and experimental characterization. Indeed typical members of the class are direct 1-dimensional analogues [43] of $BaPb_{1-x}Bi_xO_3$ and lead to our current efforts to *dope* the 1-dimensional materials *near* transitions between (e.g. charge- and spin-density-wave) ground states. More specifically, we emphasize here:
(a) The increasing appreciation of strong, competing electron-electron and electron-phonon interactions in low-dimensional materials and the consequent need to expand many-body techniques. The MX materials offer a rapidly expanding, single-crystal *class* of quasi-1-D systems which can be "tuned" (by chemistry, pressure, doping, etc.) between various ground state extremes: from strong charge-disproportionation and large lattice distortion (e.g., PtCl - 20% distortion) to weak charge-density-wave and small lattice distortion (e.g., PtI - 5% distortion), to magnetic and undistorted (e.g., NiBr);
(b) The opportunity to probe doping- and light-induced local defect states (polarons, bipolarons, kinks, excitons) and their interactions in controlled environments and the same large range of ground states;
(c) The similarities between models and theoretical issues in these materials and the recently discovered high-temperature superconductors. The MX materials are also closely connected conceptually with mixed-stack charge-transfer salts.

The MX class, then, is important in its own right, but also as a testing ground for concepts and electronic structure techniques in strongly interacting (*both* electron-electron and electron-phonon), low-D electronic materials. A joint theoretical effort is underway to give a unified understanding of these materials from the different points of view of band-structure calculations (valid in the delocalized limit), ab initio quantum chemistry calculations (valid in the high localized limit), and many-body Hamiltonians.

From a band structure point of view, the MX class is essentially a hybridized 2-band (M and X majority) system in which the antibonding band is half-filled at stoichiometry with several nonbonding levels between the bands. In the ionic limit for, e.g., M=Pt, X=Br one has a filled Pt d_{z^2} orbital and a half-filled Br p_z orbital, i.e. 6 electrons per MX unit. The structure of a single chain is shown schematically in Fig. 5. Coupling between adjacent chains is small and the ligands appear to be of secondary importance for the electronic structure along the chain. When filled non-bonding levels lying between the bonding and antibonding bands are included, as well as the extended Hubbard and strong (nonlinear) electron-phonon coupling terms, we see that we have a 1-D version of our previous description for CuO_2 planes (and chains) in the superconducting oxides. However, here we have the advantages of tunability of model parameters and of single crystal materials.

For instance, focusing on a single orbital per site and including only nearest neighbor interactions, we have investigated [44] the following two-band model for an isolated MX chain:

$$H = \sum_{l,\sigma}\{(-t_0 + \alpha\Delta_l)(c^\dagger_{l,\sigma}c_{l+1,\sigma} + c^\dagger_{l+1,\sigma}c_{l,\sigma}) + [(-1)^l 2e_o - \beta_l(\Delta_l + \Delta_{l-1})]c^\dagger_{l,\sigma}c_{l,\sigma}\}$$

$$+ \sum_l U_l n_{l\uparrow}n_{l\downarrow} + V\sum_l n_l n_{l+1} + V_{MM}\sum_{l\ even} n_l n_{l+2}$$

$$+ \frac{K}{2}\sum_l \Delta_l^2 + \frac{K_{MM}}{2}\sum_l(\Delta_{2l} + \Delta_{2l+1})^2 + \frac{1}{2}\sum_l \frac{\hat{p}_l^2}{M_l}, \quad (35)$$

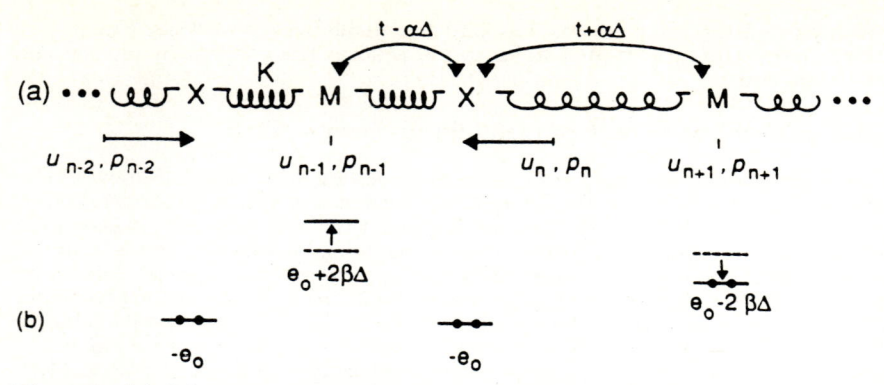

Figure 5. (a) Schematic of the MX chain showing the model parameters and a CDW distortion; (b) the corresponding $t_0 = \alpha = 0$ energy levels.

with $\beta_l = [\beta_X, \beta_M]$, $U_l = [U_X, U_M]$, relative coordinates $\Delta_l = \hat{u}_{l+1} - \hat{u}_l$, momenta $\hat{p}_l$, and displacements from uniform lattice spacing $\hat{u}_l$. This is, of course, a discrete tight-binding extended Peierls-Hubbard model with $2e_o = \epsilon_M - \epsilon_X$ ($\epsilon_{M,X}$ being on-site affinities), intra (β)- and inter (α, breathing)-site electron-phonon coupling, and an effective M-X spring (K). In addition a metal-metal spring (K_{MM} tends to preserve a uniform M-M distance, representing the "cage" effect of the 3-dimensional network.

It is not our intention to review the M-X materials here. We merely note that the full spectrum of analytic and numerical techniques used for oxide systems is also being employed [44]. This investigation has revealed a rich set of collective ground states as parameters are varied: charge-density-wave, bond-order-wave, spin-density-wave, spin-Peierls, etc. Correspondingly, defects (kinks, polarons, bipolarons, excitons) have been studied with respect to these various ground states. While expectations of superconductivity may be optimistic, interesting behavior is indicated near the transitions between ground states, where fluctuations of competing phases are soft. For this reason there is a systematic experimental program using chemistry, pressure and doping to "tune" into these regimes.

Finally, we note that simple dimerized patterns are not the only ground states of Hamiltonian (35). We have found, for example, that strong on-site electron-phonon coupling, in the presence of Coulomb interactions, drives a hierarchy of long-period (superlattice) ground state states [45]. This will also occur in higher-dimensional analogs and it is intriguing to speculate that Hamiltonians such as (35) may contain not only the secret of superconductivity but also intrinsic microscopic mechanisms for the "twinning" and superlattice structures which seem to be so characteristic of the new oxide superconductors.

3. Discussion

A microscopic theory of HTC in oxide superconductors is still not available. Given the 10 year histories of heavy fermion and organic mixed stack superconductors (see these proceedings) this is not so surprising. In this article we have indicated several strategic lines of thinking: (i) Transition metal oxides are distinguished generally by strong competitions for broken symmetry ground states of many varieties. Superconductivity is an "eye-of-the-needle" problem of operating near a transition between other broken symmetry states, where soft fluctuations can be used for pairing, without freezing into those alternative ground states. (ii) We have argued that, although charge and spin fluctuations will inevitably coexist in these small coherence length materials, charge fluctuations may play the dominant role. In $Ba_xK_{1-x}BiO_3$ this is natural, since a disproportional (charge-density-wave) reference state is close to superconductivity. In the copper-based

HTC materials, the stoichiometric materials are dominated by an antiferromagnetic spin instability (as in organic superconductors). We have argued that the role of hole doping is to drive proximity to alternative incipient charge-density-wave instabilities, possibly including a strong short range antiferromagnetic spin component. (iii) Very high superconducting transition temperatures are problematical within conventional BCS/Eliashberg mechanisms using phonons, spin *or* charge as the pairing boson. We have however argued that a synergistic *coupling* of phonons with charge (or possibly spin) is natural in these perovskite-like materials. Furthermore *anharmonic* (large-amplitude) lattice distortions are also natural here. We have argued that *coupling* of phonons to C-T channels can drive the soft phonon distortions, which can then augment underlying charge/spin pairing mechanisms. This can happen purely in the plane (in $La_{1-x}Sr_xCuO_4$). However, further enhancement is possible in 1-2-3 and other truly layered materials, using the dynamic polarizability of the material surrounding the planes. This polarizability can induce pairing in the plane using the polarization fluctuations. *Alternatively*, the C-T excitations microscopically composing the polarizability (section 2(i)) can couple to, and act as a "pump" for, soft phonons in the plane [1,22] (e.g. the buckling mode [33]) driving them into anharmonic regimes and leading to the scenario suggested above. Such a scenario relies on a combination of specific structural and chemical features in these materials, and should not be confused with simple charge fluctuation pairing.

Thus, we believe that serious attention should now be devoted to superconductivity within *strong coupling* BCS theory, but with coexisting and coupled C-T (excitonic) and phonon channels, possibly with relatively strong effects of electron-electron (Coulomb) interactions (e.g., antiferromagnetic spin fluctuations). Such theories are not constrained by the T_c limitations of pure phonon coupling but need some developments beyond familiar Eliashberg theory -- including vertex corrections, for instance.

Some preliminary steps to including multiple channels additively in the Eliashberg function ($\alpha^2(\omega)F(\omega)$) have been taken [46]. A coupling of channels should be investigated, however. In addition, it seems likely that strongly *anharmonic* phonons are necessary. Again, preliminary steps to generalize Eliashberg theory have been taken [47], but much remains to be done -- it may be necessary to go beyond Fröhlich-level Hamiltonians to include the influence of the superconductivity on phonons (indicated by certain experiments [27]) self-consistently. Inclusion of double-well anharmonicity is an interesting first step, especially if we recall that descriptions of structural phase transitions in perovskites have traditionally used "shell" models, integrating out the electronic degrees of freedom (e.g. the oxygen polarizability) to leave effective anharmonic phonons. These real phonons can be used for pairing electrons (holes) as in conventional Eliashberg theory and will result in a temperature-dependent Eliashberg function with possibly distinctive consequences for isotropic shifts, effective values for $2\Delta(o)/k_BT_c$, etc.

Acknowledgments

I am grateful to all of my collaborators on the work reported here: I. Batistic, T. Gammel, P. Lomdahl, R. Martin, K. Mueller, R. Schrieffer, Z. Tesanovic and S. Trugman.

References

1. A. R. Bishop, R. L. Martin, K. A. Mueller, Z. Tesanovic, Z. Physik B 76, 17 (1989); Solid State Comm. 68, 337 (1988).
2. Z. Schlesinger et al., preprint.
3. K. Kamoras et al., preprint.
4. K. Tanaka et al., Physica C 153-5, 1752 (1988).
5. Z. Tesanovic, private communication.
6. e.g., B. Renker et al., Z. Physik B67, 15 (1987); R. E. Cohen et al., Phys. Rev. Lett. 62, 831 (1989); M. Francois et al., Solid State Comm. 66, 1117 (1988).
7. L. Genzel et al., Phys. Rev. B (in press).
8. L. Bulaievskii et al., preprint.
9. C.-K. Loong et al., Phys. Rev. Lett. 62, 2628 (1989).

10. S. Conradson and I. Raistrick, Science 243, 1340 (1989); S. Conradson, I. Raistrick and A. R. Bishop, preprint (1989).
11. R. Zamboni, G. Ruani, A. J. Pal and C. Taliani, Solid State Comm. 70, 813 (1989).
12. A. J. Arko et al., Phys. Rev. B (in press).
13. A. Büssmann-Holder, A. Simon, H. Büttner, Phys. Rev. B (in press).
14. J. E. Hirsch, preprint (1989).
15. A. Zawadowski, Solid State Comm. 70, 439 (1989).
16. L. F. Mattheis, Phys. Rev. Lett. 58, 1028 (1987).
17. Holes in NB O_π orbitals have been suggested by, e.g., Y. Guo et al., Science 239, 896 (1988).
18. A. Shirane et al., Phys. Rev. Lett. 59, 1613 (1987).
19. J. M. Tranquada et al., Phys. Rev. Lett. 60, 156 (1988).
20. C. A. Balseiro et al., Phys. Rev. Lett. 62, 2624 (1989).
21. J. E. Hirsch et al., Phys. Rev. Lett. 60, 1668 (1988).
22. Z. Tesanovic, A. R. Bishop, R. L. Martin, C. Harris, in "Towards the Theoretical Understanding of High Temperature Superconductors," eds. S. Lundquist et al. (World Scientific, Singapore 1988).
23. C. M. Varma, S. Schmitt-Rink, E. Abrahams, Solid State Comm. 62, 681 (1987); P. B. Littlewood et al., Phys. Rev. B 39, 12371 (1989).
24. e.g., D. Scalapino et al., Phys. Rev. B 34, 8190 (1986).
25. D. Baeriswyl and A. R. Bishop, Physica Scripta T19, 239 (1987).
26. C. M. Varma in "Proceedings of the Schloss Mautendorf Meeting on HTS," ed. H. W. Weber (Plenum, NY, 1988).
27. J. B. Torrance, Phys. Rev. Lett. 61, 1127 (1988).
28. J. B. Torrance et al., preprint.
29. The cluster calculations were performed by R. L. Martin with the MESA codes designed by P. Saxe et al. (Los Alamos National Laboratory).
30. I. Batistic, A. R. Bishop, R. L. Martin, Z. Tesanovic, Phys. Rev. B 40, 6896 (1989).
31. R. J. Cava et al., Physics C156, 523 (1988).
32. J. M. Tranquada et al., Phys. Rev. B 38, 8893 (1988).
33. H. Monien and A. Zawadowski, Illinois preprint.
34. J. Hay and R. L. Martin, unpublished.
35. M. J. Rice and Y. R. Wang, Phys. Rev. B 36, 8794 (1987).
36. P. C. Pattnaik and D. M. Newns, Physica C157, 13 (1989).
37. Large c-direction dielectric constants have been reported by L. R. Testardi et al. (Nature, in press) in $YBa_2Cu_3O_{7-\delta}$; but also in the a-b plane for La_2CuO_4 by D. Reagor et al. (preprint). It is tempting to believe that a ferroelectric distortion, typical of polarizable perovskites, is nearby. However, note that for materials with mirror plane symmetry, *anti*ferroelectric fluctuations should be sought. (See section 2(ii)).
38. J. Bardeen, C. N. Cooper, J. R. Schrieffer, Phys. Rev. 108, 1175 (1957).
39. A. R. Bishop, P. S. Lomdahl, J. R. Schrieffer, S. A. Trugman, Phys. Rev. Lett. 61, 2709 (1988); D. Coffey et al., preprint (1989).
40. J. R. Schrieffer, X.-G. Wen, S.-C. Zhang, Phys. Rev. Lett. 60, 944 (1988); Phys. Rev. B (in press).
41. e.g., P. Day, in "Low Dimensional Cooperative Phenomena," ed. H. J. Keller (Plenum, NY, 1974).
42. e.g., K. Nasu, J. Phys. Soc. Jpn. 52, 3865 (1983).
43. D. Baeriswyl and A. R. Bishop, J. Phys. C21, 339 (1988).
44. A. R. Bishop et al, Synthetic Metals 29, F151 (1989); J. T. Gammel et al., Synthetic Metals 29, F161 (1989); and Los Alamos preprints.
45. I. Batistic, A. R. Bishop and J. T. Gammel, Los Alamos preprint.
46. F. Marsiglio, M. Schossmann and J. P. Carbotte, Phys. Rev. B 37, 4965 (1987); J. M. Coombes and J. P. Carbotte, Phys. Rev. B 38, 8697 (1988).
47. N. M. Plakida, V. L. Aksenov, S. L. Drechsler, Europhys. Lett. 4, 1309 (1987); J. R. Hardy and J. W. Flocken, Phys. Rev. Lett. 60, 2191 (1988), and preprint (1989).

Experimental Constraints and Theory of Layered High-Temperature Superconductors

T. Schneider and M. Frick

IBM Research Division, Zurich Research Laboratory,
CH-8803 Rüschlikon, Switzerland

Abstract. Recent experiments have revealed several key features of the normal and superconducting states of high-temperature superconductors. They impose strong constraints on theoretical models of the phenomenon. We developed a tight-binding BCS-type model. The carriers form a narrow and anisotropic band, and are subject to on-site and interlayer singlet pairing. The resulting properties agree remarkably well with experimental results, resolve conflicting interpretations of experimental results, offer an easily understandable physical picture, and point to the nature of the pairing mechanism.

1. Experimental Constraints

Recent experiments have revealed several key features of the normal and superconducting states of high-temperature superconductors. Measurements of the penetration depth [1-4], the coherence length ξ as derived from the upper critical fields [5-7], the conductivity σ [8-10], and infrared reflectivity [11-13] clearly indicate the presence of strong anisotropy, as expressed by the ratios

$$\frac{\lambda_{ab}}{\lambda_c}, \quad \frac{\xi_\perp}{\xi_\parallel}, \quad \sqrt{\frac{\sigma_\perp}{\sigma_\parallel}}, \quad \sqrt{\frac{m_\parallel}{m_\perp}} < 1, \qquad (1)$$

where m is the effective mass, $\parallel$ denotes directions parallel and $\perp$ perpendicular to the layers. Compared to the conventional superconductors, the zero-temperature correlation length is very short and the penetration depth much greater. Current estimates are

$$\frac{\lambda_{ab}(0)}{\lambda_c(0)} \simeq \frac{1400\ \text{Å}}{7000\ \text{Å}}, \quad \frac{\xi_\perp(0)}{\xi_\parallel(0)} \simeq \frac{5\ \text{Å}}{25\ \text{Å}}. \qquad (2)$$

Because $\xi_\perp$ turns out to be much smaller than the mean spacing, the interlayer interaction is small and a description in terms of a system of weakly interacting layers appears to be valid.

The temperature dependence of the penetration depth [1-4] closely follows BCS-type behavior, consistent with singlet pairing and a nodeless gap. Moreover, the relationship between transition temperature and carrier density points to an unretarded pairing interaction [14]. In contrast to conventional BCS superconductors, there is considerable evidence for an anisotropic gap from tunneling [15-18] and infrared reflectivity measurements [11-13]. As far as the normal state is concerned, Hall measurements for H parallel to the c-axis and I parallel to the a,b-plane yield a positive value for the appropriate element of the Hall tensor, i.e. hole-like, while an element for H parallel to the a,b-plane is negative [19]. Finally, photoemission [20-25] and positron annihilation [26-28] experiments provide clear evidence for anisotropic Fermi liquid states, forming narrow bands with low carrier density.

These experimental facts place strong constraints on theoretical models.

2. Model

Guided by photoemission experiments [20-25], we assume Fermi liquid states, subjected to an unretarded pairing interaction. A tight-binding description accounts for narrow bands. The Hamiltonian then reads

$$\mathcal{H} = \sum_{i,j,\sigma} t_{ij} c_{i\sigma}^+ c_{j\sigma} - \sum_{i,j} g_{ij} c_{i\uparrow}^+ c_{j\downarrow}^+ c_{j\downarrow} c_{i\uparrow}, \tag{3}$$

where t_{ij} are transfer integrals describing the hopping of the carriers between sites i and j, while g_{ij} is the strength of the pairing interaction between the carriers on these sites. For m layers per unit cell, there will be m bands which will be split according to the interlayer transfer integrals. Here, we concentrate on one layer per unit cell or, equivalently, we assume that the splitting is negligibly small, corresponding to an m-fold degenerate band. Thus, the first term describes the band structure of the carriers, while the second term represents the unretarded pairing interaction. For a square lattice (lattice constant a) within the layers, identical layers being separated by the lattice constant s, we have one layer per unit cell. Considering then nearest-neighbor hopping within the layers and between adjacent sheets, diagonalization of the first term yields

$$\varepsilon(\vec{k}) = A\{-2(\cos k_x a + \cos k_y a) + 4B \cos k_x a \cos k_y a - 2C \cos k_z s - E_F\}, \tag{4}$$

where A, AB and AC correspond to the values of the hopping matrix elements, and E_F denotes the Fermi energy given in terms of the band filling

$$\rho = \frac{1}{N} \sum_{\vec{k}} \left(\exp(\beta \varepsilon(\vec{k})) + 1 \right)^{-1}. \tag{5}$$

N is the number of sites. Guided by the photoemission results [23-25], the Hall effect measurements [19] and the anisotropy of the coherence length, we set the parameters of Eq. (4) as follows

$$A = 0.153 \text{ eV}, \quad AB = 0.069 \text{ eV}, \quad AC = 0.015 \text{ eV}. \tag{6}$$

Even though A and AC differ by only one order of magnitude, this choice is shown to lead to a highly anisotropic metallic normal state. In the superconducting state we assume BCS-type pairing. The gap equation is then given by [29]

$$\Delta(\vec{k}) = \sum_{\vec{q}} V(\vec{k}-\vec{q}) \frac{\Delta(\vec{q})}{2E(\vec{q})} \tanh \frac{\beta E(\vec{q})}{2}, \tag{7}$$

where

$$E(\vec{q}) = \left(\varepsilon^2(\vec{q}) + \Delta^2(\vec{q})\right)^{1/2} \tag{8}$$

and

$$V(\vec{q}) = \sum_{l} g_{l0} e^{i\vec{q}\cdot\vec{R}_l}, \tag{9}$$

where l labels the sites. Considering on-site and next-nearest neighbor pairing within and between the sheets, we obtain

$$V(\vec{k}) = g_0 + g_1 \cos k_z s + g_2 (\cos k_x a + \cos k_y a). \tag{10}$$

Assuming singlet pairing, where $\delta(\vec{k}) = \delta(-\vec{k})$, the solutions of the gap equation (7) adopt the form

$$\Delta(\vec{k}) = \Delta_0 + 2\Delta_1 \cos k_z s + 2\Delta_2 (\cos k_x a + \cos k_y a). \tag{11}$$

The gap thus becomes anisotropic. Of particular relevance is the behavior on or close to the Fermi surface $\varepsilon(k) = 0$, where nodes might appear. Such nodes greatly affect the thermodynamic behavior. In solving Eq. (7), we assume unretarded pairing. In this case, the k-summation extends over the full Brillouin zone and the limits of integration with respect to energy correspond to the bottom and top of the band. In the following, we concentrate on uniaxial gap anisotropy, where $g_2 = 0$. For $g_2 = 0$ and $|g_1| < g_0$, the weak-coupling solution for T_c reads

$$k_B T_c = 1.14 k_B \theta \exp(-1/N_N(0)g_0), \tag{12}$$

where

$$k_B \theta = \left(\tilde{E}_F(W - \tilde{E}_F)\right)^{1/2}. \tag{13}$$

$\tilde{E}_F = |\varepsilon(\vec{k}=0)|$ is the relative Fermi energy, W the band width and $N_N(0)$ the density of states in the normal state at the Fermi level. Weak coupling also holds for $g_0 > |g_1|$ and $k_B T_c \ll k_B \theta$. In contrast to standard BCS, where the retarded nature of the phonon-mediated pairing interaction leads to a cutoff determined by the Debye energy, unretarded pairing yields a prefactor fixed by the top and bottom of the band. Thus, all carriers are subject to pairing, and modest values for $N_N(0)\, g_0$ lead to high T_c values, determined by the band width W and the Fermi energy $\tilde{E}_F$. As noted above, the empirical relationship between T_c and the carrier density [14] points to an unretarded pairing interaction. This relation is consistent with Eqs. (5) and (12), because the carrier density $n = 2\rho/V_0$ fixes the Fermi level entering the prefactor of the weak coupling expression for T_c [Eq. (12)].

In summary, the model is of BCS-type, assumes a tight-binding band for the carriers and unretarded pairing, on-site and between nearest neighbors. The tight-binding band of the carriers includes nearest-neighbor hopping within the layers and next-nearest-neighbor hopping between the sheets. In the weak-coupling limit, T_c is related to the band width and the Fermi energy, yielding high values of T_c for modest coupling strengths. In the following, we resort to numerical solutions of the gap equation (7). Due to its nonlinearity, there are several solutions, and the one yielding the lowest free energy is chosen. It should be kept in mind, however, that our analysis was restricted to a model with one layer per unit cell. An extension to m layers per unit cell will lead to m distinct gap branches. Nevertheless, for nearly degenerate branches, the present model is still applicable. Considerable modifications are expected for the nondegenerate case, which will be treated in a forthcoming paper.

3. Model Properties

In this section, we compare the key properties of the model with current experimental facts. These results should also offer a physical explanation of such key features as the short coherence length, long penetration depth, and pronounced anisotropy. We now consider normal state properties at zero temperature. The elements of the Hall tensor are given by

$$R_{xyz} = \frac{\sigma_{xyz}}{\sigma_{xx}\sigma_{yy}} = \frac{E_y}{J_x H_z}, \quad R_{yzx} = \frac{\sigma_{yzx}}{\sigma_{yy}\sigma_{zz}} = \frac{E_y}{J_z H_x},$$
$$R_{zxy} = \frac{\sigma_{zxy}}{\sigma_{zz}\sigma_{yy}} = \frac{E_z}{J_y H_x}.$$
(14)

Assuming an isotropic relaxation time, the transport coefficients read [30]

$$\sigma_{\alpha\alpha} = e^2\tau n \frac{1}{m_{\alpha\alpha}}, \quad n = \frac{2\rho}{V_0} \tag{15}$$

and

$$\sigma_{xyz} = -2\frac{e^3\tau^2}{\hbar^4 V_0 N} \sum_{\vec{k}} \frac{\partial\varepsilon}{\partial k_x}\left(\frac{\partial\varepsilon}{\partial\vec{k}} \times \nabla_k\right)_z \frac{\partial\varepsilon}{\partial k_y} \delta(\varepsilon) \tag{16}$$

where

$$\frac{1}{m_{\alpha\alpha}} = \frac{1}{\rho N\hbar^2}\sum_{\vec{k}}\left(\frac{\partial\varepsilon}{\partial k_\alpha}\right)^2 \delta(\varepsilon) \tag{17}$$

is the effective mass, $V_0 = a^2 s$, s denotes the mean spacing of the layers, and the z-direction is perpendicular to the sheets. In this approximation, R_{xyz}, R_{yzx} and R_{zxy} are independent of the relaxation time. Moreover, it is important to observe that the free electron formula $R = -1/ne$ becomes meaningless, except in the case of a parabolic band, which is not applicable here. In fact, the Fermi surface corresponds to a corrugated column, the cross section of which is dependent on the filling, as illustrated in Fig. 1.

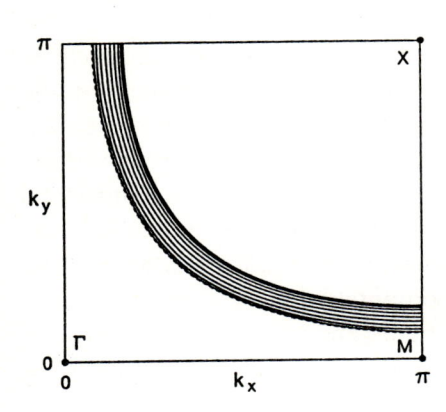

Figure 1. Cuts parallel to the (k_x, k_y) plane through the Fermi surface for different k_z values, $0 \leq k_z \leq \pi$, in steps of $\pi/(10s)$ for $\rho = 0.4$. For $k_z = 0$, the Γ, X and M points are marked. The dashed curve corresponds to $k_z = \pi$.

In the London approximation, the zero-temperature value of the penetration depth is also fully determined in terms of normal-state properties by

$$\lambda_{\alpha\alpha}^{-2} = \frac{4\pi n e^2}{c^2}\frac{1}{m_{\alpha\alpha}}. \tag{18}$$

The numerical results listed in Table I agree remarkably well with available experimental data. R_{xyz} is positive, i.e. hole-like, and has the correct order of magnitude as does the carrier density n [9]. R_{zxy} is

Table I. Normal-state properties and zero temperature penetration depth for several band fillings and the parameters listed in Eq. (6); $\widetilde{E}_F = |\varepsilon(\vec{k}=0)|$ is the relative Fermi energy, $m_{xx} = m_{yy}$ and m_{zz} are the elements of the effective mass tensor, while R_{xyz}, R_{yzx} and R_{zxy} denote independent elements of the Hall tensor in units of 10^{-3} cm^{3}C^{-1}. The other units are: energy in A [Eq. (4)] and the carrier density in 10^{21} cm^{-3} ($n = 2\rho/V_0$). As characteristic lattice constants we used $a = 3.85$ and $s = 7.76$ Å. From [31].

ρ	$\widetilde{E}_F$	$N_N(0)$	$\dfrac{m_{xx}}{m_e}$	$\dfrac{m_{zz}}{m_e}$	R_{xyz}	R_{yzx}	R_{zxy}	λ_{xx}	λ_{zz}	n
0.6	1.62	0.091	2.70	140	0.86	-0.25	-0.25	940	6738	10.43
0.5	1.51	0.175	3.64	119	0.68	-0.59	-0.59	1088	6214	8.69
0.4	1.04	0.264	5.64	100	0.58	-0.98	-0.98	1356	5698	6.96
0.3	0.75	0.461	10.84	77	0.57	-1.68	-1.68	1878	4992	5.22

negative, as observed in Y-Ba-Cu-O [19]. To obtain these magnitudes and signs, it is necessary to include the nearest-neighbor hopping term, Eq. (4), which also strongly affects the shape of the Fermi surface parallel to the sheets. While the R-values are independent of the width of the band, i.e. A in Eq. (4), this quantity enters the effective mass and the penetration depth. In fact, $m \sim A^{-1}$ and $\lambda \sim A^{1/2}$. Nevertheless, the estimates listed in Table I are in the range of the current experimental data. Moreover, the anisotropy and long penetration depth can be understood in terms of the low carrier density and the large effective mass, Eq. (18). Accordingly, our simple tight-binding parameterization leads to remarkable agreement with the available experimental facts. The narrow anisotropic band implies carriers with both hole- and electron-like character and low density, a column-like Fermi surface, and rather large effective masses.

Next, we turn to the superconducting state by assuming unretarded on-site and next-nearest-neighbor interlayer pairing. In Table II, we summarized estimates for the gap parameters and T_c for $\rho = 1/2$ as a function of g_1, the strength of the interlayer pairing interaction. These estimates, obtained from a numerical solution of Eq. (7), reveal that for $g_0 > |g_1|$ the gap becomes anisotropic but remains nodeless. In this coupling regime T_c is rather insensitive to the value of g_1. Nevertheless, an attractive g_1 enhances T_c, while in the repulsive case T_c is seen to remain unaffected. At zero temperature, however, the full nonlinearity of

Table II. Numerical estimates for Δ_0, Δ_1 and T_c as obtained from Eq. (7) for $\rho = 1/2$, $g_2 = 0$. Energies are in units of $A = 0.153$ eV. Δ_{min} denotes the minimum, Δ_{max} the maximum value of the gap. From [31]

g_0	g_1	Δ_0	Δ_1	$k_B T_c$	$\dfrac{2\Delta_{min}}{k_B T_c}$	$\dfrac{2\Delta_{max}}{k_B T_c}$	$\dfrac{\Delta_{min}}{\Delta_{max}}$
1	-0.99	0.0232	0.00067	0.0131	3.34	3.75	0.89
1	-0.9	0.0232	0.00063	0.0131	3.35	3.73	0.90
1	-0.8	0.0232	0.00059	0.0131	3.36	3.72	0.91
1	-0.6	0.0233	0.00049	0.0132	3.38	3.68	0.92
1	-0.4	0.0233	0.00036	0.0132	3.42	3.64	0.94
1	-0.2	0.0235	0.00020	0.0132	3.50	3.62	0.97
1	0	0.0235	0	0.0133	3.53	3.53	0
1	0.2	0.0236	-0.0003	0.0134	3.43	3.61	0.96
1	0.4	0.0237	-0.0006	0.0135	3.33	3.69	0.90
1	0.6	0.0239	-0.0013	0.0138	3.09	3.84	0.80
1	0.8	0.0243	-0.0026	0.0146	2.62	4.04	0.65
1	0.9	0.0244	-0.0039	0.0158	2.10	4.08	0.51
1	0.99	0.0244	-0.0061	0.0186	1.32	3.92	0.33

the gap equation (7) comes into play and the presence of interlayer pairing leads to an anisotropic gap. Thus, the standard BCS formula $2\Delta/k_B T_c \simeq 3.52$ is no longer applicable because the gap varies between Δ_{min} and Δ_{max}. Due to the column-like shape of the Fermi surface (Fig. 1) all gap values between $\Delta_{min} = \Delta_0 - 2|\Delta_1|$ and $\Delta_{max} = \Delta_0 + 2|\Delta_1|$ are present at the Fermi surface.

The transition temperatures are in the range from 23 to 33 K and are found to increase initially for fixed g_0 and g_1 by reducing ρ, owing to an increase in the density of states. Even higher values of T_c are obtained by increasing g_0.

To clarify the coupling regime and the validity of the weak-coupling solution, Eq. (12), we also studied $T_c(g_0)$ for fixed ρ and $T_c(\rho)$ for fixed g_0 by solving Eq. (7) numerically. In the range $g_0 \lesssim 2$ and for $\rho \simeq 1/2$, the weak-coupling solution, Eq. (12), turns out to be reasonable.

Next, we turn to the implications of the anisotropic gap, the coherence length, and relevant thermodynamic properties. Tunneling conductance [15-18] and infrared reflectivity [11-13] measurements yield increasing evidence for a nodeless anisotropic gap. To illustrate a fingerprint of this anisotropy, we depicted the density of states in the super-

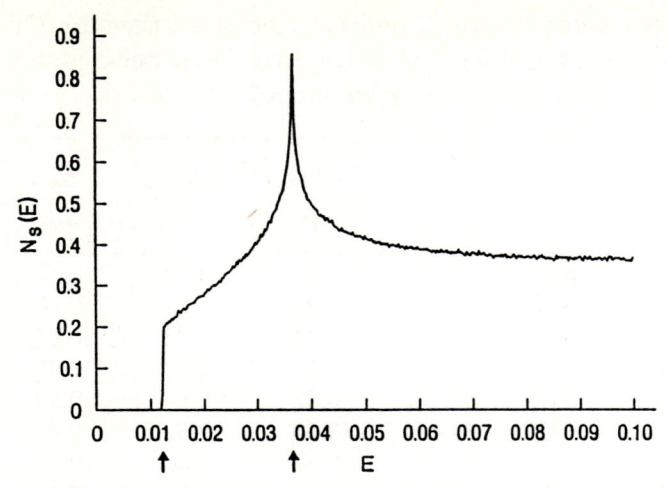

Figure 2. Density of states $N(E)$ in the superconducting state for $\Delta_0 = 0.0244$, $\Delta_1 = -0.0061$, $g_0 = 1$, $g_1 = 0.99$ and $\rho = 1/2$ (see Table II). The arrows mark the minimum ($\Delta_{min} = 0.0124$) and the maximum ($\Delta_{max} = 0.0364$) value of the gap. Energies are in units of $A = 0.153$ eV. From [31].

conducting phase in Fig. 2. The standard BCS behavior, a square-root singularity at the gap energy, is removed. The density of states is zero up to the minimum gap Δ_{min}. For larger energies, up to Δ_{max}, the structure is a characteristic of the specific gap anisotropy. In the present case, the anisotropy is uniaxial and a peak occurs at Δ_{max}, corresponding to a Van Hove singularity. In any case, there is no longer a unique gap: its energy varies between Δ_{min} and Δ_{max}. Clearly, for small anisotropy, the modifications of the standard BCS behavior will shrink to a small energy interval.

In principle, information on the gap anisotropy can be obtained from tunneling and infrared reflectivity measurements. Unfortunately, these techniques are extremely surface-sensitive. Moreover, substantial anisotropy invalidates interpretations based on standard BCS behavior. To illustrate this point, we consider specular and diffuse tunneling. For the zero-temperature voltage current characteristic in an NIS junction [29] the transfer Hamiltonian approach yields

$$I = \pm 2\pi e \sum_{\vec{k}\vec{p}} |T_{\vec{k}\vec{p}}|^2 \left(1 \mp \frac{\varepsilon(\vec{k})}{E(\vec{k})}\right) \theta(|eV| - E(\vec{k}))$$
$$\times \delta(\xi(\vec{p}) - |eV| \pm E(\vec{k})),$$
(19)

where $\vec{k}$ and $\vec{p}$ label the states on the superconducting and the normal side, respectively, and $\pm = \text{sign}(eV) \cdot \xi(\vec{p})$ describes the conduction band on the normal side, which is assumed to have a constant density of states at the Fermi level. $T_{\vec{k}\vec{p}}$ denotes the tunneling matrix element.

Assuming specular transmission, $T_{\vec{k}\vec{p}}$ is given by [29]

$$|T_{\vec{k}\vec{p}}|^2 = P \left|\frac{\partial \varepsilon(\vec{k})}{\partial k_L}\right| \left|\frac{\partial \xi(\vec{p})}{\partial p_L}\right| \delta(\vec{k}_T - \vec{p}_T) D(\varepsilon(\vec{k}), \vec{k}_T). \tag{20}$$

T and L label the transverse and longitudinal components with respect to the current in the junction; P denotes a constant prefactor. For a constant barrier height U, a WKB approximation yields

$$D(\varepsilon(\vec{k}), \vec{k}_T) = \exp\left(-2d \sqrt{\frac{2m_L}{\hbar^2}(U - \varepsilon_L(\vec{k}))}\right), \tag{21}$$

where

$$\varepsilon_L(\vec{k}) = \varepsilon(\vec{k}) - \frac{\hbar^2 \vec{k}_T^2}{2m_T}.$$

$m_{T,L}$ denotes the transverse and longitudinal effective masses of the carriers in the insulating region and are approximated by the free electron mass m.

Assuming $U \gg \varepsilon_L(\vec{k})$, Eq. (21) can be approximated by

$$D(\varepsilon(\vec{k}), \vec{k}_T) \simeq \exp\left(-2d\sqrt{\frac{2mU}{\hbar^2}}\right) \exp\left(\frac{\varepsilon(\vec{k})}{\varepsilon_0}\right) \exp\left(-\frac{k_T^2}{K^2}\right), \tag{22}$$

where

$$\varepsilon_0 = \frac{1}{d}\sqrt{\frac{\hbar^2 U}{2m}}; \quad K^2 = \frac{1}{dh}\sqrt{2mU}.$$

Reasonable values for the barrier parameters are $d \simeq 10$ Å and $U \simeq 1$ eV, which yield $\varepsilon_0 = 0.194$ eV and $K = 0.23$ Å^{-1}. The relevant voltages are of the order of the gap values on the Fermi surface of the superconductor, because only for these values can characteristics of the superconducting side be detected. Therefore, the wave vectors $\vec{k}$ contributing to the current lie around the Fermi surface in an energy shell of the order of the gap. This restriction is due to the step function in Eq. (19). Consequently, the relevant values of $\varepsilon(\vec{k})$ are much smaller than ε_0, and the exponential term $\exp(\varepsilon(\vec{k})/\varepsilon_0)$ in Eq. (22) is a constant. This approximation was proposed in Ref. [29]. Combining Eqs. (19), (20) and (22), and performing the $\vec{p}$-summation, the tunneling current for specular transmission reads

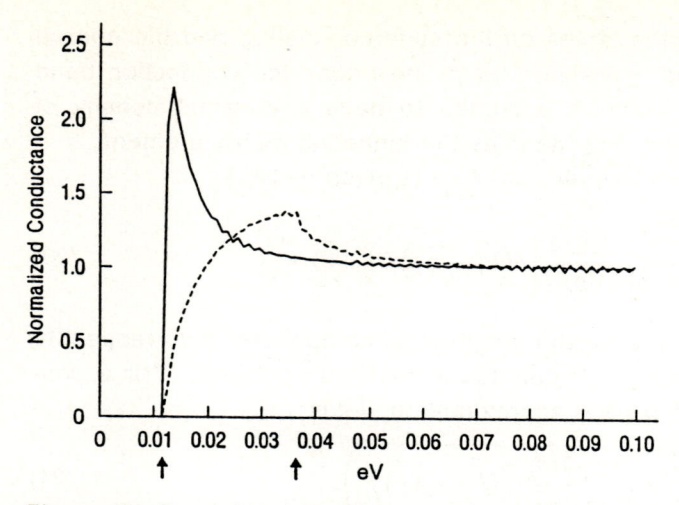

Figure 3. Specular tunneling conductance versus voltage for $C = 0.1$, $\Delta_0 = 0.0244$, $\Delta_1 = 0.0061$, $g_0 = 1$ and $g_1 = 0.99$. Solid (dashed) line: tunneling current parallel (perpendicular) to the layers. The arrows mark Δ_{min} and Δ_{max}. From [31].

$$I \propto \sum_{\vec{k}} \left(1 \mp \frac{\varepsilon(\vec{k})}{E(\vec{k})}\right) \theta(|eV| - E(\vec{k})) \left|\frac{\partial \varepsilon(\vec{k})}{\partial k_L}\right| \\ \times \exp\left(\frac{\varepsilon(\vec{k})}{\varepsilon_0}\right) \exp\left(-\frac{k_T^2}{K^2}\right). \quad (23)$$

Figure 3 shows the voltage dependence of the conductance dI/dV for tunneling currents parallel and perpendicular to the layers and with $C = 0.1$ [Eq. (4)]. In measurements, the peak position is taken as the gap. Thus, a smaller gap would be observed for a tunnel current parallel to the layers. This seems to contradict tunneling experiments on cryogenically cleaved YBCO surfaces [23-25]. In fact, the larger gap appeared for I parallel to the layers. Specular tunneling, however, requires a perfect junction. In view of this, one expects diffuse tunneling to be closer to reality.

For diffuse transmission, conservation of transverse momentum is no longer valid. Let $\vec{q}_T$ denote the transverse component of the wave vector in the insulating region. Then the $\vec{q}_T$-dependence in the expression for the D-factor has to be averaged, yielding a mean value

$$\bar{D} = \sum_{\vec{q}_T} D(\varepsilon(\vec{k}), \vec{q}_T)$$

$$= \sum_{\vec{q}_T} \exp\left(-2d\sqrt{\frac{2mU}{\hbar^2}}\right) \exp\left(\frac{\varepsilon(\vec{k})}{\varepsilon_0}\right) \exp\left(-\frac{q_T^2}{K^2}\right). \tag{24}$$

According to the arguments given above, the dependence of this term on $\varepsilon(\vec{k})$ is weak and $\bar{D}$ can be regarded as a constant. The tunneling current for diffuse transmission is thus given by

$$I \propto \sum_{\vec{k}} \left|\frac{\partial \varepsilon}{\partial k_L}\right| \left(1 \mp \frac{\varepsilon(\vec{k})}{E(\vec{k})}\right) \theta(|eV| - E(\vec{k})). \tag{25}$$

Owing to the absence of the exponential term appearing in the specular case, see Eq. (23), the current depends very little on the orientation of the junction with respect to the layers. In fact, neglecting the gradient and the ε/E-terms, the conductance becomes proportional to the density of states, which reflects the gap anisotropy as shown in Fig. 2. Thus, the standard BCS square-root singularity is removed, and the position of the maximum in the conductance determines the maximum gap value. In absolute infrared reflectivity measurements the situation is reversed: $R = 1$ up to photon energies corresponding to twice the minimum gap value. Accordingly, the ratio of gaps evaluated from infrared reflectivity and diffuse tunneling measurements then provides a measure of the gap anisotropy in terms of $\Delta_{min}/\Delta_{max}$. Hence, it is not surprising that for YBCO the infrared reflectivity estimate $2\Delta/k_B T_c = 3.5$ [13] differs markedly from the tunneling data, which converges to $2\Delta/k_B T_c = 6$ [15]. In fact, this discrepancy should be taken as clear evidence for an anisotropic gap with $\Delta_{min}/\Delta_{max} = 3.5/6 = 0.6$.

Gap anisotropy will of course also affect other properties, such as the critical field and the temperature dependence of the specific heat, nuclear magnetic relaxation rate and penetration depth.

The critical field at zero temperature is obtained from the energy difference between the superconducting and normal states in terms of

$$E_s - E_N = \sum_{\vec{k}} \left(|\varepsilon(\vec{k})| - E(\vec{k}) + \frac{\Delta^2(\vec{k})}{2E(\vec{k})}\right) = -\frac{H_c^2(0)}{8\pi}. \tag{26}$$

Numerical results for $H_c^2(0)/8\pi$ and the ratio $\gamma T_c^2/H_c^2(0)$, yielding $1/4\pi = 0.084$ in standard BCS, are given in Table III. As the gap aniso-

Table III. Zero temperature values of the critical field and the ratio $(\gamma T_c^2)/H_c^2(0)$ for several g_1 values for the parameters given in Eq. (6), $\rho = 1/2$ and $g_0 = 1$. From [31].

g_1	$\dfrac{\Delta_{min}}{\Delta_{max}}$	$k_B T_c$	$\dfrac{H_c^2(0)}{8\pi} 10^5$	$\dfrac{\gamma T_c^2}{H_c^2(0)}$
0	1	0.0133	9.34	0.087
0.4	0.9	0.0136	10.16	0.083
0.6	0.8	0.0140	10.47	0.086
0.8	0.65	0.0147	11.14	0.089
0.9	0.51	0.0160	11.40	0.103
0.99	0.33	0.0186	12.41	0.128

tropy increases with g_1 (Table II), the critical field is seen to increase with anisotropy and the ratio becomes larger and progressively deviates from the standard BCS value. γ is the Sommerfeld constant

$$\gamma = \frac{2\pi^2}{3} N_N(0) k_B^2, \qquad (27)$$

where $N_N(0)$ is 0.175.

In Table IV, we listed the normalized jump of the specific heat, $\Delta C/\gamma T_c$, for several g_1 values. The jump is seen to decrease with increasing anisotropy, and for $g_1 = 0$, one regains the standard BCS ratio 1.45. Anisotropy also affects the asymptotic low-temperature behavior, because $C \sim \exp(-\beta \Delta_{min})$.

To calculate the penetration depth at finite temperature, local electrodynamics is appropriate ($\lambda(0) \gg \xi(0)$). The finite temperature extension of Eq. (18) then reads

Table IV. Specific heat jump for the parameters listed in Eq. (6), $\rho = 1/2$ and $g_0 = 1$. From [31].

g_1	0	0.2	0.4	0.6	0.8	0.9	0.99
$\dfrac{\Delta_{min}}{\Delta_{max}}$	1	0.96	0.9	0.8	0.85	0.51	0.33
$\dfrac{\Delta C}{\gamma T_c}$	1.45	1.47	1.42	1.39	1.14	0.85	0.63

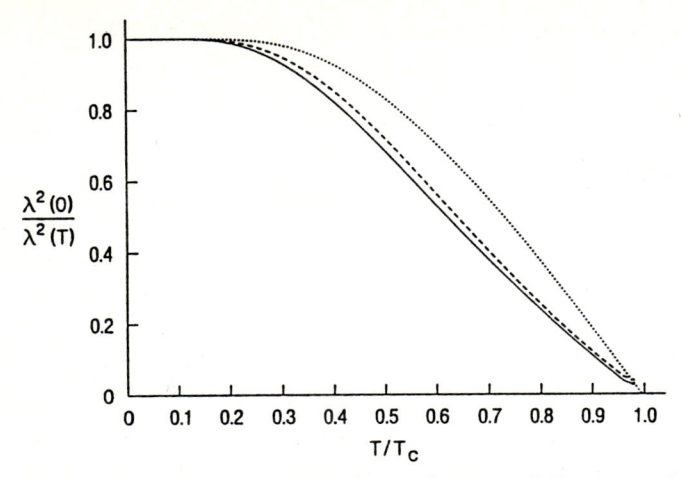

Figure 4. Temperature dependence of the penetration depth for the parameters listed in Eq. (6), and dotted line: $g_1 = 0$, λ_{xx}, λ_{zz}; dashed line: $g_1 = 0.9$, λ_{zz}; solid line: $g_1 = 0.9$, λ_{xx}. From [31].

$$\lambda_{\alpha\alpha}^{-2} = \frac{4\pi n e^2}{c^2} \sum_{\vec{k}} v_\alpha^2(\vec{k}) \left[\left(-\frac{\partial f}{\partial \varepsilon} \right) - \left(-\frac{\partial f}{\partial E} \right) \right], \qquad (28)$$

where

$$f(x) = (\exp \beta x + 1)^{-1}, \quad V_\alpha(\vec{k}) = \frac{1}{\hbar} \frac{\partial \varepsilon}{\partial k_\alpha}. \qquad (29)$$

The effect of the anisotropy of the temperature dependence is shown in Fig. 4. Above $T/T_c > 0.1$, anisotropy appears to modify the temperature dependence markedly. In particular, the amplitude of the asymptotic behavior $(\lambda_{\alpha\alpha}(0)/\lambda_{\alpha\alpha}(T))^2 = A_{\alpha\alpha}(1 - T/T_c)$ decreases from $A_{\alpha\alpha} = 2$ with increasing anisotropy. It is important to note, however, that substantial deviations from the isotropic case require close proximity to nodes in the temperature-dependent gap. In the present case, nodes appear for $g_1 = 1$, corresponding to $\Delta_{min}/\Delta_{max} \lesssim 1/3$ (Table II). Thus, the anisotropy effects seen in Fig. 4 ($g_1 = 0.9$) merely signal the close proximity to the nodes appearing for $g_1 \geq 1$. For $g_1 < 0.6$, where $\Delta_{min}/\Delta_{max} \gtrsim 0.65$, the deviations become much smaller. Experiments on YBCO clearly indicate BCS behavior [1-4], justifying first of all a BCS treatment and revealing the absence of nodes. However, as pointed out above, consistency with isotropic BCS behavior does not contradict the presence of gap anisotropy for $\Delta_{min}/\Delta_{max} \gtrsim 0.65$, which is required to resolve the conflict between the infrared and tunneling estimates for the gap.

Finally, we turn to the nuclear spin relaxation rate given by [32,33]

$$\frac{1}{T_1} \simeq \frac{1}{2} \sum_{\vec{k},\vec{k}'} \left(1 + \frac{\varepsilon(k)\varepsilon(k')}{E(k)E(k')} + \frac{\Delta(k)\Delta(k')}{E(k)E(k')}\right) \quad (30)$$
$$\times f(E(k'))(1 - f(E(k))\delta(E(k) - E(k'))).$$

For an isotropic gap, where the density of states exhibits a square-root singularity, the integral is known to diverge. This problem can be resolved by taking into account the lifetime of the quasiparticles and the anisotropy of the gap. Here we neglect the lifetime effect. Figure 5 shows a comparison between the temperature dependences resulting from a small and a large anisotropy. For $T/T_c > 1$, Korringa behavior appears, while the broad peak occurring just below T_c, reminiscent of conventional superconductors, is seen to become weaker with increasing anisotropy. Recently, the NMR relaxation of O^{17} has been measured in YBCO [34], revealing a very small enhancement just below T_c. This reduction clearly points to the presence of gap anisotropy and a finite lifetime of quasiparticles. Taking $\Delta_{min}/\Delta_{max} \simeq 0.6$, consistent with the infrared, tunneling and penetration depth measurements, an enhancement still remains (Fig. 5) and is reduced further by taking the finite lifetime of the quasiparticles into account. In fact, the enhancement

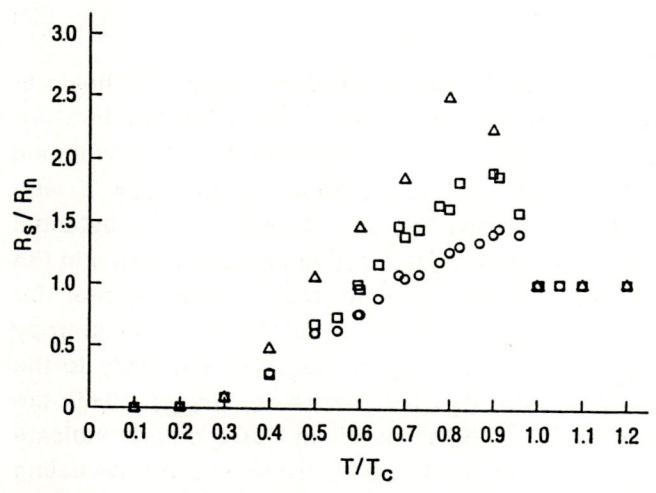

Figure 5. Temperature dependence of the nuclear spin relaxation time in terms of R_s/R_N, where $R \simeq 1/T_1$ for the parameters listed in Eq. (6), $g_0 = 1$, and circles: $g_1 = 0.9$; squares: $g_1 = 0.6$; triangles $g_1 = 0.2$. From [31].

appears close to T_c, which is so high that lifetime effects play a role. Further work is needed to disentangle the lifetime and anisotropy-induced reduction of the $1/T_1$ enhancement just below T_c.

Finally, we turn to the coherence length in the ground state. It can be obtained from the exponential decay of the correlation function of two quasiparticles, one in the origin, the other at $\vec{r}$, whose spins are antiparallel. For a spherical Fermi surface and an isotropic gap, it is given by $\xi_0 = \hbar^2 k_F/m\Delta$ [35]. The numerical estimates listed in Table V reflect the strong anisotropy determined by the small interlayer hopping matrix element. In particular, $\xi_\perp$ is found to be smaller than the spacing of the layers and the ratio $\xi_\perp/\xi_\parallel$ is in the range of values estimated from the upper critical fields [5-7]. Because the gap anisotropy does not affect the magnitude of the coherence length, a slight extension of the free electron expression

$$\xi_{\perp,\parallel} = \frac{\hbar}{\langle\Delta\rangle}\left(\frac{2\tilde{E}_F}{m_{\perp,\parallel}}\right)^{1/2} \tag{31}$$

appears to be appropriate. Thus the anisotropy comes from the effective mass, and the small magnitude is due to the large average gap and mass as well as to the small relative Fermi energy. $\xi_\perp$ and $\xi_\parallel$ are a measure of the size and shape of the Cooper pair. In the present case it is an ellipsoid with volume $V_c = (4\pi/3)\xi_\perp^2 \xi_\parallel$. Note that within the volume V_c, there are $nV_c = 94$ carriers ($n = 2\rho/V_0$, $\rho = 1/2$). Even though this number is much smaller than in conventional superconductors, where $nV_c = 10^7$, nV_c is still close to the Cooper-pairing limit ($nV_c \gg 1$) and very far from the Bose limit ($nV_c \ll 1$) [35].

Table V. Numerical estimates of the zero temperature coherence length for the parameters cited in Eq. (6). From [31].

ρ	g_0	g_1	Δ_0	Δ_1	ξ_{xx}/a	ξ_{zz}/s	$\frac{\xi_{zz}}{\xi_{xx}}\frac{a}{s}$
0.5	1	0	0.0244	0	7.6	0.39	0.05
0.5	1	0.99	0.0244	-0.0061	7.9	0.47	0.06

4. Concluding Remarks

Guided by experimental constraints, we outlined a BCS-type tight-binding model. In the highly anisotropic metallic normal state, the carriers form a narrow and anisotropic band. Normal state properties, including the elements of the Hall tensor, the effective mass, the carrier density and the penetration depth, agree remarkably well with experiment. The carriers, having hole- and electron-like character, are subject to unretarded pairing of on-site and interlayer nearest-neighbor origin. Thus, T_c is proportional to the Fermi energy and can adopt rather large values even for a weak pairing interaction. T_c turned out to be rather insensitive to the strength of interlayer pairing, giving rise to a uniaxial gap anisotropy which was found to modify the standard BCS behavior. The gap energy varies between Δ_{min} and Δ_{max}, and interpretations based on a unique ratio $2\Delta/k_B T_c$ become meaningless. Δ_{min} is accessible in infrared reflectivity measurements, while diffuse tunneling probes Δ_{max}. This interpretation offers an explanation for the apparent discrepancy of $2\Delta/k_B T_c$ values, as determined from infrared reflectivity and tunneling experiments. Moreover, we identified the effects of gap anisotropy on the temperature dependence of the nuclear spin relaxation rate T_1 and of the penetration depth. For $\Delta_{min}/\Delta_{max} \gtrsim 0.6$, the penetration depth was found to be rather insensitive to gap anisotropy. Thus, the conflict between the penetration depth measurements, which are consistent with isotropic BCS behavior, and the infrared reflectivity, as well as tunneling measurements which point to an anisotropic gap, has been resolved. A consistent picture requires that $\Delta_{min}/\Delta_{max} \gtrsim 0.6$. The missing enhancement in $1/T_1$ is hence a natural consequence of the gap anisotropy and the lifetime of quasiparticles.

The remarkable agreement with the experimental findings indicates an unretarded pairing interaction, on-site within the layers and of nearest-neighbor nature between the sheets. The origin is of electronic nature, yielding a cutoff proportional to the Fermi energy. It should be kept in mind, however, that our analysis was restricted to systems with one layer per unit cell. Accordingly, we assumed one or nearly degenerate gap branches.

Acknowledgements

The authors thank A. Aharony, A. Baratoff, S. Ciraci, A.P. Malozemoff, K.A. Müller and N. Schopohl for stimulating discussions and pertinent suggestions.

References

1. Y. J. Uemura et al., Phys. Rev. B **38**, 909 (1988).
2. L. Krusin-Elbaum et al., Phys. Rev. Lett. **62**, 217 (1989).
3. A. T. Fiory et al., Phys. Rev. Lett. **61**, 1419 (1988).
4. H. Keller et al., preprint.
5. T. T. M. Palstra et al., Phys. Rev. B **39**, 5102 (1988).
6. J. H. Kang et al., Appl. Phys. Lett. **53**, 2560 (1988).
7. U. Welp et al., Phys. Rev. Lett. **62**, 1908 (1989).
8. S. Martin et al., Phys. Rev. Lett. **60**, 3194 (1988).
9. A. Zettl et al., in *Mechanisms of High Temperature Superconductivity*, edited by H. Kamimura and A. Oshiyama (Springer-Verlag, Berlin Heidelberg, 1988), p. 249.
10. H. Takagi et al., Nature **332**, 236 (1988).
11. U. Hofman et al., preprint.
12. R. T. Collins et al., Phys. Rev. Lett. **63**, 422 (1989)
13. G. A. Thomas et al., Phys. Rev. Lett. **61**, 1313 (1988).
14. Y. J. Uemura et al., Phys. Rev. Lett. **62**, 2317 (1989).
15. M. Lee et al., in Ref. 6, p. 220.
16. J. S. Tsai et al., in Ref. 6, p. 229.
17. J. S. Tsai et al., Physica C **153-155**, 1385 (1988).
18. T. Ekino and J. Akimitsu, preprint.
19. Y. Iye, in Ref. 6, p. 263.
20. T. Takahashi et al., Nature **334**, 691 (1988).
21. J. M. Imer et al., Phys. Rev. Lett. **62**, 336 (1989).
22. F. J. Himpsel et al., Phys. Rev. B **38**, 11946 (1988).
23. F. Minami et al., Phys. Rev. B **39**, 4788 (1989).
24. T. Takahashi et al., Phys. Rev. B **39**, 6636 (1989).
25. R. Claessen et al., Phys. Rev. B **39**, 7316 (1989).
26. L. Hoffmann et al., Physica C **153-155**, 129 (1988).
27. L. C. Smedskkjaer et al., Physica C **156**, 269 (1988).
28. A. Bansil et al., Phys. Rev. Lett. **61**, 2480 (1988).
29. T. Schneider, H. de Raedt, and M. Frick, Z. Phys. B **76**, 3 (1989).
30. P. B. Allen and W. E. Pickett, Phys. Rev. B **36**, 3926 (1987).
31. T. Schneider and M. Frick, in *Strong Correlation and Superconductivity*, edited by H. Fukuyama, S. Maekawa and A. P. Malozemoff, Springer Series in Solid-State Sciences, Vol. 89 (Springer-Verlag, Berlin Heidelberg, 1989), p. 176.
32. L. C. Hebel and C. P. Schlichter, Phys. Rev. **113**, 1504 (1959).
33. M. Fibich, Phys. Rev. Lett. **14**, 561 (1965).
34. Y. Kitaoka et al., in Ref. 6, p. 148.
35. A.J. Legget, Rev. Mod. Mod. Phys. **47**, 331 (1975).

Generalized Hubbard Models for Cu-O-Based Superconductors: Field-Theoretical and Monte-Carlo Results

J. Wagner, R. Putz, G. Dopf, B. Ehlers, L. Lilly, A. Muramatsu*, and W. Hanke*

Physikalisches Institut, Universität Würzburg,
D-8700 Würzburg, Fed. Rep. of Germany
*Also at: MPI für Festkörperforschung, D-7000 Stuttgart, Fed. Rep. of Germany

Abstract: We present a summary of a) weak-coupling, b) strong-coupling and c) Monte-Carlo results we have recently obtained for a three-band Hubbard model of the copper-oxide superconductors with both intra-site (U_d, U_p) and inter-site (U_{pd}) repulsion. Within strong-coupling (hopping $t_{pd} \ll U_d$) we find that U_{pd} enhances the relative weight of s-wave versus d-wave pairing with the dominance of s-wave pairing for large transfer energies $\Delta = E_p - E_d$ and low to intermediate doping. In weak-coupling the exchange of CDW modes leads to a superconducting coupling constant for s-wave pairing which is rapidly suppressed and replaced by d-wave pairing when simultaneously onsite repulsions are included. Monte-Carlo simulations extend the parameter range to arbitrary couplings. Results on the magnetic structure factor and the possible singlet formation between Cu- and O-holes are reported for the low temperature (up to $\beta = 30$) regime.

1 Introduction

The presence of antiferromagnetism in the vicinity of the superconducting phase in the phase diagram of the high-T_c oxides indicates that antiferromagnetic (AF) correlations may play an important role in the superconducting pairing[1]. Various mechanisms which involve the AF interactions in an essential way, such as spin fluctuations[2,3], resonating valence bonds[4], spin bags[5] and others have been proposed so far. Common ingredient of all these theories is a more or less large on-site repulsion energy on the Cu-d orbitals, U_d. The corresponding large U limit of the Hubbard model in two dimensions (AF Heisenberg model) has recently been shown to be in accord with some essential magnetic properties of the Cu-O materials at half-filling[6].

On the experimental side, there is growing evidence, especially from spectroscopic data, that both Cu 3d and O 2p orbitals together with their intra- and inter-atomic correlations have to be considered[7]. Varma et al.[8] first stressed the importance of the nearest-neighbor repulsion U_{pd} and the role of Cu-O charge fluctuations, but their model did not involve U_d in an essential way[9]. Recent exact diagonalizations of very small (2 × 2) CuO$_2$ clusters indeed demonstrated that, in the presence of a strong U_d, a near-neighbor Cu-O repulsion U_{pd} can give rise to an attractive interaction between holes on the O sites.

Like in the one-band situation, as the temperature is lowered for a half-filled band, the dominant collective response is related to the particle-hole staggered spin-susceptibility at $\mathbf{Q} = (\pi, \pi)$. However, again, the nature of the pairing correlations in the particle-particle channel of the doped system is less clear. Both charge (CDW) and spin (SDW) fluctuations can play a role analogous to phonons, with their emission and absorption giving rise to an effective particle-particle interaction. If the SDW fluctuations dominate, in analogy with strong-coupling variational calculations[10] and the weak-coupling paramagnon exchange picture[2] for the one-band situation, we expect d-wave pairing in the extended model, too. In section 2 on strong coupling we will show that the additional role of the charge fluctuations depends crucially on the magnitude of the charge-transfer energy Δ : for large $\Delta = 4$ (in the following all energies are given in units of t_{pd}) the Cu-O interaction U_{pd} just serves to enhance the magnetically driven pairing with practically no influence of (virtual) charge-transfer fluctuations. On the other hand, for $\Delta = 2$ (and $U_{pd} = 1$) we find about a 15% enhancement of the s-wave (BCS) result for T_c versus doping due to the excitonic charge transfer. The dominance of s-wave pairing for $U_{pd} = 1$ is particularly interesting for low doping and the crossover to d-wave pairing for larger doping.

In the following we summarize these and other results we have obtained in both strong- and weak-coupling approaches for the multi-band Hubbard model. We also summarize some Monte-Carlo results, which aim at bridging the gap between the two perturbative regimes.

2 Strong-Coupling Expansion

The extended Hubbard model is defined for the $d_{x^2-y^2}$ orbital on Cu and the p_x and p_y orbitals on the corresponding two O atoms in the unit cell:

$$H = \sum_{i,j,\sigma} \varepsilon_{ij} c_{i\sigma}^\dagger c_{j\sigma} + \frac{1}{2} \sum_{\substack{i,j \\ \sigma,\sigma'}} U_{ij} c_{i\sigma}^\dagger c_{i\sigma} c_{j\sigma'}^\dagger c_{j\sigma'}. \tag{1}$$

Here $c_{i\sigma}^\dagger$ denotes the hole creation operator for Cu and O depending on the site index i. ε_{ij} includes on-site energies E_d (Cu) and E_p (O) with $\Delta = E_p - E_d$ and a Cu-O hopping t_{pd}, while U_{ij} describes the on-site Coulomb energies U_d and U_p and the inter-site Cu-O interaction U_{pd}.

In the following we study the competition between spin-fluctuation and charge-transfer induced pairings in the strong-coupling limit. This limit is defined by $t_{pd} \ll (U_d - \Delta, \Delta)$ and $U_{pd} < (\frac{U_d - \Delta}{2}, \Delta)$, where the small parameter is given by $t_{pd}/(U_d - \Delta, \Delta)$ and the second condition guarantees that the doped holes go on O sites with the d^9 configuration (one hole/Cu) being preserved on Cu sites. Applying standard canonical perturbation theory, as discussed in detail in Refs.[9,11] for the $(U_{pd} = 0)$-case, we can analytically derive an effective Hamiltonian up to fourth order in t_{pd}. Considering this Hamiltonian for $U_p = 0$, which is justified for small dopings, we obtain

$$H_{eff} = H_{kin} + H_{c.t.} + H_{spin} \tag{2}$$

where

$$H_{kin} = (E_p^r - \mu) \sum_{j\sigma} p_{j\sigma}^\dagger p_{j\sigma} + t_{eff} \sum_{\substack{<lj,i> \\ \sigma, l \neq j}} (-1)^{\alpha_{ij}+\alpha_{il}} p_{l\sigma}^\dagger p_{j\sigma},$$

$$H_{c.t.} = \frac{1}{2} \sum_{\substack{<ljj_1,i> \\ \sigma\sigma_1}} V_{lj}^{c.t.} (-1)^{\alpha_{ij}+\alpha_{il}} p_{l\sigma}^\dagger p_{j_1\sigma_1}^\dagger p_{j_1\sigma_1} p_{j\sigma},$$

and

$$H_{spin} = J \sum_{\substack{<lj,i> \\ \alpha\beta}} (-1)^{\alpha_{ij}+\alpha_{il}} \mathbf{S}_i p_{l\alpha}^\dagger \vec{\sigma}_{\alpha\beta} p_{j\beta} + \frac{J_{Cu}}{4} \sum_{<i,i'>} (\mathbf{S}_i + \mathbf{S}_{i'})^2. \tag{3}$$

519

The symbol $<j,i>$ means that the sum is performed over nearest neighbors only. More generally, $\sum_{<j_1\ldots j_n,i>}$ stands for a sum over all Cu-sites i followed by a sum over neighboring O-sites $j_1\ldots j_n$. The parameters entering H_{kin}, $H_{c.t.}$ and H_{spin} can be described as follows:

$$\begin{aligned}
E_p^r &= E_p + 2U_{pd} + t_{pd}^2\left(\frac{8}{\Delta+U_{pd}} - \frac{7}{\Delta} - \frac{1}{U_d-\Delta-2U_{pd}}\right), \\
t_{eff} &= \frac{t_{pd}^2}{2}\left(\frac{1}{\Delta} - \frac{1}{U_d-\Delta-2U_{pd}}\right), \\
V_{lj}^{c.t.} &= -4t_{pd}^2\left(\frac{1}{\Delta+U_{pd}} - \frac{2}{\Delta} + \frac{1}{\Delta-U_{pd}}\right)\delta_{lj} \\
&\quad + t_{pd}^2\left(-\frac{1}{\Delta} + \frac{1}{\Delta-U_{pd}} - \frac{1}{U_d-\Delta-U_{pd}} + \frac{1}{U_d-\Delta-2U_{pd}}\right), \\
J &= t_{pd}^2\left(\frac{1}{\Delta} + \frac{1}{U_d-\Delta-2U_{pd}}\right), \\
J_{Cu} &= \frac{4t_{pd}^4}{(\Delta+U_{pd})^2}\left(\frac{1}{U_d} + \frac{1}{\Delta}\right).
\end{aligned} \quad (4)$$

H_{kin} describes the hole kinetic energy on O sites with the $c(c^\dagger)$-operators of Eq. (1) being replaced by $p(p^\dagger)$ operators. The phase factor $(-1)^{\alpha_{ij}+\alpha_{il}}$ is due to the $d_{x^2-y^2}$- and $p_{x,y}$-symmetry of the Cu- and O-orbitals, respectively:

$$(-1)^{\alpha_{ij}} = \begin{cases} -1 & \text{for } j = i + \frac{\hat{x}}{2}, j = i - \frac{\hat{y}}{2} \\ +1 & \text{for } j = i - \frac{\hat{x}}{2}, j = i + \frac{\hat{y}}{2}. \end{cases} \quad (5)$$

On the other hand, $H_{c.t.}$ stands for the effective interaction between holes on O sites, which is mediated by virtual hoppings from Cu-holes onto O sites and back. The physics of this term is schematically depicted in Fig. 1. $V^{c.t.}$ (Eq. (4)) contains two terms: the first describes the energy difference originating from the virtual hopping (charge-transfer) of the Cu hole in the presence of two neighboring O holes (right part in Fig. 1) compared with the situation when the O holes are separated by a large distance (left part in Fig. 1). This term alone would correspond to a negative binding energy for statically fixed O holes. It is, however, modified by a second term in $V^{c.t.}$, which accounts for the O-hole dynamics.

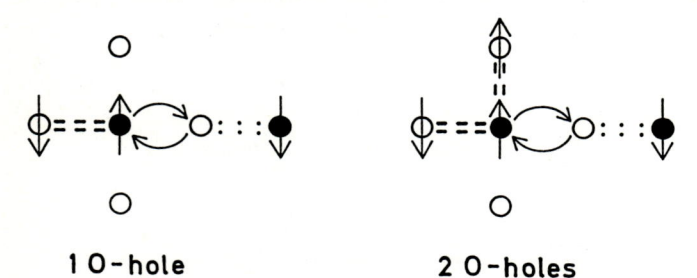

1 O-hole   2 O-holes

Figure 1: Hole hopping processes responsible for the charge-transfer interaction $H_{c.t.}$. The black (•) and the white (o) circles denote Cu- and O-sites, respectively. The dashed line (===) stands for the Cu-O repulsion in the Cu d^9 configuration, whereas (:::) denotes the corresponding repulsion in the intermediate state.

Finally, H_{spin} contains one term, proportional to J, which is the Kondo-like spin-carrier interaction. This term gives rise to the well-known magnetic semiconductor physics characterized by ferromagnetic correlations (spin polarons, double exchange). The second term in H_{spin} is the familiar antiferromagnetic coupling or superexchange[4] between the spins on Cu sites (note that we choose the energy zero to be that of the (i, i') singlet configuration). The Kondo term which is due to virtual hoppings of Cu holes to neighboring O sites and vice versa, induces a ferromagnetic polarization on Cu sites. This ferromagnetic polarization to a certain extent opposes the superexchange coupling and is responsible, for example, for the formation of "spin bags" [5].

In the next step H_{kin} is diagonalized, resulting for $t_{eff} < 0$ in a dispersionless upper and in a lower band. We then project H_{eff} onto the lower O-band. In what follows, we assume that holes are predominantly paired on O sites. This assumption is in general accord with our strong-coupling limit. We then seek for an effective coupling between the O-holes, where the Cu-spin degrees of freedom have been integrated out. This is achieved by employing a path-integral formulation which has recently been used in a similar context by Muramatsu et al.[12]. To this end, we write for the partition function of H_{spin} in Eq. (3)

$$Z = \int \mathcal{D}d^*\mathcal{D}d\mathcal{D}p^*\mathcal{D}p \, exp[S], \tag{6}$$

with

$$S = \int_0^\beta d\tau \left(d^*(\tau)\frac{\partial}{\partial \tau}d(\tau) + p^*(\tau)\frac{\partial}{\partial \tau}p(\tau) - H_{eff} \right). \tag{7}$$

H_{eff} is of identical form as in Eq. (3) but with the Fermi operators replaced by Grassmann variables $d_i^*(\tau)$, $d_i(\tau)$, $p_j^*(\tau)$, $p_j(\tau)$[13]. We completely follow Ref.[12] and introduce an auxiliary Bose field that decouples the superexchange interaction. With this the action becomes bilinear in the variables d^* and d and hence, the degrees of freedom due to the fermions on Cu-sites can be integrated out[14]; with the result for Z

$$Z = \int \mathcal{D}p^*\mathcal{D}p \, exp \left(\int_0^\beta d\tau (p^*(\tau)\frac{\partial}{\partial \tau}p(\tau) - H_{kin} - H_{c.t.} - \tilde{H}_{spin} \right). \tag{8}$$

This last step yields an effective spin Hamiltonian $\tilde{H}_{spin}$ describing only the charge carriers, where the antiferromagnetic fluctuations are incorporated in a new effective interaction. This interaction is of a rather complex, though analytical form (similar to Ref.[12], however, with U_{pd} included). It becomes much more transparent if we resort to the BCS-limit (only $\mathbf{k}$ and $-\mathbf{k}$ paired):

$$\tilde{H}_{spin}^{BCS} = \frac{1}{2N} \sum_{\substack{\mathbf{k},\mathbf{k}' \\ \alpha\beta\gamma\delta}} \frac{2J^2}{T} \frac{\vec{\sigma}_{\alpha\gamma}\vec{\sigma}_{\beta\delta} h_{\mathbf{k}}^{-2} h_{-\mathbf{k}'}^{-2}}{1 + \frac{J_{Cu}}{4T}f(\mathbf{k}+\mathbf{k}')} p_{\mathbf{k}\alpha}^\dagger p_{-\mathbf{k}\beta}^\dagger p_{-\mathbf{k}'\gamma} p_{\mathbf{k}'\delta}, \tag{9}$$

where $h_{\mathbf{k}}^{-2} = 1 - \frac{1}{2}(cos(k_x) + cos(k_y))$ and $f(\mathbf{k}) = 2 + cos(k_x) + cos(k_y)$.

Summarizing our strong-coupling procedure, the effective coupling between the charge carriers consists of a part $H_{c.t.}$ which can become attractive due to virtual Cu-O charge transfers, and a second part $\tilde{H}_{spin}$ which is the interaction mediated by the AF Cu-spin fluctuations.

For studying the superconducting state we also apply the BCS limit to $H_{c.t.}$. Resorting to standard Hartree-Fock decoupling and employing the symmetry analysis as discussed by Sigrist and Rice [15], we obtain the mean-field transition temperatures T_c plotted for various parameter regimes in Figs. 2 and 3.

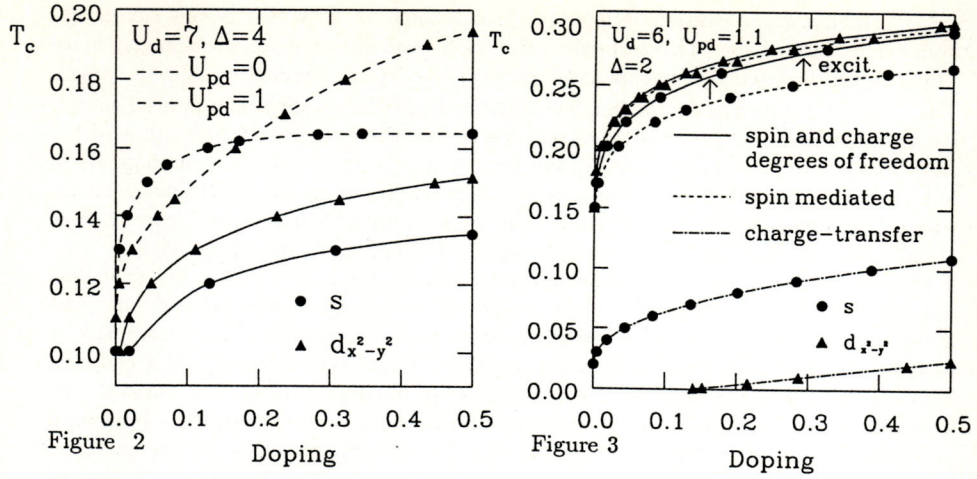

Figure 2: BCS transition temperature T_c (in units of t_{pd}) versus doping in strong-coupling for large charge transfer. Results are plotted for s and d symmetry, for two values of U_{pd}.

Figure 3: T_c versus doping for small charge transfer $\Delta = 2$. Plotted are results due to the pairing contributions in Eq. (3), i.e. $H_{c.t.}$ and H_{spin}, for s and d symmetry. Note the excitonic renormalisation (as discussed in the text) of the s-wave channel.

The results in Fig. 2 give T_c in units of t_{pd} for a large charge-transfer energy $\Delta = 4$. We observe that increasing the Cu-O repulsion U_{pd} enhances the effective pairing strength. For this relatively large Δ case, U_{pd} just serves to enhance the magnetic pairing due to the Kondo coupling J ($J = t_{pd}^2(\frac{1}{\Delta} + \frac{1}{U_d - \Delta - 2U_{pd}})$) which enters quadratically in Eq. (8), whereas the AF Cu-Cu coupling J_{Cu}, which is required for the superconducting pairing, is diminished. In this regime there is practically no influence due to the charge-transfer fluctuations ($H_{c.t.}$). This is due to a critical balance of the static and dynamic contributions to $V^{c.t.}$ (see also Eqs. (4)). Particularly interesting is the crossover from extended s-wave to d-wave pairing for $U_{pd} = 1$ as a function of doping. For $\Delta = 2$ and $U_{pd} = 1.1$ (chosen so that $t_{eff} < 0$) the enhancement of the coupling strength is due to a combined effect of $\tilde{H}_{spin}^{BCS}$ and $H_{c.t.}$. The latter term, as observed in Fig. 3, results in about a 15% enhancement of the s-wave result for T_c versus doping due to the excitonic charge-transfer. For this smaller Δ always d-wave pairing wins, with the general trend (like in Fig. 2) that U_{pd} enhances the relative weight of extended s-wave versus d-wave pairing.

3 Weak Coupling Expansion

Within a weak-coupling random-phase approximation (RPA) for the one-band Hubbard model Scalapino and coworkers[2] found that d-wave pairing is favored near a SDW metal-semiconductor transition. They examined the antiferromagnetic paramagnon exchange interaction V, which is diagrammatically illustrated in Fig. 4 and which

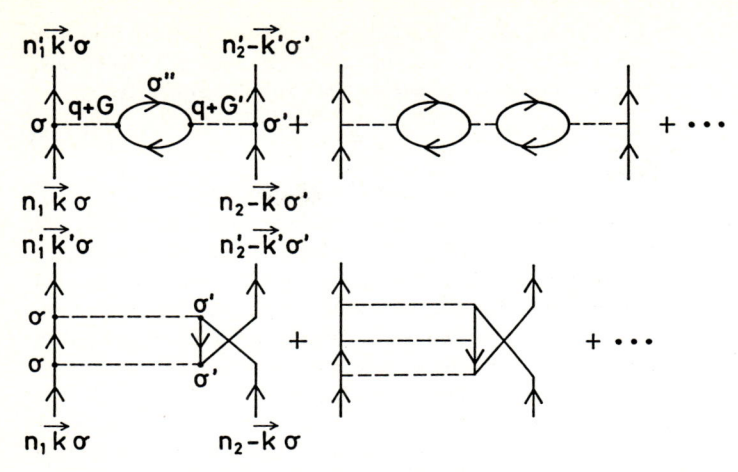

Figure 4: Diagrams contributing to the effective particle–particle interaction.

has been proposed as a possible pairing mechanism for some organic[16] and heavy-fermion[2] systems. On the basis of V, the coupling strength

$$\bar{\lambda}_\alpha = -\frac{\sum_{\mathbf{p}} \sum_{\mathbf{p'}} g_\alpha(\mathbf{p'}) V(\mathbf{p'},\mathbf{p}) g_\alpha(\mathbf{p}) \delta(\varepsilon_{\mathbf{p'}} - \mu) \delta(\varepsilon_{\mathbf{p}} - \mu)}{\sum_{\mathbf{p}} g_\alpha^2(\mathbf{p}) \delta(\varepsilon_{\mathbf{p}} - \mu)} \tag{10}$$

has been calculated for the various pairing channels. In Eq. (10) the average is taken over the Fermi surface and the factors $g_\alpha(\mathbf{p})$ describe the symmetry of the pairing (on-site s-wave: $g_s(\mathbf{p}) = 1$, extended s-wave: $g_{s\star}(\mathbf{p}) = cos(p_x) + cos(p_y)$, $g_{d_{x^2-y^2}}(p) = cos(p_x) - cos(p_y)$, etc.). However, it should be remarked, that while the RPA results on the one-band Hubbard model indicate that spin fluctuations may give rise to an attractive pairing interaction in the d-wave channel the results obtained from Monte-Carlo simulations and exact diagonalizations for the d-wave pairing susceptibility are disappointingly small[17].

We have extended these weak-coupling investigations to the multi-band Hubbard model of Eq. (1), in particular, to study the competition between magnetically (paramagnon exchange) and charge transfer (excitonic exchange) induced pairings.

The effective coupling constants $\bar{\lambda}_\alpha$, together with the self-energy renormalization coupling λ_z ($\lambda_{eff} = \bar{\lambda}_\alpha/(1 + \lambda_z)$), in the singlet and triplet channel have been obtained from the Fermi surface averages over the singlet (V_s) and triplet (V_t) interactions. Both interactions are an infinite sum of "ladder" and "bubble" contributions (Fig. 4), just as in Ref.[2], however, with the interaction lines standing not only for U_d, but also for U_p and U_{pd}. Whereas in the one-band case the corresponding Bethe-Salpeter or Dyson equations can be inverted as a scalar to derive the effective interactions V_s and V_t, in the multi-band situation we resort to the local-orbital representation of Hanke and Sham[18]: here the solution is accomplished by the inversion of low-dimensional matrices. Their dimension (11 in our case) is determined by the overlap requirements. In this local-orbital basis we obtain for the bubble contribution[19]

$$V_{\sigma\sigma'}^{Bubble}(\mathbf{k},\mathbf{k'};\omega) = M_{n_1 n_1'}^\dagger \cdot (V \cdot S^{-1} \cdot \tilde{V})_{\sigma\sigma'} \cdot M_{n_2' n_2} \tag{11}$$

where the matrix product $(\cdot)$ is with respect to the local-orbital indices s and spin indices

σ. M stands for the coupling of the "external" Bloch lines in the diagrams of Fig. 4 to the "internal" interaction lines, V is the RPA Coulomb matrix in local orbitals[18] and $S^{-1} = \chi^{(0)}(1 - V\chi^{(0)})^{-1}$, where $\chi^{(0)}$ denotes the local-orbital bubble. Similarly,

$$V^{Ladder}_{\sigma\sigma'}(\mathbf{k}, \mathbf{k}'; \omega) = M^\dagger_{n_2 n'_1} V^x_{\sigma\sigma'} S^{-1}_{x,\sigma\sigma'} V^x_{\sigma\sigma'} M_{n'_2 n_1} \tag{12}$$

where $S^{-1}_{x,\sigma\sigma'} = \chi^{(0)}(1 - V^x_{\sigma\sigma'}\chi^{(0)})^{-1}$ and V^x gives the exchange matrix[20].

Results for the various symmetry components of the coupling constant $\bar{\lambda}_\alpha$ are plotted in Fig. 5. We note that for $U_{pd} = 0$ and $U_d = 4$, similar to Ref.[2], the paramgnon exchange strongly favors singlet $d_{x^2-y^2}$-pairing ($\lambda > 0$) in the neighborhood of the RPA SDW instability. However, in contrast to Ref.[2], the s-wave coupling is repulsive in our case. The chemical potential is used to move the system away from the SDW instability. At fillings very far from half-filling ($E_F = 0$) the p-wave channel wins (not plotted).

Fig. 6 displays the CDW situation, which is plotted for $U_{pd} = 1.5$ and $U_d = 0$. In both cases we obtain two instabilities, one for $\mathbf{Q} = (\pi, \pi)$ and one for $\mathbf{Q} = 0$. Clearly the s-wave is the most atttractive, while the $d_{x^2-y^2}$ channel is slightly attractive only for (π, π). In the mixed CDW-SDW situation, with $U_d = 4$ and $U_{pd} = 1.5$, the d_{xy}-channel

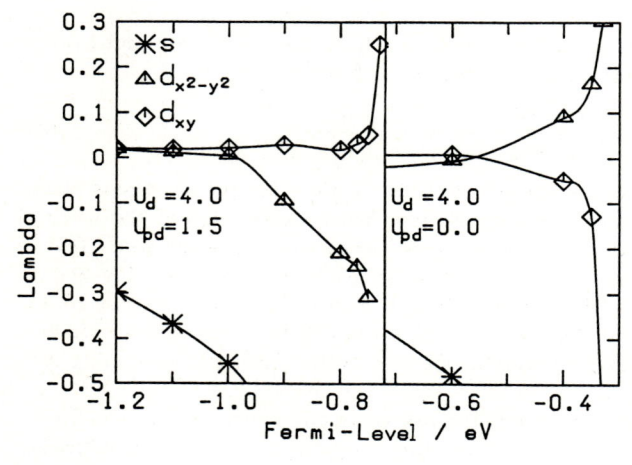

Figure 5: Superconducting coupling constants $\bar{\lambda}_\alpha$ for various symmetries as a function of E_F (doping).

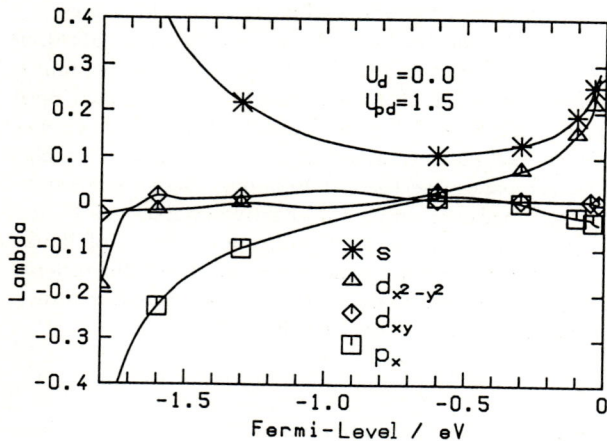

Figure 6: $\bar{\lambda}_\alpha$ as in Fig. 5, however for the CDW situation only.

wins. We found the self-energy renormalization coupling λ_z in both the CDW and SDW situations to be much larger than $\bar{\lambda}$, which would give rise to a significant reduction in T_c.

The results in Fig. 6 are in general accord with a recent calculation[21] of the particle-hole excitations and the stability of the paramagnetic state of the three-band Hamiltonian of Eq. (1). In the pure CDW case ($U_d = 0$) the authors of Ref.[21] find both a CDW instability at $\mathbf{Q} = (\pi, \pi)$ and the "charge-transfer" instability at $\mathbf{Q} = 0$, both of which are somewhat renormalized in the (U_{pd}, E_F)-plane by second-order self-energy corrections. In fact both these instabilities are responsible for the enhancement of the coupling constant λ as shown in Fig. 6. We suspect that by taking self-energy corrections into account within a conserving approximation the somewhat unphysical[22] (π, π)-instability will be suppressed.

4 Quantum Monte-Carlo Simulations

Finally, we discuss some recent Monte-Carlo results we have obtained employing an algorithm similar to the one developed in Refs. 24 and 25. This algorithm has been supplemented by a new stabilization procedure[25], which allows so far for temperatures as low as $\beta = 30$ in units of t_{pd}.

We start by considering the magnetic properties of the extended Hubbard model in Eq. (1). The results presented here correspond to a system with 4×4 elementary cells at $U_{pd} = 0$, $U_p = 0$ and, if not otherwise stated, to $U_d = 6$.

As displayed in Fig. 7, a quite drastic doping dependence is obtained for the magnetic structure factor for $\Delta = 4$, whereas for $\Delta = 1$ the structure factor practically remains constant. The $\Delta = 4$ curve corresponds to a parameter set that leads to almost complete hole localization on the Cu-site at half-filling. The strong dependence on doping in curve a) ($\Delta = 4$), which is in agreement with the experimental results, is to be contrasted with the small-Δ data in curve b), where the hole density is spread over the entire unit cell. Although, as a function of Δ, the occupation number on Cu varies appreciably already for zero doping, the local magnetic moment $<s^2>$ varies

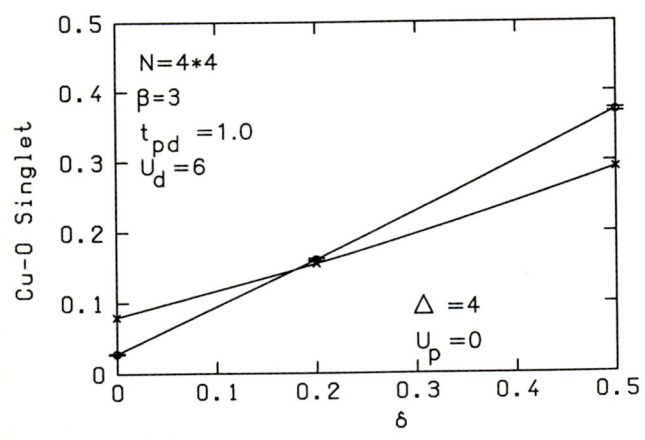

Figure 7: Magnetic structure factor as a function of doping for large and small Δ.

only between 0.535 ± 0.001 ($\Delta = 1$) and 0.835 ± 0.003 ($\Delta = 4$). Work is in progress[26] to determine, via a finite-size study, whether long-range order is present or not.

In the next step, we consider the possible formation[12] of local Cu-O singlets, which has been suggested by Zhang and Rice[27]: they argued that in the "strong-coupling" regime (by which is meant here the large-Δ case), the symmetric combination of doped holes sitting on four O-sites surrounding a Cu-site strongly couples in a singlet state to the hole on this central Cu-site.

Fig. 8 displays the square of the probability-amplitude for the singlet as a function of doping both for the interacting and the non-interacting case. Our results demonstrate that for dopings larger than 0.3 an appreciable enhancement is obtained, but not for small dopings. We have also obtained a similar result in the small-Δ regime ($\Delta = 1$, $U_d = 6$).

In Fig. 9, finally, results are plotted which demonstrate that the Monte-Carlo simulation is able to reproduce the low-temperature behavior of the system. To this

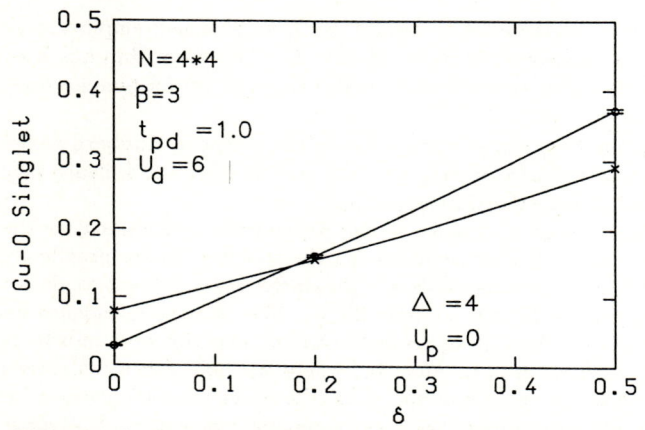

Figure 8: Square of the probability amplitude for the Cu-O singlet. Crosses and circles denote the $U_d = 0$ and $U_d = 6$ cases, respectively.

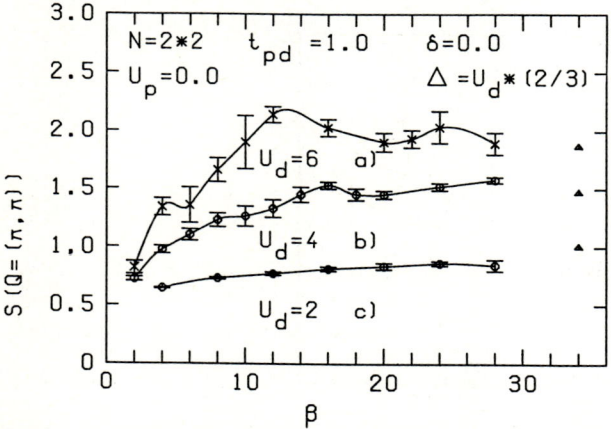

Figure 9: Magnetic structure factor as a function of $\beta = 1/k_B T$. Triangles at $\beta = 32$ are exact diagonalization results.

end, we have calculated the magnetic structure form factor for a 2 × 2 system both by Monte-Carlo and by exact diagonalization for $U_d = 6$, $U_d = 4$ and $U_d = 2$. We note that the $\beta = 30$ data of the Quantum Monte-Carlo simulation converge rather well to the exact diagonalization results, which are given by the triangles on the right-hand side of Fig. 9.

Acknowledgements

The authors would like to acknowledge support by the BMFT program No. 13N5501, by the Institute for Scientific Interchange in Torino (Italy) and by the HLRZ-Jülich, where the present calculations were performed.

References

[1] see various reviews in Physica C **153**-155 (1988)

[2] D.J. Scalapino, E. Loh and J.E. Hirsch, Phys. Rev. B **34**, 8190 (1986); Phys. Rev. B **35**, 6694 (1987) and N.E. Bickers, D.J. Scalapino and R.T. Scalettar, Int. J. Mod. Phys. B **1**, 687 (1987)

[3] J. Miyake, S. Schmitt-Rink and C. Varma, Phys. Rev. B **34**, 6554 (1986)

[4] P.W. Anderson, Science 234, 1196 (1987) and articles in Ref. 1

[5] J.R. Schrieffer, X.G. Wen and S.C. Zhang, Phys. Rev. Let. **60**, 944 (1988) and Phys. Rev. B, to be published

[6] S. Chakravarty, B.I. Halperin D.R. Nelson, Phys. Rev. B **39**, 2344 (1989)

[7] N. Nücker, J. Fink, J.C. Fuggle, P.J. Durham and W.M. Temmermann, Phys. Rev. B **37**, 5158 (1988)

[8] C.M. Varma, S. Schmitt-Rink and E. Abrahams, Sol. St. Comm. **62**, 681 (1987)

[9] V.J. Emery, G. Reiter, Phys. Rev. B **38**, 4547 (1988)

[10] C. Gros, R. Joynt and T.M. Rice, Z. Phys. B **68**, 425 (1987)

[11] J. Zaanen, A.M. Oleś, Phys. Rev. B **37**, 4923 (1988)

[12] A. Muramatsu, R. Zeyher and D. Schmeltzer, Europhys. Lett **7**,473 (1988)

[13] L.D. Fadeev amd A.A. Slavnov in "Gauge Fields, Introduction to Quantum Theory" (Benjamin/Cummings, London) 1980 and J.W. Negele and H. Orland in "Quantum Theory of Many Particle Systems" (Addison-Wesley, New York) 1988

[14] R. Rajaraman in "Solitons and Instantons" (North-Holland, Amsterdam) 1982

[15] M. Sigrist and T.M. Rice, Z. Phys. B **68**, 9 (1987)

[16] V.J. Emery, Synth. Met. **13**, 21 (1986)

[17] J.E. Hirsch, E. Loh, D.J. Scalapino, S. Tang, in Ref. 1 and H.Q. Lin, J.E. Hirsch, D.J. Scalapino, to be published

[18] W. Hanke and L.J. Sham, Phys. Rev. B **12**, 4501 (1975), Phys. Rev. B **21**, 4656 (1980)

[19] Details will be published elsewhere

[20] Note that, because $\lambda = - < \mathcal{R}e\, V(\mathbf{k},\mathbf{k}';\omega = 0) >_{E_F}$, we need only the static effective interactions.

[21] P.B. Littlewood, C.M. Varma, S. Schmitt-Rink and E. Abrahams, Phys. Rev. B **39**, 12371 (1989)

[22] B. Ehlers, L.Lilly, A. Muramatsu and W. Hanke, to be published

[23] R. Blankenbecler, et al., Phys. Rev. D **24**, 2278 (1981)

[24] J.E. Hirsch, Phys. Rev. B **31**, 4403 (1985)

[25] E.Y. Loh et al. in "Interacting Electrons in Reduced Dimensions" eds. D. Baeriswyl and D.K. Campbell (Plenum, New York) 1989, to appear

[26] G. Dopf, A. Muramatsu, W. Hanke, to be published

[27] F.C. Zhang and T.M. Rice, Phys. Rev. B **37**, 3759 (1988)

Index of Contributors

Alexander, M. 377

Barone, A. 163
Bianconi, A. 407
Bishop, A.R. 482
Blazey, K.W. 262
Borsa, F. 441
Buslaps, T. 377

Calestani, G. 407
Castrucci, P. 407
Chaudhari, P. 201
Claessen, R. 377
Corti, M. 441

Deutscher, G. 174
Dimos, D. 201
Dopf, G. 518

Ehlers, B. 518

Fabrizi, A. 407
Fink, J. 377
Fischer, Ø. 96
Flank, A.M. 407

Frick, M. 501
Fulde, P. 326

Gough, C.E. 141

Hanke, W. 518
Hervieu, M. 66

Jérome, D. 113

Katayama-Yoshida, H. 407
Keller, H. 222
Kitazawa, A. 45

Lagarde, P. 407
Lilly, L. 518

Mannhart, J. 201,208
Manzke, R. 377
Mehring, M. 467
Michel, C. 66
Morgenstern, I. 240
Müller-Heinzerling, Th. 377
Muramatsu, A. 518

Nücker, N. 377

Pflüger, J. 377
Pompa, M. 407
Portis, A.M. 278
Provost, J. 66
Putz, R. 518

Raveau, B. 66
Rega, T. 441
Rigamonti, A. 441
Romberg, H. 377

Sawatzky, G.A. 345
Scheerer, B. 377
Schneider, T. 501
Skibowski, M. 377
Steglich, F. 306
Studer, F. 66

Thomas, H. 2
Tranquada, J.M. 422

Wagner, J. 518
Waldner, F. 441

Ziolo, J. 441
Zwicknagl, G. 326